STUDENT'S SOLUTIONS MANUAL

TERI LOVELACE

PEGGY IRISH

INTERMEDIATE ALGEBRA: FUNCTIONS AND AUTHENTIC APPLICATIONS

FIFTH EDITION

Jay Lehmann

College of San Mateo

PEARSON

Boston   Columbus   Indianapolis   New York   San Francisco   Upper Saddle River
Amsterdam   Cape Town   Dubai   London   Madrid   Milan   Munich   Paris   Montreal   Toronto
Delhi   Mexico City   São Paulo   Sydney   Hong Kong   Seoul   Singapore   Taipei   Tokyo

Copyright © 2015, 2011, 2008 Pearson Education, Inc.
Publishing as Pearson, 75 Arlington Street, Boston, MA 02116.

ISBN-13: 978-0-321-86824-4
ISBN-10: 0-321-86824-2

2 3 4 5 6 V036 19 18 17 16

www.pearsonhighered.com

Contents

Chapter 1
Linear Equations and Linear Functions

Homework 1.1

1. a. (d)

   b. (c)

   c. (a)

   d. (b)

3. The more tests there are to grade, the more time it takes to grade them. Time to grade T is the dependent variable and the number of tests N is the independent variable.

5. The hotter the temperature in an oven, the faster a potato will cook. The number of minutes T is the dependent variable and the oven temperature F is the independent variable.

7. The longer the course, the more time it takes a person to run the course. The number of minutes T is the dependent variable and the length of a running course in miles L is the independent variable.

9. The greater income a person earns, the more likely it is that she owns a car. The percentage of Americans who own a car P is the dependent variable and the income I is the independent variable.

11. The larger the radius of a plate, the more spaghetti it holds. The number of ounces of spaghetti n is the dependent variable and the radius of a plate r is the independent variable.

13.

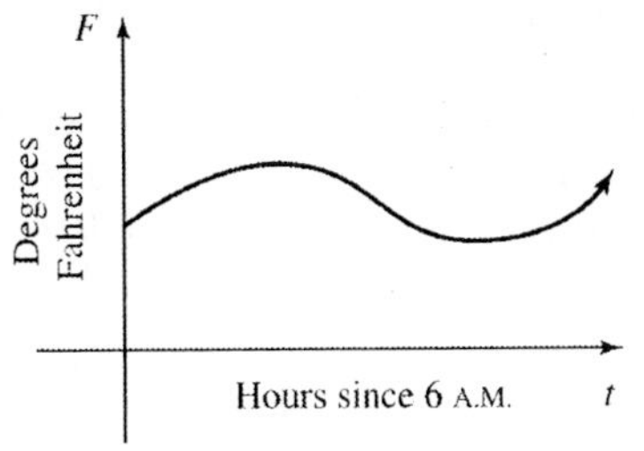

Temperature F is the dependent variable and time t is the independent variable. The graph shows that the temperature increases as the sun rises around 6 A.M. ($t = 0$). It reaches a high in the afternoon and drops slowly as the sun begins to set. Shortly after dawn of the following day, the temperature begins to rise again.

15.

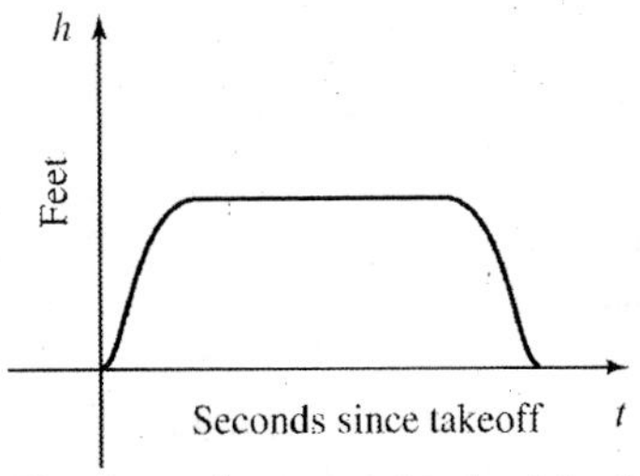

The dependent variable is altitude h and the independent variable is time t. The graph shows that the altitude of the airplane increased quickly after takeoff, and then leveled off for the flight. Near the end of the trip, the plane descended for its landing.

17.

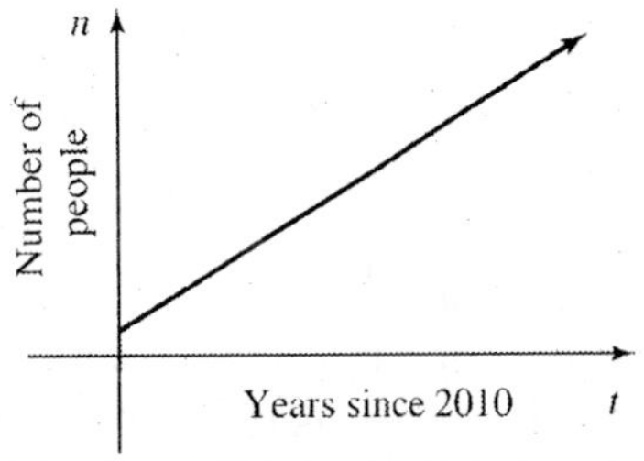

The dependent variable is the number of people undergoing laser eye surgery n and the independent variable is time t. The graph shows that the number of people having laser eye surgery is increasing steadily as time goes on from 2010.

19.

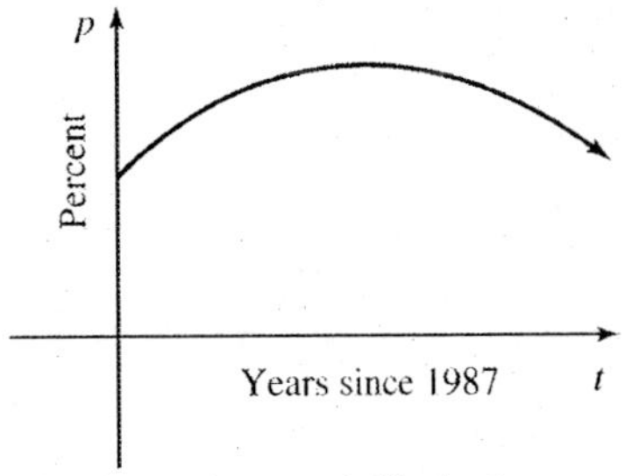

The dependent variable is the percentage of major firms that perform drug testing p and the independent variable is time t. The graph shows that the percentage of major firms performing drug testing increases over time from 1987 through 1996, and then decreases over time.

21.

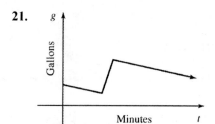

The dependent variable is amount of gas in the tank g and the independent variable is time t in minutes since the commuter left home. The graph decreases until the commuter purchases gas, rises when the gas is purchased, and then begins to decrease again as the commuter drives to work.

23.

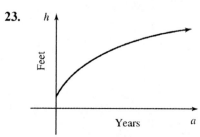

The dependent variable is height h and the independent variable is age a. The curve shows that a person's height increases from birth and levels off after a number of years. Note that height will never be 0, and that it eventually levels off at a certain age.

25.

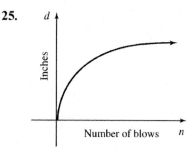

The diameter of a balloon in inches d is the dependent variable and the number of times a person has blow into the balloon n is the independent variable. The graph shows that as the number of blows into the balloon increases, the diameter increases (although at an ever slower rate).

27.

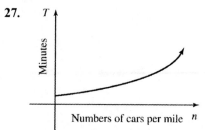

The time it takes to drive from home to school T is the dependent variable and the average number of cars per mile n is the independent variable. The graph shows that as the number of cars per mile increases, the time to drive from home to school increases.

29.

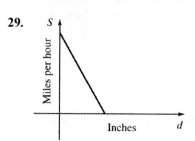

Speed S is the dependent variable and inches d is the independent variable. This graph shows that when the distance between the accelerator and the floor of the car increases, the speed of the car decreases until the car comes to a stop at $S = 0$.

31.

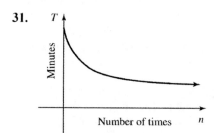

The time it takes a person to make lasagna T is the dependent variable and the number of times a person has made lasagna n is the independent variable. The graph shows that as the number of times a person has made lasagna increases, the time to make lasagna decreases.

33.

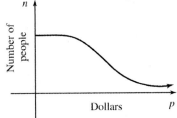

The dependent variable is *n*, the number of people that will buy the Honda Civic® CX, and the independent variable is *p*, the price at which the purchase occurs. The graph shows that more people will buy less expensive Honda Civic® CXs. As the price in dollars increases, fewer people are willing to purchase this car. The curve eventually levels off since there are only so many people who will buy the car no matter what the price.

35.

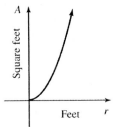

Area *A* is the dependent variable and radius *r* is the independent variable. The graph shows a rapid increase in area *A* as the circle's radius *r* increases. This is because the radius *r* is squared in the formula $A = \pi r^2$.

37. Answers may vary. The curve must pass through the origin. Example:

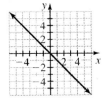

39. a. Answers may vary. Example:
Let *y* represent the height of a ball *x* seconds after it was thrown straight up in the air.

    b. Answers may vary. Example:
Let *y* represent the population of a species of bacteria *x* days after the bacteria has been introduced to an environment, assuming unlimited space and resources.

    c. Answers may vary. Example:
Let *y* represent the water level of a creek *t* days after July 10 in Georgia. The water level remained the same during a cloudy week with very little rain and it dropped again steadily as summer passed until it completely dried up.

    d. Answers may vary. Example:
Let *y* represent the temperature of a building *x* minutes after the thermostat is adjusted to reach a warmer setting.

Homework 1.2

1.

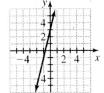

3.

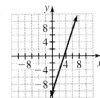

5.

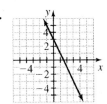

7.

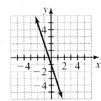

9.

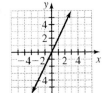

11.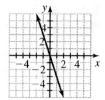

13.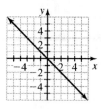

15. $9x - 3y = 0$

$$\frac{-3y}{-3} = \frac{-9x}{-3}$$

$$y = 3x$$

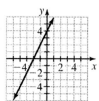

17. $3y - 6x = 12$

$$\frac{3y}{3} = \frac{12 + 6x}{3}$$

$$y = 4 + 2x$$

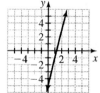

19. $8x - 2y - 10 = 0$

$$8x - 2y = 10$$

$$\frac{-2y}{-2} = \frac{-8x + 10}{-2}$$

$$y = 4x - 5$$

21. $2y - 6x - 14 = -4$

$$2y - 6x = 10$$

$$\frac{2y}{2} = \frac{10 + 6x}{2}$$

$$y = 5 + 3x$$

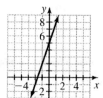

23. $8y - 7x + 3 = -4x + 5y - 9$

$$8y - 5y = -4x + 7x - 12$$

$$\frac{3y}{3} = \frac{3x - 12}{3}$$

$$y = x - 4$$

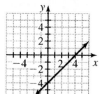

25. $-3(y - 5) = 2(3x - 6)$

$$-3y + 15 = 6x - 12$$

$$\frac{-3y}{-3} = \frac{6x - 27}{-3}$$

$$y = -2x + 9$$

27. $6x - 3(2y - 3) = y - 2(4x - 1)$

$$6x - 6y + 9 = y - 8x + 2$$

$$-6y - y = -8x - 6x - 7$$

$$\frac{-7y}{-7} = \frac{-14x - 7}{-7}$$

$$y = 2x + 1$$

29.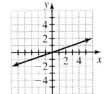

31.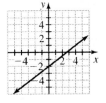

33.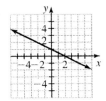

35. a.   i.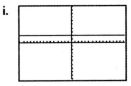

ii.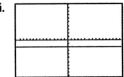

iii.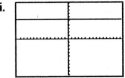

b. The graph is a horizontal line with y-intercept $(0, b)$.

37.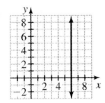

39.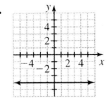

41.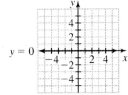

43.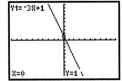

$(0, 1)$ is a point on the line.

$$y = -3x + 1$$
$$1 \overset{?}{=} -3(0) + 1$$
$$1 \overset{?}{=} 1 \quad \text{true}$$

45.
$$0.83x = 4.98y - 2$$
$$\frac{0.83x + 2}{4.98} = \frac{4.98y}{4.98}$$
$$y = \frac{(0.83x + 2)}{4.98}$$

$(0, 0.40160643)$ is a point on the line.

$$y \overset{?}{=} \frac{(0.83x + 2)}{4.98}$$
$$0.40160643 \overset{?}{=} \frac{(0.83(0) + 2)}{4.98}$$
$$0.40160643 \overset{?}{=} \frac{2}{4.98}$$
$$0.40160643 \overset{?}{=} 0.40160643 \quad \text{true}$$

47. $x - 1 + 2x = 3x - 9x + 17$
$$3x - 1 = -6x + 17$$
$$\frac{9x}{9} = \frac{18}{9}$$
$$x = 2$$

49. $-2(3w+5) = 3w-4$

$-6w-10 = 3w-4$

$\dfrac{-9w}{-9} = \dfrac{6}{-9}$

$w = -\dfrac{2}{3}$

51. $4-6(2-3x) = 2x-(4-5x)$

$4-12+18x = 2x-4+5x$

$-8+18x = 7x-4$

$\dfrac{11x}{11} = \dfrac{4}{11}$

$x = \dfrac{4}{11}$

53. $4(r-2)-3(r-1) = 2(r+6)$

$4r-8-3r+3 = 2r+12$

$r-5 = 2r+12$

$-17 = r$

55. $\dfrac{1}{2}x+\dfrac{1}{3} = \dfrac{5}{2}$

$6\left(\dfrac{1}{2}x+\dfrac{1}{3}\right) = 6\left(\dfrac{5}{2}\right)$

$3x+2 = 15$

$\dfrac{3x}{3} = \dfrac{13}{3}$

$x = \dfrac{13}{3}$

57. $-\dfrac{5}{6}b+\dfrac{3}{4} = \dfrac{1}{2}b-\dfrac{2}{3}$

$12\left(-\dfrac{5}{6}b+\dfrac{3}{4}\right) = 12\left(\dfrac{1}{2}b-\dfrac{2}{3}\right)$

$-10b+9 = 6b-8$

$\dfrac{-16b}{-16} = \dfrac{-17}{-16}$

$b = \dfrac{17}{16}$

59. $2.75x-3.95 = -6.21x+74.92$

$\dfrac{8.96x}{8.96} = \dfrac{78.87}{8.96}$

$x \approx 8.80$

61. $P = 2L+2W$

$\dfrac{P-2W}{2} = \dfrac{2L}{2}$

$L = \dfrac{P-2W}{2}$

63. $ax+by = c$

$\dfrac{by}{b} = \dfrac{c-ax}{b}$

$y = \dfrac{c-ax}{b}$

65. $y = 2x+10$

To find the *x*-intercept, let $y = 0$ and solve for *x*.

$0 = 2x+10$

$-10 = 2x$

$-5 = x$

The *x*-intercept is (−5, 0). To find the *y*-intercept, let $x = 0$ and solve for *y*.

$y = 2(0)+10 = 0+10 = 10$

The *y*-intercept is (0, 10).

67. $2x+3y = 12$

To find the *x*-intercept, let $y = 0$ and solve for *x*.

$2x+3(0) = 12$

$2x+0 = 12$

$2x = 12$

$x = 6$

The *x*-intercept is (6, 0). To find the *y*-intercept, let $x = 0$ and solve for *y*.

$2(0)+3y = 12$

$0+3y = 12$

$3y = 12$

$y = 4$

The *y*-intercept is (0, 4).

69. $y = 3x$

To find the *x*-intercept, let $y = 0$ and solve for *x*.

$0 = 3x$

$0 = x$

The *x*-intercept is (0, 0). To find the *y*-intercept, let $x = 0$ and solve for *y*.

$y = 3(0) = 0$

The *y*-intercept is (0, 0).

71. $y = 3$

Since $y = 3$ is a horizontal line, it never intersects the *x*-axis. Therefore, there is no *x*-intercept. Since the graph passes through (0, 3), this is the *y*-intercept.

73. $y = mx + b$

To find the *x*-intercept, let $y = 0$ and solve for *x*.

$0 = mx + b$

$-b = mx$

$-\dfrac{b}{m} = x$

The *x*-intercept is $\left(-\dfrac{b}{m}, 0\right)$. To find the

y-intercept, let $x = 0$ and solve for *y*.

$y = m(0) + b = b$

The *y*-intercept is $(0, b)$.

75. $a(bx + y) = c$

To find the *x*-intercept, let $y = 0$ and solve for *x*.

$a(bx + 0) = c$

$abx = c$

$x = \dfrac{c}{ab}$

The *x*-intercept is $\left(\dfrac{c}{ab}, 0\right)$. To find the

y-intercept, let $x = 0$ and solve for *y*.

$a(b \cdot 0 + y) = c$

$ay = c$

$y = \dfrac{c}{a}$

The *y*-intercept is $\left(0, \dfrac{c}{a}\right)$.

77. $ax = b(cy - d)$

To find the *x*-intercept, let $y = 0$ and solve for *x*.

$ax = b(c \cdot 0 - d)$

$ax = -bd$

$x = -\dfrac{bd}{a}$

The *x*-intercept is $\left(-\dfrac{bd}{a}, 0\right)$. To find the

y-intercept, let $x = 0$ and solve for *y*.

$a \cdot 0 = b(cy - d)$

$0 = bcy - bd$

$bd = bcy$

$\dfrac{bd}{bc} = y$

$\dfrac{d}{c} = y$

The *y*-intercept is $\left(0, \dfrac{d}{c}\right)$.

79. $\dfrac{x}{a} + \dfrac{y}{b} = 1$

To find the *x*-intercept, let $y = 0$ and solve for *x*.

$\dfrac{x}{a} + \dfrac{0}{b} = 1$

$a\left(\dfrac{x}{a}\right) = a(1)$

$x = a$

The *x*-intercept is $(a, 0)$. To find the

y-intercept, let $x = 0$ and solve for *y*.

$\dfrac{0}{a} + \dfrac{y}{b} = 1$

$b\left(\dfrac{y}{b}\right) = b(1)$

$y = b$

The *y*-intercept is $(0, b)$.

81. Answers may vary. Example:

x	*y*
−5	4
−3	3
−1	2
1	1
3	0

83. $y = -3$

85. $y \approx 3$

87. $y \approx -2.5$

89. $x \approx 6$

91. $x \approx -2$

93. $x \approx -5$

95. **a.**

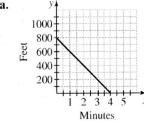

b. To find the *y*-intercept, let $x = 0$ and solve for *y*.

$y = -200(0) + 800 = 0 + 800 = 800$

The *y*-intercept is $(0, 800)$. This means that before the person begins to lower the balloon, the height is 800 feet.

c. To find the *x*-intercept, let *y* = 0 and solve for *x*.

$$0 = -200x + 800$$
$$-800 = -200x$$
$$4 = x$$

The *x*-intercept is (4, 0). This means that after 4 minutes, the balloon will be lowered to the ground (altitude = 0).

97. a.

Yes, this graph is a line with slope –4.1 and *y*-intercept 8.7.

b.

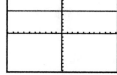

Yes, this graph is a horizontal line passing through (0, 6).

c. All graphs will be lines.
Answers may vary. Example:
$$y = 2x + 3$$

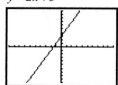

99. Substitute for *x* and *y* in the equation and solve for *b*.

$$y = 2x + b$$
$$5 = 2(7) + b$$
$$5 = 14 + b$$
$$-9 = b$$

101. The points B, C and F satisfy the equation.

103.

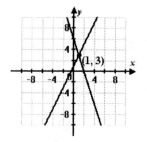

The ordered pair (1, 3) satisfies both equations, since this point lies on the intersection of the lines.

105. Answers may vary. Example:
The values of *x* must be *a*, but *y* can have any value. All such points lie on a vertical line.

107. Answers may vary. Example:
To sketch the graph of a linear equation, create a table of ordered pair solutions of the equation by substituting values of one variable and solving for the other variable. Then, plot the ordered pairs from the table and sketch the line that passes through these plotted points. The graph of an equation is all points that satisfy the equation.

Homework 1.3

1. $m_A = \dfrac{120}{4000} = 0.03$

$m_B = \dfrac{160}{6500} \approx 0.025$

Road A is steeper than road B, since the slope of road A is greater than the slope of road B.

3. $m_A = \dfrac{\left|-90\right|}{300} = \dfrac{90}{300} = 0.3$

$m_B = \dfrac{\left|-125\right|}{450} = \dfrac{125}{450} \approx 0.278$

Ski run A is steeper than ski run B, since the slope of ski run A is greater than the slope of ski run B. (In determining which is steeper, ignore the sign of the slope by considering the absolute value.)

5. $m = \dfrac{y_2 - y_1}{x_2 - x_1} = \dfrac{9 - 3}{5 - 2} = \dfrac{6}{3} = 2$

Since *m* is positive, the line is increasing.

7. $m = \dfrac{y_2 - y_1}{x_2 - x_1} = \dfrac{3 - 7}{1 - (-5)} = \dfrac{-4}{6} = -\dfrac{2}{3}$

Since *m* is negative, the line is decreasing.

9. $m = \dfrac{y_2 - y_1}{x_2 - x_1} = \dfrac{-2 - 10}{2 - (-4)} = \dfrac{-12}{2 + 4} = \dfrac{-12}{6} = -2$

Since *m* is negative, the line is decreasing.

11. $m = \dfrac{y_2 - y_1}{x_2 - x_1} = \dfrac{-4 - (-2)}{7 - 1} = \dfrac{-2}{6} = -\dfrac{1}{3}$

Since *m* is negative, the line is decreasing.

13. $m = \dfrac{y_2 - y_1}{x_2 - x_1} = \dfrac{(-2) - (-8)}{4 - (-5)} = \dfrac{6}{9} = \dfrac{2}{3}$

Since m is positive, the line is increasing.

15. $m = \dfrac{y_2 - y_1}{x_2 - x_1} = \dfrac{-1 - (-9)}{-2 - (-4)} = \dfrac{8}{2} = 4$

Since m is positive, the line is increasing.

17. $m = \dfrac{y_2 - y_1}{x_2 - x_1} = \dfrac{1 - 0}{1 - 0} = \dfrac{1}{1} = 1$

Since m is positive, the line is increasing.

19. $m = \dfrac{y_2 - y_1}{x_2 - x_1} = \dfrac{6 - 6}{7 - 2} = \dfrac{0}{5} = 0$

Since $m = 0$, the line is horizontal.

21. $m = \dfrac{y_2 - y_1}{x_2 - x_1} = \dfrac{5 - (-2)}{-6 - (-6)} = \dfrac{7}{0} =$ undefined

Since m is undefined, the line is vertical.

23. $m = \dfrac{y_2 - y_1}{x_2 - x_1} = \dfrac{-2 - 0}{0 - 5} = \dfrac{-2}{-5} = \dfrac{2}{5}$

Since m is positive, the line is increasing.

25. $m = \dfrac{y_2 - y_1}{x_2 - x_1} = \dfrac{2.6 - 5.4}{3.9 - 1.2} = \dfrac{-2.8}{2.7} \approx -1.04$

Since m is negative, the line is decreasing.

27. $m = \dfrac{y_2 - y_1}{x_2 - x_1} = \dfrac{-2.34 - (-17.94)}{21.13 - 8.94} = \dfrac{15.6}{12.19} \approx 1.28$

Since m is positive, the line is increasing.

29. The points $(3, 0)$ and $(-2, 2)$ lie on the line.

$m = \dfrac{y_2 - y_1}{x_2 - x_1} = \dfrac{2 - 0}{-2 - 3} = \dfrac{2}{-5} = -\dfrac{2}{5}$

The slope of the line is $-\dfrac{2}{5}$.

31. Since $m_1 = m_2$, the lines are parallel.

33. The lines are neither parallel nor perpendicular.

35. Since $m_2 = -\dfrac{1}{m_1}$, the lines are perpendicular.

37. The lines are neither parallel nor perpendicular.

39. Since $m_1 = 0$, l_1 is a horizontal line. Since m_2 is undefined, l_2 is a vertical line. From geometry, any horizontal line is perpendicular to any vertical line. Therefore, the lines are perpendicular.

41. No, the lines are not perpendicular because their slopes, $-\dfrac{1}{4}$ and 3, are not opposite reciprocals.

43. Answers may vary. Example:

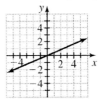

45. Answers may vary. Example:

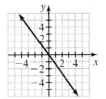

47. Answers may vary. Example:

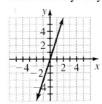

49. Answers may vary. Example:

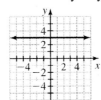

51. a. The slope is defined since it is increasing. The slope is positive.

    b. The slope is defined since it is decreasing. The slope is negative.

    c. The slope is defined since it is a horizontal line. The slope is zero.

    d. The slope of the line is undefined. The line is vertical.

53. Answers may vary. Example:

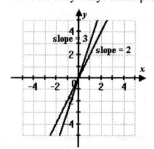

The line with slope 3 is steeper.

55. Answers may vary. Example:

The points $(-6, 2)$ and $(6, 3)$ are on the line and can be used to find the slope.

$$m = \frac{y_2 - y_1}{x_2 - x_1} = \frac{3-2}{6-(-6)} = \frac{1}{12}$$

57. Answers may vary. Example:

The points $(2, 6)$ and $(3, -6)$ are on the line and can be used to find the slope.

$$m = \frac{y_2 - y_1}{x_2 - x_1} = \frac{-6-6}{3-2} = \frac{-12}{1} = -12$$

59. Answers may vary. Example:

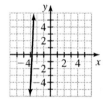

61. Answers may vary. Example:

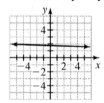

63. The student exchanged the x-coordinates with the y-coordinates. So, the correct slope is $\frac{3}{2}$ instead of $\frac{2}{3}$ since $m = \frac{y_2 - y_1}{x_2 - x_1} = \frac{8-5}{4-2} = \frac{3}{2}$.

65.

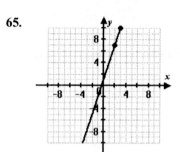

Answers may vary. Example:
Three points that lie on the line are $(1, 4)$, $(0, 1)$, and $(-1, -2)$.

67. a.   i.

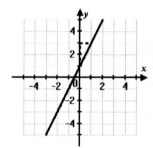

$$\frac{\text{rise}}{\text{run}} = \frac{2}{1} = 2$$

ii.

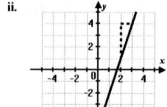

$$\frac{\text{rise}}{\text{run}} = \frac{3}{1} = 3$$

iii.

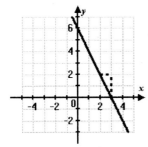

$$\frac{\text{rise}}{\text{run}} = \frac{2}{-1} = -2$$

b. The slope of the line found using $\dfrac{\text{rise}}{\text{run}}$ is the same as the coefficient of x in each of the corresponding equations.

69. a.

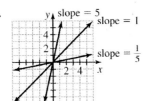

b.

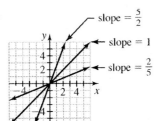

c.

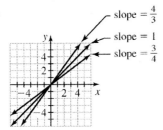

d. The graphs in parts (a)–(c) are mirror images of each other across the graph of $y = x$.

e.

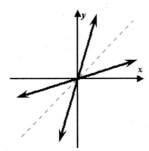

71. The line contains both points, S and Q. Answers may vary. Example:

The slope is $-\dfrac{2}{3}$ and can be written

$\dfrac{\text{rise}}{\text{run}} = \dfrac{-2}{3} = \dfrac{2}{-3}$. From point P, moving 3 units to the right and 2 units down is the equivalent of $\dfrac{\text{rise}}{\text{run}} = \dfrac{-2}{3}$. From point P, moving 3 units to the left and 2 units up is the equivalent of $\dfrac{\text{rise}}{\text{run}} = \dfrac{2}{-3}$.

73. Answers may vary. Example:
A decreasing line goes downward from left to right. Its run is positive and its rise is negative. The slope is $\dfrac{\text{rise}}{\text{run}}$, which is a negative number divided by a positive number, which is negative.

Homework 1.4

1. Since $y = 6x + 1$ is in the form $y = mx + b$, the slope is $m = 6 = \dfrac{6}{1} = \dfrac{\text{rise}}{\text{run}}$, and the y-intercept is $(0, 1)$.

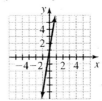

3. Since $y = -2x + 7$ is in the form $y = mx + b$, the slope is $m = -2 = \dfrac{-2}{1} = \dfrac{\text{rise}}{\text{run}}$, and the y-intercept is (0, 7).

5. Since $y = \dfrac{5}{4}x - 2$ is in the form $y = mx + b$, the slope is $m = \dfrac{5}{4} = \dfrac{\text{rise}}{\text{run}}$, and the y-intercept is (0, −2).

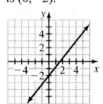

7. Since $y = -\dfrac{3}{7}x + 2$ is in the form $y = mx + b$, the slope is $m = -\dfrac{3}{7} = \dfrac{-3}{7} = \dfrac{\text{rise}}{\text{run}}$, and the y-intercept is (0, 2).

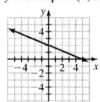

9. Since $y = -\dfrac{5}{3}x - 1$ is in the form $y = mx + b$, the slope is $m = -\dfrac{5}{3} = \dfrac{-5}{3} = \dfrac{\text{rise}}{\text{run}}$, and the y-intercept is (0, −1).

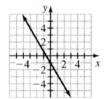

11. First, rewrite $y + x = 5$ in slope–intercept form.
$$y + x = 5$$
$$y = -x + 5$$
The slope is $m = -1 = \dfrac{-1}{1} = \dfrac{\text{rise}}{\text{run}}$, and the y-intercept is (0, 5).

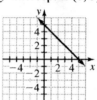

13. First, rewrite $-7x + 2y = 10$ in slope–intercept form.
$$-7x + 2y = 10$$
$$2y = 7x + 10$$
$$y = \dfrac{7}{2}x + 5$$
The slope is $m = \dfrac{7}{2} = \dfrac{\text{rise}}{\text{run}}$, and the y-intercept is (0, 5).

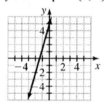

15. First, rewrite $3(x - 2y) = 9$ in slope–intercept form.
$$3(x - 2y) = 9$$
$$3x - 6y = 9$$
$$-6y = -3x + 9$$
$$y = \dfrac{1}{2}x - \dfrac{3}{2}$$
The slope is $m = \dfrac{1}{2} = \dfrac{\text{rise}}{\text{run}}$, and the y-intercept is $\left(0, -\dfrac{3}{2}\right)$.

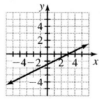

17. First, rewrite $2x - 3y + 9 = 12$ in slope–intercept form.
$$2x - 3y + 9 = 12$$
$$-3y = -2x + 3$$
$$y = \frac{2}{3}x - 1$$
The slope is $m = \frac{2}{3} = \frac{\text{rise}}{\text{run}}$, and the y-intercept is $(0, -1)$.

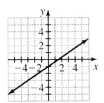

19. First, rewrite $4x - 5y + 3 = 2x - 2y - 3$ in slope–intercept form.
$$4x - 5y + 3 = 2x - 2y - 3$$
$$-3y = -2x - 6$$
$$y = \frac{2}{3}x + 2$$
The slope is $m = \frac{2}{3} = \frac{\text{rise}}{\text{run}}$, and the y-intercept is $(0, 2)$.

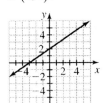

21. First, rewrite $1 - 3(y - 2x) = 7 + 3(x - 3y)$ in slope–intercept form.
$$1 - 3(y - 2x) = 7 + 3(x - 3y)$$
$$1 - 3y + 6x = 7 + 3x - 9y$$
$$6y = -3x + 6$$
$$y = -\frac{1}{2}x + 1$$
The slope is $m = -\frac{1}{2} = \frac{-1}{2} = \frac{\text{rise}}{\text{run}}$, and the y-intercept is $(0, 1)$.

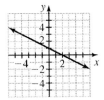

23. Rewrite $y = 4x$ as $y = 4x + 0$ to obtain slope–intercept form. The slope is $m = 4 = \frac{4}{1} = \frac{\text{rise}}{\text{run}}$, and the y-intercept is $(0, 0)$.

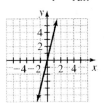

25. Since $y = -1.5x + 3$ is in the form $y = mx + b$, the slope is $m = -1.5 = \frac{-3}{2} = \frac{\text{rise}}{\text{run}}$, and the y-intercept is $(0, 3)$.

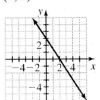

27. Rewrite $y = x$ as $y = 1x + 0$ to obtain slope–intercept form. The slope is $m = 1 = \frac{1}{1} = \frac{\text{rise}}{\text{run}}$, and the y-intercept is $(0, 0)$.

29. The linear equation $y = 4$ is a horizontal line. The slope of a horizontal line is $m = 0$, and the y-intercept is $(0, 4)$.

31. Solve for y.
$$y + 2 = 0$$
$$y = -2$$

The linear equation $y = -2$ is a horizontal line. The slope of a horizontal line is $m = 0$, and the y-intercept is $(0, -2)$.

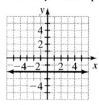

33. Solve for y.
$$ax - by = c$$
$$-by = -ax + c$$
$$y = \frac{a}{b}x - \frac{c}{b}$$

The slope is $\dfrac{a}{b}$ and the y-intercept is $\left(0, -\dfrac{c}{b}\right)$.

35. Solve for y.
$$ay = b(x - d)$$
$$y = \frac{b}{a}(x - d)$$
$$y = \frac{b}{a}x - \frac{bd}{a}$$

The slope is $\dfrac{b}{a}$ and the y-intercept is $\left(0, -\dfrac{bd}{a}\right)$.

37. Solve for y.
$$a(y + b) = x$$
$$ay + ab = x$$
$$ay = x - ab$$
$$y = \frac{1}{a}x - b$$

The slope is $\dfrac{1}{a}$ and the y-intercept is $(0, -b)$.

39. Solve for y.
$$a(x - y) = d$$
$$x - y = \frac{d}{a}$$
$$-y = -x + \frac{d}{a}$$
$$y = x - \frac{d}{a}$$

The slope is 1 and the y-intercept is $\left(0, -\dfrac{d}{a}\right)$.

41. Solve for y.
$$\frac{x}{a} + \frac{y}{a} = 1$$
$$x + y = a$$
$$y = -x + a$$

The slope is -1 and the y-intercept is $(0, a)$.

43. Set 1 is not linear. As x increases by 1, y_1 does not change by some consistent value. There is a line that comes close to every point since the changes in y_1 are roughly the same.

Set 2 is linear. As x increases by 1, y_2 decreases by -0.3. The slope is $\dfrac{-0.3}{1} = -0.3$.

Set 3 is not linear. The value of x does not change (increase or decrease) consistently, even though y_3 consistently increases by 5, so no line comes close to every point.

Set 4 is linear. As x increases by 1, y_4 decreases by 10. The slope is $-\dfrac{10}{1} = -10$.

45.

Eq. 1		Eq. 2		Eq. 3		Eq. 4	
x	y	x	y	x	y	x	y
1	12	23	69	1	47	30	15
2	15	24	53	2	41	31	24
3	18	25	37	3	35	32	33
4	21	26	21	4	29	33	42
5	24	27	5	5	23	34	51
6	27	28	-11	6	17	35	60

47. For both lines, the slope is 4. The lines are parallel.

49. The slopes are $\dfrac{3}{8}$ and $\dfrac{8}{3}$. The slopes are reciprocals, but not negative reciprocals. The lines are neither parallel nor perpendicular.

51. Solve each equation for y.

$$2x + 3y = 6 \qquad\qquad 4x + 6y = 7$$
$$3y = -2x + 6 \qquad\qquad 6y = -4x + 7$$
$$y = -\frac{2}{3}x + 2 \qquad\qquad y = -\frac{2}{3}x + \frac{7}{6}$$

Since the slopes are both $m = -\frac{2}{3}$, the lines are parallel.

53. Solve each equation for y.

$$5x - 3y = 1 \qquad\qquad 3x + 5y = -2$$
$$-3y = -5x + 1 \qquad\qquad 5y = -3x - 2$$
$$y = \frac{5}{3}x - \frac{1}{3} \qquad\qquad y = -\frac{3}{5}x - \frac{2}{5}$$

Since the slopes are negative reciprocals, the lines are perpendicular.

55. $x = -3$ and $x = 1$ are both vertical lines. All vertical lines are parallel.

57. $x = 0$ and $y = 0$ are the equations for the vertical and horizontal axes, respectively. These lines are perpendicular.

59. a. Substitute values for x in the equation $y = -3x + 18$ to solve for y.

When $x = 0$, $y = -3(0) + 18 = 0 + 18 = 18$.

When $x = 1$, $y = -3(1) + 18 = -3 + 18 = 15$.

When $x = 2$, $y = -3(2) + 18 = -6 + 18 = 12$.

When $x = 3$, $y = -3(3) + 18 = -9 + 18 = 9$.

When $x = 4$, $y = -3(4) + 18 = -12 + 18 = 6$.

When $x = 5$, $y = -3(5) + 18 = -15 + 18 = 3$.

When $x = 6$, $y = -3(6) + 18 = -18 + 18 = 0$.

Driving Time (hours) x	Amount of Gas (gallons) y
0	18
1	15
2	12
3	9
4	6
5	3
6	0

b. Each hour, the amount of gas in the tank decreases by 3 gallons. The slope of -3 in the equation shows this decrease of 3. As the value of the independent variable (driving time) increases by 1, the value of the dependent variable (amount of gas) decreases by 3 as determined by the slope.

c. Note that in one hour's time, 3 gallons of gas is used. If the person is driving at approximately 60 mph, he or she uses 3 gallons to drive 60 miles. So for 1 gallon of gas, a person can travel 20 miles $(60 \div 3)$.

61. a. Substitute values for x in the equation $y = 2x + 26$ to solve for y and complete the table.

When $x = 0$, $y = 2(0) + 26 = 0 + 26 = 26$.

When $x = 1$, $y = 2(1) + 26 = 2 + 26 = 28$.

Similar calculations yield the following table.

Time at Company (years) x	Salary (thousands of dollars) y
0	26
1	28
2	30
3	32
4	34

b. Each year the person's salary increases by $2000, which corresponds to the slope of $y = 2x + 26$ with y in thousands of dollars. As the value of the independent variable (time at company) increases by 1, the value of the dependent variable (salary) increases by 2 as determined by the slope.

63. a. Since this is a decreasing line, the slope, m, is negative ($m < 0$). The line crosses the y-axis above the origin so the y-intercept is positive ($b > 0$).

b. Since this is as increasing line, the slope, m, is positive ($m > 0$). The line crosses the y-axis below the origin so the y-intercept is negative ($b < 0$).

c. Since the line is horizontal and crosses the y-axis below the origin, the slope, m, is 0 and the y-intercept, b, is negative ($b < 0$).

d. Since this is a decreasing line that passes through the origin, the slope, m, is negative ($m < 0$), and the y-intercept, b, is 0.

65. The slope is $m = \dfrac{\text{rise}}{\text{run}} = \dfrac{-2}{1}$. The y-intercept is $(0,3)$. The equation of the line is $y = -2x + 3$.

67. Answers may vary. Example:
$$y = \frac{1}{100}x + 2$$
The slope is a small positive number, and the y-intercept is positive.

69. Answers may vary. Example:
$$y = -700x - 2$$
The slope is a large negative number, and the y-intercept is negative.

71. a.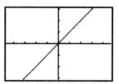

b.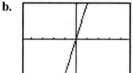

c.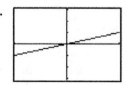

d. The slope of the graph in part (b) appears to be steeper than the graph in part (a). The graph of part (c) appears to be less steep than the graph in part (a).

e. No, you cannot make the sketch of $y = x$ appear to be a decreasing line or cross the y-axis at a point other than (0, 0) by changing the window settings. The line will always appear to be increasing and it will always lie in quadrants one and three, but it may appear to have any steepness within these constraints.

73. Since we have the slope, $m = 2$, and a point on the line, $(0, -3)$, we can substitute in the equation $y = mx + b$ to find the y-intercept, b.
$$y = mx + b$$
$$-3 = 2(0) + b$$
$$-3 = b$$
We can now write the equation of the line in slope–intercept form.
$$y = 2x + (-3)$$
$$y = 2x - 3$$

75. a.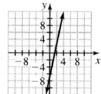

b. From the graph in part (a), we see that the y-intercept, b, is -7. Using the slope and the y-intercept, we can write the equation of the line in slope–intercept form.
$$y = 5x + (-7)$$
$$y = 5x - 7$$

c.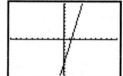

77. a.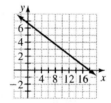

b. From the graph in part (a), we see that the y-intercept, b, is approximately $6\frac{1}{2}$. Using the slope and the y-intercept, we can write an approximate equation of the line:
$$y = -\frac{3}{8}x + \frac{13}{2}.$$ The exact equation of the line is $y = -\frac{3}{8}x + \frac{27}{4}$.

c.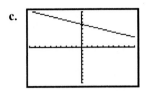

79. a. Each line is a horizontal line with a slope of 0.

b. The slope for the graph for a linear equation of the form $y = k$ is 0 when k is a constant.

81. The student is not correct.
Answers may vary. Example:
The coefficient of x is the slope of a line when the equation is written in slope–intercept form. The equation, $2x + 3y = 6$, is not written in slope intercept form. To find the slope, rewrite the equation in slope–intercept form by solving for y.

$$2x + 3y = 6$$
$$3y = -2x + 6$$
$$y = -\frac{2}{3}x + \frac{6}{3}$$
$$y = -\frac{2}{3}x + 2$$

The slope of the line is $-\frac{2}{3}$.

83. Answers may vary. Example:
The y-intercept is the point where the line crosses the y-axis. The x-coordinate of any point on the y-axis is 0. Substitute 0 for x in $y = mx + b$:

$$y = m(0) + b$$
$$y = b$$

Therefore, the y-intercept is $(0, b)$.

85. Answers may vary. Example:

$$y_1 = mx + b$$
$$y_2 = m(x + 1) + b$$
$$y_2 = mx + m + b$$
$$y_2 = y_1 + m$$

Homework 1.5

1. We are given the slope, $m = 3$, and a point on the line, $(5, 2)$. Use $y = mx + b$ to find b.

$$y = mx + b$$
$$2 = 3(5) + b$$
$$2 = 15 + b$$
$$-13 = b$$

Now, substitute for m and b in slope–intercept form to obtain the equation of the line.

$$y = 3x + (-13)$$
$$y = 3x - 13$$

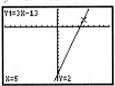

The sign of m (positive) agrees with the increasing line from the graphing calculator screen.

3. We are given the slope, $m = -2$, and a point on the line, $(3, -9)$. Use $y = mx + b$ to find b.

$$y = mx + b$$
$$-9 = -2(3) + b$$
$$-9 = -6 + b$$
$$-3 = b$$

Now, substitute for m and b in slope–intercept form to obtain the equation of the line.

$$y = -2x + (-3)$$
$$y = -2x - 3$$

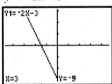

The sign of m (negative) agrees with the decreasing line from the graphing calculator screen.

5. We are given the slope, $m = \dfrac{3}{5}$, and a point on the line, (20, 7). Use $y = mx + b$ to find b.

$$y = mx + b$$
$$7 = \frac{3}{5}(20) + b$$
$$7 = 12 + b$$
$$-5 = b$$

Now, substitute for m and b in slope–intercept form to obtain the equation of the line.

$$y = \frac{3}{5}x + (-5)$$
$$y = \frac{3}{5}x - 5$$

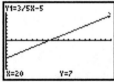

The sign of m (positive) agrees with the increasing line shown on the graphing calculator screen.

7. We are given the slope, $m = -\dfrac{1}{6}$, and a point on the line, (2, –3). Use $y = mx + b$ to find b.

$$y = mx + b$$
$$-3 = -\frac{1}{6}(2) + b$$
$$-3 = -\frac{1}{3} + b$$
$$-\frac{9}{3} = -\frac{1}{3} + b$$
$$-\frac{8}{3} = b$$

Now, substitute for m and b in slope–intercept form to obtain the equation of the line.

$$y = -\frac{1}{6}x + \left(-\frac{8}{3}\right)$$
$$y = -\frac{1}{6}x - \frac{8}{3}$$

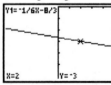

The sign of m (negative) agrees with the decreasing line shown on the graphing calculator screen.

9. We are given the slope, $m = -\dfrac{5}{2}$, and a point on the line, (–3, –4). Use $y = mx + b$ to find b.

$$y = mx + b$$
$$-4 = -\frac{5}{2}(-3) + b$$
$$-4 = \frac{15}{2} + b$$
$$-\frac{8}{2} = \frac{15}{2} + b$$
$$-\frac{23}{2} = b$$

Now, substitute for m and b in slope–intercept form to obtain the equation of the line.

$$y = -\frac{5}{2}x + \left(-\frac{23}{2}\right)$$
$$y = -\frac{5}{2}x - \frac{23}{2}$$

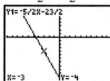

The sign of m (negative) agrees with the decreasing line shown on the graphing calculator screen.

11. We are given the slope, $m = 0$, and a point on the line, (1, 2). We know that a line with a slope of 0 is a horizontal line. Horizontal lines are of the form $y = b$. In this example, $y = 2$.

Since m is zero, the line is horizontal. This is shown on the graphing calculator screen.

13. We are given that m is undefined, and a point on the line is (3, 7). We know that a line with an undefined slope is a vertical line. Vertical lines are of the form $x = a$. In this example, $x = 3$.

Since m is undefined, the line is vertical. This is shown on the graphing calculator screen.

15. We are given the slope, $m = 1.6$, and a point on the line, $(2.1, 3.8)$. Use $y = mx + b$ to find b.

$$y = mx + b$$
$$3.8 = 1.6(2.1) + b$$
$$3.8 = 3.36 + b$$
$$0.44 = b$$

Now, substitute for m and b in slope–intercept form to obtain the equation of the line.
$$y = 1.6x + 0.44$$

To check, substitute $(2.1, 3.8)$ into the equation.
$$y = 1.6x + 0.44$$
$$3.8 \overset{?}{=} 1.6(2.1) + 0.44$$
$$3.8 \overset{?}{=} 3.8 \text{ true}$$

17. We are given the slope, $m = -3.24$, and a point on the line, $(-5.28, 1.93)$. Use $y = mx + b$ to find b.

$$y = mx + b$$
$$1.93 = -3.24(-5.28) + b$$
$$1.93 = 17.1072 + b$$
$$-15.18 \approx b$$

Substitute m and b in slope–intercept form to obtain the equation of the line.
$$y = -3.24x - 15.18$$

To check, substitute $(-5.28, 1.93)$ into the equation.
$$y = -3.24x - 15.18$$
$$1.93 \overset{?}{=} -3.24(-5.28) - 15.18$$
$$1.93 \overset{?}{=} 1.93 \text{ true}$$

19. First, find the slope.
$$m = \frac{5-3}{4-2} = \frac{2}{2} = 1$$
So, $y = 1x + b$. Since the line contains $(2, 3)$, substitute 2 for x and 3 for y and solve for b.
$$3 = 1(2) + b$$
$$3 = 2 + b$$
$$1 = b$$
So, the equation is $y = x + 1$.

21. First, find the slope.
$$m = \frac{6-(-4)}{-2-3} = \frac{6+4}{-5} = \frac{10}{-5} = -2$$
So, $y = -2x + b$. Since the line contains $(-2, 6)$, substitute -2 for x and 6 for y and solve for b.
$$6 = -2(-2) + b$$
$$6 = 4 + b$$
$$2 = b$$
So, the equation is $y = -2x + 2$.

23. First, find the slope.
$$m = \frac{-6-(-14)}{-8-(-4)} = \frac{-6+14}{-8+4} = \frac{8}{-4} = -2$$
So, $y = -2x + b$. Since the line contains $(-8, -6)$, substitute -8 for x and -6 for y and solve for b.
$$-6 = -2(-8) + b$$
$$-6 = 16 + b$$
$$-22 = b$$
So, the equation is $y = -2x - 22$.

25. First, find the slope.
$$m = \frac{1-0}{1-0} = \frac{1}{1} = 1$$
So, $y = 1x + b$. Since the line contains $(0, 0)$ substitute 0 for x and 0 for y to solve for b.
$$0 = 1(0) + b$$
$$0 = 0 + b$$
$$0 = b$$
So, the equation is $y = x$.

27. First, find the slope.
$$m = \frac{5-1}{7-2} = \frac{4}{5}$$
So, $y = \frac{4}{5}x + b$. Since the line passes through $(2, 1)$, substitute 2 for x and 1 for y to solve for b.

$$1 = \frac{4}{5}(2) + b$$
$$1 = \frac{8}{5} + b$$
$$\frac{5}{5} = \frac{8}{5} + b$$
$$-\frac{3}{5} = b$$

So, the equation is $y = \frac{4}{5}x - \frac{3}{5}$.

29. First, find the slope.

$$m = \frac{2 - (-5)}{-4 - 2} = \frac{2 + 5}{-6} = \frac{7}{-6} = -\frac{7}{6}$$

So, $y = -\frac{7}{6}x + b$. Since the line passes

through $(-4, 2)$, substitute -4 for x and 2 for y to solve for b.

$$2 = -\frac{7}{6}(-4) + b$$

$$2 = \frac{14}{3} + b$$

$$\frac{6}{3} = \frac{14}{3} + b$$

$$-\frac{8}{3} = b$$

So, the equation is $y = -\frac{7}{6}x - \frac{8}{3}$.

31. First, find the slope.

$$m = \frac{-7 - (-2)}{-5 - (-3)} = \frac{-7 + 2}{-5 + 3} = \frac{-5}{-2} = \frac{5}{2}$$

So, $y = \frac{5}{2}x + b$. Since the line passes through

$(-5, -7)$, substitute -5 for x and -7 for y to solve for b.

$$-7 = \frac{5}{2}(-5) + b$$

$$-7 = -\frac{25}{2} + b$$

$$-\frac{14}{2} = -\frac{25}{2} + b$$

$$\frac{11}{2} = b$$

So, the equation is $y = \frac{5}{2}x + \frac{11}{2}$.

33. First, find the slope.

$$m = \frac{5 - 5}{4 - 2} = \frac{0}{2} = 0$$

So, $y = 0x + b$ or $y = b$. Since this is a horizontal line passing through $(2, 5)$, the equation is $y = 5$.

35. First, find the slope.

$$m = \frac{-4 - 6}{-3 - (-3)} = \frac{-10}{0} \text{ undefined}$$

The slope is undefined. This is a vertical line passing through $(-3, -4)$, so the equation is $x = -3$.

37. First, find the slope.

$$m = \frac{-3.9 - 2.2}{-5.1 - 7.4} = \frac{-6.1}{-12.5} \approx 0.49$$

So, $y = 0.49x + b$. Since the line contains $(7.4, 2.2)$, substitute 7.4 for x and 2.2 for y to solve for b.

$$2.2 = 0.49(7.4) + b$$

$$2.2 = 3.626 + b$$

$$-1.43 \approx b$$

So, the equation is $y = 0.49x - 1.43$.

39. First, find the slope.

$$m = \frac{-6.24 - (-4.05)}{-5.97 - (-1.25)} = \frac{-2.19}{-4.72} \approx 0.46$$

So, $y = 0.46x + b$. Since the line passes through $(-5.97, -6.24)$, substitute -5.97 for x and -6.24 for y to solve for b.

$$-6.24 = 0.46(-5.97) + b$$

$$-6.24 = -2.7462 + b$$

$$-3.49 \approx b$$

So, the equation is $y = 0.46x - 3.49$.

41. The slope of the given line is 3. A line parallel to the given line also has a slope of 3 and an equation $y = 3x + b$. Since the point $(4, 5)$ lies on the parallel line, substitute 4 for x and 5 for y to solve or b.

$$5 = 3(4) + b$$

$$5 = 12 + b$$

$$-7 = b$$

The parallel line's equation is $y = 3x - 7$.

43. The slope of the given line is -2. A line parallel to this has a slope of -2 and an equation $y = -2x + b$. Substitute -3 for x and 8 for y to solve for b since the parallel line contains $(-3, 8)$.

$$8 = -2(-3) + b$$

$$8 = 6 + b$$

$$2 = b$$

The parallel line's equation is $y = -2x + 2$.

45. The slope of the given line is $\frac{1}{2}$. A line parallel to this has a slope of $\frac{1}{2}$ and an equation $y = \frac{1}{2}x + b$. Substitute 4 for x and 1 for y to solve for b since the parallel line contains $(4, 1)$.

$$1 = \frac{1}{2}(4) + b$$
$$1 = 2 + b$$
$$-1 = b$$

The parallel line's equation is $y = \frac{1}{2}x - 1$ or $y = 0.5x - 1$.

47. To find the slope, isolate y.
$$3x - 4y = 12$$
$$-4y = -3x + 12$$
$$y = \frac{3}{4}x - 3$$

The slope is $\frac{3}{4}$. A line parallel to this has the same slope and an equation $y = \frac{3}{4}x + b$.

Substitute 3 for x and 4 for y to solve for b since the parallel line contains $(3, 4)$.

$$4 = \frac{3}{4}(3) + b$$
$$4 = \frac{9}{4} + b$$
$$\frac{16}{4} = \frac{9}{4} + b$$
$$\frac{7}{4} = b$$

The parallel line's equation $y = \frac{3}{4}x + \frac{7}{4}$ or $y = 0.75x + 1.75$.

49. To find the slope, isolate y.
$$6y - x = -7$$
$$6y = x - 7$$
$$y = \frac{1}{6}x - \frac{7}{6}$$

The slope is $\frac{1}{6}$. A line parallel to this has the same slope and an equation $y = \frac{1}{6}x + b$.

Substitute -3 for x and -2 for y to solve for b since the parallel line contains $(-3, -2)$.

$$-2 = \frac{1}{6}(-3) + b$$
$$-2 = -\frac{1}{2} + b$$
$$-\frac{4}{2} = -\frac{1}{2} + b$$
$$-\frac{3}{2} = b$$

The parallel line's equation is $y = \frac{1}{6}x - \frac{3}{2}$.

51. The line $y = 6$ is horizontal and has a slope of 0. A line parallel to $y = 6$ is also horizontal. Since the parallel line contains (2, 3) and horizontal lines are of the form $y = b$, the equation of the line is $y = 3$.

53. The line $x = 2$ is vertical and has undefined slope. A line parallel to $x = 2$ is also vertical. Since vertical lines are of the form $x = a$, and the parallel line contains $(-5, 4)$, the equation of the line is $x = -5$.

55. The slope of the given line is 2. A line perpendicular to the given line must then have a slope of $-\frac{1}{2}$ and an equation $y = -\frac{1}{2}x + b$. Substitute 3 for x and 8 for y to solve for b since the line contains (3, 8).

$$8 = -\frac{1}{2}(3) + b$$
$$8 = -\frac{3}{2} + b$$
$$\frac{16}{2} = -\frac{3}{2} + b$$
$$\frac{19}{2} = b$$

The equation of the line is $y = -\frac{1}{2}x + \frac{19}{2}$ or $y = -0.5x + 9.5$.

57. The slope of the given line is –3. A line perpendicular to the given line must then have a slope of $\frac{1}{3}$ and an equation $y = \frac{1}{3}x + b$.

Substitute –1 for x and 7 for y to solve for b since the line contains (–1, 7).

$$7 = \frac{1}{3}(-1) + b$$

$$7 = -\frac{1}{3} + b$$

$$\frac{21}{3} = -\frac{1}{3} + b$$

$$\frac{22}{3} = b$$

The equation of the line is $y = \frac{1}{3}x + \frac{22}{3}$.

59. The slope of the given line is $-\frac{2}{5}$. A line perpendicular to the given line must then have a slope of $\frac{5}{2}$ and an equation $y = \frac{5}{2}x + b$.

Substitute 2 for x and 7 for y to solve for b since the line contains (2, 7).

$$7 = \frac{5}{2}(2) + b$$

$$7 = 5 + b$$

$$2 = b$$

The equation of the line is $y = \frac{5}{2}x + 2$ or

$y = 2.5x + 2$.

61. To find the slope of the given line, isolate y.

$$4x - 5y = 7$$

$$-5y = -4x + 7$$

$$y = \frac{4}{5}x - \frac{7}{5}$$

The slope of this line is $\frac{4}{5}$. A line perpendicular to this line must have a slope of $-\frac{5}{4}$ and an equation $y = -\frac{5}{4}x + b$. Substitute 10 for x and 3 for y to solve for b since the line contains $(10, 2)$.

$$3 = -\frac{5}{4}(10) + b$$

$$3 = -\frac{25}{2} + b$$

$$\frac{6}{2} = -\frac{25}{2} + b$$

$$\frac{31}{2} = b$$

The equation of the line is $y = -\frac{5}{4}x + \frac{31}{2}$ or

$y = -1.25x + 15.5$.

63. To find the slope of the given line, isolate y.

$$-2x + 3y = 5$$

$$3y = 2x + 5$$

$$\frac{3y}{3} = \frac{2x + 5}{3}$$

$$y = \frac{2}{3}x + \frac{5}{3}$$

The slope of the line is $\frac{2}{3}$. A line perpendicular to this line must have a slope of $-\frac{3}{2}$ and an equation $y = -\frac{3}{2}x + b$. Substitute –3 for x and –1 for y to solve for b since the line contains $(-3, -1)$.

$$y = -\frac{3}{2}x + b$$

$$-1 = -\frac{3}{2}(-3) + b$$

$$-1 = \frac{9}{2} + b$$

$$-\frac{2}{2} = \frac{9}{2} + b$$

$$-\frac{11}{2} = b$$

The equation of the line is $y = -\frac{3}{2}x - \frac{11}{2}$ or

$y = -1.5x - 5.5$.

65. The slope of the equation $x = 5$ is undefined. The graph of the equation is a vertical line. A line perpendicular to $x = 5$ is a horizontal line with a slope of 0. Since this perpendicular line contains (2, 3) and the y-value at this point is 3, the equation of the line is $y = 3$.

67. The slope of the equation $y = -3$ is 0. The graph of the equation is a horizontal line. A line perpendicular to $y = -3$ is a vertical line with an undefined slope. Since this perpendicular line contains $(2, 8)$ and the x-value at this point is 2, the equation of the line is $x = 2$.

69. Choose any two points to find the slope.
$$m = \frac{17 - 19}{1 - 0} = \frac{-2}{1} = -2$$
So, $y = -2x + b$. Since the point $(0, 19)$ is a solution to the equation, substitute 0 for x and 19 for y to solve for b.
$$19 = -2(0) + b$$
$$19 = 0 + b$$
$$19 = b$$
The equation describing the relationship between x and y is $y = -2x + 19$.

71. Find the slope by choosing two points on the line, such as $(2, 0)$ and $(5, 1)$.
$$m = \frac{1 - 0}{5 - 2} = \frac{1}{3}$$
So, $y = \frac{1}{3}x + b$. Since the line contains $(2, 0)$, substitute 2 for x and 0 for y to solve for b.
$$0 = \frac{1}{3}(2) + b$$
$$0 = \frac{2}{3} + b$$
$$-\frac{2}{3} = b$$
The equation for the line is $y = \frac{1}{3}x - \frac{2}{3}$.

73. Choose two points on the line to find the slope, such as $(3, 3)$ and $(5, 0)$.
$$m = \frac{3 - 0}{3 - 5} = \frac{3}{-2} = -\frac{3}{2}$$
So, $y = -\frac{3}{2}x + b$. Since the line contains $(3, 3)$, substitute 3 for x and 3 for y to solve for b.
$$3 = -\frac{3}{2}(3) + b$$
$$\frac{6}{2} = -\frac{9}{2} + b$$
$$\frac{15}{2} = b$$
The equation for the line is $y = -\frac{3}{2}x + \frac{15}{2}$.

75. a. Answers may vary. Example:
It is possible for a line to have no x-intercepts. Horizontal lines of the form $y = b$ (where b is a constant not equal to 0) have no x-intercepts.

b. Answers may vary. Example:
It is possible for a line to have exactly one x-intercept. One example is $y = x + 1$, where the x-intercept is $(-1, 0)$. Other lines of the form $y = mx + b$ have one x-intercept, as long as $m \neq 0$.

c. Answers may vary. Example:
It is not possible for a line to have exactly two x-intercepts. A line can never intersect the x-axis at exactly two points.

d. It is possible for a line to have an infinite number of x-intercepts. The line $y = 0$ lies on the x-axis and therefore, intersects the x-axis at an infinite number of points.

77. Yes, a line contains all these points. Using $(-4, 15)$ and $(-1, 9)$, note that the slope is -2.
So, $y = -2x + b$. Substitute -4 for x and 15 for y to solve for b since the line contains $(-4, 15)$.
$$15 = -2(-4) + b$$
$$15 = 8 + b$$
$$7 = b$$
The equation of the line is $y = -2x + 7$. When each of the other points is substituted into this equation, we see that the line contains all of the given points.

79. a. Answers may vary. Example:
Points on $y = 3x - 6$:

x	y
0	−6
1	−3
2	0
3	3
4	6
5	9
6	12

b. Answers may vary. Example:
Points close to but not on $y = 3x - 6$:

x	y
0	−5
1	−4
2	1
3	2
4	7
5	10
6	13

c. Answers may vary.

81. a. Answers may vary. Example:
Any equation of the form $y = -4x + b$,
where $x \neq 0$ and/or $y \neq 0$, will have a
slope of −4. One example is $y = -4x + 1$

b. Answers may vary. Example:
Any equation of the form $y = mx + \dfrac{3}{7}$
where $y \neq 0$, will have a y-intercept of
$\left(0, \dfrac{3}{7}\right)$.

c. Answers may vary. Example:
Since the line must contain the point
(−2, 8), substitute −2 for x and 8 for y to
find an equation.
$y = mx + b$
$8 = m(-2) + b$
$8 = -2m + b$
Choose any slope and then solve for b, the
y-intercept. Then, use the slope and b to
write the equation of the line. Choose
$m = 2$, for example.
$8 = 2(-2) + b$
$8 = -4 + b$
$12 = b$
An equation of a line that passes through
the point (−2, 8) is $y = 2x + 12$. Use
TRACE to verify your equation.

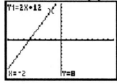

d. The line that has a slope of −4 and
y-intercept $\left(0, \dfrac{3}{7}\right)$ is $y = -4x + \dfrac{3}{7}$. Check
to see if the point (−2, 8) satisfies this
equation.
$8 \overset{?}{=} -4(-2) + \dfrac{3}{7}$
$8 \overset{?}{=} 8 + \dfrac{3}{7}$
$0 \overset{?}{=} \dfrac{3}{7}$ false
This is false. There is no such equation.

83. Answers may vary. Example:
Substitute the point into the equation for a line
and solve for b:
$y = mx + b$
$-2 = 3m + b$
$-b = 3m + 2$
$b = -3m - 2$.
Make a table of values:

m	b
−1	1
0	−2
1	−5

Substitute these values for slope and
y-intercept into the original equation to get the
equations of three lines:
$y = -x + 1$
$y = -2$
$y = x - 5$

85. Answers may vary. Example:
The student is partially correct. In the
$y = mx + b$ form of a line, the slope m is 2 and
the y-intercept b is the point where the line
crosses the y-axis, or $(0, b)$. Thus, the given
point $(3, 5)$ is not the y-intercept. You can use
$(3, 5)$ to find b by substituting it into
$y = 2x + b$ and solving for b.

87. Answers may vary. Example:
First, use the two points to find the slope of the
line. Next, use one of the points to find b, by
substituting in the equation $y = mx + b$.
Finally, substitute for m and b in the equation
$y = mx + b$. To verify that the equation
contains the two points, substitute the values
of x and y for each point to see if a true
statement results.

Homework 1.6

1. Relation 1 is not a function. The input $x = 3$ yields *two* outputs $y = 5$ and $y = 7$.
   Relation 2 could possibly be a function since each input yields only one output.
   Relation 3 could possibly be a function since each input yields only one output.
   Relation 4 is not a function. The input $x = 8$ yields *two* outputs $y = 40$ and $y = 50$.

3. No, the relation is not a function since an input yields more than one output.

5. Yes, it is possible that the relation is a function. Two inputs can yield the same output, but one input cannot yield two outputs.

7. This graph is a function since it passes the vertical line test.

9. This graph is not a function since a vertical line can intersect the graph at more than one point.

11. This graph is not a function since a vertical line can intersect the graph at more than one point.

13. This graph is a function since it passes the vertical line test.

15. The relation $y = 5x - 1$ is function since it can be put into the form $y = mx + b$ which defines a linear function.

17. First, isolate y.
$$2x - 5y = 10$$
$$-5y = -2x + 10$$
$$y = \frac{2}{5}x - 2$$
This relation is a function since it can be put into the form $y = mx + b$ which defines a linear function.

19. $y = 4$ is a horizontal line. Because every horizontal line passes the vertical line test, $y = 4$ is a function.

21. $x = -3$ is a vertical line and does not pass the vertical line test. This is not a function.

23. First, isolate y.
$$7x - 2y = 21 + 3(y - 5x)$$
$$7x - 2y = 21 + 3y - 15x$$
$$-2y - 3y = -7x - 15x + 21$$
$$\frac{-5y}{-5} = \frac{22x + 21}{-5}$$
$$y = -\frac{22}{5}x - \frac{21}{5}$$
This relation is a function since it can be put into the form $y = mx + b$ which defines a linear function.

25. Yes, any nonvertical line is a function since it passes the vertical line test.

27. No, a circle is not the graph of a function since a vertical line may intersect the circle at more than one point.

29. **a.** Answers may vary. Example:

x	y
0	$3(0) - 2 = -2$
1	$3(1) - 2 = 1$
2	$3(2) - 2 = 4$
3	$3(3) - 2 = 7$
4	$3(4) - 2 = 10$

b.

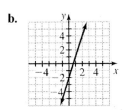

c. For each input–output pair, the output is 2 less than 3 times the input.

31. The domain is $-4 \leq x \leq 5$ and the range is $-2 \leq y \leq 3$.

33. The domain is $-5 \leq x \leq 4$ and the range is $-2 \leq y \leq 3$. The highest point on the graph appears to include a y-value of 3.

35. The domain is $-4 \leq x \leq 4$ and the range is $-2 \leq y \leq 2$. The lowest point on the graph appears to include a y-value of -2 and the highest point appears to include a y-value of 2.

37. The domain is $0 \leq x \leq 4$ and the range is $0 \leq y \leq 2$.

39. The domain is all real numbers and the range is $y \le 4$.

41. The domain is $x \ge 0$ and the range is $y \ge 0$.

43. The relation $y = \sqrt{x}$ is a function since it passes the vertical line test.

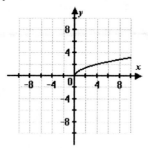

45. Consider the input $x = 16$. Substitute 16 for x and solve for y.

$$y^4 = 16$$
$$y = \pm 2$$

Since the input $x = 16$ yields *two* outputs, the relation, $y^4 = x$, is not a function.

47. Answers may vary. Example:

$y = 2^x$ is a function expressed as an equation.

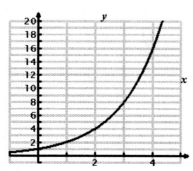

x	y
0	1
1	2
2	4
3	8
4	16

For each input–output pair, the output is 2 raised to the power of the input.

49. Answers may vary. Example:
Suppose that when $x = 2$, $y = -1$ and $y = 4$. Then suppose when $x = 6$, $y = 0$. Sketch these points.

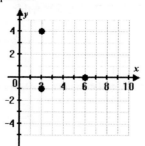

This relation is not a function since it does not pass the vertical line test.

51. Answers may vary. Example:

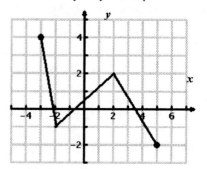

53. Answers may vary. Example:

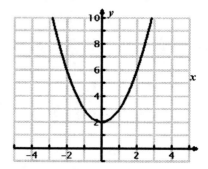

55. Answers may vary. Example:
Any nonvertical line is a linear function, since it passes the vertical line test. A vertical line is not a linear function because there is a vertical line that will intersect its graph in more than one point.

57. No, the student's conclusion is not correct. Answers may vary. Example:
In a function, *two* inputs can yield the same output, but one input cannot yield *two* outputs.

Chapter 1 Review

1. The amount of time it takes to read and reply to emails depends on the number of emails that a person receives. The amount of time to read and reply (t) is the dependent variable and the number of emails (n) is the independent variable.

2. The length of the candle (L) is the dependent variable and minutes (t) is the independent variable.

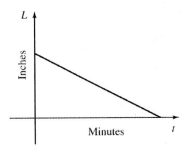

3. The amount of time (T) to cook a marshmallow is the dependent variable and the distance (d) from the campfire is the independent variable.

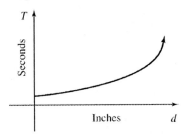

4. $3(2x-4)-2=5x-(3-4x)$
$$6x-12-2=5x-3+4x$$
$$6x-14=9x-3$$
$$\frac{-3x}{-3}=\frac{11}{-3}$$
$$x=-\frac{11}{3}$$

5. $\frac{2}{3}w-\frac{1}{2}=\frac{5}{6}w+\frac{4}{3}$
$$6\left(\frac{2}{3}w-\frac{1}{2}\right)=6\left(\frac{5}{6}w+\frac{4}{3}\right)$$
$$4w-3=5w+8$$
$$-11=w$$

6. $a(x-c)=d$
$$x-c=\frac{d}{a}$$
$$x=\frac{d}{a}+c$$

7. To find the x-intercept, substitute 0 for y.
$$3x-5y=17$$
$$3x-5(0)=17$$
$$\frac{3x}{3x}=\frac{17}{3}$$
$$x=\frac{17}{3}$$
The x-intercept is $\left(\frac{17}{3},0\right)$.

To find the y-intercept, substitute 0 for x.
$$3x-5y=17$$
$$3(0)-5y=17$$
$$\frac{-5y}{-5}=\frac{17}{-5}$$
$$y=-\frac{17}{5}$$
The y-intercept is $\left(0,-\frac{17}{5}\right)$.

8. To find the x-intercept, substitute 0 for y.
$$ax+b=cy$$
$$ax+b=(0)y$$
$$\frac{ax}{a}=\frac{-b}{a}$$
$$x=-\frac{b}{a}$$
The x-intercept is $\left(-\frac{b}{a},0\right)$.

To find the y-intercept, substitute 0 for x.
$$ax+b=cy$$
$$(0)x+b=cy$$
$$\frac{b}{c}=y$$
The y-intercept is $\left(0,\frac{b}{c}\right)$.

9. $y=-4$

10. $y\approx-\frac{2}{3}$

11. $x=-4$

12. $x=-1$

13. $m = \dfrac{-2-(-5)}{-3-2} = \dfrac{3}{-5} = -\dfrac{3}{5}$

The line is decreasing.

14. $m = \dfrac{-7-(-3)}{-9-(-1)} = \dfrac{-4}{-8} = \dfrac{1}{2}$

The line is increasing.

15. $m = \dfrac{-1-3}{4-4} = \dfrac{-4}{0}$ is undefined

The line is vertical.

16. $m = \dfrac{2.99-(-8.48)}{-5.27-3.54} = \dfrac{11.47}{-8.81} \approx -1.30$

The line with slope of -1.30 is decreasing.

17. Use the slope, -3, and the y-intercept, 10, to graph the equation.

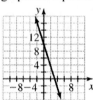

18. First, isolate y.

$y + 2x = 0$

$\quad y = -2x$

Use the slope, -2, and the y-intercept, 0, to graph the equation.

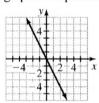

19. $y = 7$ is a horizontal line with a y-intercept of 7.

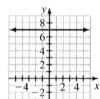

20. First, isolate y.

$3x - 2y = 12$

$\quad -2y = -3x + 12$

$\quad\quad y = \dfrac{3}{2}x - 6$

Use the slope $\dfrac{3}{2}$, and the y-intercept, -6, to graph the equation.

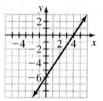

21. First, isolate y.

$-3(y+2) = 2x + 9$

$\quad -3y - 6 = 2x + 9$

$\quad\quad -3y = 2x + 15$

$\quad\quad\quad y = -\dfrac{2}{3}x - 5$

Use the slope, $-\dfrac{2}{3}$, and the y-intercept, -5, to graph the equation.

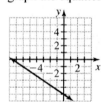

22. First, isolate y.

$3x - 2(2y-1) = 8x - 3(x+2)$

$3x - 4y + 2 = 8x - 3x - 6$

$\dfrac{-4y}{-4} = \dfrac{2x-8}{-4}$

$\quad\quad y = -\dfrac{1}{2}x + 2$

Use the slope, $-\dfrac{1}{2}$, and the y-intercept, 2, to graph the equation.

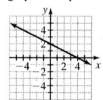

23. First, isolate y.

$$a(x - y) = c$$

$$x - y = \frac{c}{a}$$

$$-y = -x + \frac{c}{a}$$

$$y = x - \frac{c}{a}$$

The slope is 1 and the y-intercept is $\left(0, -\frac{c}{a}\right)$.

24. Solve the first equation for y to find the slope.

$$2x + 5y = 7$$

$$5y = -2x + 7$$

$$y = -\frac{2}{5}x + \frac{7}{5}$$

The slopes are $-\frac{2}{5}$ and $\frac{2}{5}$. They are not reciprocals, although they are additive inverses of each other. The lines are neither parallel nor perpendicular.

25. Solve each equation for y to find the slope.

$$3x - 8y = 7 \qquad\qquad -6x + 16y = 5$$

$$\frac{-8y}{-8} = \frac{-3x + 7}{-8} \qquad \frac{16y}{16} = \frac{6x + 5}{16}$$

$$y = \frac{3}{8}x - \frac{7}{8} \qquad\quad y = \frac{3}{8}x + \frac{5}{16}$$

The slopes are $\frac{3}{8}$ and $\frac{3}{8}$. The lines are parallel.

26. We are given the slope, $m = -4$, and a point on the line, $(-3, 7)$. Use $y = mx + b$ to find b.

$$y = mx + b$$

$$7 = -4(-3) + b$$

$$7 = 12 + b$$

$$-5 = b$$

Now, substitute for m and b in slope–intercept form to obtain the equation of the line.

$$y = -4x + (-5)$$

$$y = -4x - 5$$

27. We are given the slope, $m = -\frac{2}{3}$, and a point on the line, $(5, -4)$. Use $y = mx + b$ to find b.

$$y = mx + b$$

$$-4 = -\frac{2}{3}(5) + b$$

$$-4 = -\frac{10}{3} + b$$

$$-\frac{12}{3} = -\frac{10}{3} + b$$

$$-\frac{2}{3} = b$$

Now, substitute for m and b in slope–intercept form to obtain the equation of the line.

$$y = -\frac{2}{3}x + \left(-\frac{2}{3}\right)$$

$$y = -\frac{2}{3}x - \frac{2}{3}$$

28. First, find the slope.

$$m = \frac{6 - (-2)}{2 - (-3)} = \frac{8}{5}$$

So, $y = \frac{8}{5}x + b$. Since the line contains $(2, 6)$, substitute 2 for x and 6 for y and solve for b.

$$6 = \frac{8}{5}(2) + b$$

$$6 = \frac{16}{5} + b$$

$$\frac{30}{5} = \frac{16}{5} + b$$

$$\frac{14}{5} = b$$

So, the equation is $y = \frac{8}{5}x + \frac{14}{5}$.

29. First, find the slope.

$$m = \frac{6 - (-2)}{-4 - 2} = \frac{8}{-6} = -\frac{4}{3}$$

So, $y = -\frac{4}{3}x + b$. Since the line contains $(2, -2)$, substitute 2 for x and -2 for y and solve for b.

$$-2 = -\frac{4}{3}(2) + b$$

$$-\frac{6}{3} = -\frac{8}{3} + b$$

$$\frac{2}{3} = b$$

So, the equation is $y = -\frac{4}{3}x + \frac{2}{3}$.

30. First, find the slope.
$$m = \frac{5-(-2)}{3-3} = \frac{7}{0} \text{ undefined}$$
A line with undefined slope is a vertical line of the form $x = a$. Since (3, 5) is a point on the line and the value of x is 3 at that point, the equation of the line is $x = 3$.

31. First, find the slope.
$$m = \frac{-8.79-(-6.38)}{-3.62-2.51} = \frac{-2.41}{-6.13} \approx 0.39$$
So, $y = 0.39x + b$. Since the line contains $(-3.62, -8.79)$, substitute -3.62 for x and -8.79 for y and solve for b.
$$y = 0.39x + b$$
$$-8.79 = 0.39(-3.62) + b$$
$$-8.79 = -1.4118 + b$$
$$-7.37 \approx b$$
So, the equation is $y = 0.39x - 7.37$.

32. Line 1:
Two points on line 1 are (1, −2) and (−1, 2). Use these points to find the slope.
$$m = \frac{2-(-2)}{-1-1} = \frac{4}{-2} = -2$$
So, $y = -2x + b$. We can see from the graph that the y-intercept is 0. So, the equation of the line is $y = -2x + 0 = -2x$.
Line 2:
Two points on line 1 are (−1, 0) and (0, 2). Use these points to find the slope.
$$m = \frac{2-0}{0-(-1)} = \frac{2}{1} = 2$$
So, $y = 2x + b$. We can see from the graph that the y-intercept is 2. So, the equation of the line is $y = 2x + 2$.
Line 3:
Since line 3 is a horizontal line, it is of the form $y = b$. Since the y-intercept is −3, the equation of the line is $y = -3$.

33. The y-values decrease from 20 to 4 (for a change of −16) while the x-values increase from 2 to 6 (for a change of 4). So, the slope is $\frac{-16}{4} = \frac{-4}{1} = -4$, or the corresponding y-values decrease 4 for every x-value increase of 1. Complete the table by subtracting 4 from each y-value.

x	y
2	20
3	16
4	12
5	8
6	4
7	0

34. Line 1 is parallel to $y = 0.5x + 6$. Since parallel lines have the same slope, the slope of line 1 is 0.5. The equation of line 1 is $y = 0.5x + b$. Line 1 has the same y-intercept as the line $y = -1.5x + 3$. The y-intercept is $b = 3$. The equation of line 1 is $y = 0.5x + 3$.

35. **a.** B, C and F

b. C and E

c. C

d. A and D

36. Since the line is parallel to the line $3x - y = 6$, or $y = 3x - 6$, it will have a slope of 3. Use the slope, 3, and the point on the line, (−2, 5), to find the y-intercept.
$$5 = 3(-2) + b$$
$$5 = -6 + b$$
$$11 = b$$
The equation of the line is $y = 3x + 11$.

37. A line with an infinite number of x-intercepts is the line that lies on the x-axis. This is the horizontal line $y = 0$.

38. Relations 1 and 3 could possibly be functions, since there is only one output for each input. Relations 2 and 4 could not possibly be functions, since there are two or more outputs for a single input.

39. The graph is not a function because a vertical line can intersect the graph more than once.

40. First, isolate y.
$$5x - 6y = 3$$
$$\frac{-6y}{-6} = \frac{-5x + 3}{-6}$$
$$y = \frac{5}{6}x - 3$$

This relation is a function since it can be put into the form $y = mx + b$ which defines a linear function.

41. $x = 9$ is a vertical line. It does not pass the vertical line test and is not a function.

42. Sketch the graph of $y^2 = x$.

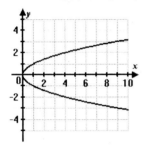

Note that this graph does not pass the vertical line test. Therefore, $y^2 = x$ is not a function.

43. The domain is all real numbers and the range is $y \le 4$.

Chapter 1 Test

1. The value in dollars of gold depends on the weight of a gold bar. The value (v) is the dependent variable and the weight (w) is the independent variable.

2. Answers may vary. Example:

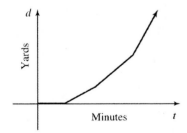

3. Answers may vary. Example:
The diameter of a balloon in inches (y) is the dependent variable and the number of seconds after the knot of the balloon was untied (x) is the independent variable.

4. $5 - 3(4x - 2) = 8 - (7x + 1)$
$$5 - 12x + 6 = 8 - 7x - 1$$
$$-12x + 11 = -7x + 7$$
$$\frac{-5x}{-5} = \frac{-4}{-5}$$
$$x = \frac{4}{5}$$

5. Line 1: Two points on line 1 are (4, 0) and (2, 5). Use these points to find the slope.
$$m = \frac{5 - 0}{2 - 4} = \frac{5}{-2} = -\frac{5}{2}$$

So, $y = -\frac{5}{2}x + b$. Use the slope, $-\frac{5}{2}$, and a point on the line to find the y-intercept.
$$y = mx + b$$
$$0 = -\frac{5}{2}(4) + b$$
$$0 = -10 + b$$
$$10 = b$$

The equation of line 1 is $y = -\frac{5}{2}x + 10$.

Line 2: Two points on line 1 are (0, 2) and (−3, 0). Use these points to find the slope.
$$m = \frac{0 - 2}{-3 - 0} = \frac{-2}{-3} = \frac{2}{3}$$

So, $y = \frac{2}{3}x + b$. We can see from the graph that the y-intercept is 2. So, the equation of the line is $y = \frac{2}{3}x + 2$.

Line 3: Since line 3 is a vertical line, it is of the form $x = a$. Since the x-intercept is −3, the equation of the line is $x = -3$.

6. a. $k > m$, since the line $y = kx + c$ is steeper than the line $y = mx + b$.

b. $b > c$, since the y-intercept of the line $y = mx + b$ is greater than the y-intercept of the line $y = kx + c$.

7. $m_A = \frac{|-85|}{270} = \frac{85}{270} \approx 0.31$

$m_B = \frac{|-140|}{475} = \frac{140}{475} \approx 0.29$

Ski run A is steeper than ski run B, since the slope of ski run A is greater than the slope of ski run B. (In determining which is steeper, ignore the sign of the slope by considering absolute value.)

8. The table is completed as follows.

x	y
4	25
5	29
6	33
7	37
8	41
9	45

9. Use the slope, $-\dfrac{1}{5}$, and the y-intercept, 4, to graph the line.

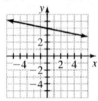

10. First, isolate y.
$$2(2x - y) = 2x + 9 + y$$
$$4x - 2y = 2x + 9 + y$$
$$-2y - y = 2x - 4x + 9$$
$$\frac{-3y}{-3} = \frac{-2x + 9}{-3}$$
$$y = \frac{2}{3}x - 3$$

Use the slope, $\dfrac{2}{3}$, and the y-intercept, -3, to graph the line.

11. $m = \dfrac{2 - (-8)}{-3 - 5} = \dfrac{10}{-8} = -\dfrac{5}{4}$

12. First, find the slope.
$$m = \frac{8 - 6}{2 - 5} = \frac{2}{-3} = -\frac{2}{3}$$
Use the slope and a point on the line, (2, 8), to find b.

$$y = mx + b$$
$$8 = -\frac{2}{3}(2) + b$$
$$8 = -\frac{4}{3} + b$$
$$\frac{24}{3} = -\frac{4}{3} + b$$
$$\frac{28}{3} = b$$

The equation of the line is $y = -\dfrac{2}{3}x + \dfrac{28}{3}$. To find points on the line, substitute values for x to find y. For example, let $x = 8$.
$$y = -\frac{2}{3}(8) + \frac{28}{3} = -\frac{16}{3} + \frac{28}{3} = \frac{12}{3} = 4$$
The point, (8, 4) is on the line. Similar calculations will show that (11, 2) and (14, 0), among others, also lie on the line. Answers may vary.

13. Use the slope, $-\dfrac{3}{7}$, and the point $(-2, 5)$, to find the y-intercept, b.
$$y = mx + b$$
$$5 = -\frac{3}{7}(-2) + b$$
$$5 = \frac{6}{7} + b$$
$$\frac{35}{7} - \frac{6}{7} = b$$
$$\frac{29}{7} = b$$

The equation of the line is $y = -\dfrac{3}{7}x + \dfrac{29}{7}$.

14. First, find the slope.
$$m = \frac{7 - (-5)}{-3 - 2} = -\frac{12}{5}$$
Use the slope and a point on the line, (2, −5), to find b.
$$y = mx + b$$
$$-5 = -\frac{12}{5}(2) + b$$
$$-\frac{25}{5} = -\frac{24}{5} + b$$
$$-\frac{1}{5} = b$$

The equation of the line is $y = -\dfrac{12}{5}x - \dfrac{1}{5}$.

15. All of the points except (0, 2) lie on a line. Choose two of the points to find the slope.

$$m = \frac{5 - (-3)}{-2 - 2} = \frac{8}{-4} = -2$$

Use the slope and a point on the line to find the y-intercept, b.

$$y = mx + b$$
$$5 = -2(-2) + b$$
$$5 = 4 + b$$
$$1 = b$$

The equation of the line is $y = -2x + 1$.

16. First, find the slope of the line by isolating y.

$$3x - 5y = 20$$
$$-5y = -3x + 20$$
$$y = \frac{3}{5}x - 4$$

The slope of the line is $\frac{3}{5}$. The slope of a line perpendicular to this line is $-\frac{5}{3}$. Use the slope, $-\frac{5}{3}$, and a point on the line, (4, −1), to find the y-intercept.

$$y = mx + b$$
$$-1 = -\frac{5}{3}(4) + b$$
$$-1 = -\frac{20}{3} + b$$
$$-\frac{3}{3} = -\frac{20}{3} + b$$
$$\frac{17}{3} = b$$

The equation of the line is $y = -\frac{5}{3}x + \frac{17}{3}$.

17. To find the x-intercept, set y = 0 and solve for x.

$$2(0) + 5 = 4(x - 1) + 3$$
$$0 + 5 = 4x - 4 + 3$$
$$6 = 4x$$
$$\frac{6}{4} = x$$
$$\frac{3}{2} = x$$

To find the y-intercept, set x = 0 and solve for y.

$$2y + 5 = 4(0 - 1) + 3$$
$$2y + 5 = -4 + 3$$
$$2y = -6$$
$$y = -3$$

The x-intercept is $\left(\frac{3}{2}, 0\right)$. The y-intercept is $(0, -3)$.

18. a. Answers may vary. Example:

x	y
0	$2(0) - 4 = -4$
1	$2(1) - 4 = -2$
2	$2(2) - 4 = 0$
3	$2(3) - 4 = 2$
4	$2(4) - 4 = 4$

b.

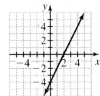

c. For each input–output pair, the output is 4 less than 2 times the input.

19. Answers may vary. Any graph that does not pass the vertical line test is not the graph of a function.

20. Sketch the graph of $y = \pm\sqrt{x}$.

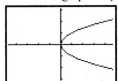

Note that this graph does not pass the vertical line test. Therefore, $y = \pm\sqrt{x}$ is not a function.

21. The graph of $y = -2x + 5$ is a nonvertical line. Any nonvertical line is a function since it passes the vertical line test.

22. The domain is $-3 \le x \le 5$ and the range is $-3 \le y \le 4$. The relation is a function since the graph can be intersected by any vertical line only once.

Chapter 2
Modeling with Linear Functions

Homework 2.1

1. a.

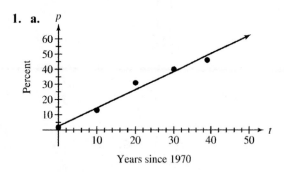

Years since 1970

b. Draw a line through the points. See the graph in part (a).

c. In 2018, 60% of dentistry degrees will be earned by women.

d. In 1985, about 20% of dentistry degrees were earned by women.

3. a.

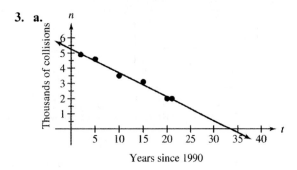

Years since 1990

b. Draw a line through the points. See the graph in part (a).

c. In 2007, there were about 2.6 thousand collisions. We performed interpolation because we used a part of the model whose t-coordinates are between the t-coordinates of two data points.

d. If the graph is extended, there will be 1.0 thousand collisions in 2017. We performed extrapolation because we used a part of the model whose t-coordinates are not between the t-coordinates of two data points.

e. The n-intercept of the model is when $n = 5.3$, or $(0, 5.3)$. This means that in 1990, there were 5.3 thousand collisions.

f. If the graph is extended, the t-intercept of the model occurs when $t = 34$, or $(34, 0)$. This means that there will be no collisions in 2024. Model breakdown has likely occurred.

5. a.

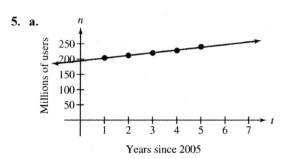

Years since 2005

b. Draw a line through the points. See the graph in part (a).

c. The n-intercept is when $n = 194$. This means that there were 194 million Internet users in 2005. We performed extrapolation because we used a part of the model whose n-coordinates are not between the n-coordinates of two data points.

d. The number of Internet users is increasing by 9 million people per year.

e. In 2019 the model predicts that everyone in the U.S. will be an Internet user. We performed extrapolation because we used a part of the model whose t-coordinates are not between the t-coordinates of two data points. Model breakdown has likely occurred.

7. a.

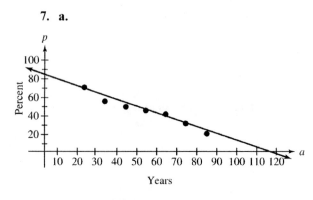

Years

b. Draw a line through the points. See the graph in part (a).

c. The model predicts that at age 27 years, 65% of Americans believe same-sex couples should be recognized by the law as valid.

d. The *p*-intercept of the model is when $p = 85$, or $(0, 85)$. This means that 85% of newborns believe marriages between same-sex couples should be recognized by the law as valid. Model breakdown has occurred.

e. The *a*-intercept of the model is when $a = 117$, or $(117, 0)$. This means that no 117-year-old Americans believe marriages between same-sex couples should be recognized by law as valid. Model breakdown has occurred.

9. a.

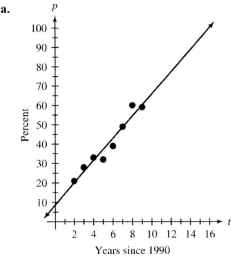

b. Draw a line through the points. See the graph in part (a).

c. About 100% of Americans will be satisfied in 2006.

d.

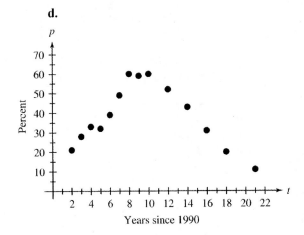

e. The error is 69 percentage points.
Answers may vary. Example:
Up to 2000 the percent of Americans who were satisfied was steadily increasing. After 2000 the percent of Americans who are satisfied was steadily decreasing.

11. a. To find the *t*-intercept, estimate the number of years since 2002 when $p = 0$. This estimate is approximately $(2, 0)$. This point represents that in 2004 $(t = 2)$, the profit was 0 dollars.

b. To find the *p*-intercept, estimate when $t = 0$. This estimate is approximately $(0, -68)$. This point represents that the profit was -68 million dollars in 2002.

13. Answers may vary. Example:
A linear model is a linear function that describes the relationship between two quantities for an authentic situation. Every linear model is a linear function. However, not every linear function is a linear model. Functions are used both to describe situations and to describe certain mathematical relationships between two variables.

15. Answers may vary.

17. Answers may vary. Example:
The linear model that comes closest to all data points is the most accurate model, and allows for the best estimates and predictions based on the given data.

Homework 2.2

1. Use the points $(9, 22.2)$ and $(11, 92.0)$ to write the equation of the line. First find the slope.
$$m = \frac{92.0 - 22.2}{11 - 9} = 34.9$$
So, $n = 34.9t + b$. Substitute 9 for *t* and 22.2 for *n* since the line contains $(9, 22.2)$ and then solve for *b*.
$$n = mt + b$$
$$22.2 = 34.9(9) + b$$
$$22.2 = 314.1 + b$$
$$-291.9 = b$$
The equation of the line is $n = 34.9t - 291.9$.

3. Use $(7, 912)$ and $(10, 613)$ to write the equation of the line. First, find the slope.
$$m = \frac{613 - 912}{10 - 7} = \frac{-299}{3} \approx -99.67$$

So, $p = -\dfrac{299}{3}t + b$. Substitute 7 for t and 912

for p since the line contains $(7, 912)$ and then solve for b.

$$p = mt + b$$
$$912 = -\dfrac{299}{3}(7) + b$$
$$\dfrac{2736}{3} = -\dfrac{2093}{3} + b$$
$$\dfrac{4829}{3} = b$$
$$1609.67 \approx b$$

The equation of the line is
$p = -99.67t + 1609.67$

5. Use the points $(7, 11)$ and $(10, 16)$ to write the equation of the line. First find the slope.

$$m = \dfrac{16 - 11}{10 - 7} = \dfrac{5}{3} \approx 1.67$$

So, $p = \dfrac{5}{3}t + b$. Substitute 7 for t and 11 for p

since the line contains $(7, 11)$ and then solve for b.

$$p = mt + b$$
$$11 = \dfrac{5}{3}(7) + b$$
$$\dfrac{33}{3} = \dfrac{35}{3} + b$$
$$-\dfrac{2}{3} = b$$
$$-0.67 \approx b$$

The equation of the line is $p = 1.67t - 0.67$.

7. Use the points $(6, 27)$ and $(20, 54.5)$ to find the equation of the line.

$$m = \dfrac{54.5 - 27}{20 - 6} = \dfrac{27.5}{14} = \dfrac{55}{28} \approx 1.96$$

So, $L = \dfrac{55}{28}a + b$. Substitute 6 for a and 27 for

L since the line contains $(6, 27)$ and then solve for b.

$$L = ma + b$$
$$27 = \dfrac{55}{28}(6) + b$$
$$27 = \dfrac{55}{14}(3) + b$$
$$\dfrac{378}{14} = \dfrac{165}{14} + b$$
$$\dfrac{213}{14} = b$$
$$15.21 \approx b$$

The equation of the line is $L = 1.96a + 15.21$.

9. Use $(3, 5)$ and $(7, 15)$ to write the equation of the line. First, find the slope.

$$m = \dfrac{15 - 5}{7 - 3} = \dfrac{10}{4} = 2.5$$

So, $y = 2.5x + b$. Substitute 3 for x and 5 for y since the line contains $(3, 5)$ and then solve for b.

$$y = mx + b$$
$$5 = 2.5(3) + b$$
$$5 = 7.5 + b$$
$$-2.5 = b$$

The equation of the line is $y = 2.5x - 2.5$.
(Your equation may be slightly different if you chose different points, or if you used the linear regression feature of your graphing calculator.) Use the graphing calculator to check your results.

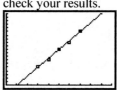

11. **a.**

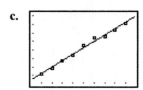

   b. Use the points $(70, 10.7)$ and $(110, 40.8)$ to write the equation of the line. First find the slope.

$$m = \dfrac{40.8 - 10.7}{110 - 70} \approx 0.75$$

So, $p = 0.75t + b$. Substitute 70 for t and 10.7 for p, since the line contains $(70, 10.7)$ and then solve for b.

$$p = mt + b$$
$$10.7 = 0.75(70) + b$$
$$10.7 = 52.5 + b$$
$$-41.8 = b$$

The equation of the line is $p = 0.75t - 41.8$.
(Your equation may be slightly different if you chose different points, or if you used the linear regression feature of your graphing calculator.)

   c.

Notice that the points (70, 10.7) and
(110, 40.8) are filled in, showing that the
line goes through these points.

13. a.

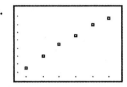

b. Use the points (2, 30) and (4, 56) to find
the equation of the line.
$$m = \frac{56 - 30}{4 - 2} = 13$$
So, $p = 13n + b$. Substitute 2 for n and 30
for p since the line contains (2, 30) and
then solve for b.
$$p = mn + b$$
$$30 = 13(2) + b$$
$$30 = 26 + b$$
$$4 = b$$
The equation of the line is $p = 13n + 4$.
(Your equation may be slightly different if
you chose different points, or if you used
the linear regression feature of your
graphing calculator.)

c.

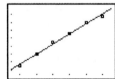

Notice that the points (2, 30) and (4, 56)
are filled in, showing that the line goes
through these points.

15. a.

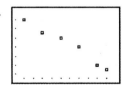

b. Use the points (9, 54) and (18, 43) to find
the equation of the line.
$$m = \frac{43 - 54}{18 - 9} \approx -1.22$$
So, $p = -1.22t + b$. Substitute 9 for t and 54
for p since the line contains (9, 54) and
then solve for b.
$$p = mt + b$$
$$54 = -1.22(9) + b$$
$$54 = -10.98 + b$$
$$64.98 = b$$

The equation of the line is
$p = -1.22t + 64.98$. (Your equation may be
slightly different if you chose different
points, or if you used the linear regression
feature of your graphing calculator.)

c.

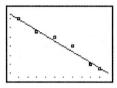

17. a.

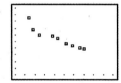

b. Use the points (69, 51.7) and (83, 47.99) to
find the equation of the line.
$$m = \frac{47.99 - 51.7}{83 - 69} \approx -0.27$$
So, $r = -0.27t + b$. Substitute 69 for t and
51.7 for r since the line contains (69, 51.7)
and then solve for b.
$$r = mt + b$$
$$51.7 = -0.27(69) + b$$
$$51.7 = -18.63 + b$$
$$70.33 = b$$
The equation of the line is
$r = -0.27t + 70.33$. (Your equation may be
slightly different if you chose different
points, or if you used the linear regression
feature of your graphing calculator.)

c.

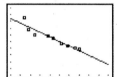

19. a.

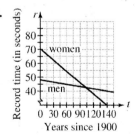

b. Yes, the model predicts that the women's
record time will equal the men's record
time. This will happen in the year 2003
with a record time of approximately
42.6 seconds.

c. Yes, the model predicts that the women's record time will be less than the men's record time in years 2004 and after.

21. a.

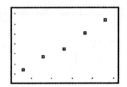

b. Use the points (23, 34) and (43, 84) to find the equation of the line.

$$m = \frac{84 - 34}{43 - 23} = 2.5$$

So, $p = 2.5x + b$. Substitute 23 for x and 34 for p since the line contains (23, 34) and then solve for b.

$$p = mx + b$$
$$34 = 2.5(23) + b$$
$$34 = 57.5 + b$$
$$-23.5 = b$$

The equation of the line is $p = 2.5x - 23.5$. (Your equation may be slightly different if you chose different points, or if you used the linear regression feature of your graphing calculator.)

c.

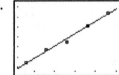

23. Student B made the best choice of points. The graph that best fits all the points goes though the points (3, 9.0) and (4, 11.0).

25.

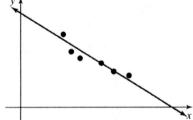

In this model, m will be the same as the original model. However, the value of b will increase.

27. Answers may vary. Example:
To find an equation of a linear model, first make a scattergram of the points. Then determine if there is a line that comes close to all the data points. If so, choose two points (not necessarily data points) that you can use to find the equation of a linear model. Find an equation of the line between the two points you found. To verify that the linear function models the situation well, use a graphing calculator to verify that the graph of your equation comes close to the points of the scattergram.

Homework 2.3

1. Substituting 5 wherever there is an x in $f(x)$:
$$f(5) = 6(5) - 4 = 30 - 4 = 26$$

3. Substituting $\frac{2}{3}$ wherever there is an x in $f(x)$:
$$f\left(\frac{2}{3}\right) = 6\left(\frac{2}{3}\right) - 4 = 4 - 4 = 0$$

5. Substituting $a + 2$ wherever there is an x in $f(x)$:
$$f(a + 2) = 6(a + 2) - 4 = 6a + 12 - 4 = 6a + 8$$

7. Substituting 2 wherever there is an x in $g(x)$:
$$g(2) = 2(2)^2 - 5(2) = 2(4) - 10 = 8 - 10 = -2$$

9. Substituting -3 wherever there is an x in $g(x)$:
$$g(-3) = 2(-3)^2 - 5(-3)$$
$$= 2(9) + 15$$
$$= 18 + 15$$
$$= 33$$

11. Substituting 2 wherever there is an x in $h(x)$:
$$h(2) = \frac{3(2) - 4}{5(2) + 2} = \frac{6 - 4}{10 + 2} = \frac{2}{12} = \frac{1}{6}$$

13. Substituting $a - 3$ wherever there is an x in $h(x)$:
$$h(a - 3) = \frac{3(a - 3) - 4}{5(a - 3) + 2}$$
$$= \frac{3a - 9 - 4}{5a - 15 + 2}$$
$$= \frac{3a - 13}{5a - 13}$$

15. Substituting -2 wherever there is an x in $g(x)$:
$$g(-2) = -3(-2)^2 + 2(-2)$$
$$= -3(4) - 4$$
$$= -12 - 4$$
$$= -16$$

17. Substituting 5 wherever there is an x in $f(x)$:
$$f(5) = -2(5) + 7 = -10 + 7 = -3$$

19. Since $h(x) = -4$, for all x, $h(7) = -4$.

21. Substituting $5a$ wherever there is an x in $f(x)$:
$$f(5a) = -4(5a) - 7 = -20a - 7$$

23. Substituting $\dfrac{a}{2}$ wherever there is an x in $f(x)$:
$$f\left(\frac{a}{2}\right) = -4\left(\frac{a}{2}\right) - 7 = -2a - 7$$

25. Substituting $a + 4$ wherever there is an x in $f(x)$:
$$\begin{aligned} f(a+4) &= -4(a+4) - 7 \\ &= -4a - 16 - 7 \\ &= -4a - 23 \end{aligned}$$

27. Substituting $a - h$ wherever there is an x in $f(x)$:
$$f(a-h) = -4(a-h) - 7 = -4a + 4h - 7$$

29. To find x when $f(x) = 6$, substitute 6 for $f(x)$ and solve for x.
$$\begin{aligned} 6 &= -3x + 7 \\ 3x &= 1 \\ x &= \frac{1}{3} \end{aligned}$$

31. To find x when $f(x) = \dfrac{5}{2}$, substitute $\dfrac{5}{2}$ for $f(x)$ and solve for x.
$$\begin{aligned} \frac{5}{2} &= -3x + 7 \\ \frac{-9}{2} &= -3x \\ \frac{3}{2} &= x \end{aligned}$$

33. To find x when $f(x) = a$, substitute a for $f(x)$ and solve for x.
$$\begin{aligned} a &= -3x + 7 \\ a - 7 &= -3x \\ \frac{7-a}{3} &= x \end{aligned}$$

35. Substituting 10.91 wherever there is an x in $f(x)$:
$$\begin{aligned} f(10.91) &= -5.95(10.91) + 183.22 \\ &\approx -64.91 + 183.22 \\ &\approx 118.31 \end{aligned}$$

37. To find x when $f(x) = 99.34$, substitute 99.34 for $f(x)$ and solve for x.
$$\begin{aligned} 99.34 &= -5.95x + 183.22 \\ -83.88 &= -5.95x \\ 14.10 &\approx x \end{aligned}$$

39. The third row of the chart indicates that $f(2) = 4$.

41. The second and fourth rows of the chart indicate that $f(x) = 2$ when $x = 1$ or $x = 3$.

43. Since the line includes the point $(-6, 4)$, $f(-6) = 4$.

45. Since the line includes a point at approximately $(2.5, 1.2)$, $f(2.5) \approx 1.2$.

47. Since the line includes the point $(6, 0)$, $x = 6$ when $f(x)$ or $y = 0$.

49. Since the line includes the point $(-3, 3)$, $x = -3$ when $f(x)$ or $y = 3$.

51. Since the line includes the point $(4.5, \frac{1}{2})$,
$x = 4.5$ when $f(x)$ or $y = \dfrac{1}{2}$.

53. The domain of f is all the x-coordinates of the points in the graph. In this case f has a domain of all real numbers.

55. Sine the curve includes the point $(-2, 1)$, $g(-2) = 1$.

57. The domain of g is all of the x-coordinates of the points in the graph. In this case g has a domain of $-4 \le x \le 5$.

59. Since the curve includes the point $(1, -3)$, $h(1) = -3$.

61. The domain of h is all of the x-coordinates of the points in the graph. In this case h has a domain of $-5 \le x \le 4$.

63. $f(x)$ or $y = 5x - 8$
To find the y-intercept, set $x = 0$ and solve for y.
$$y = 5(0) - 8 = 0 - 8 = -8$$
The y-intercept is $(0, -8)$.
To find the x-intercept, set $y = 0$ and solve for x.

$0 = 5x - 8$

$8 = 5x$

$\dfrac{8}{5} = x$

The x-intercept is $\left(\dfrac{8}{5}, 0\right)$.

65. $f(x)$ or $y = 3x$

To find the y-intercept, set $x = 0$ and solve for y.

$y = 3(0) = 0$

The y-intercept is $(0, 0)$.

To find the x-intercept, set $y = 0$ and solve for x.

$0 = 3x$

$0 = x$

The x-intercept is $(0, 0)$.

67. $f(x)$ or $y = 5$ is the equation of a horizontal line. The y-intercept of the line will be $(0, 5)$. There is no x-intercept.

69. $f(x)$ or $y = \dfrac{1}{2}x - 3$

To find the y-intercept, set $x = 0$ and solve for y.

$y = \dfrac{1}{2}(0) - 3 = -3$

The y-intercept is $(0, -3)$.

To find the x-intercept, set $y = 0$ and solve for x.

$0 = \dfrac{1}{2}x - 3$

$3 = \dfrac{1}{2}x$

$6 = x$

The x-intercept is $(6, 0)$.

71. $f(x)$ or $y = 2.58x - 45.21 f(x)$

To find the y-intercept, set $x = 0$ and solve for y.

$y = 2.58(0) - 45.21 = 45.21$

The y-intercept is $(0, -45.21)$.

To find the x-intercept, set $y = 0$ and solve for x.

$0 = 2.58x - 45.21$

$45.21 = 2.58x$

$17.52 \approx x$

The x-intercept is $(17.52, 0)$.

73. a. Answers may vary. Example:

x	$g(x)$
-2	10
-1	7
0	4
1	1
2	-2

b.

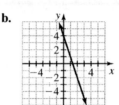

c. Multiply the input by -3 and add 4 to find the output.

75. a. $f(t) = 0.76t - 42.04$

b. $f(118) = 0.76(118) - 42.04$

$= 89.68 - 42.04$

$= 47.64$

When $t = 118$, $p = 47.64$. This means that in the year $1900 + 118 = 2018$, 47.6% of births will be outside marriage.

c. $47 = 0.76t - 42.04$

$89.04 = 0.76t$

$117.16 \approx t$

When $f(t) = 47$, $t \approx 117.16$. This means that in the year $1900 + 117 = 2017$, 47% of births will be outside marriage.

d. $100 = 0.76t - 42.04$

$142.04 = 0.76t$

$186.9 \approx t$

In the year $1900 + 186.9 \approx 2087$, all births will be outside marriage according to this model. Model breakdown has likely occurred.

e. Find p when $t = 1997 - 1900 = 97$.

$f(97) = 0.76(97) - 42.04$

$= 73.72 - 42.04$

$= 31.68$

This means that in the year 1997, the percentage of births outside marriage will be about 31.7%. Since the actual percentage was 32.4%, the error was $31.7\% - 32.4\% = -0.7\%$.

77. a. $f(t) = -1.19t + 64.86$

b. $f(28) = -1.19(28) + 64.86$
$$= -33.32 + 64.86$$
$$= 31.54$$
When $t = 28$, $p \approx 31.54$. This means that in $1990 + 28 = 2018$, baseball will be the favorite sport of about 32% of Americans.

c. $28 = -1.19t + 64.86$
$$-36.86 = -1.19t$$
$$30.97 \approx t$$
This means that in $1990 + 30.97 \approx 2021$, baseball will be the favorite sport of 28 percent of Americans.

d. Since the model $f(t) = -1.19t + 64.86$ is in slope–intercept form, the p-intercept is $(0, 64.86)$.
This means that in $1990 + 0 = 1990$, baseball was the favorite sport of about 65% of Americans.

e. To find the t-intercept of the model, we must find t when $f(t) = 0$.
$$0 = -1.19t + 64.86$$
$$-64.86 = -1.19t$$
$$54.50 \approx t$$
This means that in $1990 + 54.50 \approx 2045$, baseball will be the favorite sport of 0 percent of Americans. Model breakdown has likely occurred.

79. a. $f(n) = 12.74n + 4.40$

b. To find the p-intercept of the model, we must find p when $n = 0$.
$$p = -12.74(0) + 4.40 = 4.40$$
This means the cost of a ski rental package for 0 days is $4.40. Model breakdown has occurred.

c. To find the cost for seven days, we must find $f(n)$ when $n = 7$.
$$f(7) = 12.74(7) + 4.40$$
$$= 89.18 + 4.40$$
$$= 93.58$$

d.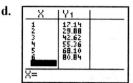

As n increases by 1, p increases by 12.74. The price increases by $12.74 for each additional day. This is equal to the slope of the equation.

e. $g(1) = \dfrac{12.74(1) + 4.40}{1} = 17.14$

$g(2) = \dfrac{12.74(2) + 4.40}{2} = 14.94$

$g(3) = \dfrac{12.74(3) + 4.40}{3} \approx 14.21$

$g(4) = \dfrac{12.74(4) + 4.40}{4} = 13.84$

$g(5) = \dfrac{12.74(5) + 4.40}{5} = 13.62$

$g(6) = \dfrac{12.74(6) + 4.40}{6} \approx 13.47$

The least value is $13.47. This represents the cost per day for a six-day ski rental package.

f. The rate of increase for packages over 6 days is greater than the rate of increase for packages 6 days and shorter. Even so, the cost per day will reach its lowest value for packages over 6 days.

81. a.

b. Use the points $(1, 452)$ and $(10, 666)$ to find the equation of the line.
$$m = \frac{666 - 452}{10 - 1} = \frac{214}{9} \approx 23.78$$
So, $y = 23.78x + b$. Substitute 1 for x and 452 for y since the line contains $(1, 452)$ and then solve for b.
$$y = mx + b$$
$$452 = 23.78(1) + b$$
$$452 = 23.78 + b$$
$$428.22 = b$$

The equation of the line is
$y = 23.78x + 428.22$. (Your equation may
be slightly different if you chose different
points, or if you used the linear regression
feature of your graphing calculator.)

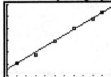

Graphing the line with the scatterplot, we
see that the model fits the data very well.

c. Since $2001 - 1990 = 11$, find $f(11)$.
$$f(11) = 23.78(11) + 428.22$$
$$= 261.58 + 428.22$$
$$= 689.8$$
According to the model, approximately
690 million boardings were made in 2001.
The actual number of boardings was
622 million. This is an error of 68 million
boardings.

d. 68 million boardings is equivalent to
$\dfrac{68}{4} = 17$ million round trips. If on average
each round trip is $340, the airlines lost
$17 \cdot \$340 = \$5780$ million or $5.78 billion.

83. a.

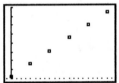

The data points lie close to the line that
passes through (0, 32) and (100, 212). To
find the equation for f, start by finding the
slope using these points.
$$m = \frac{212 - 32}{100 - 0} = \frac{180}{100} = 1.8$$
So, $f(C) = 1.8C + b$. To solve for b,
substitute 0 for C and 32 for $f(C)$ since the
line contains (0, 32).
$32 = 1.8(0) + b$
$32 = b$
The equation for f is $f(C) = 1.8C + 32$.

b. $f(24) = 1.8(24) + 32 = 43.2 + 32 = 75.2$
If it is $24°C$, it is $75.2°F$.

c. $64 = 1.8C + 32$
$32 = 1.8C$
$C \approx 17.78$
If it is $64°F$, then it is $17.78°C$.

d. $f(-273.15) = 1.8(-273.15) + 32$
$$= -491.67 + 32$$
$$= -459.67$$
If it is $-273.15°C$, it is $-459.67°F$.

85. a. $f(x) = 2.48x - 23.64$

b. $100 = 2.48x - 23.64$
$123.64 = 2.48x$
$x \approx 50$
This means that the cutoff score would
have to be 50 (out of 50) to ensure that all
students succeed in the intermediate
algebra course.

c. $0 = 2.48x - 23.64$
$23.64 = 2.48x$
$x \approx 10$
This means that for scores 10 and under,
no students would succeed in the
intermediate algebra course.

d. For the $16 - 20$ range, we use a score of 18
to represent the group. Find the percentage
for $x = 18$.
$p = 2.48(18) - 23.64 = 44.64 - 23.64 = 21$
This means that 21% of the students who
score in the $16 - 20$ range pass the
intermediate algebra course. If 145 students
scored in this range, we could expect 21%
or $0.21 \cdot 145 \approx 30$ students to pass.
It would not make sense for CSM to lower
the placement score cutoff to 16. While
21% of students might pass, based on the
model, the vast majority, 79%, would not
pass.

e. First calculate the percentages for each of
the score groups. Use the average score in
each range to represent the group.
$p = 2.48(23) - 23.64 = 33.4$
$p = 2.48(28) - 23.64 = 45.8$
$p = 2.48(33) - 23.64 = 58.2$
$p = 2.48(38) - 23.64 = 70.6$
$p = 2.48(43) - 23.64 = 83$
$p = 2.48(48) - 23.64 = 95.4$
Next, multiply each of the percentages by
the number of students in that category.

$0.334 \cdot 94 = 31.4$ students
$0.458 \cdot 44 = 20.2$ students
$0.582 \cdot 19 = 11.1$ students
$0.706 \cdot 12 = 8.5$ students
$0.83 \cdot 9 = 7.5$ students
$0.954 \cdot 4 = 3.8$ students
Add these results to obtain the total number of students, 82.5. This means that of students who scored at least 21 points on the placement test, approximately 83 students will pass the class.

87. If we let t be the number of years after 1980 and let E be the public school per-student expenditures (in thousands of dollars), we can make a linear model of the given information. To start, we can use the two points (0, 2.2) and (28, 10.8) to find the slope of the model:
$$m = \frac{10.8 - 2.2}{28 - 0} = \frac{8.6}{28} \approx 0.31$$
So $E(t) = 0.31t + b$. To solve for b, substitute 0 for t and 2.2 for E, since the line contains (0, 2.2).
$2.2 = 0.31(0) + b$
$2.2 = b$
The equation for E is $E(t) = 0.31t + 2.2$. To find E for 2017, or when $t = 37$, substitute 37 for t and solve for E.
$E(37) = 0.31(37) + 2.2$
$\qquad = 11.47 + 2.2$
$\qquad = 13.67$
According to the model, in 2017 the public school per-student expenditures will be about $13.7 thousand.

89. If we let t be the number of years after 2000 and let D be the number of blood donations to the American Red Cross (in millions of pints), we can make a linear model of the given information. To start, we can use the two points (9, 6.6) and (12, 5.9) to find the slope of the model:
$$m = \frac{5.9 - 6.6}{12 - 9} = \frac{-0.7}{3} \approx -0.23$$
So $D(t) = -0.23t + b$. To solve for b, substitute 9 for t and 6.6 for D, since the line contains (9, 6.6).
$6.6 = -0.23(9) + b$
$6.6 = -2.07 + b$
$8.67 = b$
The equation for D is $D(t) = -0.23t + 8.67$.

To find when the number of blood donations will be 4.5 million pints, or when $D = 4.5$, substitute 4.5 for D and solve for t.

$4.5 = -0.23t + 8.67$
$-4.17 = -0.23t$
$18.13 \approx t$
According to the model, in 2018 the blood donations will be 4.5 million pints.

91. **a.** If we let t be the number of years after 1990 and let R be the revenue of Kodak (in billions of dollars), we can make a linear model of the given information. To start, we can use the two points (0, 19.0) and (21, 6.0) to find the slope of the model:
$$m = \frac{6.0 - 19.0}{21 - 0} = \frac{-13.0}{21} \approx -0.62$$
So $R(t) = -0.62t + b$. To solve for b, substitute 0 for t and 19.0 for R, since the line contains (0, 19.0).
$19.0 = -0.62(0) + b$
$19.0 = b$
The equation for R is $R(t) = -0.62t + 19.0$.
To find the revenue of Kodak in 2013, or when $t = 23$, substitute 23 for t and solve for R.
$R(23) = -0.62(23) + 19.0$
$\qquad = -14.26 + 19.0$
$\qquad = 4.74$
According to the model, in 2013 the revenue of Kodak was $4.74 billion.

b. To find when the revenue will be 0, or when $R = 0$, substitute 0 for R and solve for t.
$0 = -0.62t + 19.0$
$-19.0 = -0.62t$
$30.6 \approx t$
According to the model, Kodak's revenue will be 0 in 2021.

93. If we let t be the amount of time a fourth-grader studies history per week and s be the average score a fourth-grader gets on the National Assessment of Educational Progress test in U.S. history, we can make a linear model of the given information. To start, we can use the two points (45, 195) and (150, 211) to find the slope of the model:
$$m = \frac{211 - 195}{150 - 45} \approx 0.15$$
So $s = 0.15t + b$. To solve for b, substitute 45 for t and 195 for s, since the line contains (45, 195).
$195 = 0.15(45) + b$
$195 = 6.75 + b$
$188.25 = b$

The equation for s is $s = 0.15t + 188.25$.
To find the average score for fourth graders who study history about 200 minutes per week, substitute 200 for t and solve for s.
$s = 0.15(200) + 188.25 = 30 + 188.25 = 218.25$
According to the model, when a fourth grader studies history for about 200 minutes per week, they will score 218 points.

95. **a.** From the information given, we have the data points $(0, 640)$ and $(4, 0)$. Use these points to find the slope of the equation for f:
$m = \dfrac{640 - 0}{0 - 4} = \dfrac{640}{-4} = -160$
So, $f(t) = -160t + b$. To solve for b, substitute 0 for t and 640 for $f(t)$, since the line contains $(0, 640)$.
$640 = -160(0) + b$
$640 = b$
The equation for f is $f(t) = -160t + 640$.

b.

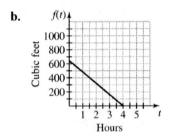

c. Since it takes 4 hours to pump out the water, the domain is $0 \le t \le 4$. Since the water level is at 640 cubic feet before starting to pump, the range is
$0 \le f(t) \le 640$.

97. The student mistakenly substituted 5 for x and solved for $f(x)$. The student should have substituted 5 for $f(x)$ and solved for x.
$5 = x + 2$
$3 = x$

99. **a.** To find $f(3)$, $f(5)$ and $f(8)$, substitute the correct value for x in $f(x)$:
$f(3) = 4(3) = 12$
$f(5) = 4(5) = 20$
$f(8) = 4(8) = 32$
To check $f(3 + 5) = f(3) + f(5)$, substitute for the above values:
$f(3 + 5) \overset{?}{=} f(3) + f(5)$
$f(8) \overset{?}{=} 12 + 20$
$32 \overset{?}{=} 32$ true
This is a true statement, so
$f(3 + 5) = f(3) + f(5)$ is a true statement.

b. To find $f(2)$, $f(3)$ and $f(5)$, substitute the correct value for x in $f(x)$:
$f(2) = (2)^2 = 4$
$f(3) = (3)^2 = 9$
$f(5) = (5)^2 = 25$
To check $f(2 + 3) = f(2) + f(3)$, substitute for the above values:
$f(2 + 3) \overset{?}{=} f(2) + f(3)$
$f(5) \overset{?}{=} 4 + 9$
$25 \overset{?}{=} 13$ false
This is not a true statement, so
$f(2 + 3) = f(2) + f(3)$ is not a true statement.

c. To find $f(9)$, $f(16)$ and $f(25)$, substitute the correct value for x in $f(x)$:
$f(9) = \sqrt{9} = 3$
$f(16) = \sqrt{16} = 4$
$f(25) = \sqrt{25} = 5$
To check $f(9 + 16) = f(9) + f(16)$, substitute for the above values:
$f(9 + 16) \overset{?}{=} f(9) + f(16)$
$f(25) \overset{?}{=} 3 + 4$
$5 \overset{?}{=} 7$ false
This is not a true statement, so
$f(9 + 16) = f(9) + f(16)$ is not a true statement.

d. No, it is not true that $f(a + b) = f(a) + f(b)$ for every function f. This is shown by two of these three functions.

101. No, the student is not correct.
Answers may vary. Example:
If $t = -3$, then the year is $2010 - 3 = 2007$, which is a valid year.

103. When $f(3) = 5$, the input for f is 3 and the output is 5. Possible equations for f may vary. Example:
$f(x) = 2x - 1$
$f(x) = \dfrac{1}{3}x + 4$
$f(x) = \dfrac{5}{3}x$

Homework 2.4

1. The average rate of change of the number of shredder models the company makes is given by dividing the change in the number of models offered by the change in years.

$$m = \frac{\text{change in models}}{\text{change in years}}$$
$$= \frac{41-2}{2012-1990}$$
$$= \frac{39}{22}$$
$$\approx 1.77$$

This means that the average rate of change of the number of shredder models was about 1.77 shredder models per year.

3. The average rate of change of the number of trips by Americans to Canada is given by dividing the change in trips by the change in years.

$$m = \frac{\text{change in trips}}{\text{change in years}} = \frac{11-44}{2011-2000} = \frac{-33}{11} = -3$$

This means that the average rate of change of the number of trips by Americans to Canada was –3 million trips per year, or a decrease of 3 million trips per year.

5. The average rate of change of the number of executions from the death penalty is given by dividing the change in the number of executions by the change in years.

$$m = \frac{\text{change in number}}{\text{change in years}}$$
$$= \frac{43-98}{2011-1999}$$
$$= \frac{-55}{12}$$
$$\approx -4.58$$

This means that the average rate of change of the number of executions from the death penalty is –4.58 executions per year, or a decrease of 4.58 executions per year.

7. The average rate of change of the cost per credit hour of classes is given by dividing the change in cost by the change in credits.

$$m = \frac{\text{change in cost}}{\text{change in credits}}$$
$$= \frac{1176-882}{12-9}$$
$$= \frac{294}{3}$$
$$= 98$$

This means that the average cost of classes at Triton College is $98 per credit hour.

9. **a.** Yes, there is a linear relationship between t and d. For every hour that passes, the car has gone 70 miles. The slope of the graph is 70 miles per hour, the speed of the car.

   b. $d = 70t$

11. **a.** Yes there is a linear relationship between t and n; the rate of increase is constant. The slope of the model is 5.3, which means the number of households that pay bills online has increased by 5.3 million per year.

   b. The n-intercept of the model is (0, 66). This means that at $t = 0$, or 2010, the number of U.S. households that paid bills online was 66 million.

   c. $n = 5.3t + 66$

   d. households
   $$= \frac{\text{households}}{\text{year}} \cdot \text{year} + \text{households}$$
   $$= \text{households} + \text{households}$$
   $$= \text{households}$$
   This unit analysis shows that this model uses the correct units.

   e. Since $2016 - 2010 = 6$, the year 2016 corresponds to $t = 6$.
   $$n = 5.3(6) + 66 = 31.8 + 66 = 97.8$$
   The model predicts that 97.8 million households will pay their bills online in 2016.

13. **a.** The slope of the graph is –0.7, which means the unemployment rate has decreased by 0.7 percentage point per year.

   b. $g(t) = -0.7t + 7.8$

   c. To find this quantity, we substitute 6 for $g(t)$ and solve for t.
   $$6 = -0.7t + 7.8$$
   $$-1.8 = -0.7t$$
   $$2.57 \approx t$$
   The model predicts that the unemployment rate will be 6% in $2012 + 2.57 \approx 2015$. This result is one year earlier than Smith's prediction.

d. Since $2017 - 2012 = 5$ years since $t = 0$, we can substitute 5 for t and solve for $g(t)$.
$$g(5) = -0.7(5) + 7.8 = -3.5 + 7.8 = 4.3$$
So in 2017, the unemployment rate will be 4.3%.

e. The t-intercept of the model is given by solving for t when $g(t) = 0$.
$$0 = -0.7t + 7.8$$
$$-7.8 = -0.7t$$
$$11.14 \approx t$$
The model predicts that no one will be unemployed in 2023. Model breakdown has occurred.

15. a. The slope of the graph is 45, or \$45 per course a student is taking.

b. $f(n) = 45n + 15.50$

c. $\text{dollars} = \dfrac{\text{dollars}}{\text{course}} \text{course} + \text{dollars}$
$$= \text{dollars} + \text{dollars}$$
$$= \text{dollars}$$
This unit analysis shows that this model uses the correct units.

d. To find $f(4)$, substitute 4 for n in the equation we found and compute.
$$f(4) = 45(4) + 15.50 = 180 + 15.50 = 195.50$$
$f(4)$ means that the cost of an ID plus textbook rental for 4 courses is \$195.50.

e. $240.50 = 45(n) + 15.50$
$$225 = 45n$$
$$5 = n$$
This means that a student who pays \$240.50 for an ID and textbook rental is taking 5 courses.

17. a. Since 0.05 gallons of gas are being used each mile, the slope is −0.05. This means that for every mile that is driven, the amount of gas in the tank decreases by 0.05 gallons.

b. An equation for g is in the form $g(x) = mx + b$. Since at the beginning of the trip, there are 15.3 gallons of gas in the tank, when $x = 0$, $b = 15.3$. So, $f(t) = -0.05t + 15.3$.

c. To find the x-intercept, let $g(x) = 0$ and solve for x.
$$0 = -0.05x + 15.3$$
$$-15.3 = -0.05x$$
$$306 = x$$
The x-intercept is (306, 0). This means that when 306 miles have been driven, the gas tank will be empty.

d. Since the car can be driven between 0 and 306 miles on the 15.3-gallon tank of gas, the domain is $0 \le x \le 306$. Since the gas tank is filled with 15.3 gallons and empty with 0 gallons, the range is $0 \le y \le 15.3$.

e. Find x when $g(x) = 1$.
$$1 = -0.05x + 15.3$$
$$-14.3 = -0.05x$$
$$286 = x$$
The car can be driven 286 miles before refueling.

19. a. Let s the sales of energy drinks in the United States (in millions of gallons) in the year that is t years since 2012.
$$s = 33.3t + 400$$

b. $\text{gallons} = \dfrac{\text{gallons}}{\text{year}} \text{year} + \text{gallons}$
$$= \text{gallons} + \text{gallons}$$
$$= \text{gallons}$$
This unit analysis shows that this model uses the correct units.

c. The slope is 33.3. This means sales are increasing by 33.3 million gallons per year.

d. To find this quantity, we substitute 600 for s and solve for t.
$$600 = 33.3t + 400$$
$$200 = 33.3t$$
$$6.0 \approx t$$
In the year $2012 + 6 = 2018$, sales will reach 600 million gallons.
$$\frac{600}{290} \approx 2.1$$
The average annual consumption per person will be about 2.1 gallons in 2018.

21. a. Let c be cab fare (in dollars) for traveling d miles.
$$c = 2d + 5.75$$

b. To find this quantity, we substitute 31.75 for c and solve for d.

$$31.75 = 2d + 5.75$$
$$26 = 2d$$
$$13 = d$$

The ride was 13 miles.

23. Let p be percentage of aluminum cans that were recycled in the United States in the year that is t years since 2011.

$$p = 3.6t + 65.1$$

To find this quantity, we substitute 75 for p and solve for t.

$$75 = 3.6t + 65.1$$
$$9.9 = 3.6t$$
$$2.75 = t$$

In $2011 + 2.75 \approx 2014$, 75% of aluminum cans will be recycled. This result suggests that the goal will be met.

25. Let b be the monthly bill for local calling and using m long-distance minutes.

$$b = 0.05m + 46.99$$

To find this quantity, we substitute 65.49 for b and solve for m.

$$65.49 = 0.05m + 46.99$$
$$18.50 = 0.05m$$
$$370 = m$$

The total number of long-distance minutes used was $500 + 370 = 870$ minutes.

27. The slope of this model is 1.82. This means that each year after 1980 the average salary of professors at four-year public colleges and universities increases by $1820.

29. The slope of this model is –0.27. This means that each year after 1900 the record time for the women's 400-meter run decreases by 0.27 seconds.

31. The slope of this model is 1.8. This means that a 1 degree Celsius change corresponds to a 1.8 degree Fahrenheit change.

33. a. Create a scattergram using the data in the table.

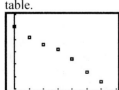

Use the first and last data points, (0, 70) and (60, 48) to find the slope.

$$m \approx \frac{70 - 48}{0 - 60} = \frac{22}{-60} \approx -0.37$$

So, the equation is of the form $f(t) = -0.37t + b$. Substitute 0 for t and 70 for $f(t)$ and solve for b.

$$70 = -0.37(0) + b$$
$$70 = 0 + b$$
$$70 = b$$

The equation is $f(t) = -0.37t + 70$.

(Your equation may be slightly different if you chose different points, or if you used the linear regression feature of your graphing calculator.)

b. The slope of f is –0.37. This means that the percentage of the world's population that lives in rural areas decreases by 0.37 of one percent each year.

c. In 2020, $t = 2020 - 1950 = 70$.

$$f(70) = -0.37(70) + 70$$
$$= -25.9 + 70$$
$$= 44.1$$

In 2020, about 44.1 percent of the world's population will live in rural areas. If the population will be 7.6 billion in 2020, then about $7.6(0.441) \approx 3.4$ billion people will live in rural areas.

d. 46% of the world's population will live in rural areas when $f(t) = 46$.

$$46 = -0.37t + 70$$
$$-24 = -0.37t$$
$$64.86 \approx t$$

So 46% of the world's population will live in rural areas in $1950 + 64.86 \approx 2015$.

e. The t-intercept is the value of t when $f(t) = 0$.

$$0 = -0.37t + 70$$
$$-70 = -0.37t$$
$$189.19 \approx t$$

This means that in the year $1950 + 189 = 2139$, no one in the world will live in a rural area. It is highly improbable that there will ever be a time that no one lives in a rural area. Also, it is highly unlikely that a mathematical model based on 58 years of data will be valid for a span of more than 180 years. Model breakdown has likely occurred.

35. a. Create a scattergram using the data in the table.

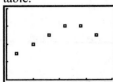

The data cannot be modeled well by a linear function because a line does not lie near all of the points.

b. Use the data points $(8, 0.5)$ and $(17, 0.5)$ to find the slope.
$$m = \frac{0.5 - 0.5}{17 - 8} = 0$$
So, the slope is zero which means an average rate of change of 0 billion barrels per year. If the data could be modeled well by a linear function, an average rate of change of 0 would mean that the oil production did not trend upward or downward during the period from 1992–2007. But the data shows an upward trend from 1992–2001 and a downward trend from 2004–2007, so the data cannot be modeled well by a linear model for this period.

c. Create a scattergram using the data in the table.

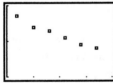

The data is approximately linear, so it can be modeled using a linear model. Use the data points, $(2, 2.7)$ and $(17, 1.8)$ to find the slope.
$$m = \frac{1.8 - 2.7}{17 - 2} = \frac{-0.9}{15} = -0.06$$
The equation is of the form
$U(t) = -0.06t + b$. Substitute 2 for t and
2.7 for $U(t)$ and solve for b.
$$2.7 = -0.06(2) + b$$
$$2.7 = -0.12 + b$$
$$2.82 = b$$
The equation is $U(t) = -0.06t + 2.82$.

(Your equation may be slightly different if you chose different points, or if you used the linear regression feature of your graphing calculator.)

d. The rate of change of the total U.S. oil production was –0.06 billion barrels of oil per year

e. If $U(t) = 1.1$ we can solve for t.
$$1.1 = -0.06t + 2.82$$
$$-1.72 = -0.06t$$
$$28.67 \approx t$$
The model predicts that the total U.S. oil production will be 1.1 billion barrels in $1990 + 28.67 \approx 2018$.

37. a. Create a scattergram using the data in the table.

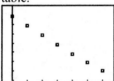

Use the first and last data points, $(0, 29.92)$ and $(6, 23.98)$ to find the slope.
$$m = \frac{29.92 - 23.98}{0 - 6} = -0.99$$
So the equation is of the form
$f(a) = -0.99a + b$. Substitute 0 for a and 29.92 for $f(a)$ and solve for b.
$$29.92 = -0.99(0) + b$$
$$29.92 = b$$
The equation is $f(a) = -0.99a + 29.92$.
(Your equation may be slightly different if you chose different points, or if you used the linear regression feature of your graphing calculator.)
Graphing the equation on the same graph as the scattergram, we see that the equation fits the data very well.

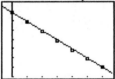

b. The slope of the model is –0.99. This means that for every increase of one thousand feet above sea level, the pressure measured in inches of mercury will decrease by 0.99 inches.

c. To find the average rate of change in pressure, divide the differences in the pressures by the differences in the elevation.

i. $m = \dfrac{25.84 - 28.86}{4 - 1} \approx -1.01$

This is a slightly steeper rate of change than we found in part (b).

ii. $m = \dfrac{24.89 - 27.82}{5 - 2} \approx -0.98$

This is a slightly shallower rate of change than we found in part (b).

iii. Since we used the points (0, 29.92) and (6, 23.98), we have already found this average rate of change to be −0.99.

d. To find the pressure at 14,440 feet, substitute 14.44 for a and solve for $f(a)$.
$f(a) = -0.99(14.44) + 29.92$
$f(a) \approx -14.30 + 29.92$
$f(a) \approx 15.62$
This means that at 14,440 feet the pressure is 15.62 inches of mercury.

39. a.

t	$f(t)$
0	0
1	500
2	1000
3	1500
4	1900
5	2300

b.

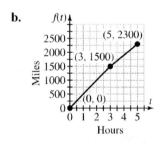

c. No, it will take some amount of time to decelerate from 500 mph to 400 mph. Answers may vary. Example: The given speeds represent average speeds over a time interval.

d. Answers may vary. Example: The constant rate of change property does not apply in this exercise. Since the rate is 500 miles per hour for the first few hours and 400 miles per hour for the next 2 hours, the rate of change is not constant. Since we do not have a constant rate of

change, we cannot say that the relationship between the variables is linear.

41. The points (2, 20) and (8, 10) are on the line and can be used to find the slope.

$m = \dfrac{y_2 - y_1}{x_2 - x_1} = \dfrac{10 - 20}{8 - 2} = \dfrac{-10}{6} = -\dfrac{5}{3} \approx -1.67$

The annual revenue is decreasing by about $1.67 million per year.

43. The points (0, 15) and (4, 20) are on the line and can be used to find the slope.

$m = \dfrac{y_2 - y_1}{x_2 - x_1} = \dfrac{20 - 15}{4 - 0} = \dfrac{5}{4} = 1.25$

The price increases $1.25 per ingredient.

45. Answers may vary. Example:

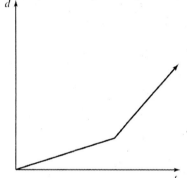

47. Answers may vary. Example:

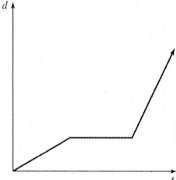

49. Answers may vary. Example: The rate of change is the quotient of the change in the dependent variable and the change in the independent variable. If both changes are positive, the quotient is positive.

51. Answers may vary. Example: Slope is a measurement of how much the dependent variable changes when the independent variable changes per unit.

Chapter 2 Review

1. $f(3) = 3(3)^2 - 7 = 3(9) - 7 = 27 - 7 = 20$

2. $f(-3) = 3(-3)^2 - 7 = 3(9) - 7 = 27 - 7 = 20$

3. $g(2) = \dfrac{2(2)+5}{3(2)+6} = \dfrac{4+5}{6+6} = \dfrac{9}{12} = \dfrac{3}{4}$

4. $h\left(\dfrac{3}{5}\right) = -10\left(\dfrac{3}{5}\right) - 3 = -2(3) - 3 = -6 - 3 = -9$

5. $h(a+3) = -10(a+3) - 3$
   $\quad\quad\quad = -10a - 30 - 3$
   $\quad\quad\quad = -10a - 33$

6. $-6 = 2x + 3$
   $-6 - 3 = 2x + 3 - 3$
   $-9 = 2x$
   $\dfrac{-9}{2} = x$

7. $\dfrac{2}{3} = 2x + 3$

   $\dfrac{2}{3} - 3 = 2x + 3 - 3$

   $\dfrac{-7}{3} = 2x$

   $\dfrac{1}{2}\left(\dfrac{-7}{3}\right) = \dfrac{1}{2}(2x)$

   $\dfrac{-7}{6} = x$

8. $a + 7 = 2x + 3$
   $a + 7 - 3 = 2x + 3 - 3$
   $a + 4 = 2x$
   $\dfrac{a+4}{2} = \dfrac{2x}{2}$
   $\dfrac{a}{2} + 2 = x$

9. Since the graph includes the point (2, 0), $f(2) = 0$.

10. Since the graph includes the point (0, 1), $f(0) = 1$.

11. Since the graph includes a point close to $(-3, 3.5)$, $f(-3) \approx 3.5$.

12. Since the graph includes the point $(-2, 3)$, $x = -2$ when $f(x) = 3$.

13. Since the graph includes the point (2, 0), $x = 2$ when $f(x) = 0$.

14. Since the graph includes the point (4, –1), $x = 4$ when $f(x) = -1$.

15. The domain of f is $-5 \leq x \leq 6$.

16. The range of f is $-2 \leq y \leq 4$.

17. Since x is 0 when $f(x) = 1$, $f(0) = 1$.

18. Since x is 2 when $f(x) = 4$, $f(2) = 4$.

19. When $f(x) = 0$, $x = 4$.

20. When $f(x) = 2$, $x = 1$.

21. $f(x)$ or $y = -7x + 3$
    To find the x-intercept, set $y = 0$ and solve for x.
    $0 = -7x + 3$
    $-3 = -7x$
    $x = \dfrac{-3}{-7} = \dfrac{3}{7}$
    The x-intercept is $\left(\dfrac{3}{7}, 0\right)$.

    To find the y-intercept, set $x = 0$ and solve for y.
    $y = -7(0) + 3 = 0 + 3 = 3$
    The y-intercept is (0, 3).

22. $f(x)$ or $y = 4$ is the equation of a horizontal line. The y-intercept of the line is (0, 4). There is no x-intercept.

23. $f(x)$ or $y = -\dfrac{4}{7}x + 2$
    To find the y-intercept, set $x = 0$ and solve for y.
    $y = -\dfrac{4}{7}(0) + 2 = 0 + 2 = 2$
    The y-intercept is (0, 2).
    To find the x-intercept, set $y = 0$ and solve for x.
    $0 = -\dfrac{4}{7}x + 2$
    $-2 = -\dfrac{4}{7}x$
    $-2\left(-\dfrac{7}{4}\right) = x$
    $x = \dfrac{7}{2}$
    The x-intercept is $\left(\dfrac{7}{2}, 0\right)$.

24. To find the *x*-intercept, set *y* equal to zero and solve for *x*;

$$2.56x - 9.41(0) = 78.25$$
$$2.56x = 78.25$$
$$x \approx 30.57$$

The *x*-intercept is (30.57, 0).

To find the *y*-intercept, set *x* equal to zero and solve for *y*;

$$2.56(0) - 9.41y = 78.25$$
$$-9.41y = 78.25$$
$$y \approx -8.32$$

The *y*-intercept is (0, −8.32).

25.

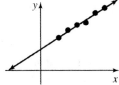

The new line has an increased slope since it is steeper than the model. The *y*-intercept is lower in the new line as it intersects the *y*-axis at a lower point than the model.

26. a. The student's car has 13 gallons of gas in the tank when no hours have been driven, so when $t = 0$, $b = 13$. Since the car uses 1.8 gallons of gas per hour, the slope is −1.8. An equation for *f* is

$$f(t) = -1.8t + 13 .$$

b. The slope of *f* is −1.8. This means that the amount of gasoline decreases at a constant rate of 1.8 gallons per hour of driving.

c. To find the *A*-intercept of *f*, let $t = 0$.

$$A = f(t) = -1.8(0) + 13 = 0 + 13 = 13$$

The *A*-intercept is (0, 13). This represents the amount of gas in a full tank before any time has been spent driving ($t = 0$). At that time, the car has 13 gallons of gas in the tank.

d. $\text{gallons} = \text{gallons} - \dfrac{\text{gallons}}{\text{hour}} \text{hour}$
$= \text{gallons} - \text{gallons}$
$= \text{gallons}$

This unit analysis shows that this model uses the correct units.

e. To find the *t*-intercept of *f*, let $f(t) = 0$.

$$0 = -1.8t + 13$$
$$0 + 1.8t = -1.8t + 13 + 1.8t$$
$$1.8t = 13$$
$$t = \frac{13}{1.8}$$
$$t \approx 7.22$$

The *t*-intercept is (7.22, 0). This means that the student can drive for 7.22 hours before running out of gas.

f. Since the car can be driven for 7.22 hours before running out of gas, the domain is $0 \le t \le 7.22$. Since the gas tank has between 0 and 13 gallons of gas, the range is $0 \le A \le 13$.

27. a. The slope of *C* is 37. The median annual compensation is increasing by \$37 thousand per year.

b. In 2010, $t = 0$ and the median annual compensation was \$870 thousand. Therefore, $b = 870$. Since the slope is 37, the equation for *C* is $C(t) = 37t + 870$.

c. To find $C(8)$, substitute 8 for *t*.

$$C(8) = 37(8) + 870 = 296 + 870 = 1166$$

The median annual compensation will be \$1166 thousand (\$1.166 million) in 2018.

d. Substitute 1000 for $C(t)$ and solve for *t*.

$$1000 = 37t + 870$$
$$130 = 37t$$
$$3.51 \approx t$$

The median annual compensation will be \$1000 thousand (\$1 million) in 2014.

28. The rate of change of annual sales of new light-duty vehicles is given by the difference in the sales divided by the difference in years.

$$m = \frac{\text{change in sales}}{\text{change in years}}$$
$$= \frac{5.65 - 7.78}{2010 - 2006}$$
$$= \frac{-2.13}{4}$$
$$\approx -0.53$$

The rate of change of annual sales is −0.53 million vehicles per year (−530 thousand vehicles per year).

29. If we let t be the number of years after 1990 and p be the average U.S. personal income, we can make a linear model of the given information. To start, we can use the two points (0, 14,420) and (18, 27,589) to find the slope of the model:

$$m = \frac{27,589 - 14,420}{18 - 0} = \frac{13,169}{18} \approx 731.6$$

So $p = 731.6t + b$. To solve for b substitute 0 for t and 14,420 for p since the line contains (0, 14,420).

$$14,420 = 488(0) + b$$
$$14,420 = b$$

The equation for p is $p = 731.6t + 14,420$.

To find the average U.S. personal income in 2010, we can substitute 20 for t and solve for p.

$$p = 731.6(20) + 14,420$$
$$= 14,632 + 14,420$$
$$= 29,052$$

According to the model, in 2010, the average U.S. personal income was $29,052. The actual average personal income for 2010 was $26,059, so the result from the model is an overestimate. Answers may vary. Example:
The average personal income in the years 2008–2010 may not have followed the linear model due to the poor economy.

30. a.

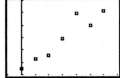

From the given information, we can select the data points (0, 32.5) and (12, 55.5). Use these points to find the slope of the equation for f.

$$m = \frac{55.5 - 32.5}{12 - 0} = \frac{23}{12} \approx 1.92$$

So, $f(t) = 1.92t + b$. Since the line contains (0, 32.5), substitute 0 for t and 32.5 for $f(t)$ and solve for b.

$$32.5 = 1.92(0) + b$$
$$32.5 = b$$

The linear model is $f(t) = 1.92t + 32.5$.

(Your equation may be slightly different if you chose different points, or if you used the linear regression feature of your graphing calculator.)

b. The slope of the model is 1.92. This means that for every year between 2000 and 2012, the standard mileage rate increases by 1.92 cents per mile.

c. The M-intercept is given by letting t equal zero and solving for M.
$$M = 1.92(0) + 32.5 = 32.5$$
This means that during 2000 the standard mileage rate was 32.5 cents per mile.

d. To predict when the standard mileage rate will be 69 cents per mile, substitute 69 for M and solve for t.
$$69 = 1.92t + 32.5$$
$$36.5 = 1.92t$$
$$19.0 \approx t$$
The model predicts that the standard mileage rate will be about 69 cents per mile in 2000 + 19 = 2019.

e. 2017 corresponds to $t = 17$.
$$M = 1.92(17) + 32.5 = 32.64 + 32.5 = 65.14$$
The model predicts that the standard mileage rate in 2017 will be 65 cents per mile, or $0.65 per mile.
If someone will drive 12,500 miles in 2017, their deduction will be given by the standard mileage rate times the distance they traveled.
$$12,500(0.65) = 8125$$
This person will be able to deduct $8125.

f. The estimate of 54.5 is the average of 50.5 and 58.5.

31. a.

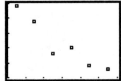

From the given information, we can select the data points (3, 34) and (9, 19). Use these points to find the slope of the equation for f.

$$m = \frac{19 - 34}{9 - 3} = \frac{-15}{6} = -2.5$$

So, $f(t) = -2.5t + b$. To solve for b, substitute 3 for t and 34 for $f(t)$ since the line contains (3, 34).

$$34 = -2.5(3) + b$$
$$34 = -7.5 + b$$
$$41.5 = b$$

The equation for f is $f(t) = -2.5t + 41.5$.

(Your equation may be slightly different if you chose different points, or if you used the linear regression feature of your graphing calculator.)

b. The slope is –2.5. This means that the percentage of Americans who think the First Amendment goes too far in the rights it guarantees decreases by 2.5 percentage points per year.

c. Set $f(t)$ equal to 3 and solve for t.
$$3 = -2.5t + 41.5$$
$$-38.5 = -2.5t$$
$$15.4 = t$$
According to the model, in $2000 + 15.4 \approx 2015$, 3% of Americans will think the First Amendment goes too far in the rights it guarantees.

d. $f(8) = -2.5(8) + 41.5 = -20 + 41.5 = 21.5$
The model predicts that in $2000 + 8 = 2008$, 21.5 % of Americans thought the First Amendment went too far in the rights it guarantees.

e.
$$8 = -2.5t + 41.5$$
$$-33.5 = -2.5t$$
$$13.4 = t$$
The model predicts that in $2000 + 13.4 \approx 2013$, 8% of Americans thought the First Amendment went too far in the rights it guarantees.

f. To find the t-intercept, set $f(t)$ equal to 0 and solve for t.
$$0 = -2.5t + 41.5$$
$$-41.5 = -2.5t$$
$$16.6 = t$$
According to the model, in $2000 + 16.6 \approx 2017$, there will be no Americans (0%) who think the First Amendment went too far in the rights it guarantees. Model breakdown has occurred.

Chapter 2 Test

1. Since the line includes the point $(-3, -2)$, $f(-3) = -2$.

2. Since the line includes the point $(3, 0)$, $f(3) = 0$.

3. Since the line includes the point $(0, -1)$, $f(0) = -1$.

4. Since the line includes the point $\left(-5, \dfrac{-8}{3}\right)$, $f(-5) = -\dfrac{8}{3}$ or approximately –2.7.

5. Since the line includes the point $(-6, -3)$, $x = -6$ when $f(x) = -3$.

6. Since the line includes the point $(-3, -2)$, $x = -3$ when $f(x) = -2$.

7. Since the line includes the point $(3, 0)$, $x = 3$ when $f(x) = 0$.

8. Since the line includes the point $(4.5, 0.5)$, $x = 4.5$ when $f(x) = 0.5$.

9. The domain of f is all the x-coordinates of the points in the graph. In this case f has a domain of $-6 \leq x \leq 6$.

10. The range of f is all the y-coordinates of the points in the graph. In this case f has a range of $-3 \leq y \leq 1$.

11. To find $f(-3)$, substitute -3 for x.
$$f(-3) = -4(-3) + 7 = 12 + 7 = 19$$

12. To find $f(a - 5)$, substitute $a - 5$ for x.
$$f(a - 5) = -4(a - 5) + 7$$
$$= -4a + 20 + 7$$
$$= -4a + 27$$

13. To find x when $f(x) = 2$, substitute a for $f(x)$.
$$2 = -4x + 7$$
$$-5 = -4x$$
$$\frac{5}{4} = x$$

14. To find x when $f(x) = a$, substitute 2 for $f(x)$.
$$a = -4x + 7$$
$$a - 7 = -4x$$
$$\frac{a - 7}{-4} = x \quad \text{or} \quad x = \frac{-a + 7}{4}$$

15. To find the x-intercept, let $f(x) = 0$ and solve for x.

$$0 = 3x - 7$$
$$7 = 3x$$
$$\frac{7}{3} = x$$

The x-intercept is $\left(\frac{7}{3}, 0\right)$.

To find the y-intercept, let $x = 0$ and solve for y or $f(x)$ since $y = f(x)$.

$$y = 3(0) - 7 = -7$$

The y-intercept is $(0, -7)$.

16. To find the x-intercept, let $g(x) = 0$ and solve for x.

$$0 = -2x$$
$$0 = x$$

The x-intercept is $(0, 0)$.

To find the y-intercept, let $x = 0$ and solve for y or $g(x)$ since $y = g(x)$.

$$y = -2(0) = 0$$

The y-intercept is $(0, 0)$.

17. To find the x-intercept, let $k(x) = 0$ and solve for x.

$$0 = \frac{1}{3}x - 8$$
$$8 = \frac{1}{3}x$$
$$24 = x$$

The x-intercept is $(24, 0)$.

To find the y-intercept, let $x = 0$ and solve for y or $k(x)$ since $y = k(x)$.

$$y = \frac{1}{3}(0) - 8 = -8$$

The y-intercept is $(0, -8)$.

18. a. Use the first three steps of the modeling process to find the equation. Begin by creating a scattergram using the data in the table.

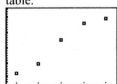

A line close to the data points passes through (16, 56.8) and (20, 81.8). Use those points to find the slope of the line.

$$m = \frac{56.8 - 81.8}{16 - 20} = 6.25$$

So $f(a) = 6.25a + b$. To solve for b, substitute 16 for a and 56.8 for $f(a)$ since the line contains (16, 56.8).

$$56.8 = 6.25(16) + b$$
$$56.8 = 100 + b$$
$$-43.2 = b$$

The equation for f is $f(a) = 6.25a - 43.2$. (Your equation may be slightly different if you chose different points, or if you used the linear regression feature of your graphing calculator.)

The graph of the model fits the data well.

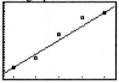

b. The slope of the model is 6.25. This means that the percentage of teenagers who have driver's licenses increases by 6.25 percentage points per year of age.

c.
$$0 = 6.25a - 43.2$$
$$43.2 = 6.25a$$
$$7 \approx a$$

According to the model, 0% of 7 year olds have a driver's license. While this might be true, it implies that some percentage of 8–15 year olds do have a driver's license, something that probably doesn't happen.

d. Let $a = 21$ and solve for $f(a)$.

$$f(21) = 6.25(21) - 43.2$$
$$= 131.25 - 43.2$$
$$= 88.05$$

This means that 88.05 percent of 21-year-old adults have a driver's license.

e. Let $f(a) = 100$ and solve for a.

$$100 = 6.25a - 43.2$$
$$143.2 = 6.25a$$
$$23 \approx a$$

This means that 100 percent of 23-year-old adults have a driver's license. Model breakdown has likely occurred.

19. a. Use the first three steps of the modeling process to find the equation. Begin by creating a scattergram using the data in the table.

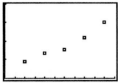

A line close to the data points passes through (8, 2.7) and (16, 4.4). Use those points to find the slope of the line.

$$m = \frac{4.4 - 2.7}{16 - 8} = \frac{1.7}{8} \approx 0.21$$

So $f(t) = 0.21t + b$. To solve for b, substitute 8 for t and 2.7 for $f(t)$ since the line contains (8, 2.7).

$$2.7 = 0.21(8) + b$$
$$2.7 = 1.68 + b$$
$$1.02 = b$$

The equation for f is $f(t) = 0.21t + 1.02$. (Your equation may be slightly different if you chose different points, or if you used the linear regression feature of your graphing calculator.)

b. The slope of the model is 0.21. This means the number of farmers markets is increasing by 0.21 thousand (210) markets per year.

c. $f(8) = 0.21(8) + 1.02 = 1.68 + 1.02 = 2.70$

This means in 1998, there were about 2.7 thousand farmers markets.

d. Let $f(t) = 8$ and solve for t.
$$8 = 0.21t + 1.02$$
$$6.98 = 0.21t$$
$$33.24 \approx t$$
This means in $1990 + 33.24 \approx 2023$, there will be 8 thousand farmers markets.

e. To find the t-intercept, set $f(t)$ equal to 0 and solve for t.
$$0 = 0.21t + 1.02$$
$$-1.02 = 0.21t$$
$$-4.86 \approx t$$
According to the model, in $1990 - 4.86 \approx 1985$, there were no farmers markets. Model breakdown has occurred.

20. Use the points (8, 239.1) and (20, 613.8) to find the slope of the line.
$$m = \frac{613.8 - 239.1}{20 - 8} = \frac{374.7}{12} \approx 31.23$$

Let t be the number of years after 1990. So $f(t) = 31.23t + b$. To solve for b, substitute 8 for t and 239.1 for $f(t)$ since the line contains (8, 239.1).
$$239.1 = 31.23(8) + b$$
$$239.1 = 249.84 + b$$
$$-10.74 = b$$
The equation for f is $f(t) = 31.23t - 10.74$. Predicting for 2017 using this equation, let $t = 27$ for $f(t)$.
$$f(27) = 31.23(27) - 10.74$$
$$= 843.21 - 10.74$$
$$= 832.47$$
This means that in 2017, total ad spending for the NCAA basketball tournament can be predicted to be about $832.5 million.

21. a. Yes, the rate of increase has remained constant, showing a linear relationship. The slope in this case is 1.7, or an increase of 1.7 states per year.

b. The n-intercept is the n value when $t = 0$, or during 2012. In 2012, $n = 29$, meaning there were 29 states with ethanol plants in 2012.

c. Based on the first two parts, the equation of the model is $n(t) = 1.7t + 29$.

d. Since $2018 - 2012 = 6$, $t = 6$ corresponds to 2018.
$$f(6) = 1.7(6) + 29 = 10.2 + 29 = 39.2$$
The model predicts that in 2012, 39 states will have ethanol plants.

e. To find when all states have ethanol plants, substitute 50 for n, and solve for t.
$$50 = 1.7t + 29$$
$$21 = 1.7t$$
$$12.35 \approx t$$
The model predicts that in $2012 + 12.35 \approx 2024$, all states will have ethanol plants.

22. Answers may vary. Example:

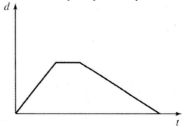

Chapter 3
Systems of Linear Equations

Homework 3.1

1.

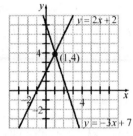

Verify that $(1,4)$ satisfies both equations.

$y = 2x + 2$ $y = -3x + 7$

$4 \overset{?}{=} 2(1) + 2$ $4 \overset{?}{=} -3(1) + 7$

$4 \overset{?}{=} 2 + 2$ $4 \overset{?}{=} -3 + 7$

$4 \overset{?}{=} 4$ true $4 \overset{?}{=} 4$ true

3.

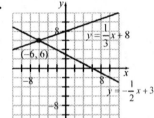

Verify that $(-6, 6)$ satisfies both equations.

$y = -\dfrac{1}{2}x + 3$ $y = \dfrac{1}{3}x + 8$

$6 \overset{?}{=} -\dfrac{1}{2}(-6) + 3$ $6 \overset{?}{=} \dfrac{1}{3}(-6) + 8$

$6 \overset{?}{=} 3 + 3$ $6 \overset{?}{=} -2 + 8$

$6 \overset{?}{=} 6$ true $6 \overset{?}{=} 6$ true

5. Write $y = 3(x-1)$ in slope–intercept form.

$y = 3(x-1)$

$y = 3x - 3$

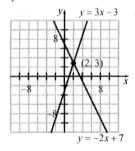

Verify that $(2,3)$ satisfies both equations.

$y = 3(x-1)$ $y = -2x + 7$

$3 \overset{?}{=} 3(2-1)$ $3 \overset{?}{=} -2(2) + 7$

$3 \overset{?}{=} 3(1)$ $3 \overset{?}{=} -4 + 7$

$3 \overset{?}{=} 3$ true $3 \overset{?}{=} 3$ true

7. Write both equations in slope–intercept form.

$x + 4y = 20$ $2x - 4y = -8$

$4y = 20 - x$ $-4y = -8 - 2x$

$y = 5 - \dfrac{1}{4}x$ $y = 2 + \dfrac{1}{2}x$

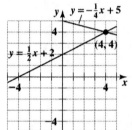

Verify that $(4,4)$ satisfies both equations.

$x + 4y = 20$ $2x - 4y = -8$

$(4) + 4(4) \overset{?}{=} 20$ $2(4) - 4(4) \overset{?}{=} -8$

$4 + 16 \overset{?}{=} 20$ $8 - 16 \overset{?}{=} -8$

$20 \overset{?}{=} 20$ true $-8 \overset{?}{=} -8$ true

9. Write both equations in slope–intercept form.

$5(y-2) = 21 - 2(x+3)$ $y = 3(x-1) + 8$

$5y - 10 = 21 - 2x - 6$ $y = 3x - 3 + 8$

$5y = -2x + 25$ $y = 3x + 5$

$y = -\dfrac{2}{5}x + 5$

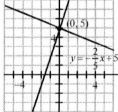

Verify that $(0,5)$ satisfies both equations.

$5(y-2) = 21 - 2(x+3)$ $y = 3(x-1) + 8$

$5(5-2) \overset{?}{=} 21 - 2(0+3)$ $5 \overset{?}{=} 3(0-1) + 8$

$5(3) \overset{?}{=} 21 - 2(3)$ $5 \overset{?}{=} 3(-1) + 8$

$15 \overset{?}{=} 15$ true $5 \overset{?}{=} 5$ true

11. Write $4y - 12 = -8x$ in slope–intercept form.

$$4y - 12 = -8x$$
$$4y = -8x + 12$$
$$y = -2x + 3$$

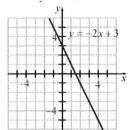

The system is dependent. The solution set is the set of numbers represented by points on the line $y = 2x + 3$.

13. Write both equations in slope–intercept form.

$$4x - 6y = 24 \qquad\qquad 6x - 9y = 18$$
$$-6y = -4x + 24 \qquad -9y = -6x + 18$$
$$y = \frac{2}{3}x - 4 \qquad\qquad y = \frac{2}{3}x - 2$$

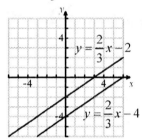

The system is inconsistent. There is no solution. The solution set is the empty set.

15. Write both equations in slope–intercept form.

$$\frac{1}{2}x - \frac{1}{2}y = 1 \qquad\qquad \frac{1}{4}x + \frac{1}{2}y = 2$$
$$x - y = 2 \qquad\qquad x + 2y = 8$$
$$-y = -x + 2 \qquad\qquad 2y = -x + 8$$
$$y = x - 2 \qquad\qquad y = -\frac{1}{2}x + 4$$

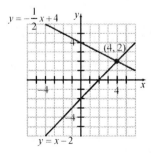

Verify that $(4, 2)$ satisfies both equations.

$$\frac{1}{2}x - \frac{1}{2}y = 1 \qquad\qquad \frac{1}{4}x + \frac{1}{2}y = 2$$
$$\frac{1}{2}(4) - \frac{1}{2}(2) \stackrel{?}{=} 1 \qquad \frac{1}{4}(4) + \frac{1}{2}(2) \stackrel{?}{=} 2$$
$$2 - 1 \stackrel{?}{=} 1 \qquad\qquad 1 + 1 \stackrel{?}{=} 2$$
$$1 \stackrel{?}{=} 1 \ \text{ true} \qquad\qquad 2 \stackrel{?}{=} 2 \ \text{ true}$$

17.

The approximate solution is $(-1.12, -3.69)$.

19. Write both equations in slope–intercept form.

$$2x + 5y = 7 \qquad\qquad 3x - 4y = -13$$
$$5y = 7 - 2x \qquad\qquad -4y = -3x - 13$$
$$y = \frac{7}{5} - \frac{2}{5}x \qquad\qquad y = \frac{3}{4}x + \frac{13}{4}$$

 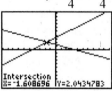

The approximate solution is $(-1.61, 2.04)$.

21. Write $2(2x - y) = 2$ in slope–intercept form.

$$2(2x - y) = 2$$
$$4x - 2y = 2$$
$$-2y = -4x + 2$$
$$y = 2x - 1$$

The equations are identical.

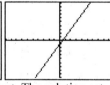

The system is dependent. The solution set contains all ordered pairs (x, y) such that $y = 2x - 1$.

23. Write $0.2y - x = 1$ in slope–intercept form.

$$0.2y - x = 1$$
$$0.2y = x + 1$$
$$y = 5x + 5$$

The system is inconsistent. The solution set is the empty set.

25. Write both equations in slope–intercept form.

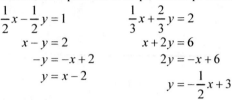

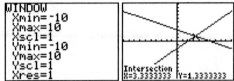

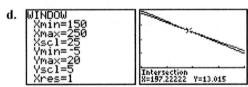

The approximate solution is $(3.33, 1.33)$.

27. a. In 2010, $t = 40$.

$$W(40) = -0.153(40) + 43.19 = 37.07$$

$$M(40) = -0.135(40) + 39.64 = 34.24$$

The women's time estimate is 37.07 seconds and the men's time estimate is 34.24 seconds. The error for the women's time estimate is $37.07 - 38.050 = -0.98$ second and the error for the men's time estimate is $34.24 - 34.910 = -0.67$ second.

b. The absolute value of the slope of W is more than the absolute value of the slope of M. This shows that the winning times of women decrease at a greater rate than the winning times of men.

c. Since the winning time for women is decreasing at a faster rate than for men, the winning time for men may equal the winning time for women in an upcoming year since these times are only a few seconds apart in the year 2010 and they are getting closer as each year passes.

d.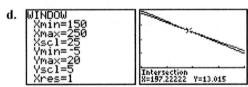

Men and women will have a winning time of 13.02 seconds in the year $1970 + 197 = 2167$.

29. a. Start by plotting the data. Then find the regression lines for the data.

Paid by check:

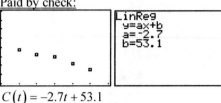

$C(t) = -2.7t + 53.1$

Paid online:

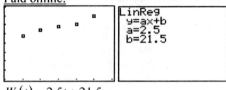

$W(t) = 2.5t + 21.5$

b.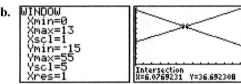

The percentage of bills paid by checks was equal to the percentage of bills paid by online payments in $2000 + 6.08 \approx 2006$. The percentage was about 37%.

c. In 2018, $t = 18$.

$$C(18) = -2.7(18) + 53.1 = 4.5$$

$$W(18) = 2.5(18) + 21.5 = 66.5$$

The total percentage of bills that will be paid by checks and online payments in 2018 is $4.5 + 66.5 = 71\%$.

31. a. Start by plotting the data. Then find the regression lines for the data.

Internet access:

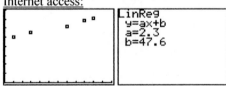

$I(t) = 2.3t + 47.6$

Broadband internet access:

$B(t) = 6.83t + 1.4$

b.

```
WINDOW
Xmin=0
Xmax=12
Xscl=1
Ymin=0
Ymax=80
Yscl=10
Xres=1
```

```
Intersection
X=10.191176  Y=71.039706
```

All households with Internet access had broadband Internet access in $2000 + 10.19 \approx 2010$.

c. Answers may vary. Example:
After 2010, the percentage of households with broadband Internet access is greater than the percentage of households with Internet access, which is not possible.

33. The solution of this system is estimated to be $(-1.9, -2.8)$.

35. To estimate the solution, identify the slopes of each line to extend the graphs until they intersect. The blue line has a slope of $-\dfrac{1}{2}$, while the red line has a slope of $\dfrac{1}{4}$. The blue line goes through the point (6, 1), while the red line goes through the point (6, –2). If both lines are extended, they intersect at (10, –1).

37. Using the tables of each line, notice that when $x = 3$, both equations are equal to –4. The intersection of these two lines is the point (3, –4).

39. Using the tables of each function notice that f has a slope of –3 and g has a slope of 5. Every time x increases by one, f decreases by 3 and g increases by 5. Using the table you can see that $f(3) = 21$, $f(4) = 18$, $g(3) = 17$ and $g(4) = 22$. This shows that f and g intersect between $x = 3$ and $x = 4$, or approximately at $x = 3.5$. Since the slope of f is –3, the y-value should be 21 – 3(0.5) = 19.5. The approximate solution is (3.5, 19.5).

41. $f(-4) = 0$

43. $f(x) = 3$ when $x = 5$.

45. $f(x) = g(x)$ when $x = -1$.

47. a. Points B and E satisfy $y = ax + b$ since these points lie on the graph of this equation.

b. Points E and F satisfy $y = cx + d$ since these points lie on the graph of this equation.

c. Point E satisfies both equations because it lies at the intersection of the graphs of the two equations.

d. Points A, C, and D do not satisfy either equation because they do not lie on the graph of either equation.

49. a. Answers may vary. Example:

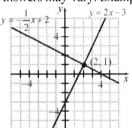

b. Answers may vary. Example:

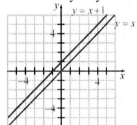

c. Answers may vary. Example:

$$y = \frac{3}{2}x$$

$$y = \frac{1}{2}(3x + 2) - 1$$

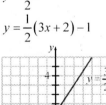

51. Graph all three equations.

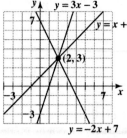

The solution of the system is $(2,3)$.

53. Answers may vary. Example:

$$y = -\frac{3}{4}x$$
$$y = x + 7$$
$$y = -2x - 5$$

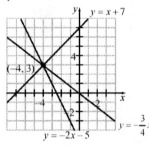

55. The student did not check whether $(1, 2)$ satisfies $y = -2x + 9$, which it does not. The student should examine the graph of the lines for their point of intersection. The common point to both graphs is $(2, 5)$.

57. Answers may vary. Example:
The graph of a linear equation is a visual representation of all the ordered pairs that satisfy the equation. Therefore, any intersection of two graphs shows an ordered pair that satisfies both equations, and is therefore a solution to the system of equations.

Homework 3.2

1. Substitute $x - 5$ for y in $x + y = 9$.

$$x + (x - 5) = 9$$
$$2x - 5 = 9$$
$$2x = 14$$
$$x = 7$$

Let $x = 7$ in $y = x - 5$.

$$y = 7 - 5 = 2$$

The solution is $(7, 2)$.

3. Substitute $4y + 7$ for x in $2x - 3y = -1$.

$$2(4y + 7) - 3y = -1$$
$$8y + 14 - 3y = -1$$
$$5y + 14 = -1$$
$$5y = -15$$
$$y = -3$$

Let $y = -3$ in $x = 4y + 7$.

$$x = 4(-3) + 7 = -12 + 7 = -5$$

The solution is $(-5, -3)$.

5. Substitute $2(x - 5)$ for y in $3x - 5y - 29 = 0$.

$$3x - 5(2(x - 5)) - 29 = 0$$
$$3x - 5(2x - 10) = 29$$
$$3x - 10x + 50 = 29$$
$$-7x = -21$$
$$x = 3$$

Let $x = 3$ in $y = 2(x - 5)$.

$$y = 2(3 - 5) = 2(-2) = -4$$

The solution is $(3, -4)$.

7. Substitute $99x$ for y in $y = 100x$.

$$99x = 100x$$
$$-x = 0$$
$$x = 0$$

Let $x = 0$ in $y = 99x$.

$$y = 99(0) = 0$$

The solution is $(0, 0)$.

9. Substitute $4x + 5$ for y in $y = 2x - 1$.

$$(4x + 5) = 2x - 1$$
$$2x + 5 = -1$$
$$2x = -6$$
$$x = -3$$

Let $x = -3$ in $y = 2x - 1$.

$$y = 2(-3) - 1 = -6 - 1 = -7$$

The solution is $(-3, -7)$.

11. Substitute $0.2x + 0.6$ for y in $2y - 3x = -4$.

$$2(0.2x + 0.6) - 3x = -4$$
$$0.4x + 1.2 - 3x = -4$$
$$-2.6x = -5.2$$
$$x = 2$$

Let $x = 2$ in $y = 0.2x + 0.6$.

$$y = 0.2(2) + 0.6 = 0.4 + 0.6 = 1$$

The solution is $(2, 1)$.

13. First solve $4x + 3y = 2$ for y.

$$4x + 3y = 2$$
$$3y = -4x + 2$$
$$y = -\frac{4}{3}x + \frac{2}{3}$$

Now substitute $-\frac{4}{3}x + \frac{2}{3}$ for y in

$2x - y = -4$.

$$2x - \left(-\frac{4}{3}x + \frac{2}{3}\right) = -4$$
$$2x + \frac{4}{3}x - \frac{2}{3} = -4$$
$$\frac{10}{3}x - \frac{2}{3} = -4$$
$$\frac{10}{3}x = -\frac{10}{3}$$
$$x = -1$$

Let $x = -1$ in $2x - y = -4$

$$2(-1) - y = -4$$
$$-2 - y = -4$$
$$-y = -2$$
$$y = 2$$

The solution is $(-1, 2)$.

15. Substitute $\frac{1}{2}x - 5$ for y in $2x + 3y = -1$.

$$2x + 3\left(\frac{1}{2}x - 5\right) = -1$$
$$2x + \frac{3}{2}x - 15 = -1$$
$$\frac{7}{2}x = 14$$
$$x = 4$$

Let $x = 4$ in $y = \frac{1}{2}x - 5$.

$$y = \frac{1}{2}(4) - 5 = 2 - 5 = -3$$

The solution is $(4, -3)$.

17. Add the two equations.

$$-x + 3y = -25$$
$$\underline{x - 5y = 39}$$
$$-2y = 14$$
$$y = -7$$

Substitute $y = -7$ into $x - 5y = 39$.

$$x - 5(-7) = 39$$
$$x + 35 = 39$$
$$x = 4$$

The solution is $(4, -7)$.

19. Multiply the first equation by -1 and add the equations.

$$-3x + 4y = 6$$
$$\underline{5x - 4y = -2}$$
$$2x = 4$$
$$x = 2$$

Substitute $x = 2$ into $5x - 4y = -2$ and solve for y.

$$5(2) - 4y = -2$$
$$10 - 4y = -2$$
$$-4y = -12$$
$$y = 3$$

The solution is $(2, 3)$.

21. Multiply the first equation by 2 and add the equations.

$$4x + 2y = 4$$
$$\underline{5x - 2y = -13}$$
$$9x = -9$$
$$x = -1$$

Substitute $x = -1$ into $5x - 2y = -13$ and solve for y.

$$5(-1) - 2y = -13$$
$$-5 - 2y = -13$$
$$-2y = -8$$
$$y = 4$$

The solution is $(-1, 4)$.

23. Multiply the first equation by 2 and add the equations.

$$6x - 4y = 14$$
$$\underline{-6x - 5y = 4}$$
$$-9y = 18$$
$$y = -2$$

Substitute $y = -2$ into $3x - 2y = 7$ and solve for x.

$$3x - 2(-2) = 7$$
$$3x + 4 = 7$$
$$3x = 3$$
$$x = 1$$

The solution is $(1, -2)$.

25. Multiply the first equation by 2 and the second equation by 5, and then add the equations.

$$6x + 10y = 6$$
$$\underline{35x - 10y = -170}$$
$$41x = -164$$
$$x = -4$$

Substitute $x = -4$ into $3x + 5y = 3$ and solve for y.

$3(-4) + 5y = 3$

$-12 + 5y = 3$

$5y = 15$

$y = 3$

The solution is (–4, 3).

27. Multiply the first equation by –3 and the second equation by 2, and then add the equations.

$-24x + 27y = 129$

$24x + 30y = 42$

$57y = 171$

$y = 3$

Substitute $y = 3$ into $8x - 9y = -43$ and solve for x.

$8x - 9(3) = -43$

$8x - 27 = -43$

$8x = -16$

$x = -2$

The solution is $(-2, 3)$.

29. Multiply the first equation by 5 and the second equation by 4, and then add the equations.

$20x - 35y = -145$

$-20x - 8y = 16$

$-43y = -129$

$y = 3$

Substitute $y = 3$ into $4x - 7y = -29$ and solve for x.

$4x - 7(3) = -29$

$4x - 21 = -29$

$4x = -8$

$x = -2$

The solution is (–2, 3).

31. Write the first equation in standard form. Multiply the first equation by 4 and the second equation by 3, and then add the equations.

$-8x + 12y = -24$

$15x - 12y = 3$

$7x = -21$

$x = -3$

Substitute $x = -3$ into $3y = 2x - 6$ and solve for y.

$3y = 2(-3) - 6$

$3y = -6 - 6$

$3y = -12$

$y = -4$

The solution is $(-3, -4)$.

33. Multiply the second equation by 2 and add the equations.

$0.9x + 0.4y = 1.9$

$0.6x - 0.4y = 2.6$

$1.5x = 4.5$

$x = 3$

Substitute $x = 3$ into $0.9x + 0.4y = 1.9$ and solve for y.

$0.9(3) + 0.4y = 1.9$

$2.7 + 0.4y = 1.9$

$0.4y = -0.8$

$y = -2$

The solution is $(3, -2)$.

35. Use the distributive property to simplify both equations.

$3(2x - 1) + 4(y - 3) = 1$

$6x - 3 + 4y - 12 = 1$

$6x + 4y = 16$

$4(x + 5) - 2(4y + 1) = 18$

$4x + 20 - 8y - 2 = 18$

$4x - 8y = 0$

The system can be rewritten as

$6x + 4y = 16$

$4x - 8y = 0$

Multiply the first equation by 2, and then add the equations.

$12x + 8y = 32$

$4x - 8y = 0$

$16x = 32$

$x = 2$

Substitute $x = 2$ into $4x - 8y = 0$ and solve for y.

$4(2) - 8y = 0$

$8 - 8y = 0$

$-8y = -8$

$y = 1$

The solution is $(2, 1)$.

37. Multiply the first equation by 3 and add the equations.

$\frac{3}{5}x + \frac{9}{2}y = 21$

$\frac{2}{5}x - \frac{9}{2}y = -16$

$x = 5$

Substitute $x = 5$ into $\frac{1}{5}x + \frac{3}{2}y = 7$ and solve for y.

$$\frac{1}{5}(5) + \frac{3}{2}y = 7$$

$$1 + \frac{3}{2}y = 7$$

$$\frac{3}{2}y = 6$$

$$y = 4$$

The solution is $(5, 4)$.

39. Multiply the first equation by 3 and the second equation by -4, then add the equations.

$$2x + \frac{3}{2}y = \frac{1}{2}$$

$$\underline{-2x - 5y = -11}$$

$$-\frac{7}{2}y = -\frac{21}{2}$$

$$y = 3$$

Substitute $y = 3$ into $\frac{2}{3}x + \frac{1}{2}y = \frac{1}{6}$ and solve for x.

$$\frac{2}{3}x + \frac{1}{2}(3) = \frac{1}{6}$$

$$\frac{2}{3}x + \frac{3}{2} = \frac{1}{6}$$

$$\frac{2}{3}x = -\frac{4}{3}$$

$$x = -2$$

The solution is $(-2, 3)$.

41. Substitute $2x + 5$ for y in $6x - 3y = -3$.

$$6x - 3(2x + 5) = -3$$

$$6x - 6x - 15 = -3$$

$$-15 = -3 \quad \text{false}$$

This is a contradiction. The system is inconsistent. The lines are parallel so the solution set is the empty set.

43. Multiply the first equation by 3 and the second equation by 2, then add the equations.

$$39x + 30y = -21$$

$$\underline{34x - 30y = 94}$$

$$73x = 73$$

$$x = 1$$

Substitute $x = 1$ into $13x + 10y = -7$ and solve for y.

$$13(1) + 10y = -7$$

$$13 + 10y = -7$$

$$10y = -20$$

$$y = -2$$

The solution is $(1, -2)$.

45. Multiply the first equation by 3 and add the equations.

$$12x - 15y = 9$$

$$\underline{-12x + 15y = -9}$$

$$0 = 0 \quad \text{true}$$

This is an identity. The system is dependent. The solution set is the set of ordered pairs (x, y) such that $4x - 5y = 3$.

47. Multiply the first equation by 5 and add the equations.

$$20x - 15y = 5$$

$$\underline{-20x + 15y = -3}$$

$$0 = -2 \quad \text{false}$$

This is a false statement, the system is inconsistent. There is no solution to this system.

49. Substitute $-2x + 4$ for y in $y = -4x + 10$ and solve for x.

$$(-2x + 4) = -4x + 10$$

$$-2x = -4x + 6$$

$$2x = 6$$

$$x = 3$$

Substitute $x = 3$ into $y = -2x + 4$ and solve for y.

$$y = -2(3) + 4 = -6 + 4 = -2$$

The solution is $(3, -2)$.

51. Use the distributive property to simplify both equations.

$$2(x + 3) - (y + 5) = -6$$

$$2x + 6 - y - 5 = -6$$

$$2x - y = -7$$

$$5(x - 2) + 3(y - 4) = -34$$

$$5x - 10 + 3y - 12 = -34$$

$$5x + 3y = -12$$

The system can be rewritten as

$$2x - y = -7$$

$$5x + 3y = -12$$

Multiply the first equation by 3, then add the two equations.

$$6x - 3y = -21$$

$$\underline{5x + 3y = -12}$$

$$11x = -33$$

$$x = -3$$

Substitute $x = -3$ into $2x - y = -7$

$$2(-3) - y = -7$$

$$-6 - y = -7$$

$$-y = -1$$

$$y = 1$$

The solution is $(-3, 1)$.

53. Substitute $\frac{1}{2}x + 3$ for y in $2y - x = 6$.

$$2\left(\frac{1}{2}x + 3\right) - x = 6$$
$$x + 6 - x = 6$$
$$6 = 6 \quad \text{true}$$

This is an identity. The system is dependent. The solution set is the set of ordered pairs (x, y) such that $y = \frac{1}{2}x + 3$.

55. Multiply the first equation by 2 and the second equation by 5, then add the equations.

$$\frac{5}{3}x + \frac{1}{2}y = 6$$
$$-\frac{5}{3}x + \frac{25}{2}y = 20$$
$$\overline{\phantom{-\frac{5}{3}x + }13y = 26}$$
$$y = 2$$

Substitute $y = 2$ into $\frac{5}{6}x + \frac{1}{4}y = 3$ and solve for x.

$$\frac{5}{6}x + \frac{1}{4}(2) = 3$$
$$\frac{5}{6}x + \frac{1}{2} = 3$$
$$\frac{5}{6}x = \frac{5}{2}$$
$$x = 3$$

The solution is $(3, 2)$.

57. First simplify the equations by multiplying by their LCD.

$$2(x + 2y) - 3(x - y) = 13$$
$$2x + 4y - 3x + 3y = 13$$
$$-x + 7y = 13$$
$$2(x + 3y) + (x + y) = 17$$
$$2x + 6y + x + y = 17$$
$$3x + 7y = 17$$

The system can be rewritten as
$$-x + 7y = 13$$
$$3x + 7y = 17$$

Multiply the first equation by -1 and add both equations.

$$x - 7y = -13$$
$$3x + 7y = 17$$
$$\overline{4x = 4}$$
$$x = 1$$

Substitute $x = 1$ into $3x + 7y = 17$ and solve for y.

$$3(1) + 7y = 17$$
$$3 + 7y = 17$$
$$7y = 14$$
$$y = 2$$

The solution is $(1, 2)$.

59. Substitute $2.58x - 8.31$ for y into $y = -3.25x + 7.86$ and solve for x.

$$(2.58x - 8.31) = -3.25x + 7.86$$
$$2.58x = -3.25x + 16.17$$
$$5.83x = 16.17$$
$$x \approx 2.77$$

Substitute $x = 2.77$ into $y = 2.58x - 8.31$ and solve for y.

$$y = 2.58(2.77) - 8.31 \approx 7.15 - 8.31 \approx -1.16$$

The approximate solution is $(2.77, -1.16)$.

61. Multiply the first equation by -3 and add the equations.

$$-3y = 2.31x - 14.52$$
$$y = -2.31x - 1.49$$
$$\overline{-2y = -16.01}$$
$$y \approx 8.01$$

Substitute $y = 8.01$ into $y = -0.77x + 4.84$ and solve for y.

$$8.01 = -0.77x + 4.84$$
$$3.17 = -0.77x$$
$$-4.12 \approx x$$

The approximate solution is $(-4.12, 8.01)$.

63. The graph shows that $y = \frac{1}{2}x + \frac{5}{2}$ and $y = 2x + 7$ intersect at the point $(-3, 1)$. So the solution of the given equation is $x = -3$.

65. The graph shows that $y = \frac{1}{2}x + \frac{5}{2}$ and $y = 3$ intersect as the point $(1, 3)$. So the solution of the given equation is $x = 1$.

67. The graph shows that $y = 2x + 7$ and $y = -3$ intersect at the point $(-5, -3)$. So the solution of the given equation is $x = -5$.

69. The graph shows that $y = \frac{1}{3}x + \frac{5}{3}$ and $y = x - 1$ intersect at the point $(4, 3)$. So the solution of the given equation is $x = 4$.

71. The graph shows that $y = \frac{1}{3}x + \frac{5}{2}$ and $y = 2$ intersect at the point $(1, 2)$. So the solution to the given equation is $x = 1$.

73. The graph shows that $y = \dfrac{1}{3}x + \dfrac{5}{3}$ and

$y = -3x - 5$ intersect at the point $(-2, 1)$. So the solution to the given system of equations is $(-2, 1)$.

75.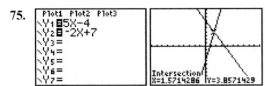

The solution to the equation is $x \approx 1.57$.

77.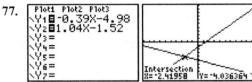

The solution to the equation is $x \approx -2.42$.

79.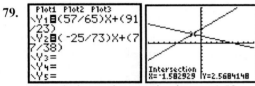

The solution to the equation is $x \approx -1.58$.

81. Notice that for $y = \dfrac{1}{2}x + \dfrac{7}{2}$ and $y = \dfrac{4}{5}x + 2$
when $x = 5$ they are both equal to 6. So the solution to the given equation is $x = 5$.

83. Notice that for $y = \dfrac{11}{10}x + \dfrac{17}{10}$ and $y = 5$ when
$x = 3$ they are both equal to 5. So the solution to the given equation is $x = 3$.

85. Notice that for $y = \dfrac{4}{5}x + 2$ and $y = \dfrac{11}{10}x + \dfrac{17}{10}$
when $x = 1$ they are both equal to 2.8. So the solution to the given system of equations is $(1, 2.8)$.

87. The function f has a slope of 4 and an f-intercept of $(0, 3)$. Therefore, $f(x) = 4x + 3$.
The function g has a slope of -6 and a g-intercept of $(0, 50)$. Therefore,
$g(x) = -6x + 50$. Write the system that describes the functions f and g.
$y = 4x + 3$
$y = -6x + 50$
Solve using substitution.

Substitute $4x + 3$ for y in $y = -6x + 50$ and solve for x.
$$4x + 3 = -6x + 50$$
$$10x = 47$$
$$x = 4.7$$
Substitute $x = 4.7$ into $y = 4x + 3$ and solve for y.
$$y = 4(4.7) + 3 = 18.8 + 3 = 21.8$$
The solution is $(4.7, 21.8)$.

89. Elimination:
Rewrite the second equation so the variables are on the left side of the equation.
$$3x + y = 11$$
$$2x + y = 9$$
Multiply the second equation by -1 and add the equations.
$$3x + y = 11$$
$$\underline{-2x - y = -9}$$
$$x = 2$$
Substitute $x = 2$ into $y = -2x + 9$ and solve for y.
$$y = -2(2) + 9 = 5$$
The solution is $(2, 5)$.

Substitution:
Substitute $-2x + 9$ for y in the first equation.
$$3x + (-2x + 9) = 11$$
$$3x - 2x + 9 = 11$$
$$x = 2$$
Substitute $x = 2$ into $y = -2x + 9$ and solve for y.
$$y = -2(2) + 9 = -4 + 9 = 5$$
The solution is $(2, 5)$.

Graphing:

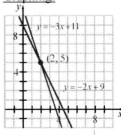

91. a. Multiply the first equation by 3 and the second equation by –2, then add the equations.

$$\begin{array}{r} 6x + 12y = 30 \\ -6x + 14y = -4 \\ \hline 26y = 26 \\ y = 1 \end{array}$$

Substitute $y = 1$ into $2x + 4y = 10$

$$2x + 4(1) = 10$$
$$2x + 4 = 10$$
$$2x = 6$$
$$x = 3$$

The solution is (3, 1).

b. Multiply the first equation by 7 and the second equation by 4, then add the equations.

$$\begin{array}{r} 14x + 28y = 70 \\ 12x - 28y = 8 \\ \hline 26x = 78 \\ x = 3 \end{array}$$

Substitute $x = 3$ into $2x + 4y = 10$

$$2(3) + 4y = 10$$
$$6 + 4y = 10$$
$$4y = 4$$
$$y = 1$$

The solution is (3, 1).

c. The results are the same if the x terms or the y terms are eliminated. It doesn't matter which variable you choose to eliminate first.

93. The student is not correct. A system is inconsistent (i.e. the solution set is the empty set) when the lines are parallel. Parallel lines must have the same slope. The slopes are different yet they are close enough that the lines look parallel around the origin. To find the correct solution, substitute $2x + 3$ for y in $y = 2.01x + 1$ and solve for x.

$$2x + 3 = 2.01x + 1$$
$$-0.01x = -2$$
$$x = 200$$

Substitute $x = 200$ into $y = 2x + 3$ and solve for y.

$$y = 2(200) + 3 = 403$$

The solution is $(200, 403)$.

95. The coordinates for A are $(0,0)$ since it lies at the origin. The coordinates for B are the same as the coordinates of the y-intercept of ℓ_1, which is $(0,3)$. The coordinates for C are the same as the point of intersection of ℓ_1 and ℓ_2. Solve the following system.

$$\ell_1 : y = 2x + 3$$
$$\ell_2 : 3y + x = 30$$

Substitute $2x + 3$ for y in the second equation.

$$3(2x + 3) + x = 30$$
$$6x + 9 + x = 30$$
$$7x = 21$$
$$x = 3$$

Substitute $x = 3$ into the first equation and solve for y.

$$y = 2(3) + 3 = 6 + 3 = 9$$

The solution is $(3,9)$ so the coordinates of C are $(3,9)$. The coordinates for D are the same as the coordinates of the point of intersection of ℓ_2 and ℓ_3. Solve the following system.

$$\ell_2 : 3y + x = 30$$
$$\ell_3 : y + 3x = 26$$

Multiply the second equation by -3 and add the equations.

$$\begin{array}{r} 3y + x = 30 \\ -3y - 9x = -78 \\ \hline -8x = -48 \\ x = 6 \end{array}$$

Substitute $x = 6$ into $3y + x = 30$ and solve for y.

$$3y + (6) = 30$$
$$3y = 24$$
$$y = 8$$

The solution is $(6,8)$ so the coordinates of D are $(6,8)$. The coordinates for E are the same as the coordinates of the point of intersection of ℓ_3 and ℓ_4. Solve the following system.

$$\ell_3 : y + 3x = 26$$
$$\ell_4 : y = 2x - 10$$

Substitute $2x - 10$ in for y in $y + 3x = 26$.

$$2x - 10 + 3x = 26$$
$$5x = 36$$
$$x = 7.2$$

Substitute $x = 7.2$ into $y = 2x - 10$ and solve for x.

$$y = 2(7.2) - 10 = 4.4$$

The solution is $(7.2, 4.4)$ so the coordinates of E are $(7.2, 4.4)$. The coordinates of F are the same as the coordinates of the x-intercept of ℓ_4. Let $y = 0$ in ℓ_4 and solve for x.

$$y = 2x - 10$$
$$0 = 2x - 10$$
$$2x = 10$$
$$x = 5$$

The coordinates of F are $(5, 0)$.

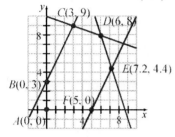

97. a. You may use more than one method to solve the system. For example, you may use substitution as follows.
Solve the first equation for y.

$$ax + by = c$$
$$by = -ax + c$$
$$y = \frac{-ax + c}{b}$$

Substitute this result for y in the second equation and solve for x.

$$kx + p\left(\frac{-ax + c}{b}\right) = d$$
$$kx - \frac{apx}{b} + \frac{cp}{b} = d$$
$$bkx - apx + cp = bd$$
$$(bk - ap)x + cp = bd$$
$$(kd - ap)x = bd - cp$$
$$x = \frac{bd - cp}{bk - ap} \quad \text{or} \quad \frac{cp - bd}{ap - bk}$$

Substitute this result for x in $ax + by = c$ and solve for y.

$$a\left(\frac{cp - bd}{ap - bk}\right) + by = c$$
$$\frac{acp - abd}{ap - bk} + by = c$$
$$by = c - \frac{acp - abd}{ap - bk}$$
$$by = \frac{acp - bck}{ap - bk} - \frac{acp - abd}{ap - bk}$$
$$by = \frac{-bck + abd}{ap - bk}$$
$$by = \frac{abd - bck}{ap - bk}$$
$$y = \frac{ad - ck}{ap - bk} \quad \text{or}$$
$$y = \frac{ck - ad}{bk - ap}$$

Therefore, the solution is

$$\left(\frac{cp - bd}{ap - bk}, \frac{ad - ck}{ap - bk}\right), \text{ assuming}$$

$ap - bk \neq 0$.

b. Substitute 3 for a, 5 for b, 2 for c, 4 for k, 3 for p, and 4 for d in the solution from part (a).

$$x = \frac{2(3) - 5(4)}{3(3) - 5(4)} = \frac{-14}{-11} = \frac{14}{11}$$
$$y = \frac{3(4) - 2(4)}{3(3) - 5(4)} = \frac{4}{-11} = -\frac{4}{11}$$

The solution is $\left(\frac{14}{11}, -\frac{4}{11}\right)$.

99. Answers may vary.

101. Answers may vary.

Homework 3.3

1. Solve the system
    $$y = -0.153t + 43.19$$
    $$y = -0.135t + 39.64$$
    Substitute $y = -0.153t + 43.19$ for y in the second equation and solve for t.
    $$-0.153t + 43.19 = -0.135t + 39.64$$
    $$-0.018t = -3.55$$
    $$t \approx 197.22$$
    Substitute this result into the first equation and solve for y.
    $$y = -0.153(197.22) + 43.19 \approx 13.02$$
    According to the models, the winning times for women and men will both be 13.02 seconds in the year $1970 + 197.22 \approx 2167$. Model breakdown has likely occurred.

3. **a.** The slope of I is 2.30. The slope of B is 6.83. The percentage of households that have Internet access is increasing by 2.30 percentage points per year. The percentage of households that have broadband Internet access is increasing by 6.83 percentage points per year.

    b. Answers may vary. Example: Since the rate of increase in the percentage of households that have broadband Internet access is greater than the rate of increase in the percentage of households that have Internet access, some households must be switching from another form of Internet access to broadband Internet access.

    c. Substitute $I(t) = 2.30t + 47.6$ for $B(t)$ in the second equation and solve for t.
    $$2.30t + 47.6 = 6.83t + 1.4$$
    $$-4.53t = -46.2$$
    $$t \approx 10.2$$
    The model predicts that in $2000 + 10.2 \approx 2010$, all households with Internet access will have broadband Internet access.

5. **a.** Find the regression lines for the data.
    Milk:
    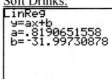
    $$M(t) = -0.28t + 50.87$$
    Soft Drinks:

    LinReg
    y=ax+b
    a=.8190651558
    b=-31.99730878

    $$S(t) = 0.82t - 32.00$$

    b. Substitute $M(t) = -0.28t + 50.87$ for $S(t)$ in the second equation and solve for t.
    $$-0.28t + 50.87 = 0.82t - 32.00$$
    $$82.87 = 1.10t$$
    $$75.34 \approx t$$
    During 1975, the per-person milk and soft drink consumptions were the same.
    $$M(75.34) = -0.28(75.34) + 50.87 \approx 29.8$$
    $$S(75.34) = 0.82(75.34) - 32.00 \approx 29.8$$
    The annual consumption per-person in 1975 was about 29.8 gallons.

    c.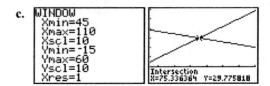

    d. Answers may vary.

    e. Answers may vary.

7. **a.** Find the regression lines for the data.
    Knowledge:

    LinReg
    y=ax+b
    a=.0144186047
    b=-.7465116279

    $$K(a) = 0.014a - 0.747$$

    Memory:

    LinReg
    y=ax+b
    a=-.028255814
    b=1.664534884

    $$M(a) = -0.028a + 1.66$$

b. Solve the system
$$K(a) = 0.014a - 0.747$$
$$M(a) = -0.028a + 1.66$$
Substitute $K(a) = 0.014a - 0.747$ for $M(a)$ in the second equation and solve for a.
$$0.014a - 0.747 = -0.028a + 1.66$$
$$0.042a = 2.407$$
$$a \approx 57.31$$
The scores will be roughly equal when a person is 57 years old.

c. Find the intersection point using a graphing utility.

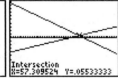

This confirms our original solution of 57 years old.

9. a. Solve the system
$$y = 13.5t + 229$$
$$y = -5t + 365$$
Substitute $13.5t + 229$ for y in the second equation and solve for t.
$$13.5t + 229 = -5t + 365$$
$$18.5t = 136$$
$$t \approx 7.35$$
Substitute this result into the first equation and solve for y.
$$y = 13.5(7.35) + 229 \approx 328.23$$
According to the models, the two newspapers had equal circulations of roughly 328 thousand in 1997.

b. Since the circulations were roughly equal in 1997, competition heated up as each newspaper tried to overtake the other.

c. $D(10) = 13.5(10) + 229 = 135 + 229 = 364$
$R(10) = -5(10) + 365 = -50 + 365 = 315$
$826 - (364 + 315) = 826 - 679 = 147$
According to the models, the combined increase due to bonus copies was roughly 147 thousand bonus copies.

d. $D(11) = 13.5(11) + 229 = 377.5$
$R(11) = -5(11) + 365 = 310$
$377.5 + 310 = 687.5$
According to the models, the combined circulation of the two newspapers was roughly 688 thousand copies in 2001.

e. The estimate in part (d) was an overestimate. After joining revenue streams, the competition for subscribers ceased (or, at least, reduced if there were other competitors). The end of bonus copies, or just the merger in general, may have caused some subscribers to cancel subscriptions.

11. a. For both $F(t)$ and $D(t)$ we are given the F- and D- intercepts and the slope in terms of the price of the cars in 2011 and their depreciation rates, respectively.
$$F(t) = -1414t + 14,290$$
$$D(t) = -3740t + 30,450$$

b. Substitute $-1414t + 14,290$ for D in $D(t) = -3740t + 30,450$ and solve for t.
$$-1414t + 14,290 = -3740t + 30,450$$
$$2326t = 16,160$$
$$t \approx 6.948$$
So both cars will have the same value around $2012 + 6.948 \approx 2019$. To find that value, substitute $t = 6.948$ into $F(t)$.
$$F(6.948) = -1414(6.948) + 14,290 \approx 4466$$
So in 2012 they will both be worth about $4466.

c. Find the intersection point using a graphing utility.

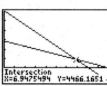

This confirms our original solution of (6.948, 4466).

13. a. Since the Nutrisystem program fees increase by a constant $91 each week, the function N is linear and its slope is 91. The N-intercept is $(0,0)$, since there is no start-up fee at $t = 0$. So, an equation for $N(t)$ is $N(t) = 91t$.

Since the Weight Watchers program fees increase by a constant $87 each week ($12 fee + $75 food), the function W is linear and its slope is 87. The W-intercept is $(0,20)$ since the start-up fee is $20 at $t = 0$. An equation for $W(t)$ is

$W(t) = 87t + 20$.

b. $N(t) = 91t$

$$\text{dollars} = \frac{\text{dollars}}{\text{week}} \text{weeks}$$

dollars = dollars

The units of $N(t)$ are correct.

$W(t) = 87t + 20$

$$\text{dollars} = \frac{\text{dollars}}{\text{week}} \text{weeks} + \text{dollars}$$

dollars = dollars + dollars

dollars = dollars

The units of $W(t)$ are correct.

c. Solve the system

$y = 91t$

$y = 87t + 20$

Substitute $87t + 20$ for y into the first equation.

$87t + 20 = 91t$

$\quad 20 = 4t$

$\quad\quad 5 = t$

Substitute this result into the first equation and solve for y.

$y = 91(5) = 455$

In 5 weeks, the total cost at both Nutrisystem and Weight Watchers is $455.

d. Find the intersection point using a graphing utility.

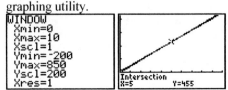

e. After 7 weeks, the cost of the Nutrisystem program is $N(7) = 91(7) = \$637$ and the cost of the Weight Watchers program is $W(7) = 87(7) + 20 = \$629$. The Weight Watchers program is less expensive for the first 7 weeks.

15. The percentage of U.S. electricity generated from natural gas $N(t)$ and the percentage of U.S. electricity generated from coal $C(t)$ can be represented by functions in terms of t, the years after 2010.

$N(t) = 0.9t + 24$

$C(t) = -0.7t + 44$

To predict when the percentage of U.S. electricity generated from natural gas will be equal to that from coal, set $N(t) = C(t)$ and solve for t.

$0.9t + 24 = -0.7t + 44$

$\quad\quad 1.6t = 20$

$\quad\quad\quad t = 12.5$

In $2010 + 12.5 \approx 2023$, the percentage of U.S. electricity generated from natural gas will be equal to that from coal.

To find this percentage, substitute this result into the first equation and solve for $N(t)$.

$N(12.5) = 0.9(12.5) + 24 \approx 35.3$

In 2023, 35.3% of U.S. electricity will be generated from natural gas and 35.3% will be generated from coal.

17. a. For each state we are given the personal incomes in 2000 and the personal incomes in 2010. Since the personal income in both states increased linearly we can write a linear model. To do this we first need to find the rate of increase for both states.

$$m_H = \frac{37.6 - 24.5}{2010 - 2000} = \frac{13.1}{10} = 1.31$$

$$m_C = \frac{38.8 - 28.9}{2010 - 2000} = \frac{9.9}{10} = 0.99$$

So $H(t) = 1.31t + 24.5$ and

$C(t) = 0.99t + 28.9$.

b. Substitute $1.31t + 24.5$ for $C(t)$ in
$C(t) = 0.99t + 28.9$ and solve for t.

$$1.31t + 24.5 = 0.99t + 28.9$$
$$0.32t = 4.4$$
$$t = 13.75$$

So during $2000 + 13.75 \approx 2014$, personal income in Hawaii will equal the personal income in Colorado.

To find this income, substitute this result into the first equation and solve for $H(t)$.

$$H(13.75) = 1.31(13.75) + 24.5 \approx 42.5$$

In 2014, the average personal income in both Hawaii and Colorado will be approximately $42.5 thousand.

c. Find the intersection point using a graphing utility.

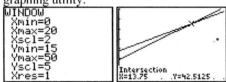

d. $H(17) = 1.31(17) + 24.5 = 46.77$
$C(17) = 0.99(17) + 28.9 = 45.73$

$$\frac{H(17) + C(17)}{2} = \frac{46.77 + 45.73}{2} = 46.25$$

The average is $46.25 thousand. This result is likely to be an overestimate. Answers may vary.

19. First find the slope of General Motor's market share, $G(t)$, and then find the slope of Ford's market share, $F(t)$.

$$m_G = \frac{19.4 - 22.1}{2011 - 2008} = \frac{-2.7}{3} = -0.90$$

$$m_F = \frac{16.5 - 14.2}{2011 - 2008} = \frac{2.3}{3} \approx 0.77$$

If we let t be the years after 2008, then $G(0) = 22.1$ and $F(0) = 14.2$, and we find that

$$G(t) = -0.90t + 22.1$$
$$F(t) = 0.77t + 14.2$$

Substitute $-0.90t + 22.1$ for $F(t)$ in $F(t) = 0.77t + 14.2$ and solve for t.

$$-0.90t + 22.1 = 0.77t + 14.2$$
$$7.9 = 1.67t$$
$$4.73 \approx t$$

So their market share will be equal around $2008 + 4.73 \approx 2013$. To find the market share, substitute $t = 4.73$ in $G(t)$.

$$G(4.73) = -0.90(4.73) + 22.1 \approx 17.8$$

So General Motors and Ford will both have a market share of about 17.8% in 2013.

21. a. Due to rounding, the solution is not exact.

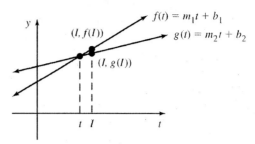

b. $f(I)$ will be larger than $g(I)$. Notice in the previous graph the slope of f is larger than the slope of g.

Homework 3.4

1. The total number of seats is 5000. To find the total revenue multiply the number of each type of seat by how much they cost and add these numbers together. Let the $27 seats be x and the $40 seats be y. This gives the following system of equations.

$$x + y = 5000$$
$$27x + 40y = 150,600$$

Substitute $5000 - x$ for y in the second equation and solve for x.

$$27x + 40(5000 - x) = 150,600$$
$$27x + 200,000 - 40x = 150,600$$
$$-13x + 200,000 = 150,600$$
$$-13x = -49,400$$
$$x = 3800$$

So about 3800 $27 tickets should be sold and 1200 $40 tickets should be sold.

3. Let the number of *Plans* albums sold be represented by x and the number of *Codes and Keys* albums sold be represented by y. This gives the following system of equations.

$$x + y = 836$$
$$9.99x + 10.99y = 9059.64$$

Substitute $836 - x$ for y in the second equation and solve for x.

$$9.99x + 10.99(836 - x) = 9059.64$$
$$9.99x + 9187.64 - 10.99x = 9059.64$$
$$-x = -128$$
$$x = 128$$

So 128 *Plans* albums and $836 - 128 = 708$ *Codes and Keys* albums were sold.

5. Let the price of main-level seats be p and let the price of the balcony seats be m. These two prices are related by $m = p - 15$. This gives the following system of equations.
$$m = p - 15$$
$$500m + 1800p = 84,500$$
Substitute $p - 15$ for m in the second equation and solve for p.
$$500(p - 15) + 1800p = 84,500$$
$$500p - 7500 + 1800p = 84,500$$
$$2300p = 92,000$$
$$p = 40$$
So the main-level seats should cost \$40 and the balcony seats should cost \$25.

7. A system of equations can be written based on the given information: there will be three times as many full-time students as part-time students, each full-time student takes 14 units, each part-time student takes 3 units, and each unit costs \$13.
$$3p = f$$
$$13(14f + 3p) = 877,500$$
Substitute $3p$ for f in the second equation and solve for p.
$$13(14(3p) + 3p) = 877,500$$
$$13(42p + 3p) = 877,500$$
$$13(45p) = 877,500$$
$$585p = 877,500$$
$$p = 1500$$
There need to be 1500 part-time students and 4500 full-time students in order for the revenue to be \$877,500.

9. **a.** The total revenue will be given by multiplying the number of tickets by their respective price, and adding these amounts. The total number of tickets will be 20,000 so $x + y = 20,000$. We only want R to be in terms of x so $y = 20,000 - x$.
$$f(x) = 50x + 75(20,000 - x)$$
$$f(x) = -25x + 1,500,000$$

   b.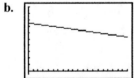

   The slope of the graph is –25. This represents the fact that as more \$50 tickets are sold, the revenue decreases by \$25 per ticket compared to selling only \$75 tickets.

c. $f(16,000)$
$$= 50(16,000) + 75(20,000 - 16,000)$$
$$= 800,000 + 75(4000)$$
$$= 800,000 + 300,000$$
$$= 1,100,000$$
$f(16,000)$ represents the amount of revenue made when 16,000 \$50 tickets are sold.

d. The total revenue must cover the \$475,000 production cost and make \$600,000 in profit. This means that the total revenue must be \$1,075,000.
$$1,075,000 = 50x + 75(20,000 - x)$$
$$1,075,000 = 50x + 1,500,000 - 75x$$
$$1,075,000 = -25x + 1,500,000$$
$$-425,000 = -25x$$
$$17,000 = x$$
So 17,000 of the \$50 tickets and 3000 of the \$75 tickets must be sold.

11. **a.** The total revenue will be given by multiplying the number of tickets by their respective prices, and adding these amounts. The total number of tickets will be 12,000 so $x + y = 12,000$. We only want R to be in terms of x so $y = 12,000 - x$.
$$f(x) = 45x + 70(12,000 - x)$$
$$= 45x + 840,000 - 70x$$
$$= -25x + 840,000$$

   b.

X	Y₁
0	840000
2000	790000
4000	740000
6000	690000
8000	640000
10000	590000
12000	540000

X=0

   In each case $f(x)$ is the total revenue when x of the \$45 tickets are sold.

   c. The possible revenues from the concert range from \$540,000 to \$840,000.

   d. $602,500 = -25x + 840,000$
$$-237,500 = -25x$$
$$9500 = x$$
   So 9500 of the \$45 tickets and 2500 of the \$70 tickets should be sold.

13. a. Since a first-class ticket is $242 more than a coach ticket, $y = 242 + x$. To find the total revenue of a single flight, multiply the number of tickets sold by their respective prices, and then add these amounts.
$$R = 8y + 126x$$
$$f(x) = 8(242 + x) + 126x$$
$$= 1936 + 8x + 126x$$
$$= 134x + 1936$$

b. The slope of the graph of f is 134. As the price of a coach ticket increases by a dollar, the total revenue increases by $134 on a full plane.

c.
$$14,130 = 134x + 1936$$
$$12,194 = 134x$$
$$91 = x$$
United Airlines should charge $91 for a coach ticket, and $333 for a first-class ticket.

15. Let x be the amount of money put into the American Funds New Perspective F account, and y be the amount of money put into the Oppenheimer Global Y account. Both amounts of money add to $15,000, and their interest after one year will be $1410.
$$x + y = 15,000$$
$$0.09x + 0.11y = 1410$$
Substitute $15,000 - y$ for x in the second equation and solve for y.
$$0.09(15,000 - y) + 0.11y = 1410$$
$$1350 - 0.09y + 0.11y = 1410$$
$$1350 + 0.02y = 1410$$
$$0.02y = 60$$
$$y = 3000$$
So $3000 should go into the Oppenheimer Global Y account and $12,000 should go into the American Funds New Perspective F account.

17. Let x be the amount of money put into the GMO Growth III account, and y be the amount of money put into the Gartmore Destinations Mod Agg Svc account. Both amounts of money add to $8,500, and their interest after one year will be $303.
$$x + y = 8500$$
$$0.023x + 0.066y = 303$$

Substitute $8500 - y$ for x in the second equation and solve for y.
$$0.023(8500 - y) + 0.066y = 303$$
$$195.5 - 0.023y + 0.066y = 303$$
$$0.043y = 107.5$$
$$y = 2500$$
So $2500 should go into the Gartmore Destinations Mod Agg Svc account and $6000 should go into the GMO Growth III account.

19. Let x be the amount of money put into the Lord Abbett Developing Growth B account, and y be the amount of money put into the Bridgeway Micro-Cap Limited account. The problem states that $x = 2y$. The interest for any account is the amount deposited into the account multiplied by its interest rate in decimal form.
$$I = 0.082x + 0.215y$$
$$f(y) = 0.082(2y) + 0.215y$$
$$= 0.164y + 0.215y$$
$$= 0.379y$$
$$758 = 0.379y$$
$$2000 = y$$
So $2000 should be invested in the Bridgeway Micro-Cap Limited account and $4000 in the Lord Abbett Developing Growth B account.

21. Let x be the amount of money put into the Dreyfus Premier Worldwide Growth R account, and y be the amount of money put into the Oppenheimer Global Opportunities Y account.
$$x = 3y$$
$$0.045x + 0.149y = 426$$
Substitute $3y$ for x in the second equation and solve for y.
$$0.045(3y) + 0.149y = 426$$
$$0.135y + 0.149y = 426$$
$$0.284y = 426$$
$$y = 1500$$
So $1500 should go into the Oppenheimer Global Opportunities Y account and $4500 in the Dreyfus Premier Worldwide Growth R account.

23. a. Since the total amount of money to be invested is $10,000, the sum of x and y must be $10,000. The interest for any account is the amount deposited into the account, multiplied by its return as a decimal.

$$x + y = 10000$$
$$I = 0.0287x + 0.081y$$
$$f(x) = 0.0287x + 0.081(10,000 - x)$$
$$= 0.0287x + 810 - 0.081x$$
$$= -0.0523x + 810$$

b.

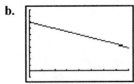

The slope is –0.0523, this is the decrease in the total amount of interest for every dollar put into the Charter One Bank CD. The more money put into the mutual fund, the higher the interest. The more money put into the CD the lower the interest.

c. $$400 = -0.0523x + 810$$
$$-410 = -0.0523x$$
$$7839.39 \approx x$$

So about $7839.39 should go into the CD, while about $2160.61 should go into the mutual fund to earn a total of $400 in one year.

25. a. Since the total amount of money to be invested is $9000, the sum of x and y must be $9000. The interest for any account is the product of the amount deposited into the account and its interest rate in decimal form.

$$x + y = 9000$$
$$I = 0.025x + 0.0945y$$
$$f(x) = 0.025x + 0.0945(9,000 - x)$$
$$= 0.025x + 850.5 - 0.0945x$$
$$= -0.695x + 850.5$$

b. When $x = 500$, this means that $500 has been invested in the CD. $f(500)$ is the amount of interest earned per year when $500 has been invested in the CD, and the rest has been invested in the mutual fund.

$$f(500) = -0.0695(500) + 850.5$$
$$= -34.75 + 850.5$$
$$= 815.75$$

c. $f(x) = 500$ is when x amount of dollars has been invested into the CD, and both investments return $500 in interest per year.

$$500 = -0.0695x + 850.5$$
$$-350.5 = -0.0695x$$
$$5043.17 = x$$

So when about $5043.17 is invested in the CD and about $3956.83 is invested in the mutual fund, the total interest earned is $500.

d. When $x = 10,000$, this means that $10,000 has been invested in the CD. $f(10,000)$ is the amount of interest earned per year when $10,000 has been invested in the CD.

$$f(10,000) = -0.0695(10,000) + 850.5$$
$$= -695 + 850.5$$
$$= 155.5$$

Since the total amount of money available to be invested is $9000, this implies that negative $1000 has been invested in the mutual fund. This is impossible, so model breakdown has occurred.

27. a. Since the total amount of money to be invested is $8000, the sum of x and y must be $8000. The interest for any account is the product of the amount deposited into the account and its interest rate in decimal form.

$$x + y = 8000$$
$$I = 0.015x + 0.116y$$
$$f(x) = 0.015x + 0.116(8000 - x)$$
$$= 0.015x + 928 - 0.116x$$
$$= -0.101x + 928$$

b. The minimum principal of $2500 relates to the function above by restricting the domain. Now $2,500 \le x \le 8,000$.

$$f(2500) = -0.101(2500) + 928$$
$$= -252.5 + 928$$
$$= 675.5$$
$$f(8000) = -0.101(8000) + 928$$
$$= -808 + 928$$
$$= 120$$

The total interest will be between $120 and $675.50, inclusive.

c. $$400 = -0.101x + 928$$
$$-528 = -0.101x$$
$$5227.72 = x$$

So $5227.72 should be invested in the CD and $2772.28 should be invested in the mutual fund.

29. a. Since the total amount of money to be invested is $6000, the sum of x and y must be $6000. The interest for any account is the product of the amount deposited into the account and its interest rate in decimal form.

$$x + y = 6000$$
$$I = 0.0285x + 0.09y$$
$$f(x) = 0.0285x + 0.09(6000 - x)$$
$$= 0.0285x + 540 - 0.09x$$
$$= -0.0615x + 540$$

b. The I-intercept of the model is when $x = 0$, or when no money has been invested in the CD and all of it is invested in the mutual fund.

$$I = -0.0615(0) + 540 = 540$$

This means that if no money was invested in the CD, the interest per year would be $540.

c. The x-intercept is when $I = 0$, or when there is no interest being earned.

$$0 = -0.0615x + 540$$
$$-540 = -0.0615x$$
$$8780.49 = x$$

The x-intercept is (8780.49, 0), which means that when $8780.49 is invested in the CD, no interest is earned. However, only $6000 is being invested, so model breakdown has occurred.

d. The slope is –0.0615, or a decrease of 0.0615 dollars of interest for every dollar invested in the CD.

31. Let x be the amount of 10% alcohol solution and y be the amount of 30% alcohol solution. The sum of both amounts needs to be 10 ounces, and should make a mixture of 22% alcohol.

$$x + y = 10$$
$$0.10x + 0.30y = 0.22(10)$$

Substitute $10 - y$ for x in the second equation and solve for y.

$$0.10(10 - y) + 0.30y = 0.22(10)$$
$$1 - 0.10y + 0.30y = 2.2$$
$$0.20y = 1.2$$
$$y = 6$$

So 6 ounces of the 30% solution and 4 ounces of the 10% solution are needed.

33. Let x be the amount of 5% antifreeze solution and y be the amount of 20% antifreeze solution. The sum of both amounts needs to be 3 gallons, and should make a 15% antifreeze solution.

$$x + y = 3$$
$$0.05x + 0.20y = 0.15(3)$$

Substitute $3 - y$ for x in the second equation and solve for y.

$$0.05(3 - y) + 0.20y = 0.15(3)$$
$$0.15 - 0.05y + 0.20y = 0.45$$
$$0.15 + 0.15y = 0.45$$
$$0.15y = 0.30$$
$$y = 2$$

So 2 gallons of the 20% solution and 1 gallon of the 5% solution are needed.

35. Let x be the amount of 10% acid solution and y be the amount of 25% acid solution. The sum of both amounts needs to be 6 cups, and should make a 15% acid solution.

$$x + y = 6$$
$$0.1x + 0.25y = 0.15(6)$$

Substitute $6 - x$ for y in the second equation, and solve for x.

$$0.1x + 0.25(6 - x) = 0.15(6)$$
$$0.1x + 1.5 - 0.25x = 0.9$$
$$-0.15x + 1.5 = 0.9$$
$$-0.15x = -0.6$$
$$x = 4$$

So 4 cups of the 10% solution and 2 cups of the 25% solution are needed.

37. Let x be the amount of water and y be the amount of 25% alcohol solution. The sum of both amounts should total 5 liters and make a 20% alcohol solution. Assume that water is a 0% alcohol solution.

$$x + y = 5$$
$$0x + 0.25y = 0.2(5)$$
$$0.25y = 1$$
$$y = 4$$

So 4 liters of the 25% alcohol solution needs to be added to 1 liter of water.

39. Answers may vary. Example:
Since the amounts of the two solutions are the same, the resulting alcohol percentage is the average of the two original percentages.

$$\frac{15 + 25}{2} = \frac{40}{2} = 20\%$$

41. Answers may vary. Example:
A political fundraiser has two types of admission, a $2500 dinner and a $9500 meet-and-greet. The organizers want to make $579,000, and they want to sell 30 fewer meet-and-greets than dinners. How many of each type of admission should they sell?
Substitute the first equation into the second, and solve for y.

$$9500(y-30)+2500y = 579,000$$
$$9500y-285,000+2500y = 579,000$$
$$12,000y = 864,000$$
$$y = 72$$

$$x = y-30 = 72-30 = 42$$

They need to sell 72 of the $2500 dinners and 42 of the $9500 meet-and-greets.

43. Answers may vary. Example:
To have a correct balance of investments, a banker wants to put three times as much money in Fund X as in Fund Y. Fund X has an annual return of 5%, while Fund Y has an annual return of 12%. If the banker wants $135 in interest, how much money does she need to put into each fund?
Substitute the first equation into the second equation and solve for y.

$$0.05(3y)+0.12y = 135$$
$$0.15y+0.12y = 135$$
$$0.27y = 135$$
$$y = 500$$

So about $500 needs to be put into Fund Y and about $1500 needs to be put into Fund X.

Homework 3.5

1.

	In Words	Inequality	Graph	Interval Notation
	Numbers greater than 3	$x > 3$		$(3, \infty)$
	Numbers less than or equal to -4	$x \le -4$		$(-\infty, -4]$
	Numbers less than 5	$x < 5$		$(-\infty, 5)$
	Numbers greater than or equal to -1	$x \ge -1$		$[-1, \infty)$

---

3. $x+2 \ge 5$
$x+2-2 \ge 5-2$
$x \ge 3$
Interval: $[3, \infty)$

5. $-4x \ge 12$
$\dfrac{-4x}{-4} \le \dfrac{12}{-4}$
$x \le -3$
Interval: $(-\infty, -3]$

7. $2w+7 < 11$
$2w+7-7 < 11-7$
$2w < 4$
$\dfrac{2w}{2} < \dfrac{4}{2}$
$w < 2$
Interval: $(-\infty, 2)$

9.
$$9x < 4 + 5x$$
$$9x - 5x < 4 + 5x - 5x$$
$$4x < 4$$
$$\frac{4x}{4} < \frac{4}{4}$$
$$x < 1$$

Interval: $(-\infty, 1)$

11.
$$2.1x - 7.4 \le 4.36$$
$$2.1x - 7.4 + 7.4 \le 4.36 + 7.4$$
$$2.1x \le 11.76$$
$$\frac{2.1x}{2.1} \le \frac{11.76}{2.1}$$
$$x \le 5.6$$

Interval: $(-\infty, 5.6]$

13.
$$2b - 3 > 7b + 22$$
$$2b - 3 + 3 > 7b + 22 + 3$$
$$2b > 7b + 25$$
$$2b - 7b > 7b + 25 - 7b$$
$$-5b > 25$$
$$\frac{-5b}{-5} < \frac{25}{-5}$$
$$b < -5$$

Interval: $(-\infty, -5)$

15. $3 - 2(x - 4) > 4x + 1$
$$3 - 2x + 8 > 4x + 1$$
$$-2x + 11 > 4x + 1$$
$$-2x + 11 - 11 > 4x + 1 - 11$$
$$-2x > 4x - 10$$
$$-2x - 4x > 4x - 10 - 4x$$
$$-6x > -10$$
$$\frac{-6x}{-6} < \frac{-10}{-6}$$
$$x < \frac{5}{3}$$

Interval: $\left(-\infty, \frac{5}{3}\right)$

17.
$$6.2a + 61.31 < 5(3.1 - 2.7a) + 0.5$$
$$6.2a + 61.31 < 15.5 - 13.5a + 0.5$$
$$6.2a + 61.31 < 16 - 13.5a$$
$$6.2a + 61.31 - 61.31 < 16 - 13.5a - 61.31$$
$$6.2a < -45.31 - 13.5a$$
$$6.2a + 13.5a < -45.31 - 13.5a + 13.5a$$
$$19.7a < -45.31$$
$$\frac{19.7a}{19.7} < \frac{-45.31}{19.7}$$
$$a < -2.3$$

Interval: $(-\infty, -2.3)$

19. $7(x + 1) - 8(x - 2) \le 0$
$$7x + 7 - 8x + 16 \le 0$$
$$-x + 23 \le 0$$
$$-x \le -23$$
$$\frac{-x}{-1} \ge \frac{-23}{-1}$$
$$x \ge 23$$

Interval: $[23, \infty)$

21. $5r - 4(2r - 6) - 1 \ge 3(3r - 1) + r$
$$5r - 8r + 24 - 1 \ge 9r - 3 + r$$
$$-3r + 23 \ge 10r - 3$$
$$-3r + 23 - 23 \ge 10r - 3 - 23$$
$$-3r \ge 10r - 26$$
$$-3r - 10r \ge 10r - 10r - 26$$
$$-13r \ge -26$$
$$\frac{-13r}{-13} \le \frac{-26}{-13}$$
$$r \le 2$$

Interval: $(-\infty, 2]$

23.
$$-\frac{2}{3}x > 4$$
$$-\frac{3}{2} \cdot -\frac{2}{3}x < -\frac{3}{2} \cdot 4$$
$$x < -6$$

Interval: $(-\infty, -6)$

25.
$$\frac{2}{3} - \frac{3}{4}t \le \frac{5}{2}$$
$$\frac{2}{3} - \frac{3}{4}t - \frac{2}{3} \le \frac{5}{2} - \frac{2}{3}$$
$$-\frac{3}{4}t \le \frac{11}{6}$$
$$-\frac{4}{3} \cdot -\frac{3}{4}t \ge -\frac{4}{3} \cdot \frac{11}{6}$$
$$t \ge -\frac{22}{9}$$
Interval: $\left[-\frac{22}{9}, \infty\right)$

27.
$$-\frac{1}{2}x - \frac{5}{6} \ge \frac{1}{3} + \frac{3}{2}x$$
$$-\frac{1}{2}x - \frac{5}{6} + \frac{5}{6} \ge \frac{1}{3} + \frac{3}{2}x + \frac{5}{6}$$
$$-\frac{1}{2}x \ge \frac{3}{2}x + \frac{7}{6}$$
$$-\frac{1}{2}x - \frac{3}{2}x \ge \frac{3}{2}x + \frac{7}{6} - \frac{3}{2}x$$
$$-2x \ge \frac{7}{6}$$
$$x \le -\frac{7}{12}$$
Interval: $\left(-\infty, -\frac{7}{12}\right]$

29.
$$\frac{4c-5}{6} \le \frac{3c+7}{4}$$
$$\left(\frac{4c-5}{6}\right)(12) \le \left(\frac{3c+7}{4}\right)(12)$$
$$8c - 10 \le 9c + 21$$
$$8c - 10 + 10 \le 9c + 21 + 10$$
$$8c \le 9c + 31$$
$$8c - 9c \le 9c + 31 - 9c$$
$$-c \le 31$$
$$\frac{-c}{-1} \ge \frac{31}{-1}$$
$$c \ge -31$$
Interval: $[-31, \infty)$

31.
$$\frac{3x+1}{6} - \frac{5x-2}{9} > \frac{2}{3}$$
$$\left(\frac{3x+1}{6} - \frac{5x-2}{9}\right)(18) > \left(\frac{2}{3}\right)(18)$$
$$(3x+1)(3) - (5x-2)(2) > (2)(6)$$
$$9x + 3 - 10x + 4 > 12$$
$$-x + 7 > 12$$
$$-x + 7 - 7 > 12 - 7$$
$$-x > 5$$
$$\frac{-x}{-1} < \frac{5}{-1}$$
$$x < -5$$
Interval: $(-\infty, -5)$

33.
$$4 < x + 3 < 8$$
$$4 - 3 < x + 3 - 3 < 8 - 3$$
$$1 < x < 5$$
Interval: $(1, 5)$

35.
$$-15 \le 2x - 5 \le 7$$
$$-15 + 5 \le 2x - 5 + 5 \le 7 + 5$$
$$-10 \le 2x \le 12$$
$$\frac{-10}{2} \le \frac{2x}{2} \le \frac{12}{2}$$
$$-5 \le x \le 6$$
Interval: $[-5, 6]$

37.
$$-17 < 3 - 4x \le 15$$
$$-17 - 3 < 3 - 4x - 3 \le 15 - 3$$
$$-20 < -4x \le 12$$
$$\frac{-20}{-4} > \frac{-4x}{-4} \ge \frac{12}{-4}$$
$$5 > x \ge -3$$
Interval: $[-3, 5)$

39.
$$\frac{1}{3} \le 4 - \frac{2}{3}x < 2$$
$$\left(\frac{1}{3}\right)(3) \le \left(4 - \frac{2}{3}x\right)(3) < (2)(3)$$
$$1 \le 12 - 2x < 6$$
$$1 - 12 \le 12 - 2x - 12 < 6 - 12$$
$$-11 \le -2x < -6$$
$$\frac{-11}{-2} \ge \frac{-2x}{-2} > \frac{-6}{-2}$$
$$5.5 \ge x > 3$$
Interval: $(3, 5.5]$

41.
$$F < D$$
$$-1414t + 14,290 < -3740t + 30,450$$
$$-1414t + 14,290 - 14,290 < -3740t + 30,450 - 14,290$$
$$-1414t < -3740t + 16,160$$
$$-1414t + 3740t < -3740t + 16,160 + 3740t$$
$$2326t < 16,160$$
$$\frac{2326t}{2326} < \frac{16,160}{2326}$$
$$t < 6.95$$
The value of the Ford Fusion is less than the value of the Cadillac DTS for the years before 2019 ($2012 + 6.95 \approx 2019$).

43. a. Since U-Haul's charge increases at a constant rate of $0.69 per mile, the equation is linear with slope 0.69. The U-intercept is 19.95 since U-Haul charges a flat fee of $19.95.
$U(d) = 0.69d + 19.95$.

Using the same method, an equation can be made for Penske's charge.
$P(d) = 0.39d + 29.95$.

b.
$$U < P$$
$$0.69d + 19.95 < 0.39d + 29.95$$
$$0.69d + 19.95 - 19.95 < 0.39d + 29.95 - 19.95$$
$$0.69d < 0.39d + 10$$
$$0.69d - 0.39d < 0.39 + 10 - 0.39d$$
$$0.3d < 10$$
$$\frac{0.3d}{0.3} < \frac{10}{0.3}$$
$$d < \frac{100}{3}$$
U-Haul will be cheaper for miles driven less than 33.33 miles.

45. The percentage of residents of Ohio who smoke can be modeled by $O(t) = -0.85t + 20.3$, where t is the number of years after 2009. The percentage of residents of Wisconsin who smoke can be modeled by $W(t) = -0.61t + 18.8$.
$$O < W$$
$$-0.85t + 20.3 < -0.61t + 18.8$$
$$-0.85t + 20.3 - 20.3 < -0.61t + 18.8 - 20.3$$
$$-0.85t < -0.61t - 1.5$$
$$-0.85t + 0.61t < -0.61t - 1.5 + 0.61t$$
$$-0.24t < -1.5$$
$$\frac{-0.24t}{-0.24} > \frac{-1.5}{-0.24}$$
$$t > 6.25$$
By this model, the percentage of residents of Ohio who smoke will be less than the percentage of residents of Wisconsin who smoke after the year $2009 + 6.25 \approx 2015$.

47. a.

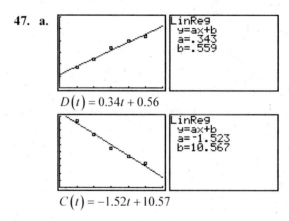

$D(t) = 0.34t + 0.56$

$C(t) = -1.52t + 10.57$

b.
$$D(t) > C(t)$$
$$0.34t + 0.56 > -1.52t + 10.57$$
$$1.86t + 0.56 > 10.57$$
$$1.86t > 10.01$$
$$t > 5.38$$
The revenue from downloaded music will be more than the revenue from music CDs. in $2005 + 5.38 \approx 2010$ and after.

c. The sum of the slopes is
$0.34 + (-1.52) = -1.18$.

The total revenue from downloaded music and music CDs is decreasing by $1.18 billion per year.

49. a. $W(38) = 0.115(38) + 77.44$
$= 4.37 + 77.44$
≈ 81.8
$M(38) = 0.208(38) + 69.86$
$= 7.904 + 69.86$
≈ 77.8
$81.8 - 77.8 = 4.0$
Women born in 2018 will live approximately 4.0 years longer, on average, than men born in 2018.

b. $W < M$
$0.115t + 77.44 < 0.208t + 69.86$
$77.44 < 0.093t + 69.86$
$7.58 < 0.093t$
$81.5 < t$

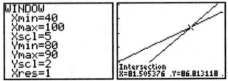

After $t = 81.5$, or $1980 + 81.5 \approx 2062$, $M(t)$ is greater than $W(t)$.

c.   i.   The woman should marry a younger man since the life expectancies increase as year of birth increases.

ii.   The woman's average life expectancy is 77.4 years. As time passes t increases, as does the age of the woman. So after t years, she wants to find a man of life expectancy $M(t)$, but she will be t years older.
$77.4 < M(t) + t$
$77.4 < 0.208t + 69.86 + t$
$77.4 < 1.208t + 69.86$
$7.54 < 1.208t$
$6.24 < t$
She should marry a man born during or after $1980 + 6.24 \approx 1986$.

51.   The student made a mistake. It is not necessary to switch the direction of the inequality when dividing by a positive number.
$3x + 7 > 1$
$3x + 7 - 7 > 1 - 7$
$3x > -6$
$\dfrac{3x}{3} > \dfrac{-6}{3}$
$x > -2$

53. a.   Solve the inequality for x.
$3(x - 2) + 1 \geq 7 - 4x$
$3x - 6 + 1 \geq 7 - 4x$
$3x - 5 \geq 7 - 4x$
$3x - 5 + 5 \geq 7 - 4x + 5$
$3x \geq 12 - 4x$
$3x + 4x \geq 12 - 4x + 4x$
$7x \geq 12$
$\dfrac{7x}{7} \geq \dfrac{12}{7}$
$x \geq \dfrac{12}{7}$
Any three numbers that are greater than or equal to $\dfrac{12}{7}$ are possible solutions.

b.   From part (a), any number less than $\dfrac{12}{7}$ is not a solution.

55.   Answers may vary. Example:
Since the inequality symbols are facing different directions in the two inequalities, m must be a negative number. Let's choose $m = -1$.
$mx < c$
$(-1)x < c$
$x > -c$
We want $x > 2$, so c should be -2.
Check:
$mx < c$
$(-1)x < (-2)$
$\dfrac{-x}{-1} > \dfrac{-2}{-1}$
$x > 2$
So, when $m = -1$ and $c = -2$, the solution set of $mx < c$ is $x > 2$.

57. a. $x + 1 = -2x + 10$
$-9 = -3x$
$3 = x$

b. $x + 1 < -2x + 10$
$-9 < -3x$
$3 > x$

c. $x + 1 > -2x + 10$
$-9 > -3x$
$3 < x$

d.

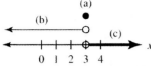

Answers may vary. Example:
The first equation's solution is a solid circle on 3. The second equation has solutions to the left of 3. The third equation has solutions to the right of 3. The solutions to the second and third equations are symmetrical around $x = 3$.

e.

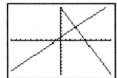

Notice that when x is less than 3 $g(x)$ is above, or larger than, $f(x)$. When x is more than 3, $g(x)$ is below, or smaller than, $f(x)$. When $x = 3$, $g(x) = f(x)$.

59. False. When $x = -4$, the graph of f is below the graph of g.

61. False. When $x = -1$, the graph of f intersects the graph of g.

63. The graph of f is above the graph of g for all values of x such that $x < 2.8$. So $f(x) > g(x)$ when $x < 2.8$.

65. Answers may vary. Example:

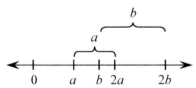

67. Answers may vary.

Homework 3.6

1. $y \geq 2x - 4$

Graph the line $y = 2x - 4$ with a solid line and shade the region above it.

3. $y < -\dfrac{1}{2}x + 3$

Graph the line with a dashed line and shade the region below it.

5. $y \leq -2x + 6$

Graph the line $y = -2x + 6$ with a solid line and shade the region below it.

7. $y > x$

Graph the line $y = x$ with a dashed line and shade the region above it.

9. $y < -\dfrac{1}{3}x$

Graph the line $y = -\dfrac{1}{3}x$ with a dashed line and shade the region below it.

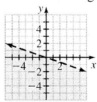

11. $2x + 5y < 10$

$$5y < -2x + 10$$

$$y < -\frac{2}{5}x + 2$$

Graph the line $y = -\frac{2}{5}x + 2$ with a dashed

line and shade the region below it.

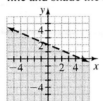

13. $4x - 6y - 6 \geq 0$

$$-6y \geq -4x + 6$$

$$y \leq \frac{2}{3}x - 1$$

Graph the line $y = \frac{2}{3}x - 1$ with a solid line

and shade the region below it.

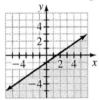

15. $3(x - 2) + y \leq -2$

$$3x - 6 + y \leq -2$$

$$y \leq -3x + 4$$

Graph the line $y = -3x + 4$ with a solid line

and shade the region below it.

17. $y \leq 2$

Graph the line $y = 2$ with a solid line and

shade the region below it.

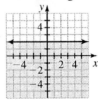

19. $x > -3$

Graph the line $x = -3$ with a dashed line and

shade the region above it.

21. $y \geq \frac{1}{3}x - 2$

$$y > -x + 3$$

Graph the line $y = \frac{1}{3}x - 2$ with a solid line

and the line $y = -x + 3$ with a dashed line.

The solution region of the system is the

intersection of the solution regions of

$y \geq \frac{1}{3}x - 2$ and $y > -x + 3$.

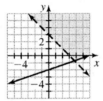

23. $y \leq x - 4$

$$y \geq -3x$$

Graph the line $y = x - 4$ with a solid line and

the line $y = -3x$ with a solid line. The

solution region of the system is the

intersection of the solution regions of

$y \leq x - 4$ and $y \geq -3x$.

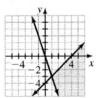

25. $y \leq -3x + 9$

$$y \geq 2x - 3$$

$$x \geq 0$$

$$y \geq 0$$

Graph the lines $y = -3x + 9$, $y = 2x - 3$,

$x = 0$, and $y = 0$ with solid lines. The

solution region of the system is the

intersection of the solution regions of

$y \leq -3x + 9$, $y \geq 2x - 3$, $x \geq 0$ and $y \geq 0$.

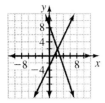

27. $y < -x + 5$
  $y \leq x + 5$
  $y > \dfrac{1}{2}x + 1$

Graph the lines $y = -x + 5$ and $y = \dfrac{1}{2}x + 1$

with dashed lines, and the line $y = x + 5$ with a solid line. The solution region of the system is the intersection of the solution regions of

$y < -x + 5$, $y \leq x + 5$ and $y > \dfrac{1}{2}x + 1$.

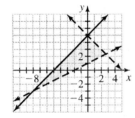

29. $y \leq -3$
  $y \geq -5$

Graph the lines $y = -3$ and $y = -5$ with solid lines. The solution region of the system is the intersection of the solution regions of $y \leq -3$
and $y \geq -5$.

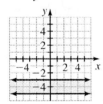

31. $2x - 4y \leq 8$ $\qquad$ $3x + 5y \leq 10$
  $-4y \leq -2x + 8$ $\qquad$ $5y \leq -3x + 10$
  $y \geq \dfrac{1}{2}x - 2$ $\qquad$ $y \leq -\dfrac{3}{5}x + 2$

Graph the lines $y = \dfrac{1}{2}x - 2$ and $y = -\dfrac{3}{5}x + 2$

with solid lines. The solution region of the system is the intersection of the solution

regions of $y \geq \dfrac{1}{2}x - 2$ and $y \leq -\dfrac{3}{5}x + 2$.

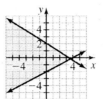

33. $x - 2y > 6$ $\qquad$ $x + 3y \leq 3$
  $-2y > -x + 6$ $\qquad$ $3y \leq -x + 3$
  $y < \dfrac{1}{2}x - 3$ $\qquad$ $y \leq -\dfrac{1}{3}x + 1$

Graph the line $y = \dfrac{1}{2}x - 3$ with a dashed line

and the line $y = -\dfrac{1}{3}x + 1$ with a solid line. The

solution region of the system is the intersection of the solution regions of

$y < \dfrac{1}{2}x - 3$ and $y \leq -\dfrac{1}{3}x + 1$.

35. $1 + y \geq \dfrac{1}{2}(x - 4)$ $\qquad$ $3 - y > 2(x - 1)$
$\qquad\qquad\qquad\qquad\qquad$ $-y > 2x - 2 - 3$
  $y \geq \dfrac{1}{2}x - 2 - 1$ $\qquad$ $-y > 2x - 5$
$\qquad\qquad\qquad\qquad\qquad$ $y < -2x + 5$
  $y \geq \dfrac{1}{2}x - 3$

Graph the line $y = \dfrac{1}{2}x - 3$ with a solid line

and the line $y = -2x + 5$ with a dashed line. The solution region of the system is the intersection of the solution regions of

$y \geq \dfrac{1}{2}x - 3$ and $y < -2x + 5$.

37. $5y \le 2x + 20$

$y \le \dfrac{2}{5}x + 4$

$5y \ge 2x + 5$

$y \ge \dfrac{2}{5}x + 1$

$x \ge 3$

$x \le 5$

Graph the lines $y = \dfrac{2}{5}x + 4$, $y = \dfrac{2}{5}x + 1$, $x = 3$, and $x = 5$ with solid lines. The solution region of the system is the intersection of the solution regions of $y \le \dfrac{2}{5}x + 4$, $y \ge \dfrac{2}{5}x + 1$, $x \ge 3$, and $x \le 5$.

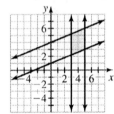

39. a. Answers may vary. Example:
By entering the sets of data into a graphing calculator, using the linear regression feature, and rounding to the second decimal place, we find the following equations.
$B(t) = 0.196t + 70.00$
$T(t) = 0.167t + 51.73$

b. The life expectancies of U.S. males from 0 years to 20 years must be less than or equal to $T(t)$ and greater than or equal to $B(t)$. Also, limiting the years to 1980 to 2020 means that t must be greater than or equal to 0 and less than or equal to 40. Therefore, the system of equations is:
$L \le 0.196t + 70.00$
$L \ge 0.167t + 51.73$
$t \ge 0$
$t \le 40$

c.

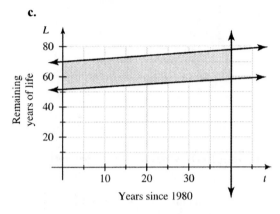

Years since 1980

d. $T(35) \le L \le B(35)$
$0.167(35) + 51.73 \le L \le 0.196(35) + 70.00$
$5.845 + 51.73 \le L \le 6.86 + 70.00$
$57.575 \le L \le 76.86$
The life expectancies of U.S. males from 0 years through 20 years old in 2015 is between 57.6 years and 76.9 years.

41. a. Answers may vary. Example:
By entering the sets of data into a graphing calculator, using the linear regression feature, and rounding to the second decimal place, we find the following equations.
$B(w) = 0.44w + 84.21$
$I(w) = 0.44w + 89.21$

b. The ski lengths for beginning to intermediate skiers must be greater than or equal to $B(w)$ and less than or equal to $I(w)$. Also, w must be greater than or equal to 130 and less than or equal to 150. Therefore, the system of equations is:
$L \ge 0.44w + 84.21$
$L \le 0.44w + 89.21$
$w \ge 130$
$w \le 150$

c.

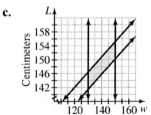

Pounds

d. $B(140) \le L \le I(140)$
$0.44(140) + 84.21 \le L \le 0.44(140) + 89.21$
$61.6 + 84.21 \le L \le 61.6 + 89.21$
$145.81 \le L \le 150.81$

43. Answers may vary. Example:
Since the coefficient of the y variable is negative, the direction of the inequality will need to be switched. Therefore, the graph of $2x - 3y < 6$ will actually be above the line $2x - 3y = 6$.

45. Answers may vary. Example:
$y > x$

$(3,4)$ is a solution but $(4,3)$ is not a solution.

47. The intersection of the solution regions of $y \geq 2x + 1$ and $y \leq 2x + 1$ is the solid line $y = 2x + 1$.

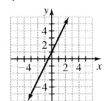

49. a. i. $y = -2x - 3$

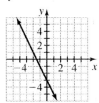

  ii. $y < -2x - 3$

Graph the line $y = -2x - 3$ with a dashed line and shade the region below it.

  iii. $y \leq -2x - 3$

Graph the line $y = -2x - 3$ with a solid line and shade the region below it.

  iv. $y > -2x - 3$

Graph the line $y = -2x - 3$ with a dashed line and shade the region above it.

  v. $y \geq -2x - 3$

Graph the line $y = -2x - 3$ with a solid line and shade the region above it.

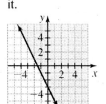

  b. Answers may vary. Example:
Each process begins with graphing the equation of the (related) line. A dashed line is used for $<$ and $>$. A solid line is used for $\leq$ and $\geq$. The area above the line is shaded for $>$ and $\geq$. The area below the line is shaded for $<$ and $\leq$.

51. Answers may vary. Example:
To graph an inequality in two variables, first isolate the y variable on the left side of the inequality. If you multiply or divide both sides of the equation by a negative number during this process, reverse the direction of the inequality symbol. Then, substitute an equal sign for the inequality symbol, and graph the line given by this equation. If the inequality symbol is $<$ or $>$, use a dotted line; if it is $\leq$ or $\geq$, use a solid line. Finally, if the inequality symbol is $<$ or $\leq$, shade below the line; if the symbol is $>$ or $\geq$, shade above the line.

Chapter 3 Review

1. $y = -\dfrac{3}{2}x + 1$

   $y = \dfrac{1}{4}x - 6$

   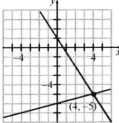

   The solution is $(4, -5)$.

2. $3x - 5y = -1$ $\qquad$ $y = -2(x - 4)$
   $-5y = -3x - 1$ $\qquad$ $y = -2x + 8$
   $y = \dfrac{3}{5}x + \dfrac{1}{5}$

   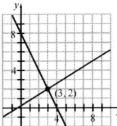

   The solution is $(3, 2)$.

3. Solve using elimination.
   Multiply the first equation by 3 and the second equation by -4, then add the equations.
   $12x - 15y = -66$
   $\underline{-12x - 8y = 20}$
   $-23y = -46$
   $y = 2$
   Substitute $y = 2$ into $3x + 2y = -5$ and solve for x.
   $3x + 2(2) = -5$
   $3x + 4 = -5$
   $3x = -9$
   $x = -3$
   The solution is $(-3, 2)$.

4. Solve using elimination.
   Multiply the first equation by -2 and add the equations.
   $-6x + 14y = -10$
   $\underline{6x - 14y = -1}$
   $0 = -11$ false
   This is a contradiction. The system is inconsistent. The solution set is the empty set.

5. Solve using elimination.
   Rewrite the second equation so all the variables are on the left side of the equation and the constant is on the right side.
   $-4x - 5y = 3$
   $8x + 10y = -6$
   Multiply the first equation by 2 and add the equations.
   $-8x - 10y = 6$
   $\underline{8x + 10y = -6}$
   $0 = 0$ true
   This is an identity. The system is dependent. The solution set contains all ordered pairs (x, y) such that $y = -\dfrac{4}{5}x - \dfrac{3}{5}$.

6. Solve using substitution.
   Substitute $4.2x - 7.9$ for y in the second equation.
   $y = -2.8x + 0.5$
   $4.2x - 7.9 = -2.8x + 0.5$
   $4.2x - 7.9 + 7.9 = -2.8x + 0.5 + 7.9$
   $4.2x = -2.8x + 8.4$
   $4.2x + 2.8x = -2.8x + 8.4 + 2.8x$
   $7x = 8.4$
   $x = 1.2$
   Substitute this result into the first equation and solve for y.
   $y = 4.2(1.2) - 7.9 \approx -2.86$
   The solution is $(1.2, -2.86)$.

7. Solve using substitution.
   Substitute $4.9x$ for y in the second equation and solve for x.
   $-3.2y = x$
   $-3.2(4.9x) = x$
   $-15.68x = x$
   $-16.68x = 0$
   $x = 0$
   Substitute this result into the first equation and solve for y.
   $y = 4.9(0) = 0$
   The solution is $(0, 0)$.

8. Solve using substitution.

Substitute $\frac{1}{2}x - 4$ for y in the first equation.

$$3x - 5\left(\frac{1}{2}x - 4\right) = 21$$

$$3x - \frac{5}{2}x + 20 = 21$$

$$\frac{1}{2}x + 20 = 21$$

$$\frac{1}{2}x = 1$$

$$x = 2$$

Substitute $x = 2$ into the second equation and solve for y.

$$y = \frac{1}{2}(2) - 4 = 1 - 4 = -3$$

The solution is $(2, -3)$.

9. Solve using elimination. Multiply both sides of the first equation by 2.

$$\frac{6}{5}x - \frac{4}{3}y = 8$$

$$-\frac{6}{5}x + \frac{8}{3}y = -4$$

$$\frac{4}{3}y = 4$$

$$y = 3$$

Substitute $y = 3$ into $\frac{3}{5}x - \frac{2}{3}y = 4$ and solve for x.

$$\frac{3}{5}x - \frac{2}{3}(3) = 4$$

$$\frac{3}{5}x - 2 = 4$$

$$\frac{3}{5}x = 6$$

$$x = 10$$

The solution is $(10, 3)$.

10. First use the distributive property to simplify each equation.

$$2(3x - 4) + 3(2y - 1) = -5$$

$$6x - 8 + 6y - 3 = -5$$

$$6x + 6y - 11 = -5$$

$$6x + 6y = 6$$

$$-3(2x + 1) + 4(y + 3) = -7$$

$$-6x - 3 + 4y + 12 = -7$$

$$-6x + 4y + 9 = -7$$

$$-6x + 4y = -16$$

Solve using elimination.

$$6x + 6y = 6$$

$$-6x + 4y = -16$$

Add the two equations together.

$$6x + 6y = 6$$

$$\underline{-6x + 4y = -16}$$

$$10y = -10$$

$$y = -1$$

Substitute $y = -1$ into $6x + 6y = 6$ and solve for x.

$$6x + 6y = 6$$

$$6x + 6(-1) = 6$$

$$6x - 6 = 6$$

$$6x = 12$$

$$x = 2$$

The solution is $(2, -1)$.

11. Elimination:
Rewrite the second equation so that all the variables are on the left side of the equation and the constant is on the right side.

$$2x - 5y = 15$$

$$2x + y = 9$$

Multiply the second equation by -1 and add the equations.

$$2x - 5y = 15$$

$$\underline{-2x - y = -9}$$

$$-6y = 6$$

$$y = -1$$

Substitute $y = -1$ into $2x - 5y = 15$ and solve for x.

$$2x - 5(-1) = 15$$

$$2x + 5 = 15$$

$$2x = 10$$

$$x = 5$$

The solution is $(5, -1)$.

Substitution:
Substitute $-2x + 9$ for y in the first equation and solve for x.

$$2x - 5y = 15$$

$$2x - 5(-2x + 9) = 15$$

$$2x + 10x - 45 = 15$$

$$12x = 60$$

$$x = 5$$

Substitute this result into the second equation and solve for y.

$$y = -2x + 9 = -2(5) + 9 = -10 + 9 = -1$$

The solution is $(5, -1)$.

Graphically:

$2x - 5y = 15$ $y = -2x + 9$

$-5y = -2x + 15$

$y = \dfrac{2}{5}x - 3$

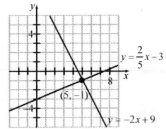

The solution is $(5, -1)$.

12. **a.** Answers may vary. Example:

$x + 2y = 4$

$3x + 6y = 12$

Solving this system yields an identity such as $0 = 0$. The solution set is the set of ordered pairs (x, y) such that $x + 2y = 4$.

b. Answers may vary. Example:

$x + 2y = 4$

$3x + 6y = 10$

Solving this system yields a contradiction such as $0 = -2$. The solution set is the empty set.

c. Answers may vary. Example:

$x + y = 10$

$3x - 3y = -6$

The point $(4, 6)$ satisfies both equations.

13. The two lines intersect at the point $(1, -2)$, so the solution is 1.

14. The red line intersects the line $y = 2$ at the point $(-5, 2)$, so the solution is -5.

15. Answers may vary. Example:
Notice that the functions cross between 0 and 1, and between 2 and 3. Approximate solutions may include (0.5, 15.5) and (2.5, 21.5).

16. $2(5) + 3(3) = a$ $6(5) - 4(3) = b$

$10 + 9 = a$ $30 - 12 = b$

$19 = a$ $18 = b$

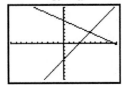

17. The coordinates of point A are the same as the coordinates of the origin, $(0, 0)$.

The coordinates of point B is the same as the y-intercept of $y = 3x + 4$, $(0, 4)$.

Use substitution to find the intersection of l_1 and l_2.

$3(3x + 4) + 2x = 34$

$9x + 12 + 2x = 34$

$11x = 22$

$x = 2$

Substitute to find the other coordinate.

$y = 3(2) + 4 = 10$

C is the point $(2, 10)$.

Substitute $-4x + 28$ into l_2.

$3(-4x + 28) + 2x = 34$

$-12x + 84 + 2x = 34$

$-10x = -50$

$x = 5$

Substitute to find the other coordinate.

$y + 4(5) = 28$

$y + 20 = 28$

$y = 8$

D is the point $(5, 8)$.

Substitute $3x - 14$ for y in l_3.

$(3x - 14) + 4x = 28$

$7x - 14 = 28$

$7x = 42$

$x = 6$

$y = 3(6) - 14 = 18 - 14 = 4$

E is the point $(6, 4)$.

Find the x-intercept of l_4.

$0 = 3x - 14$

$14 = 3x$

$\dfrac{14}{3} = x$

F is the point $\left(\dfrac{14}{3}, 0\right)$.

18. Graph each of the equations.

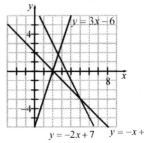

The three graphs do not share a common point. The solution set is the empty set.

19.
$$3x - 8 \le 13$$
$$3x - 8 + 8 \le 13 + 8$$
$$3x \le 21$$
$$\frac{3x}{3} \le \frac{21}{3}$$
$$x \le 7$$
Interval: $(-\infty, 7]$

20.
$$29.19 - 3.6a \ge 3.9(a + 2.1)$$
$$29.19 - 3.6a \ge 3.9a + 8.19$$
$$29.19 - 3.6a - 29.19 \ge 3.9a + 8.19 - 29.19$$
$$-3.6a \ge 3.9x - 21$$
$$-3.6x - 3.9x \ge 3.9x - 21 - 3.9x$$
$$-7.5x \ge -21$$
$$\frac{-7.5x}{-7.5} \le \frac{-21}{-7.5}$$
$$x \le 2.8$$
Interval: $(-\infty, 2.8]$

21.
$$-5(2x + 3) \ge 2(3x - 4)$$
$$-10x - 15 \ge 6x - 8$$
$$-10x - 15 + 15 \ge 6x - 8 + 15$$
$$-10x \ge 6x + 7$$
$$-10x - 6x \ge 6x + 7 - 6x$$
$$-16x \ge 7$$
$$\frac{-16x}{-16} \le \frac{7}{-16}$$
$$x \le -\frac{7}{16}$$
Interval: $\left(-\infty, -\frac{7}{16}\right]$

22.
$$\frac{2x - 1}{4} - \frac{4x + 3}{6} > \frac{5}{3}$$
$$\left(\frac{2x - 1}{4} - \frac{4x + 3}{6}\right)(12) > \left(\frac{5}{3}\right)(12)$$
$$3(2x - 1) - 2(4x + 3) > 5(4)$$
$$6x - 3 - 8x - 6 > 20$$
$$-2x - 9 > 20$$
$$-2x > 29$$
$$x < -\frac{29}{2}$$
Interval: $\left(-\infty, -\frac{29}{2}\right)$

23.
$$1 \le 2x + 5 < 11$$
$$1 - 5 \le 2x + 5 - 5 < 11 - 5$$
$$-4 \le 2x < 6$$
$$-2 \le x < 3$$
Interval: $[-2, 3)$

24. a. Solve the inequality for x.
$$7 - 2(3x + 5) < 4x + 1$$
$$7 - 6x - 10 < 4x + 1$$
$$-6x - 3 < 4x + 1$$
$$-6x - 3 + 3 < 4x + 1 + 3$$
$$-6x < 4x + 4$$
$$-6x - 4x < 4x + 4 - 4x$$
$$-10x < 4$$
$$\frac{-10x}{-10} > \frac{4}{-10}$$
$$x > -\frac{2}{5}$$

Any three numbers that are greater than $-\frac{2}{5}$ are possible solutions.

b. From part (a), any number less than or equal to $-\frac{2}{5}$ is not a solution.

25. The student's work is incorrect. When dividing both sides of an inequality by a negative number, the direction of the inequality must be switched.
$$5 - 3x \le 11$$
$$5 - 3x - 5 \le 11 - 5$$
$$\frac{-3x}{-3} \ge \frac{6}{-3}$$
$$x \ge -2$$

26. $f(4) = 1$

27. $g(x) = 0$ when $x = -5$.

28. $f(x) = g(x)$ when $x = -2$.

29. The graph of $f(x)$ is above the graph of $g(x)$ for values of x that are greater than -2. Thus, $f(x) > g(x)$ when $x > -2$.

30. Answers may vary. Example:
Since the inequality symbols are facing
different directions in the two inequalities, a
must be a negative number. Let's choose
$a = -1$.
$$ax + b > c$$
$$(-1)x + b > c$$
$$-x > c - b$$
$$x < b - c$$
We want $x < 5$, so $b - c$ should be 5. Let's pick
$b = 2$ and $c = -3$.
Check:
$$ax + b > c$$
$$(-1)x + 2 > -3$$
$$-x > -5$$
$$x < 5$$
So, when $a = -1$, $b = 2$ and $c = -3$, the solution
set of $ax + b > c$ is $x < 5$.

31. $y \le 2x - 6$

Graph the line $y = 2x - 6$ with a solid line and
shade the region below it.

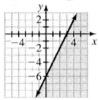

32. $4x - 2y < 8$
$$-2y < 8 - 4x$$
$$y > 2x - 4$$

Graph the line $y = 2x - 4$ with a dashed line
and shade the region above it.

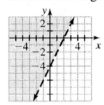

33. $-2(y + 3) + 4x \ge -8$
$$-2y - 6 + 4x \ge -8$$
$$-2y \ge -4x - 2$$
$$y \le 2x + 1$$

Graph the line $y = 2x + 1$ with a solid line and
shade the region below it.

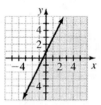

34. $y < -2$

Graph the line $y = 2$ with a dashed line and
shade the region below it.

35. $y \ge \dfrac{2}{5}x + 1$

$$y < -\dfrac{1}{4}x + 2$$

Graph the line $y = \dfrac{2}{5}x + 1$ with a solid line and

the line $y = -\dfrac{1}{4}x + 2$ with a dashed line. The

solution region of the system is the
intersection of the solution regions for

$y \ge \dfrac{2}{5}x + 1$ and $y < -\dfrac{1}{4}x + 2$.

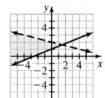

36. $3x - 4y \ge 12$ $6y - 2x \le 12$
$$-4y \ge -3x + 12$$ $$6y \le 2x + 12$$
$$y \le \dfrac{3}{4}x - 3$$ $$y \le \dfrac{1}{3}x + 2$$

Graph the lines $y = \dfrac{3}{4}x - 3$ and $y = \dfrac{1}{3}x + 2$

with solid lines. The solution region of the
system is the intersection of the solution

regions for $y \le \dfrac{3}{4}x - 3$ and $y \le \dfrac{1}{3}x + 2$.

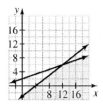

37. $x - y < 3$, so

$$y > x - 3$$

$x + y < 5$, so

$$y < -x + 5$$

$$x \geq 0$$

$$y \geq 0$$

Graph the lines $y = x - 3$ and $y = -x + 5$ with dashed lines, and graph the lines $x = 0$ and $y = 0$ with solid lines. The solution region of the system is the intersection of the solution regions of $y > x - 3$, $y < -x + 5$, $x \geq 0$, and $y \geq 0$.

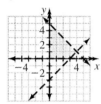

38. $2x - 5y > 10$

$$-5y > -2x + 10$$

$$y < \frac{2}{5}x - 2$$

Answers may vary. Example:
Three possible solutions:
$(-5, -5), (0, -4),$ and $(5, -2)$

$-5 \overset{?}{<} \frac{2}{5}(-5) - 2$ $-4 \overset{?}{<} \frac{2}{5}(0) - 2$

$-5 \overset{?}{<} -2 - 2$ $-4 \overset{?}{<} 0 - 2$

$-5 \overset{?}{<} -4$ true $-4 \overset{?}{<} -2$ true

$-2 \overset{?}{<} \frac{2}{5}(5) - 2$

$-2 \overset{?}{<} 2 - 2$

$-2 \overset{?}{<} 0$ true

Three possible non-solutions:
$(-5, 0), (0, 1),$ and $(5, 2)$

$0 \overset{?}{<} \frac{2}{5}(-5) - 2$ $1 \overset{?}{<} \frac{2}{5}(0) - 2$

$0 \overset{?}{<} -2 - 2$ $1 \overset{?}{<} 0 - 2$

$0 \overset{?}{<} -4$ false $1 \overset{?}{<} -2$ false

$2 \overset{?}{<} \frac{2}{5}(5) - 2$

$2 \overset{?}{<} 2 - 2$

$2 \overset{?}{<} 0$ false

39. a. Start by plotting the data sets, then find the regression line for each set.
Households with Phone Landlines:

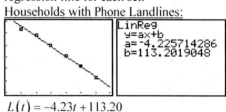

$L(t) = -4.23t + 113.20$

Households with Only Wireless:

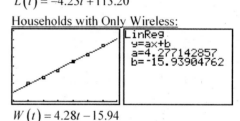

$W(t) = 4.28t - 15.94$

b. The slopes are –4.23 aand 4.28. The percentage of households with phone landlines is decreasing by 4.23 percentage points per year; the percentage of households with only wireless phones is increasing by 4.28 percentage points per year.

c.
$$L(t) = W(t)$$
$$-4.23t + 113.20 = 4.28t - 15.94$$
$$113.20 = 8.51t - 15.94$$
$$129.14 = 8.51t$$
$$15.18 \approx t$$

The percentage of households with phone landlines will be equal to the percentage of households with only wireless phones in $2000 + 15.18 \approx 2015$.

$$L(15.18) = -4.23(15.18) + 113.20$$
$$\approx -64.21 + 113.20$$
$$\approx 49.0$$

About 49% of households will have landlines and about 49% will have only wireless phones in 2015.

d. The percentage found in part (c) is not 50% because some households do not have phones.

e.
$$L(t) < W(t)$$
$$-4.23t + 113.20 < 4.28t - 15.94$$
$$113.20 < 8.51t - 15.94$$
$$129.14 < 8.51t$$
$$15.18 < t$$
The percentage of households with phone landlines will be less than the percentage of households with only wireless phones after 2015.

40. a. Start by plotting the data sets, then find the regression line for each set of ratings.
World Series:

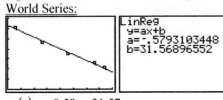

$$w(t) = -0.58t + 31.57$$

Prime Time:

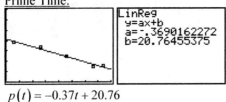

$$p(t) = -0.37t + 20.76$$

b. $-0.58t + 31.57 = -0.37t + 20.76$
$$31.57 = 0.21t + 20.76$$
$$10.81 = 0.21t$$
$$51.48 \approx t$$
The ratings will be equal approximately 51 years after 1970, in the year 2021.
$$w(51.48) = -0.58(51.48) + 31.57$$
$$\approx -29.86 + 31.57$$
$$\approx 1.7$$
The rating is approximately 1.7.

41. a. Since U-Haul's charge increases at a constant rate of $0.69 per mile, the equation is linear with slope 0.69. The U-intercept is 29.95 since U-Haul charges a flat fee of $29.95.
$$U(x) = 0.69x + 29.95 .$$
Similar work yields an equation for Rent A Wreck's charge.
$$R(x) = 0.22x + 75.00 .$$

b.
$$U(x) = R(x)$$
$$0.69x + 29.95 = 0.22x + 75$$
$$0.69x + 29.95 - 29.95 = 0.22x + 75 - 29.95$$
$$0.69x = 0.22x + 45.05$$
$$0.69x - 0.22x = 0.22x + 45.05 - 0.22x$$
$$0.47x = 45.05$$
$$x \approx 95.85$$
The two charges will be the same when the number of miles driven is roughly 95.85 miles, for a charge of approximately $96.09.

c.
$$R < U$$
$$0.22x + 75 < 0.69x + 29.95$$
$$0.22x + 75 - 75 < 0.69x + 29.95 - 75$$
$$0.22x < 0.69x - 45.05$$
$$0.22x - 0.69x < 0.69 - 45.05 - 0.69x$$
$$-0.47x < -45.05$$
$$\frac{-0.47x}{-0.47} > \frac{-45.05}{-0.47}$$
$$x > 95.85$$
Rent A Wreck will be cheaper for miles driven more than 95.85 miles.

42. Let $P(t)$ represent the average price (in dollars) of a home in a community and $S(t)$ represent the amount of money (in dollars) a family has saved at t years since 2012. Since the average price of a home increases at a constant $9000 per year, the function P is linear and its slope is 9000. The P-intercept is $(0, 250,000)$ since the price of a home is $250,000 in year $t = 0$.
$$P(t) = 9000t + 250,000 .$$
Similar work yields the equation for the function S.
$$S(t) = 2760t + 12,000 .$$ (The slope of S is 2760 because the family plans to save $230 each month, which is $2760 each year.)
In order to predict when the family will be able to pay a 10% down payment on an average-priced house, solve the following system for t.
$$y = 0.1(P(t)) = 0.1(9000t + 250,000)$$
$$y = S(t) = 2760t + 12,000$$
Substitute $0.1(9000t + 250,000)$ for y in the second equation and solve for t.

$$0.1(9000t + 250,000) = 2760t + 12,000$$
$$900t + 25,000 = 2760t + 12,000$$
$$-1860t = -13,000$$
$$t \approx 6.99$$

The family will be able to pay a 10% down payment in about 7 years, in $2012 + 7 = 2019$.

43. Let A represent the number of $55 tickets sold, and B represent the number of $70 tickets sold.

There are 20,000 total tickets available, so $A + B = 20,000$.

Selling A tickets at $55 and B tickets at $70, the theater wants to make $1,197,500.
$55A + 70B = 1,197,500$.

Use substitution or elimination to solve for A and B.
$$55(A + B) = 55(20,000)$$
$$55A + 70B = 1,197,500$$

Subtract the second equation from the first.
$$55A + 55B = 1,100,000$$
$$-(55A + 70B = 1,197,500)$$
$$\overline{\quad -15B = -97,500}$$
$$B = 6500$$

Now use this information to solve for A.
$$A + B = 20,000$$
$$A + 6500 = 20,000$$
$$A = 20,000 - 6500$$
$$A = 13,500$$

The theater must sell 13,500 tickets at $55 and 6500 tickets at $70 in order to generate a total revenue of $1,197,500.

44. a. To formulate a function $f(x)$ for total interest, substitute $(8000 - x)$ for y.
$$x + y = 8000$$
$$x + y - x = 8000 - x$$
$$y = 8000 - x$$

Write the equation for the total interest earned (in dollars) from investing $8000 for one year.
$$I = f(x) = 0.068x + 0.13y$$
$$f(x) = 0.068x + 0.13(8000 - x)$$
$$f(x) = -0.062x + 1040$$

b. $f(575) = -0.062(575) + 1040 = 1004.35$

This means that if $575 is invested in Hartford Global Leaders Y (and $7425 in Mutual Discovery Z), the total interest will be $1004.35.

c.
$$575 = -0.062x + 1040$$
$$575 - 1040 = -0.062x + 1040 - 1040$$
$$-465 = -0.062x$$
$$7500 = x$$

This means that if $7500 is invested in Hartford Global Leaders Y (and $500 in Mutual Discovery Z), the total interest will be $575.

Chapter 3 Test

1. Solve using substitution.
Substitute $3x - 1$ for y in the second equation and solve for x.
$$3x - 2y = -1$$
$$3x - 2(3x - 1) = -1$$
$$3x - 6x + 2 = -1$$
$$-3x + 2 = -1$$
$$-3x = -3$$
$$x = 1$$

Substitute $x = 1$ into the first equation and solve for y.
$$y = 3(1) - 1 = 3 - 1 = 2$$

The solution is $(1, 2)$.

2. Solve using elimination.
First write the second equation so that all the variables are on the left side of the equation and the constant is on the right side.
$$2x - 5y = 3$$
$$6x - 15y = 9$$

Multiply the first equation by -3 and add the equations.
$$-6x + 15y = -9$$
$$\underline{6x - 15y = 9}$$
$$0 = 0 \quad \text{true}$$

This is an identity. The system is dependent. The solution set is the set of ordered pairs (x, y) such that $2x - 5y = 3$.

3. Solve using elimination.
Multiply the first equation by 3 and the second equation by -2, then add the equations.
$$12x - 18y = 15$$
$$\underline{-12x + 18y = 4}$$
$$0 = 19 \quad \text{false}$$

This is a contradiction. The system is inconsistent. The solution set is the empty set.

4. Solve using elimination.
Multiply the second equation by 3 and add the equations.

$$\frac{2}{5}x - \frac{3}{4}y = 8$$
$$\frac{9}{5}x + \frac{3}{4}y = 3$$
$$\frac{11}{5}x = 11$$
$$x = 5$$

Substitute $x = 5$ into $\frac{9}{5}x + \frac{3}{4}y = 3$ and solve for y.

$$\frac{3}{5}(5) + \frac{1}{4}y = 1$$
$$3 + \frac{1}{4}y = 1$$
$$\frac{1}{4}y = 1 - 3$$
$$\frac{1}{4}y = -2$$
$$y = -8$$

The solution is $(5, -8)$.

5. First use the distributive property to simplify each equation.

$$-4(x + 2) + 3(2y - 1) = 21$$
$$-4x - 8 + 6y - 3 = 21$$
$$-4x + 6y - 11 = 21$$
$$-4x + 6y = 32$$
$$2x - 3y = -16$$

$$5(3x - 2) - (4y + 3) = -59$$
$$15x - 10 - 4y - 3 = -59$$
$$15x - 4y - 13 = -59$$
$$15x - 4y = -46$$

These equations can be written as a system.

$$2x - 3y = -16$$
$$15x - 4y = -46$$

Solve using elimination.
Multiply the first equation by 4 and the second equation by -3, then add the equations.

$$8x - 12y = -64$$
$$-45x + 12y = 138$$
$$-37x = 74$$
$$x = -2$$

Substitute $x = -2$ into $2x - 3y = -16$ and solve for y.

$$2(-2) - 3y = -16$$
$$-4 - 3y = -16$$
$$-3y = -12$$
$$y = 4$$

The solution is $(-2, 4)$.

6. Answers may vary. Example:
$$x - 2y = 1$$
$$3x + y = 17$$

7. Elimination:
First rewrite the second equation so that all the variables are on the left side and the constant is on the right side.

$$4x - 3y = 9$$
$$-2x + y = -5$$

Multiply the second equation by 2 and add the equations.

$$4x - 3y = 9$$
$$-4x + 2y = -10$$
$$-y = -1$$
$$y = 1$$

Substitute $y = 1$ into $4x - 3y = 9$ and solve for x.

$$4x - 3(1) = 9$$
$$4x - 3 = 9$$
$$4x = 12$$
$$x = 3$$

The solution is $(3, 1)$.

Substitution:
Substitute $2x - 5$ for y in the first equation.

$$4x - 3y = 9$$
$$4x - 3(2x - 5) = 9$$
$$4x - 6x + 15 = 9$$
$$-2x + 15 = 9$$
$$-2x = -6$$
$$x = 3$$

Substitute $x = 3$ into $y = 2x - 5$ and solve for y.

$$y = 2(3) - 5 = 1$$

The solution is $(3, 1)$.

Graphically:

$$4x - 3y = 9 \qquad\qquad y = 2x - 5$$

$$y = \frac{4}{3}x - 3$$

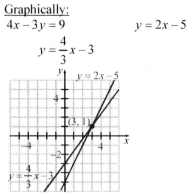

The solution is $(3,1)$.

8. If the solution set is the empty set, the system consists of two parallel lines. Therefore, m in $y = mx + b$ is 5, because this is the slope in $y = 5x - 13$ and parallel lines have the same slope. The y-intercept in $y = mx + b$ is any number other than -13 since $y = 5x - 13$ and $y = mx + b$ intersect the y-axis at different points. So $b \ne -13$.

9.
$$2 - 10x \ge 3x + 14$$
$$2 - 10x - 2 \ge 3x + 14 - 2$$
$$-10x \ge 3x + 12$$
$$-10x - 3x \ge 3x + 12 - 3x$$
$$-13x \ge 12$$
$$\frac{-13x}{-13} \le \frac{12}{-13}$$
$$x \le -\frac{12}{13}$$

Interval: $\left(-\infty, -\frac{12}{13}\right]$

10.
$$3(x + 4) + 1 < 5(x - 2)$$
$$3x + 12 + 1 < 5x - 10$$
$$3x + 13 < 5x - 10$$
$$3x + 13 - 13 < 5x - 10 - 13$$
$$3x < 5x - 23$$
$$3x - 5x < 5x - 23 - 5x$$
$$-2x < -23$$
$$\frac{-2x}{-2} > \frac{-23}{-2}$$
$$x > \frac{23}{2}$$

Interval: $\left(\frac{23}{2}, \infty\right)$

11.
$$2.6(t - 3.1) > 4.7t - 9.74$$
$$2.6t - 8.06 > 4.7t - 9.74$$
$$2.6t - 8.06 + 8.06 > 4.7t - 9.74 + 8.06$$
$$2.6t > 4.7t - 1.68$$
$$2.6x - 4.7t > 4.7t - 1.68 - 4.7t$$
$$-2.1t > -1.68$$
$$\frac{-2.1t}{-2.1} < \frac{-1.68}{-2.1}$$
$$t < 0.8$$

Interval: $(-\infty, -0.8)$

12.
$$-\frac{5}{3}w + \frac{1}{6} \le \frac{7}{4}w$$
$$-\frac{5}{3}w + \frac{1}{6} + \frac{5}{3}w \le \frac{7}{4}w + \frac{5}{3}w$$
$$\frac{1}{6} \le \frac{41}{12}w$$
$$\frac{12}{41} \cdot \frac{1}{6} \le \frac{12}{41} \cdot \frac{41}{12}x$$
$$\frac{2}{41} \le x$$
$$x \ge \frac{2}{41}$$

Interval: $\left[\frac{2}{41}, \infty\right)$

13. The graph of f is lower than the graph of g for all values of x that are greater than 13. Thus, $f(x) < g(x)$ for $x > 13$.

14. $f(5) = 3$

15. $g(x) = 3$ when $x = -4$.

16. $f(x) = g(x)$ when $x = 2$.

17. The graph of f is below the graph of g for $x < 2$ and the two graphs are equal for $x = 2$. Therefore, $f(x) \le g(x)$ when $x \le 2$.

18.

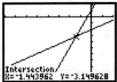

The point of intersection is $(-1.44, -3.15)$, so the value of x for which this statement is true is -1.44.

19. a. Solve the inequality for x.
$$3x - 11 < 7 - 6x$$
$$3x - 11 + 11 < 7 - 6x + 11$$
$$3x < 18 - 6x$$
$$3x + 6x < 18 - 6x + 6x$$
$$9x < 18$$
$$\frac{9x}{9} < \frac{18}{9}$$
$$x < 2$$

Answers may vary. Any number less than 2 will satisfy the inequality.

b. Answers may vary. From part (a), any number greater than or equal to 2 does not satisfy the inequality.

20. $y \le -\dfrac{2}{5}x + 3$

Graph the line $y = -\dfrac{2}{5}x + 3$ with a solid line and shade the region below it.

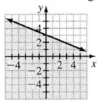

21. $4x - 3y < 6$
$$-3y < 6 - 4x$$
$$y > \frac{4}{3}x - 2$$

Graph the line $y = \dfrac{4}{3}x - 2$ with a dashed line and shade the region above it.

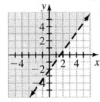

22.
$$3x + 2y > 6 \qquad\qquad 2x - y \ge 4$$
$$2y > -3x + 6 \qquad\qquad -y \ge -2x + 4$$
$$y > -\frac{3}{2}x + 3 \qquad\qquad y \le 2x - 4$$

Graph the line $y = -\dfrac{3}{2}x + 3$ with a dashed line, and the line $y = 2x - 4$ with a solid line.

The solution region of the system is the intersection of the solution regions for

$$y \le -\frac{3}{2}x + 3 \text{ and } y \le 2x - 4 .$$

23. a. Start by plotting the data sets, then find the regression line for each set.

Clarity in the Winter (in feet):

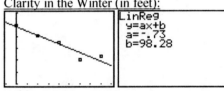

$$w(t) = -0.73t + 98.28$$

Clarity in the Summer (in feet):

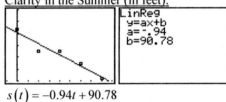

$$s(t) = -0.94t + 90.78$$

b.
$$w(t) = s(t)$$
$$-0.73t + 98.28 = -0.94t + 90.78$$
$$98.28 = -0.21t + 90.78$$
$$7.50 = -0.21t$$
$$-35.71 \approx t$$

The clarity in the winter was equal to the clarity in the summer in $1970 + (-35.71) \approx 1934$.

$$w(-35.71) = -0.73(-35.71) + 98.28$$
$$\approx 26.07 + 98.28$$
$$\approx 124.4$$

This clarity was about 124.4 feet.

c. $w(48) = -0.73(48) + 98.28$
$= -35.04 + 98.28$
$= 63.24$
$s(48) = -0.94(48) + 90.78$
$= -45.12 + 90.78$
$= 45.66$

Average clarity $= \dfrac{w(48) + s(48)}{2}$
$= \dfrac{63.24 + 45.66}{2}$
$= \dfrac{108.9}{2}$
≈ 54.5

The average clarity throughout all of 2018 will be about 54.5 feet.

24. a. For each retailer we are given the revenue from books, CDs, and DVDs in 2007 and in 2010. Since the revenues increased or decreased linearly we can write a linear model. To do this we first need to find the rate of increase or decrease for both retailers.

$m_A = \dfrac{7.0 - 4.6}{2010 - 2007} = \dfrac{2.4}{3} = 0.8$

$m_B = \dfrac{2.3 - 3.7}{2010 - 2007} = \dfrac{-1.4}{3} \approx -0.47$

The equation for Amazon will be of the form $y = 0.8x + b$. Substitute $2007 - 2000 = 7$ for x and 4.6 for y and solve for b.

$4.6 = 0.8(7) + b$
$4.6 = 5.6 + b$
$-1.0 = b$

The equation for Amazon is $A(t) = 0.8t - 1.0$.

The equation for Borders will be of the form $y = -0.47x + b$. Substitute $2007 - 2000 = 7$ for x and 3.7 for y and solve for b.

$3.7 = -\dfrac{1.4}{3}(7) + b$
$3.7 = -3.27 + b$
$6.97 \approx b$

The equation for Borders is $B(t) = -0.47t + 6.97$.

b. $0.8t - 1.0 = -0.47t + 6.97$
$1.27t - 1.0 = 6.97$
$1.27t = 7.97$
$t \approx 6.28$

The annual revenue of Amazon from books, CDs, and DVDs was equal to the annual revenue of Borders in $2000 + 6.28 \approx 2006$.

$A(6.28) = 0.8(6.28) - 1.0 = 5.024 - 1 \approx 4.0$

The revenue was about \$4.0 billion.

c. $0.8t - 1.0 < -0.47t + 6.97$
$1.27t - 1.0 < 6.97$
$1.27t < 7.97$
$t < 6.28$

The annual revenue of Amazon from books, CDs, and DVDs was less than the annual revenue of Borders in the years before $2000 + 6.28 \approx 2006$.

d. The slope of the graph of B is -0.47. The revenue of Borders decreased by \$0.47 billion (470 million) per year. Answers may vary.

25. Let x be the number of gallons of the 10% solution, and y be the number of gallons of the 20% solution.

$x + y = 10$
$x(0.10) + y(0.20) = 10(0.16)$
$(10 - y)(0.10) + 0.2y = 1.6$
$1 - 0.1y + 0.2y = 1.6$
$0.1y = 0.6$
$y = 6$

Substitute $y = 6$.
$x + 6 = 10 = 4$
4 gallons of a 10% antifreeze solution and 6 gallons of a 20% antifreeze solution must be mixed to make 10 gallons of a 16% antifreeze solution.

26. a. Add the revenues from the two types of tickets to find an equation of the total revenue R.
$R = 35x + 50y$

The total number of tickets sold for a sellout performance is 10,000:
$x + y = 10,000$
$y = 10,000 - x$

Substitute $10,000 - x$ into $R = 35x + 50y$.

$$R = 35x + 50y$$
$$= 35x + 50(10,000 - x)$$
$$= 35x + 500,000 - 50x$$
$$= -15x + 500,000$$

So, $f(x) = -15x + 500,000$.

b.

The equation $f(x) = -15x + 500,000$ is of the form $y = mx + b$. The slope is -15. The more $35 tickets sold (so, the fewer $50 tickets sold), the lower the total revenue will be (by $15 for each additional $35 ticket sold).

c.
$$390,500 = -15x + 500,000$$
$$-109,500 = -15x$$
$$7300 = x$$
$$y = 10,000 - x = 10,000 - 7300 - 2700$$

For the revenue to be $390,500, 7300 $35 tickets and 2700 $50 tickets must be sold.

Cumulative Review of Chapters 1–3

1.

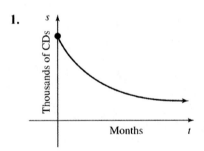

2. $5x - 3y = 15$
$$-3y = -5x + 15$$
$$y = \frac{5}{3}x - 5$$

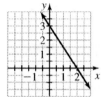

3. $3(x - 4) = -2(y + 5) + 4$
$$3x - 12 = -2y - 10 + 4$$
$$3x - 12 = -2y - 6$$
$$2y = -3x + 6$$
$$y = -\frac{3}{2}x + 3$$

4. $m = \dfrac{y_2 - y_1}{x_2 - x_1} = \dfrac{-1 - 2}{3 - (-4)} = \dfrac{-3}{7} = -\dfrac{3}{7}$

5. $m = -\dfrac{3}{5}$
$$y = mx + b$$
$$-3 = -\frac{3}{5}(2) + b$$
$$-3 = -\frac{6}{5} + b$$
$$b = -\frac{9}{5}$$

Write the equation of the line.
$$y = -\frac{3}{5}x - \frac{9}{5} \text{ or } y = -0.6x - 1.8$$

6. First find the slope.
$$m = \frac{3 - (-2)}{-2 - (-5)} = \frac{3 + 2}{-2 + 5} = \frac{5}{3}$$

Next use the slope and either point (we will use the first point) to find b.
$$m = \frac{5}{3}; (-5, -2)$$
$$y = mx + b$$
$$-2 = \frac{5}{3}(-5) + b$$
$$-2 = -\frac{25}{3} + b$$
$$b = \frac{19}{3}$$
$$y = \frac{5}{3}x + \frac{19}{3}$$

7. $2x - 5y = 20$

$-5y = -2x + 20$

$y = \dfrac{2}{5}x - 4$

Since the slope of the given line is $\dfrac{2}{5}$, our line

must have slope $-\dfrac{5}{2}$ since perpendicular lines

have opposite-reciprocal slopes.

$m = -\dfrac{5}{2}; (-5, 3)$

$y = mx + b$

$3 = -\dfrac{5}{2}(-5) + b$

$3 = \dfrac{25}{2} + b$

$b = -\dfrac{19}{2}$

Write the equation of the line.

$y = -\dfrac{5}{2}x - \dfrac{19}{2}$ or $y = -2.5x - 9.5$

8. The lines $y = 2x + 3$ and $y = 2x + 3.1$ are parallel. A line between these lines and parallel to both is $y = 2x + 3.05$. We can select any three points on this line. Answers may vary. Example:

$(0, 3.05)$

$(1, 5.05)$

$(2, 7.05)$

---

9. Determine the slopes from the given data and use them to obtain the remaining values.

$f(x)$: $m = \dfrac{58-97}{3-0} = \dfrac{-39}{3} = -13$		$g(x)$: $m = \dfrac{43-4}{9-6} = \dfrac{39}{3} = 13$		$h(x)$: $m = \dfrac{-22-23}{6-1} = \dfrac{-45}{5} = -9$		$k(x)$: $m = \dfrac{-16-(-28)}{14-10} = \dfrac{12}{4} = 3$	
Equation 1		**Equation 2**		**Equation 3**		**Equation 4**	
x	$f(x)$	x	$g(x)$	x	$h(x)$	x	$k(x)$
0	97	4	−22	1	23	10	−28
1	84	5	−9	2	14	11	−25
2	71	6	4	3	5	12	−22
3	**58**	7	17	4	−4	13	−19
4	45	8	30	5	−13	14	**−16**
5	32	9	43	6	−22	15	−13

10. $f(5) = 32$

11. $g(x) = 30$ when $x = 8$.

12. $f(x) = -\dfrac{3}{2}x + 7$

$f(-4) = -\dfrac{3}{2}(-4) + 7 = 6 + 7 = 13$

13. $f(x) = -\dfrac{3}{2}x + 7$

$\dfrac{5}{3} = -\dfrac{3}{2}x + 7$

$-\dfrac{16}{3} = -\dfrac{3}{2}x$

$x = \left(-\dfrac{16}{3}\right)\left(-\dfrac{2}{3}\right)$

$x = \dfrac{32}{9}$

14. The x-intercept is found by solving $f(x) = 0$.

$$0 = -\frac{3}{2}x + 7$$

$$\frac{3}{2}x = 7$$

$$\frac{2}{3} \cdot \frac{3}{2}x = \frac{2}{3} \cdot 7$$

$$x = \frac{14}{3}$$

The x-intercept is $\left(\frac{14}{3}, 0\right)$.

15. Since the equation is in the form $y = mx + b$, the y-intercept is $(0, b)$ or $(0, 7)$.

16. $f(x) = -\frac{3}{2}x + 7$

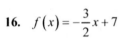

17. $g(3) = -3$

18. $f(x) = 1$ when $x = 0$.

19. The y-intercept of f is $(0, 1)$.

20. To find an equation for f, first determine the slope by finding two points on the graph. Two possible points are $(0, 1)$ and $(3, 3)$. Use the points to find the slope of the line.

$$m = \frac{3 - 1}{3 - 0} = \frac{2}{3}$$

Since the y-intercept is $(0, 1)$, the equation of the line is $y = \frac{2}{3}x + 1$.

21. $f(x) = g(x)$ when $x = -3$.

22. The graph of f is at or below the graph of g for all values of x that are less than or equal to -3. Therefore, $f(x) \le g(x)$ when $x \le -3$.

23.
$$-5x - 3(2x + 4) = 8 - 2x$$
$$-5x - 6x - 12 = 8 - 2x$$
$$-11x - 12 = 8 - 2x$$
$$-11x - 12 + 12 = 8 - 2x + 12$$
$$-11x = 20 - 2x$$
$$-11x + 2x = 20 - 2x + 2x$$
$$-9x = 20$$
$$x = -\frac{20}{9}$$

24.
$$\frac{7}{8} - \frac{1}{4}b = \frac{1}{2} + \frac{3}{8}b$$
$$\frac{7}{8} - \frac{1}{4}b - \frac{7}{8} = \frac{1}{2} + \frac{3}{8}b - \frac{7}{8}$$
$$-\frac{1}{4}b = \frac{3}{8}b - \frac{3}{8}$$
$$-\frac{1}{4}b - \frac{3}{8}b = \frac{3}{8}b - \frac{3}{8} - \frac{3}{8}b$$
$$-\frac{5}{8}b = -\frac{3}{8}$$
$$b = \frac{3}{5}$$

25.
$$\frac{b}{c} - d = \frac{k}{c}$$
$$\frac{b}{c} - d + d = \frac{k}{c} + d$$
$$\frac{b}{c}(c) = \frac{k}{c}(c) + d(c)$$
$$b = k + cd$$

26. To find the x-intercept, substitute 0 for y.
$$5x - 3y + 2 = 0$$
$$5x - 3(0) + 2 = 0$$
$$5x = -2$$
$$x = -\frac{2}{5}$$

The x-intercept is $\left(-\frac{2}{5}, 0\right)$.

To find the y-intercept, put the equation in slope–intercept form.
$$5x - 3y + 2 = 0$$
$$5x - 3y = -2$$
$$-3y = -5x - 2$$
$$y = \frac{5}{3}x + \frac{2}{3}$$

The equation is of the format $y = mx + b$, and so the y-intercept is $\left(0, \frac{2}{3}\right)$.

27. The domain of the relation is $-5 \le x \le 5$.

28. The range of the relation is $-2 \le y \le 3$.

29. Yes, the graph is that of a function. The graph passes the vertical line test.

30.
$$2x + 4y = -8$$
$$2x = -4y - 8$$
$$x = -2y - 4$$
$$5x - 3y = 19$$
$$5(-2y - 4) - 3y = 19$$
$$-10y - 20 - 3y = 19$$
$$-13y = 39$$
$$y = -\frac{39}{13} = -3$$
$$x = -2(-3) - 4$$
$$x = 2$$
The solution to the system is (2, –3).

31. Solve by substitution.
$$3x - 7y = 14$$
$$3x - 7\left(\frac{3}{7}x - 2\right) = 14$$
$$3x - 3x + 14 = 14$$
$$14 = 14$$
The system is dependent. The solution set is the set of ordered pairs (x, y) such that $3x - 7y = 14$.

32. The green and blue lines intersect at the point $(-2, -1)$. Therefore, the solution is –2.

33. The green line intersects the line $y = -3$ at the point $(6, -3)$. Therefore, the solution is 6.

34. The red and green lines intersect at the point $(2, -2)$, so that is the solution of the system.

35. $-2(4x + 5) \ge 3(x - 7) + 1$
$$-8x - 10 \ge 3x - 21 + 1$$
$$-8x + 10 \ge 3x$$
$$10 \ge 11x$$
$$\frac{10}{11} \ge x$$
Interval: $\left(-\infty, \dfrac{10}{11}\right]$

36. a. $-3x + 6 = 0$
$$-3x = -6$$
$$x = 2$$

b. $-3x + 6 < 0$
$$-3x < -6$$
$$x > 2$$

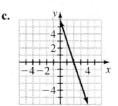

c.

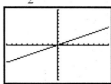

d. $4x - 1 = -3x + 6$
$$7x - 1 = 6$$
$$7x = 7$$
$$x = 1$$
$$y = 4(1) - 1$$
$$y = 3$$
The solution of the system of equations is (1, 3).

37. a. Answers may vary. Example:
$$4x = 8$$
$$x = 2$$

b. Answers may vary. Example:
$$4x = 8y$$
$$8y = 4x$$
$$y = \frac{1}{2}x$$
The solution set is every point on the line $y = \frac{1}{2}x$.

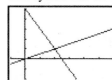

c. Answers may vary. Example:
$$4x = 8y$$
$$y = -2x + 5$$
This system has the solution $(2, 1)$.

d. Answers may vary. Example:

$$4x < 8$$
$$x < 2$$

The solution is $x < 2$ and the solution interval is $(-\infty, 2)$.

e. Answers may vary. Example:

$$y < -\frac{1}{2}x + 3$$

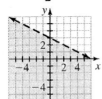

f. Answers may vary. Example:

$$y \le x - 4$$
$$y \ge -3x$$

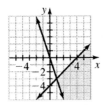

38. a. Start by plotting the data set, and then find the regression line for the data.

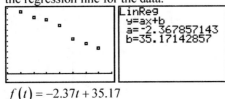

$$f(t) = -2.37t + 35.17$$

b. $f(4) = -2.37(4) + 35.17$
$$= -9.48 + 35.17$$
$$\approx 25.69$$

This means that in the year 2004, about 25.7 thousand foreign children were adopted by American families.

c. $f(t) = -2.37t + 35.17$
$$4 = -2.37t + 35.17$$
$$-31.17 = -2.37t$$
$$13.15 \approx t$$

This means that in the year 2013, 4 thousand foreign children were adopted by American families.

d. The equation $f(t) = -2.37t + 35.17$ is of the form $y = mx + b$. The slope is -2.37. This means that each year there is a decrease of 2.37 thousand foreign children adopted by American families.

e. $0 = -2.37t + 35.17$
$$-35.17 = -2.37t$$
$$14.84 \approx t$$

The t-intercept is $(14.84, 0)$.

In $2000 + 14.84 \approx 2015$, American families will not adopt any foreign children. Model breakdown has likely occurred.

39. a. At $t = 0$ (2009), the number of bicyclists younger than 16 who were hit and killed by motor vehicles was 82. The slope is -9. These values can be substituted into an equation of the form $y = mx + b$.
$$f(t) = -9t + 82.$$

b. The slope, -9, means that each year 9 fewer bicyclists younger than 16 are hit and killed by motor vehicles.

c. The n-intercept is 82, which is how many bicyclists younger than 16 were hit and killed by motor vehicles in the year 2009 $(t = 0)$.

d. $0 = -9t + 82$
$$9t = 82$$
$$t \approx 9.11$$

This means that in the year $2009 + 9.11 \approx 2018$, no bicyclists younger than 16 will be hit and killed by motor vehicles. Model breakdown has likely occurred.

e. According to this model, in every year after 2018 there will be fewer than 0 bicyclists younger than 16 hit and killed by motor vehicles, which is impossible.

40. a. For each retailer we are given the DSI for 2010, which is represented by $t = 0$. We are also given the rates of change, or slopes, of the DSIs. Since the DSIs increased linearly we can write linear models. The equation for Home Depot is
$$H(t) = 2.2t + 86.8.$$

The equation for Lowes is
$$L(t) = 3.2t + 95.9.$$

b. The slope of $H(t)$ is 2.2 and the slope of $L(t)$ is 3.2. The DSI of Home Depot is increasing by 2.2 days per year; the DSI of Lowes is increasing by 3.2 days per year. A negative slope would be better. Answers may vary.

c.
$$H(t) = L(t)$$
$$2.2t + 86.8 = 3.2t + 95.9$$
$$-1.0t + 86.8 = 95.9$$
$$-1.0t = 9.1$$
$$t = -9.1$$
The DSI of Home Depot was equal to the DSI of Lowes in $2010 + (-9.1) \approx 2001$.
$$H(-9.1) = 2.2(-9.1) + 86.8$$
$$= -20.02 + 86.8$$
$$\approx 66.8$$
The DSI was about 66.8 days.

d.
$$H(t) < L(t)$$
$$2.2t + 86.8 < 3.2t + 95.9$$
$$-1.0t + 86.8 < 95.9$$
$$-1.0t < 9.1$$
$$t > -9.1$$
The DSI of Home Depot was less than the DSI of Lowes after $2010 + (-9.1) \approx 2001$.

41. a. Start by plotting the data sets, then find the regression lines for each set of data.
Women:

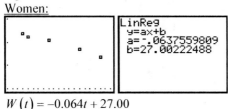

$$W(t) = -0.064t + 27.00$$
Men:

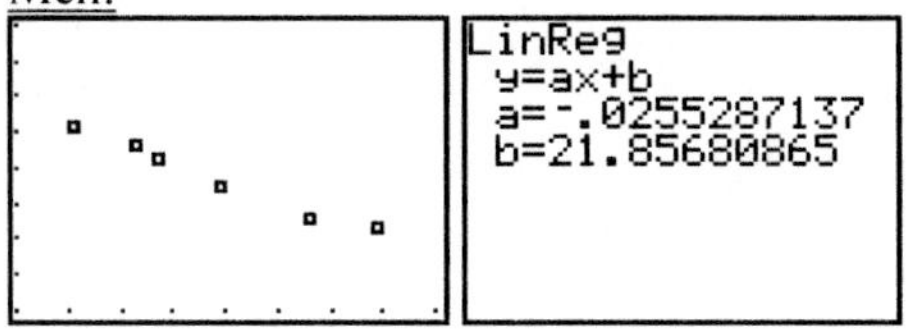

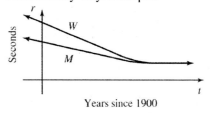

$$M(t) = -0.026t + 21.86$$

b. $W(118) = -0.064(118) + 27.00 \approx 19.45$
$M(118) = -0.026(118) + 21.86 \approx 18.79$
This means that in the year 2018, the women's record time will be 19.45 seconds and the men's record time will be 18.79 seconds.

c. The absolute value of W's slope is greater than the absolute value of M's slope. Women's record times are decreasing at a greater rate than men's record times.

d. Answers may vary. Example:
Over time, the record times of men and women will become closer together until women's times are equal to men's.

e. $-0.064t + 27.00 = -0.026t + 21.86$
$-0.038t + 27.00 = 21.86$
$-0.038t = -5.14$
$t \approx 135.26$
Find the time associated with this year.
$W(135.26) = -0.064(135.26) + 27.00$
≈ 18.34
The record times will be equal in about $1900 + 135.26 \approx 2035$, at a time of about 18.34 seconds.

f. $-0.064t + 27.00 > -0.026t + 21.86$
$-0.038t + 27.00 > 21.86$
$-0.038t > -5.14$
$t < 135.26$
This means that women's record times are greater than men's record times for years before $1900 + 135.26 \approx 2035$.

g.
$$W(t) = 0$$
$$-0.064(t) + 27.00 = 0$$
$$-0.064t = -27.00$$
$$t \approx 421.9$$
$$M(t) = 0$$
$$-0.026t + 21.86 = 0$$
$$-0.026t = -21.86$$
$$t \approx 840.77$$
This means the women's record time will be 0 seconds in $1900 + 421.9 \approx 2322$, and the men's record time will be 0 seconds in $1900 + 840.77 \approx 2741$. Model breakdown has occurred.

h. Answers may vary. Example:

(Graph: vertical axis labeled "Seconds", horizontal axis labeled t "Years since 1900", with lines W and M.)

42. Start by setting up a system of equations for U, amount invested in a UBS Global Equity Y account (at 7.2% interest), and F, amount invested in a Fidelity Worldwide account (at 9.4% interest).

$$U = 2F$$
$$0.072U + 0.094F = 595$$

Now solve for either U or F.

$$0.072(2F) + 0.094F = 595$$
$$0.144F + 0.094F = 595$$
$$0.238F = 595$$
$$F = 2500$$

Substitute this value into the original equation to solve for the other variable.

$$U = 2F$$
$$U = 2(2500) = 5000$$

To earn \$595 in one year, the person will have to invest \$2500 into the Fidelity Worldwide account, and \$5000 into the UBS Global Equity Y account.

Chapter 4
Exponential Functions

1. $2^{-1} = \dfrac{1}{2}$

3. $5^0 = 1$

5. $-4^2 = -(4 \cdot 4) = -16$

7. $(-4)^2 = (-4)(-4) = 16$

9. $\left(2^3\right)^2 = 2^6 = 2 \cdot 2 \cdot 2 \cdot 2 \cdot 2 \cdot 2 = 64$

11. $2^{-1} + 3^{-1} = \dfrac{1}{2} + \dfrac{1}{3} = \dfrac{3}{6} + \dfrac{2}{6} = \dfrac{5}{6}$

13. $\dfrac{7^{902}}{7^{900}} = 7^{902-900} = 7^2 = 49$

15. $13^{500} 13^{-500} = 13^{500+(-500)} = 13^0 = 1$

17. $\left(25^3 - 411^5 + 89^2\right)^0 = 1$

19. $b^7 b^{-9} = b^{7-9} = b^{-2} = \dfrac{1}{b^2}$

21. $\left(7b^{-3}\right)\left(-2b^{-5}\right) = \left(7(-2)\right)\left(b^{-3}b^{-5}\right)$
$$= -14b^{-3-5}$$
$$= -14b^{-8}$$
$$= \dfrac{-14}{b^8} \text{ or } -\dfrac{14}{b^8}$$

23. $\left(-9b^{-7}c^5\right)\left(-8b^6c^{-5}\right)$
$$= \left(-9(-8)\right)\left(b^{-7}b^6\right)\left(c^5 c^{-5}\right)$$
$$= 72b^{-7+6}c^{5-5}$$
$$= 72b^{-1}c^0$$
$$= \dfrac{72}{b}$$

25. $\left(3b^2c^4\right)^3 \left(2b^3c^5\right)^2 = 3^3 b^{2\cdot3} c^{4\cdot3} 2^2 b^{3\cdot2} c^{5\cdot2}$
$$= 3^3 b^6 c^{12} 2^2 b^6 c^{10}$$
$$= 27(4)b^{6+6}c^{12+10}$$
$$= 108b^{12}c^{22}$$

27. $3\left(b^5c\right)^{-2} = 3b^{5(-2)}c^{-2} = 3b^{-10}c^{-2} = \dfrac{3}{b^{10}c^2}$

29. $\left(2b^4c^{-2}\right)^5 \left(3b^{-3}c^{-4}\right)^{-2}$
$$= (2)^5 \left(b^4\right)^5 \left(c^{-2}\right)^5 (3)^{-2} \left(b^{-3}\right)^{-2} \left(c^{-4}\right)^{-2}$$
$$= 32b^{4\cdot5}c^{-2\cdot5} \dfrac{1}{3^2} b^{-3(-2)} c^{-4(-2)}$$
$$= 32b^{20}c^{-10} \dfrac{1}{9} b^6 c^8$$
$$= 32\left(\dfrac{1}{9}\right)b^{20}b^6 c^{-10}c^8$$
$$= \dfrac{32b^{26}}{9c^2}$$

31. $\dfrac{b^{-10}}{b^{15}} = b^{-10-15} = b^{-25} = \dfrac{1}{b^{25}}$

33. $\dfrac{2b^{-12}}{5b^{-9}} = \dfrac{2b^{-12-(-9)}}{5} = \dfrac{2b^{-12+9}}{5} = \dfrac{2b^{-3}}{5} = \dfrac{2}{5b^3}$

35. $\dfrac{-12b^{-6}c^5}{14b^4c^5} = \dfrac{-6b^{-6-4}c^{5-5}}{7}$
$$= \dfrac{-6b^{-10}c^0}{7}$$
$$= -\dfrac{6}{7b^{10}}$$

37. $\dfrac{15b^{-7}c^{-3}d^8}{-45c^2b^{-6}d^8} = \dfrac{b^{-7-(-6)}c^{-3-2}d^{8-8}}{-3}$
$$= \dfrac{b^{-1}c^{-5}d^0}{-3}$$
$$= \dfrac{1}{-3bc^5} \text{ or } -\dfrac{1}{3bc^5}$$

39. $\dfrac{\left(-5b^{-3}c^4\right)\left(4b^{-5}c^{-1}\right)}{80b^2c^{17}} = \dfrac{-20b^{-3+(-5)}c^{4+(-1)}}{80b^2c^{17}}$

$= -\dfrac{b^{-8}c^3}{4b^2c^{17}}$

$= -\dfrac{b^{-8-2}c^{3-17}}{4}$

$= -\dfrac{b^{-10}c^{-14}}{4}$

$= -\dfrac{1}{4b^{10}c^{14}}$

41. $\dfrac{\left(24b^3c^{-6}\right)\left(49b^{-1}c^{-2}\right)}{\left(28b^2c^4\right)\left(14b^{-5}c\right)}$

$= \dfrac{24 \cdot 49 b^{3+(-1)}c^{-6+(-2)}}{28 \cdot 14 b^{2+(-5)}c^{4+1}}$

$= \dfrac{3b^2c^{-8}}{b^{-3}c^5}$

$= 3b^{2-(-3)}c^{-8-5}$

$= 3b^5c^{-13}$

$= \dfrac{3b^5}{c^{13}}$

43. $\dfrac{\left(3b^5c^{-2}\right)^3}{2^{-1}b^{-3}c} = \dfrac{3^3b^{5\cdot3}c^{-2\cdot3}}{2^{-1}b^{-3}c}$

$= \dfrac{27b^{15}c^{-6}}{2^{-1}b^{-3}c}$

$= 27(2)b^{15-(-3)}c^{-6-1}$

$= 54b^{18}c^{-7}$

$= \dfrac{54b^{18}}{c^7}$

45. $\dfrac{\left(2b^{-4}c\right)^{-3}}{\left(2b^2c^{-5}\right)^2} = \dfrac{2^{-3}b^{-4\cdot-3}c^{-3}}{2^2b^{2\cdot2}c^{-5\cdot2}}$

$= \dfrac{2^{-3}b^{12}c^{-3}}{2^2b^4c^{-10}}$

$= \dfrac{b^{12-4}c^{-3-(-10)}}{8 \cdot 4}$

$= \dfrac{b^8c^7}{32}$

47. $\left(\dfrac{6b^5c^{-2}}{7b^2c^4}\right)^2 = \dfrac{6^2b^{5\cdot2}c^{-2\cdot2}}{7^2b^{2\cdot2}c^{4\cdot2}}$

$= \dfrac{36b^{10}c^{-4}}{49b^4c^8}$

$= \dfrac{36b^{10-4}c^{-4-8}}{49}$

$= \dfrac{36b^6c^{-12}}{49}$

$= \dfrac{36b^6}{49c^{12}}$

49. $\left(\dfrac{5b^4c^{-3}}{15b^{-2}c^{-1}}\right)^{-4} = \left(\dfrac{b^4c^{-3}}{3b^{-2}c^{-1}}\right)^{-4}$

$= \dfrac{b^{4(-4)}c^{-3(-4)}}{3^{-4}b^{-2(-4)}c^{-1(-4)}}$

$= \dfrac{3^4b^{-16}c^{12}}{b^8c^4}$

$= \dfrac{81c^{12-4}}{b^{16+8}}$

$= \dfrac{81c^8}{b^{24}}$

51. $\left(\dfrac{7b^4c^{-5}}{14b^7c^{-2}}\right)^0 = 1$

53. $b^{-1}c^{-1} = \dfrac{1}{bc}$

55. $\dfrac{1}{b^{-1}} + \dfrac{1}{c^{-1}} = b + c$

57. $b^{4n}b^{3n} = b^{4n+3n} = b^{7n}$

59. $\dfrac{b^{7n-1}}{b^{2n+3}} = b^{(7n-1)-(2n+3)} = b^{7n-1-2n-3} = b^{5n-4}$

61. $f(3) = 2(3)^3 = 2(27) = 54$

63. $f(-4) = 2(3)^{-4} = \dfrac{2}{3^4} = \dfrac{2}{81}$

65. $g(a+2) = 4^{a+2} = 4^a \cdot 4^2 = 16\left(4^a\right)$

67. $g(2a) = 4^{2a} = \left(4^2\right)^a = 16^a$

69. a.

x	$f(x)$
-3	$f(-3) = 2^{-3} = \dfrac{1}{2^3} = \dfrac{1}{8} = 0.125$
-2	$f(-2) = 2^{-2} = \dfrac{1}{2^2} = \dfrac{1}{4} = 0.25$
-1	$f(-1) = 2^{-1} = \dfrac{1}{2} = 0.5$
0	$f(0) = 2^0 = 1$
1	$f(1) = 2^1 = 2$
2	$f(2) = 2^2 = 4$
3	$f(3) = 2^3 = 8$
4	$f(4) = 2^4 = 16$

b.

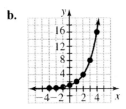

c. $2^{\frac{1}{2}} \approx 1.4$

71. $3.965 \times 10^2 = 3.965 \cdot 100 = 396.5$

73. $2.39 \times 10^{-1} = 2.39 \cdot \dfrac{1}{10} = 0.239$

75. $5.2 \times 10^2 = 5.2 \cdot 100 = 520$

77. $9.113 \times 10^{-5} = 9.113 \cdot \dfrac{1}{100,000} = 0.00009113$

79. $-6.52 \times 10^{-4} = -6.52 \cdot \dfrac{1}{10,000} = -0.000652$

81. $9 \times 10^5 = 9 \cdot 100,000 = 900,000$

83. $-8 \times 10^0 = -8 \cdot 1 = -8$

85. $54,260,000 = 5.426 \times 10,000,000$
$\qquad = 5.426 \times 10^7$

87. $23,587 = 2.3587 \times 10,000 = 2.3587 \times 10^4$

89. $0.00098 = \dfrac{9.8}{10,000} = 9.8 \times 10^{-4}$

91. $0.0000346 = \dfrac{3.46}{100,000} = 3.46 \times 10^{-5}$

93. $-42,215 = -4.2215 \times 10,000 = -4.2215 \times 10^4$

95. $-0.00244 = \dfrac{-2.44}{1000} = -2.44 \times 10^{-3}$

97. $6.3\text{E}^-6 = \dfrac{6.3}{1,000,000} = 0.0000063$

$1.3\text{E}^-4 = \dfrac{1.3}{10,000} = 0.00013$

$3.2\text{E}6 = 3.2 \cdot 1,000,000 = 3,200,000$

$6.4\text{E}7 = 6.4 \cdot 10,000,000 = 64,000,000$

99. $3.6 \times 10^9 = 3.6 \cdot 1,000,000,000$
$\qquad = 3,600,000,000$ years

101. $6.3 \times 10^{-8} = 6.3 \cdot \dfrac{1}{100,000,000}$
$\qquad = 0.000000063$ mole per liter

103. $10,080,000 = 1.008 \times 10,000,000$
$\qquad = 1.008 \times 10^7$ gallons

105. $0.00000047 = \dfrac{4.7}{10,000,000} = 4.7 \times 10^{-7}$ meter

107. a.

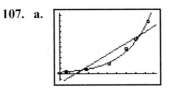

The exponential function $n = 0.30(1.078)^t$ describes the function better.

b. $n = 0.30(1.078)^{57} \approx 21.7$

The model predicts the number of bald eagle pairs to be 21.7 thousand in 2017.

c. $n = 0.21(57) - 1.66 = 11.34 - 1.66 \approx 10.3$

The model predicts the number of bald eagle pairs to be 10.3 thousand in 2017.

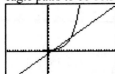

Answers may vary. Example:
The exponential model grows at a much faster rate than the linear model.

109. Student B is correct. Student A did not distribute the -1 exponent to the base 5 correctly. It should have been 5^{-1} and not -5. Student A mistakenly multiplied 5 by -1.

111. The student moved 3 to the denominator with b^{-2}.

$$\frac{3b^{-2}c^4}{d^7} = \frac{3c^4}{b^2 d^7}$$

113. $-2^2 = -4$

$2(-1) = -2$

$\left(\dfrac{1}{2}\right)^2 = \dfrac{1}{4}$

$\dfrac{1}{2}$

$2^{-1} = \dfrac{1}{2}$

$\left(\dfrac{1}{2}\right)^{-1} = \dfrac{2}{1} = 2$

$(-2)^2 = 4$

$2^2 = 4$

The "ties" are $2^{-1} = \dfrac{1}{2}$ and $(-2)^2 = 2^2$.

115. a. $5^0 = 4^0 = 3^0 = 2^0 = 1^0 = 1$

It is reasonable to assume that $0^0 = 1$.

b. $0^5 = 0^4 = 0^3 = 0^2 = 0^1 = 0$

It is reasonable to assume that $0^0 = 0$.

c. Answers may vary. Example:
It is a good idea to leave 0^0 meaningless, since depending on the reasoning used, we get different values for 0^0.

117. Answers may vary. Example:

b^m means that the base b is multiplied by itself m times. Also, b^n means that the base b is multiplied by itself n times. So, if we multiply m factors of b by n factors of b, we will have a total of $m + n$ factors of b.

$\left(b^m\right)^n$ means that there are n factors of b^m.

Since each b^m has m factors of b, we have a total of $m \cdot n$ or mn factors of b.

119. $f(3) = 2(3) = 6$

121. $g(3) = 2^3 = 8$

123. $y = 3x + 1$
$y = 2x - 4$
$3x + 1 = 2x - 4$
$x + 1 = -4$
$x = -5$
$y = 3(-5) + 1 = -15 + 1 = -14$
The solution is $(-5, -14)$.
This is a linear system in two variables.

125. $3x + 1 = 2x - 4$
$x = -5$
This is a linear equation in one variable.

Homework 4.2

1. $16^{1/2} = 4$, since $4^2 = 16$.

3. $1000^{1/3} = 10$, since $10^3 = 1000$.

5. $49^{1/2} = 7$, since $7^2 = 49$.

7. $125^{1/3} = 5$, since $5^3 = 125$.

9. $8^{4/3} = \left(8^{1/3}\right)^4 = 2^4 = 16$

11. $9^{3/2} = \left(9^{1/2}\right)^3 = 3^3 = 27$

13. $32^{2/5} = \left(32^{1/5}\right)^2 = 2^2 = 4$

15. $4^{5/2} = \left(4^{1/2}\right)^5 = 2^5 = 32$

17. $27^{-1/3} = \dfrac{1}{27^{1/3}} = \dfrac{1}{3}$

19. $-36^{-1/2} = -\dfrac{1}{36^{1/2}} = -\dfrac{1}{6}$

21. $4^{-5/2} = \dfrac{1}{4^{5/2}} = \dfrac{1}{\left(4^{1/2}\right)^5} = \dfrac{1}{2^5} = \dfrac{1}{32}$

23. $(-27)^{-4/3} = \dfrac{1}{(-27)^{4/3}}$

$\qquad = \dfrac{1}{\left((-27)^{1/3}\right)^4}$

$\qquad = \dfrac{1}{(-3)^4}$

$\qquad = \dfrac{1}{81}$

25. $2^{1/4}2^{3/4} = 2^{\frac{1}{4}+\frac{3}{4}} = 2^{\frac{4}{4}} = 2^1 = 2$

27. $\left(3^{1/2}2^{3/2}\right)^2 = 3^{\left(\frac{1}{2}\right)(2)}2^{\left(\frac{3}{2}\right)(2)} = 3^1 \cdot 2^3 = 3 \cdot 8 = 24$

29. $\dfrac{7^{1/3}}{7^{-5/3}} = 7^{\frac{1}{3}-\left(-\frac{5}{3}\right)} = 7^{\frac{6}{3}} = 7^2 = 49$

31. $f\left(\dfrac{3}{4}\right) = 81^{3/4} = \left(81^{1/4}\right)^3 = 3^3 = 27$

33. $g\left(\dfrac{1}{3}\right) = 4(27)^{1/3} = 4 \cdot 3 = 12$

35. $g\left(-\dfrac{1}{3}\right) = 4(27)^{-1/3} = \dfrac{4}{27^{1/3}} = \dfrac{4}{3}$

37. $h\left(\dfrac{3}{2}\right) = -2(4)^{3/2}$

$\qquad = -2\left(4^{1/2}\right)^3$

$\qquad = -2(2)^3$

$\qquad = -2 \cdot 8$

$\qquad = -16$

39.

x	$f(x)$
$-\dfrac{3}{4}$	$f\left(-\dfrac{3}{4}\right) = 16^{-3/4} = \dfrac{1}{16^{3/4}}$ $= \dfrac{1}{\left(16^{1/4}\right)^3} = \dfrac{1}{2^3} = \dfrac{1}{8}$
$-\dfrac{1}{2}$	$f\left(-\dfrac{1}{2}\right) = 16^{-1/2} = \dfrac{1}{16^{1/2}} = \dfrac{1}{4}$
$-\dfrac{1}{4}$	$f\left(-\dfrac{1}{4}\right) = 16^{-1/4} = \dfrac{1}{16^{1/4}} = \dfrac{1}{2}$
0	$f(0) = 16^0 = 1$
$\dfrac{1}{4}$	$f\left(\dfrac{1}{4}\right) = 16^{1/4} = 2$
$\dfrac{1}{2}$	$f\left(\dfrac{1}{2}\right) = 16^{1/2} = 4$
$\dfrac{3}{4}$	$f\left(\dfrac{3}{4}\right) = 16^{3/4} = \left(16^{1/4}\right)^3 = 2^3 = 8$
1	$f(1) = 16^1 = 16$

41. $b^{7/6}b^{5/6} = b^{\frac{7}{6}+\frac{5}{6}} = b^{12/6} = b^2$

43. $b^{3/5}b^{-13/5} = b^{\frac{3}{5}+\left(-\frac{13}{5}\right)} = b^{-10/5} = b^{-2} = \dfrac{1}{b^2}$

45. $\left(16b^8\right)^{1/4} = (16)^{1/4}\left(b^8\right)^{1/4} = 2b^{8\left(\frac{1}{4}\right)} = 2b^2$

47. $4\left(25b^8c^{14}\right)^{-1/2} = \dfrac{4}{\left(25b^8c^{14}\right)^{1/2}}$

$\qquad = \dfrac{4}{(25)^{1/2}\left(b^8\right)^{1/2}\left(c^{14}\right)^{1/2}}$

$\qquad = \dfrac{4}{5\left(b^{8\left(\frac{1}{2}\right)}\right)\left(c^{14\left(\frac{1}{2}\right)}\right)}$

$\qquad = \dfrac{4}{5b^4c^7}$

49. $\left(b^{3/5}c^{-1/4}\right)\left(b^{2/5}c^{-7/4}\right)=b^{\frac{3}{5}+\frac{2}{5}}c^{-\frac{1}{4}+\left(-\frac{7}{4}\right)}$

$=b^{\frac{5}{5}}c^{-\frac{8}{4}}$

$=b^1c^{-2}$

$=\dfrac{b}{c^2}$

51. $\left(5bcd\right)^{1/5}\left(5bcd\right)^{4/5}=\left(5bcd\right)^{\frac{1}{5}+\frac{4}{5}}$

$=\left(5bcd\right)^{\frac{5}{5}}$

$=5bcd$

53. $\left[\left(3b^5\right)^3\left(3b^9c^8\right)\right]^{1/4}$

$=\left(3b^5\right)^{3/4}\left(3b^9c^8\right)^{1/4}$

$=\left(3\right)^{3/4}\left(b^5\right)^{3/4}\left(3\right)^{1/4}\left(b^9\right)^{1/4}\left(c^8\right)^{1/4}$

$=3^{3/4}b^{15/4}3^{1/4}b^{9/4}c^2$

$=3^{\frac{3}{4}+\frac{1}{4}}b^{\frac{15}{4}+\frac{9}{4}}c^2$

$=3b^6c^2$

55. $\dfrac{b^{-2/5}c^{1/8}}{b^{18/5}c^{-5/8}}=\dfrac{c^{\frac{11}{8}-\left(-\frac{5}{8}\right)}}{b^{\frac{2}{5}+\frac{18}{5}}}=\dfrac{c^{16/8}}{b^{20/5}}=\dfrac{c^2}{b^4}$

57. $\left(\dfrac{9b^3c^{-2}}{25b^{-5}c^4}\right)^{-1/2}=\left(\dfrac{9b^3b^5}{25c^2c^4}\right)^{-1/2}$

$=\left(\dfrac{9b^{3+5}}{25c^{2+4}}\right)^{-1/2}$

$=\left(\dfrac{9b^8}{25c^6}\right)^{-1/2}$

$=\left(\dfrac{25c^6}{9b^8}\right)^{1/2}$

$=\dfrac{\left(25\right)^{1/2}\left(c^6\right)^{1/2}}{\left(9\right)^{1/2}\left(b^8\right)^{1/2}}$

$=\dfrac{5c^3}{3b^4}$

59. $32^{1/5}b^{3/7}b^{2/5}=2b^{\frac{3}{7}+\frac{2}{5}}=2b^{\frac{15+14}{35}}=2b^{29/35}$

61. $\dfrac{b^{5/6}}{b^{1/4}}=b^{\frac{5}{6}-\frac{1}{4}}=b^{\frac{10}{12}-\frac{3}{12}}=b^{7/12}$

63. $\dfrac{\left(9b^5\right)^{3/2}}{\left(27b^4\right)^{2/3}}=\dfrac{\left(9\right)^{3/2}\left(b^5\right)^{3/2}}{\left(27\right)^{2/3}\left(b^4\right)^{2/3}}$

$=\dfrac{27b^{15/2}}{9b^{8/3}}$

$=3b^{\frac{45-16}{6}}$

$=3b^{29/6}$

65. $\left(\dfrac{8b^{2/3}}{2b^{4/5}}\right)^{3/2}=\left(\dfrac{4b^{2/3}}{b^{4/5}}\right)^{3/2}$

$=\dfrac{\left(4\right)^{3/2}\left(b^{2/3}\right)^{3/2}}{\left(b^{4/5}\right)^{3/2}}$

$=\dfrac{\left(4^{1/2}\right)^3 b}{b^{6/5}}$

$=2^3b^{1-\frac{6}{5}}$

$=8b^{-1/5}$

$=\dfrac{8}{b^{1/5}}$

67. $\dfrac{\left(8bc^3\right)^{1/3}}{\left(81b^{-5}c^3\right)^{3/4}}=\dfrac{\left(8\right)^{1/3}b^{1/3}\left(c^3\right)^{1/3}}{\left(81\right)^{3/4}\left(b^{-5}\right)^{3/4}\left(c^3\right)^{3/4}}$

$=\dfrac{2b^{1/3}c}{27b^{-15/4}c^{9/4}}$

$=\dfrac{2b^{\frac{1}{3}+\frac{15}{4}}c^{1-\frac{9}{4}}}{27}$

$=\dfrac{2b^{\frac{4}{12}+\frac{45}{12}}c^{-\frac{5}{4}}}{27}$

$=\dfrac{2b^{49/12}}{27c^{5/4}}$

69. $b^{2/5}\left(b^{8/5}+b^{3/5}\right)=b^{2/5}b^{8/5}+b^{2/5}b^{3/5}$

$=b^{\frac{2}{5}+\frac{8}{5}}+b^{\frac{2}{5}+\frac{3}{5}}$

$=b^{10/5}+b^{5/5}$

$=b^2+b$

71. a.

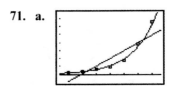

The exponential function $A=2.73\left(1.8\right)^t$ describes the situation better.

b. $A = 2.73(1.8)^{18} \approx 107,416$

About 107,416 megawatts of solar power will be installed in 2018.

c.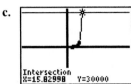

Intersection
X=15.82990   Y=30000

The model predicts that the amount of solar power installed will be 30,000 megawatts in 2016.

73. Answers may vary. Example:

$5^{1/n}$ may be represented by $\sqrt[n]{5}$. So, when $n = 2$, $\sqrt[n]{5} = \sqrt[2]{5}$, which is equivalent to $\sqrt{5}$.

75. The student did not compute $36^{1/2}$ correctly.

$$\left(36x^{36}\right)^{1/2} = 36^{1/2}\left(x^{36}\right)^{1/2} = 6x^{18}$$

77. Answers may vary. Example:

$9^{3/2}$ can be written as either $\left(9^{1/2}\right)^3$ or $\left(9^3\right)^{1/2}$. Student A's answer is easier because 3^3 is easier to simplify than $729^{1/2}$.

79. a. $(-9)^{1/2}$, $(-81)^{1/4}$, and $(-1)^{1/6}$ are not real numbers.

b. If b is negative and n is even, $b^{1/n}$ is not a real number. Answers may vary. Example: An even power of any real number is nonnegative.

81. Answers may vary. Example:

We can compute either $b^{m/n} = \left(b^m\right)^{1/n}$ or $b^{m/n} = \left(b^{1/n}\right)^m$.

83. $f\left(\dfrac{1}{3}\right) = 8 \cdot \dfrac{1}{3} = \dfrac{8}{3}$

85. $g\left(\dfrac{1}{3}\right) = 8^{1/3} = 2$

87. $f\left(-\dfrac{1}{3}\right) = 8 \cdot -\dfrac{1}{3} = -\dfrac{8}{3}$

89. $g\left(-\dfrac{1}{3}\right) = 8^{-1/3} = \dfrac{1}{8^{1/3}} = \dfrac{1}{2}$

91.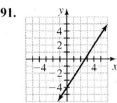

$f(x) = \dfrac{3}{2}x - 4$ is a linear function.

93. $5 = \dfrac{3}{2}x - 4$

$9 = \dfrac{3}{2}x$

$x = 6$

$f(x) = \dfrac{3}{2}x - 4$ is a linear function.

Homework 4.3

1. $y = 3^x$

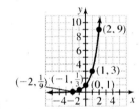

3. $y = 10^x$

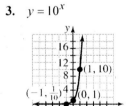

5. $y = 3(2)^x$

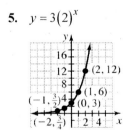

7. $y = 6(3)^x$

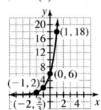

9. $y = 15\left(\dfrac{1}{3}\right)^x$

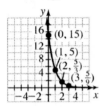

11. $y = 12\left(\dfrac{1}{2}\right)^x$

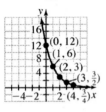

13. $f(x) = 2^x, g(x) = -2^x$

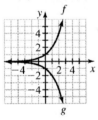

15. $f(x) = 4(3)^x, g(x) = -4(3)^x$

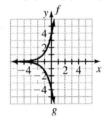

17. $f(x) = 8\left(\dfrac{1}{2}\right)^x, g(x) = -8\left(\dfrac{1}{2}\right)^x$

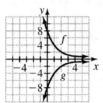

19.

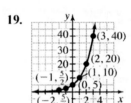

Domain: All real numbers
Range: $y > 0$

21.

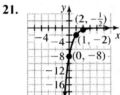

Domain: All real numbers
Range: $y < 0$

23. a. Answers may vary. Example:

x	$f(x)$
0	4
1	8
2	16
3	32
4	64

b. Answers may vary. Example:

c. Answers may vary. Example:
For each input–output pair, the output is 4 times 2 raised to the power equal to the input.

25.

x	$f(x)$	$g(x)$	$h(x)$	$k(x)$
0	162	3	2	800
1	54	12	10	400
2	18	48	50	200
3	6	192	250	100
4	2	768	1250	50

27.

x	$f(x)$	$g(x)$	$h(x)$	$k(x)$
0	5	160	162	3
1	10	80	54	12
2	20	40	18	48
3	40	20	6	192
4	80	10	2	768

29. $f(-3) = 8$

31. $f(0) = 1$

33. $x = -2$

35. $x = 0$

37. $f(3) = 24$

39. $f(5) = 96$

41. $x = 0$

43. $x = 3$

45. **a.** Make a scattergram of the data.

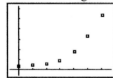

It is better to use an exponential function to model the data as the average ticket price appears to increase at an exponential rate, as opposed to a linear rate.

b.

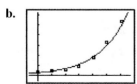

Yes, the graph of f comes close to the data points.

c. $f(68) = 1.22(1.051)^{68} \approx 35.92$

The average ticket price in 2018 will be about \$35.92.

d.

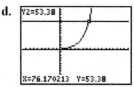

The average ticket prices to all major league games will reach \$53.38 in 2026.

47. $y = 7^x$

No x-intercept; y-intercept: $(0, 1)$

49. $y = 3\left(\dfrac{1}{5}\right)^x$

No x-intercept; y-intercept: $(0, 3)$

51. $f(2) = 2^2 + 3^2 = 4 + 9 = 13$

53. $f(-2) = 2^{-2} + 3^{-2} = \dfrac{1}{4} + \dfrac{1}{9} = \dfrac{9}{36} + \dfrac{4}{36} = \dfrac{13}{36}$

55. $3 = 3^x$
$3^1 = 3^x$
$1 = x$ or $x = 1$

57. $1 = 3^x$
$3^0 = 3^x$
$0 = x$ or $x = 0$

59. $f(x) = 2^{3x}, g(x) = 8^x$

$g(x) = 8^x = \left(2^3\right)^x = 2^{3x} = f(x)$

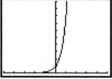

The graphs coincide. f and g are equivalent functions.

61. $f(x) = 2^{x+3}$, $g(x) = 8(2)^x$

$$g(x) = 8(2)^x = 2^3(2)^x = 2^{3+x} = 2^{x+3} = f(x)$$

The graphs coincide. f and g are equivalent functions.

63. $f(x) = \dfrac{6^x}{3^x}$, $g(x) = 2^x$

$$g(x) = 2^x = \left(\frac{6}{3}\right)^x = \frac{6^x}{3^x} = f(x)$$

The graphs coincide. f and g are equivalent functions.

65. $f(x) = \dfrac{3^{2x}}{3^x}$, $g(x) = 3^x$

$$f(x) = \frac{3^{2x}}{3^x} = 3^{2x-x} = 3^x = g(x)$$

The graphs coincide. f and g are equivalent functions.

67. $f(x) = x^{1/2}$, $g(x) = \sqrt{x}$

$$f(x) = x^{1/2} = \sqrt{x} = g(x)$$

The graphs coincide. f and g are equivalent functions.

69. $f(x) = 2^x$, $g(x) = 8^{x/3}$

$$g(x) = 8^{x/3} = \left(2^3\right)^{x/3} = 2^{3(x/3)} = 2^x = f(x)$$

The graphs coincide. f and g are equivalent functions.

71. a. $a < 0$ and $b > 1$
negative output and exponential growth

b. $a > 0$ and $b > 1$
positive output and exponential growth

c. $a > 0$ and $0 < b < 1$
positive output and exponential decay

d. $a < 0$ and $0 < b < 1$
negative output and exponential decay

73. Answers may vary. Example:
The family of exponential curves,

$f(x) = ab^x$, where $b = \dfrac{1}{2}$ and the coefficient

is an integer between -4 and 4, inclusive, excluding 0.

$f(x) = -4\left(\dfrac{1}{2}\right)^x$ $j(x) = 4\left(\dfrac{1}{2}\right)^x$

$g(x) = -3\left(\dfrac{1}{2}\right)^x$ $k(x) = 3\left(\dfrac{1}{2}\right)^x$

$h(x) = -2\left(\dfrac{1}{2}\right)^x$ $l(x) = 2\left(\dfrac{1}{2}\right)^x$

$i(x) = -1\left(\dfrac{1}{2}\right)^x$ $m(x) = 1\left(\dfrac{1}{2}\right)^x$

75. Since the y-intercept is $(0, 3)$, we know that

the equation is of the form $f(x) = 3(b)^x$.

Since the point $(1, 6)$ lies on the graph, we know that when x is 1, $f(x)$ is 6. Therefore, $6 = 3b$, and $b = 2$. The equation is

$$f(x) = 3(2)^x.$$

77. Answers may vary. Example:

$$f(x) = 2\left(\frac{2}{3}\right)^x \quad h(x) = 2(2)^x$$

$$g(x) = 4\left(\frac{9}{8}\right)^x \quad k(x) = 3\left(\frac{1}{2}\right)^x$$

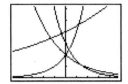

79. a. $f(0) = 100$, y-intercept $(0, 100)$

  $g(0) = 5$, y-intercept $(0, 5)$

  b. Since $g(x)$ has a larger base (3 versus 2), $g(x)$ will increase faster than $f(x)$.

  c. Eventually $g(x)$ will be much greater than $f(x)$. For every increase in x of one, $g(x)$ increases by a factor of three, while $f(x)$ increases by a factor of two.

  d.

X	Y₁	Y₂		X	Y₁	Y₂
0	100	5		7	12800	10935
1	200	15		8	25600	32805
2	400	45		9	51200	98415
3	800	135		10	102400	295245
4	1600	405		11	204800	885735
5	3200	1215		12	409600	2.66E6
6	6400	3645		13	819200	7.97E6
X=0				X=7		

81. a. $f(3 + 4) = 2^{3+4} = 2^7 = 128$

  $f(3) + f(4) = 2^3 + 2^4 = 8 + 16 = 24$
  $f(3 + 4) \neq f(3) + f(4)$
  The statement is not true.

  b. $f(x + y) = 2^{x+y} = 2^x \cdot 2^y$

  $f(x) + f(y) = 2^x + 2^y$
  $2^x \cdot 2^y \neq 2^x + 2^y$
  The statement is not true.

83. Answers may vary. Example:
If $b > 1$, then f is increasing because b increases as it is multiplied by itself x times. If $0 < b < 1$, then f is decreasing because b decreases toward 0 as it is multiplied by itself x times.

85. Answers may vary. Example:
If $f(x) = -ab^x$, then $-f(x) = ab^x$. Since

$g(x)$ also equals ab^x, $g(x) = -f(x)$. This

means that for the same x-value, f and g return opposite y-values. Therefore, functions f and g would reflect each other across the x-axis at each x.

87.

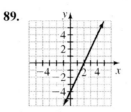

89.

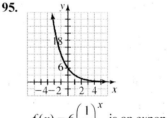

91. $0 = 8 + 4x$
  $-8 = 4x$
  $x = -2$
  $y = 8 + 4(0) = 8$

The x-intercept is $(-2, 0)$ and the y-intercept is $(0, 8)$.

93. The given values fit $f(x) = 13 - 4x$, which is linear. So f might be linear.

The given values fit $g(x) = 4(3)^x$, which is exponential. So g might be exponential.

The given values fit $h(x) = 48\left(\frac{1}{2}\right)^x$, which is

exponential. So h might be exponential.
k is neither linear nor exponential.

95.

$f(x) = 6\left(\frac{1}{2}\right)^x$ is an exponential function.

97. $f(-2) = 6\left(\frac{1}{2}\right)^{-2} = 6(2)^2 = 6 \cdot 4 = 24$

  $f(x) = 6\left(\frac{1}{2}\right)^x$ is an exponential function.

Homework 4.4

1. The *y*-intercept of *f* is (0, 4). As the value of *x* increases by 1, the value of *y* is multiplied by 2. By the base multiplier property, we know that *f* is an exponential function with base 2.
$f(x) = 4(2)^x$
The *y*-intercept of *g* is (0, 36). As *x* increases by 1, the value of *y* is multiplied by $\frac{1}{3}$.

$$g(x) = 36\left(\frac{1}{3}\right)^x$$

The *y*-intercept of *h* is (0, 5). As *x* increases by 1, the value of *y* is multiplied by 10.
$h(x) = 5(10)^x$
The *y*-intercept of *k* is (0, 250). As *x* increases by 1, the value of *y* is multiplied by $\frac{1}{5}$.

$$k(x) = 250\left(\frac{1}{5}\right)^x$$

3. The *y*-intercept of *f* is (0, 100). As *x* increases by 1, the value of *y* is multiplied by $\frac{1}{2}$. *f* is exponential with base $\frac{1}{2}$.

$$f(x) = 100\left(\frac{1}{2}\right)^x$$

The *y*-intercept of *g* is (0, 100). As *x* increases by 1, the value of *y* decreases by 50. *g* is linear with slope of −50.
$g(x) = -50x + 100$
The *y*-intercept of *h* is (0, 2). As *x* increases by 1, the value of *y* increases by 4. *h* is linear with a slope of 4.
$h(x) = 4x + 2$
The *y*-intercept of *k* is (0, 2). As *x* increases by 1, the value of *y* is multiplied by 3. *k* is exponential with base 3.
$k(x) = 2(3)^x$

5. $b^2 = 16$
$$b = \pm(16)^{1/2}$$
$$b = \pm 4$$

7. $b^3 = 27$
$$b = 27^{1/3}$$
$$b = 3$$

9. $3b^5 = 96$
$$b^5 = 32$$
$$b = 32^{1/5}$$
$$b = 2$$

11. $35b^4 = 15$
$$b^4 \approx 0.429$$
$$b \approx \pm(0.429)^{1/4}$$
$$b \approx \pm 0.81$$

13. $3.6b^3 = 42.5$
$$b^3 \approx 11.81$$
$$b \approx 11.81^{1/3}$$
$$b \approx 2.28$$

15. $32.7b^6 + 8.1 = 392.8$
$$32.7b^6 = 384.7$$
$$b^6 \approx 11.765$$
$$b \approx \pm(11.765)^{1/6}$$
$$b \approx \pm 1.51$$

17. $\frac{1}{4}b^3 - \frac{1}{2} = \frac{9}{4}$
$$4\left(\frac{1}{4}b^3 - \frac{1}{2}\right) = 4\left(\frac{9}{4}\right)$$
$$b^3 - 2 = 9$$
$$b^3 = 11$$
$$b = 11^{1/3}$$
$$b \approx 2.22$$

19. $\dfrac{b^6}{b^2} = 81$
$$b^{6-2} = 81$$
$$b^4 = 81$$
$$b = \pm(81)^{1/4}$$
$$b = \pm 3$$

21. $\dfrac{b^8}{b^3} = \dfrac{79}{5}$
$$b^{8-3} = \frac{79}{5}$$
$$b^5 = \frac{79}{5}$$
$$b = \left(\frac{79}{5}\right)^{1/5}$$
$$b \approx 1.74$$

23. $b^n + c = d$

$$b^n = d - c$$

$$b = (d - c)^{1/n}$$

25. $ab^n - c = d$

$$ab^n = c + d$$

$$b^n = \frac{c + d}{a}$$

$$b = \left(\frac{c + d}{a}\right)^{1/n}$$

27. $\dfrac{b^m}{b^n} = d$

$$b^{m-n} = d$$

$$b = d^{1/(m-n)}$$

29. $\dfrac{ab^m}{b^n} + c = d$

$$\frac{ab^m}{b^n} = d - c$$

$$\frac{b^m}{b^n} = \frac{d - c}{a}$$

$$b^{m-n} = \frac{d - c}{a}$$

$$b = \left(\frac{d - c}{a}\right)^{1/(m-n)}$$

31. (0, 4) and (1, 8)
The y-intercept is (0, 4), and the equation if of the form $y = 4b^x$. Substitute 1 for x and 8 for y.

$$8 = 4b^1$$

$$4b = 8$$

$$b = 2$$

The equation is $y = 4(2)^x$.

33. (0, 3), (5, 100)
The y-intercept is (0, 3), and the equation is of the form $y = 3b^x$. Substitute 5 for x and 100 for y to find b.

$$100 = 3b^5$$

$$3b^5 = 100$$

$$b^5 = \frac{100}{3}$$

$$b = \left(\frac{100}{3}\right)^{1/5}$$

$$b \approx 2.02$$

The equation is $y = 3(2.02)^x$.

35. (0, 87), (6, 14)
The y-intercept is (0, 87), and the equation is of the form $y = 87b^x$. Substitute 6 for x and 14 for y to find b.

$$14 = 87b^6$$

$$b^6 = \frac{14}{87}$$

$$b = \left(\frac{14}{87}\right)^{1/6}$$

$$b \approx 0.74$$

The equation is $y = 87(0.74)^x$.

37. (0, 5.5), (2, 73.9)
The y-intercept is (0, 5.5), and the equation is of the form $y = 5.5b^x$. Substitute 2 for x and 73.9 for y to find b.

$$73.9 = 5.5b^2$$

$$5.5b^2 = 73.9$$

$$b^2 = \frac{73.9}{5.5}$$

$$b = \left(\frac{73.9}{5.5}\right)^{1/2}$$

$$b \approx 3.67$$

The equation is $y = 5.5(3.67)^x$.

39. (0, 7.4), (3, 1.3)
The y-intercept is (0, 7.4), and the equation is of the form $y = 7.4b^x$. Substitute 3 for x and 1.3 for y to find b.

$$1.3 = 7.4b^3$$

$$7.4b^3 = 1.3$$

$$b^3 = \frac{1.3}{7.4}$$

$$b = \left(\frac{1.3}{7.4}\right)^{1/3}$$

$$b \approx 0.56$$

The equation is $y = 7.4(0.56)^x$.

41. (0, 39.18), (15, 3.66)
The y-intercept is (0, 39.18), and the equation is of the form $y = 39.18b^x$. Substitute 15 for x and 3.66 for y to find b.

$$3.66 = 39.18b^{15}$$

$$39.18b^{15} = 3.66$$

$$b^{15} = \frac{3.66}{39.18}$$

$$b = \left(\frac{3.66}{39.18}\right)^{1/15}$$

$$b \approx 0.85$$

The equation is $y = 39.18(0.85)^x$.

43. (1, 4), and (2, 12)

We can form a system of equations since both points satisfy $y = a(b)^x$.

$4 = ab^1$
$12 = ab^2$

Combine the two equations.

$\dfrac{12}{4} = \dfrac{ab^2}{ab^1}$

$3 = b$

Use b and one of the points to solve for a.
Substitute 1 for x and 4 for y.

$4 = a(3)^1$

$a(3)^1 = 4$

$3a = 4$

$a \approx 1.33$

The equation of the curve is $y = 1.33(3)^x$.

45. (3, 4) and (5, 9)

We can form a system of equations since both points satisfy $y = a(b)^x$.

$4 = ab^3$
$9 = ab^5$

Combine the two equations.

$\dfrac{9}{4} = \dfrac{ab^5}{ab^3}$

$\dfrac{9}{4} = b^2$

$b = \left(\dfrac{9}{4}\right)^{1/2}$

$b = \dfrac{3}{2} = 1.5$

Use b and one of the points to solve for a.
Substitute 3 for x and 4 for y.

$4 = a(1.5)^3$

$a = \dfrac{4}{(1.5)^3}$

$a \approx 1.19$

The equation of the curve is $y = 1.19(1.5)^x$.

47. (10, 329) and (30, 26)

We can form a system of equations since both points satisfy $y = a(b)^x$.

$329 = ab^{10}$

$26 = ab^{30}$

Combine the two equations.

$\dfrac{26}{329} = \dfrac{ab^{30}}{ab^{10}}$

$\dfrac{26}{329} = b^{20}$

$b = \left(\dfrac{26}{329}\right)^{1/20}$

$b \approx 0.88$

Use b and one of the points to solve for a.
Substitute 10 for x and 329 for y.

$329 \approx a(0.88)^{10}$

$a \approx \dfrac{329}{(0.88)^{10}}$

$a \approx 1181.32$

The equation of the curve is

$y = 1181.32(0.88)^x$.

49. (5, 8.1) and (9, 2.4)

We can form a system of equations since both points satisfy $y = a(b)^x$.

$8.1 = ab^5$

$2.4 = ab^9$

Combine the two equations.

$\dfrac{2.4}{8.1} = \dfrac{ab^9}{ab^5}$

$\dfrac{2.4}{8.1} = b^4$

$b \approx 0.74$

Use b and one of the points to solve for a.
Substitute 5 for x and 8.1 for y.

$8.1 \approx a(0.74)^5$

$a \approx \dfrac{2.4}{(0.74)^5}$

$a \approx 36.50$

The equation of the curve is

$y = 36.50(0.74)^x$.

51. (13, 24.71) and (21, 897.35)

We can form a system of equations since both points satisfy $y = a(b)^x$.

$897.35 = ab^{21}$

$24.71 = ab^{13}$

Combine the two equations.

$\dfrac{897.35}{24.71} = \dfrac{ab^{21}}{ab^{13}}$

$\dfrac{897.35}{24.71} = b^8$

$b \approx 1.57$

Use b and one of the points to solve for a.
Substitute 13 for x and 24.71 for y.

$$24.71 \approx a(1.57)^{13}$$
$$a \approx \frac{24.71}{(1.57)^{13}}$$
$$a \approx 0.070$$

The equation of the curve is $y = 0.070(1.57)^x$.

53. (2, 73.8) and (7, 13.2)
    We can form a system of equations since both
    points satisfy $y = a(b)^x$.

    $$13.2 = ab^7$$
    $$73.8 = ab^2$$
    Combine the two equations.
    $$\frac{13.2}{73.8} = \frac{ab^7}{ab^2}$$
    $$\frac{13.2}{73.8} = b^5$$
    $$b \approx 0.71$$
    Use b and one of the points to solve for a.
    Substitute 2 for x and 73.8 for y.
    $$73.8 \approx a(0.71)^2$$
    $$a \approx \frac{73.8}{(0.71)^2}$$
    $$a \approx 146.40$$
    The equation of the curve is
    $y = 146.40(0.71)^x$.

55. The points (0, 4) and (1, 2) lie on the graph.
    Use these points to determine the equation of
    the exponential curve. The y-intercept is (0, 4),
    and the equation is of the form $y = 4b^x$.
    Substitute 1 for x and 2 for y to find b.
    $$2 = 4b^1$$
    $$4b^1 = 2$$
    $$b = \frac{2}{4} = \frac{1}{2}$$
    The equation is $y = 4\left(\frac{1}{2}\right)^x$.

57. The points (1, 2) and (3, 5) lie on the
    exponential curve. Use the points to form a
    system of equations since both points satisfy
    the equation $y = a(b)^x$.

    $$5 = ab^3$$
    $$2 = ab^1$$
    Combine the two equations.

$$\frac{5}{2} = \frac{ab^3}{ab^1}$$
$$\frac{5}{2} = b^2$$
$$b \approx 1.58$$
Use b and one of the points to solve for a.
Substitute 1 for x and 2 for y.
$$2 \approx a(1.58)^1$$
$$a \approx \frac{2}{1.58}$$
$$a \approx 1.27$$
The equation of the curve is $y = 1.27(1.58)^x$.

59. Both the exponential equations have the
    coefficient 6. Therefore they both have the same
    y-intercept (0, 6). Thus, the solution is (0, 6).

61. a. i. Yes. Answers may vary. Example:
        The equation $y = 2(2)^x$ contains the
        point (0, 2).

     ii. No. Answers may vary. Example:
         For (2, 0) to be a solution to an
         exponential equation, the equation
         must be of the form $0 = a(b)^2$.
         However, since $a \neq 0$ and $b > 0$ in all
         exponential equations, (2, 0) can never
         be found on an exponential curve.

    b. No. Answers may vary. Example:
       Create a system of equations based on
       $y = a(b)^x$.
       $$1 = ab^3$$
       $$-1 = ab^2$$
       Combine the two equations.
       $$\frac{1}{-1} = \frac{ab^3}{ab^2}$$
       $$-1 = b$$
       Since $b = -1$, the equation is not
       exponential.

63. Answers may vary. Example:
    The base multiplier property states that when
    $y = ab^x$, if the value of x increases by 1, the
    value of y is multiplied by the base b. This
    makes sense because increasing x by 1 is the
    same as multiplying by another factor b, due to
    the nature of exponents. For example, if
    $y = 5(3)^x$, y is 45 when x is 2. If you increase x
    by 1, y becomes 135, which is the same as the
    product of 45 and 3.

65. $\dfrac{b^7}{b^2} = b^{7-2} = b^5$

67. $\dfrac{b^7}{b^2} = 76$

$b^5 = 76$

$b = (76)^{1/5}$

$b \approx 2.38$

69. $\dfrac{8b^3}{6b^{-1}} = \dfrac{4b^3}{3b^{-1}} = \dfrac{4b^{3-(-1)}}{3} = \dfrac{4b^4}{3}$

71. $\dfrac{8b^3}{6b^{-1}} = \dfrac{3}{7}$

$\dfrac{4b^4}{3} = \dfrac{3}{7}$

$b^4 = \dfrac{9}{28}$

$b = \pm\left(\dfrac{9}{28}\right)^{1/4}$

$b \approx \pm 0.75$

73. L is a linear function of the form $y = mx + b$. The y-intercept is $(0, 2)$, so $b = 2$ and the equation is of the form $y = mx + 2$. Substitute 1 for x and 6 for y to find m.

$6 = m(1) + 2$

$6 = m + 2$

$4 = m$

The equation for the linear function is $L(x) = 4x + 2$.

E is an exponential function of the form $y = ab^x$. The y-intercept is $(0, 2)$, so $a = 2$ and the equation is of the form $y = 2b^x$.

Substitute 1 for x and 6 for y to find b.

$6 = 2b^1$

$\dfrac{6}{2} = b^1$

$3 = b$

The equation for the exponential function is $E(x) = 2(3)^x$.

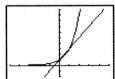

75. A linear function has the form $y = mx + b$. Substitute for x and y and solve.

$6 = 7m + b$

$3 = 5m + b$

$3 = 2m$

$m = \dfrac{3}{2}$

Solve for b.

$6 = 7\left(\dfrac{3}{2}\right) + b$

$6 = \dfrac{21}{2} + b$

$b = -\dfrac{9}{2}$

The equation for the linear function is

$y = \dfrac{3}{2}x - \dfrac{9}{2}$.

An exponential function has the form $y = ab^x$. Substitute for x and y and solve.

$\dfrac{6}{3} = \dfrac{ab^7}{ab^5}$

$2 = b^2$

$b \approx \pm 1.41$

Solve for a.

$6 = a(1.41)^7$

$a \approx 0.54$

The equation for the exponential function is

$y = 0.54(1.41)^x$.

So, the function could be either linear or exponential.

77. a. $L(x) = 2x + 100$, $E(x) = 3(2)^x$

To find the y-intercept, set $x = 0$.

$L(0) = 2(0) + 100 = 100$

The y-intercept of L is $(0, 100)$.

$E(0) = 3(2)^0 = 3(1) = 3$

The y-intercept of E is $(0, 3)$.

b. L is linear. By the slope addition property, when x increases by 1, the value of y increases by 2. E is exponential. By the base multiplier property, when x increases by 1, the value of y is multiplied by 2.

c. The exponential function, E, will eventually dominate. L increases by a fixed amount for every change in x, while E increases by an increasing amount for every change in x.

d.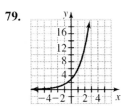

79.

$f(x) = 3(2)^x$ is an exponential function.

81. $f(-3) = 3(2)^{-3} = \dfrac{3}{2^3} = \dfrac{3}{8} = 0.375$

$f(x) = 3(2)^x$ is an exponential function.

Homework 4.5

1. a. Complete a table for $f(t)$ based on the assumption that the total number of Americans who have heard the rumor triples each day.

t (days)	$f(t)$ (Americans)
0	40
1	120
2	360
3	1080
4	3240

For every increase of 1 in t, $f(t)$ is multiplied by 3. $f(t)$ is exponential with base 3 and y-intercept 40. The equation is $f(t) = 40(3)^t$.

b. $f(10) = 40(3)^{10} = 2{,}361{,}960$ Americans

c. $f(15) = 40(3)^{15} = 573{,}956{,}280$ Americans
Model breakdown has occurred, because this number exceeds the U.S. population.

3. a. Complete a table for $f(t)$ based on the assumption that the total number of web pages doubles each year.

t (years)	$f(t)$ (trillions of web pages)
0	30
1	60
2	120
3	240
4	480

For every increase in t of 1, $f(t)$ is multiplied by 2. $f(t)$ is exponential with base 2 and y-intercept 30. The equation is $f(t) = 30(2)^t$.

b. Since 2018 is 6 years after 2012, find $f(6)$.
$f(6) = 30(2)^6 = 1920$
In the year 2018, 1920 trillion, or 1.92 quadrillion, web pages will be indexed.

c. First convert pages to inches.
1.92×10^{15} pages $\cdot \dfrac{2 \text{ inches}}{500 \text{ pages}}$
$= 7.68 \times 10^{12}$ inches
Now, convert inches to miles.
7.680×10^{12} inches $\times \dfrac{\text{foot}}{12 \text{ inches}} \times \dfrac{\text{mile}}{5280 \text{ feet}}$
$= 121{,}212{,}121.2$ miles
The height of the pile would be about 121 million miles.

5. a. Complete a table for $f(t)$, assuming that each year the market share is about 2.08 times the previous year's market share.

t (years)	$f(t)$ (percent of market share)
0	8.3
1	17.26
2	35.91
3	74.69
4	155.36

For every increase in t of 1, $f(t)$ is multiplied by 2.08. $f(t)$ is exponential with base 2.08 and y-intercept 8.3. The equation is $f(t) = 8.3(2.08)^t$.

b. The M-intercept is the value of a in the equation $f(t) = ab^t$. In this case, the M-intercept is (0, 8.3). This means that in 2010, the market share of eBooks was 8.3%.

c. The base, b, of the model is 2.08. Each year, the market share is 2.08 times that of the previous year.

d. Since 2013 is 3 years after 2010, find $f(3)$.
$f(3) = 8.3(2.08)^3 \approx 74.7$
The market share in 2013 will be 74.7%.
74.7% of $5 billion is $3.735 billion.

7. a. Complete a table for $D(t)$, assuming that the number of subscribers increases by 50% per year.

t (years)	$D(t)$ (thousands of subscribers)
0	2.5
1	$2.5(1.5) = 3.75$
2	$3.75(1.5) = 5.625$
3	$5.625(1.5) \approx 8.44$
4	$8.44(1.5) \approx 12.66$

For every increase in t of 1, $D(t)$ is multiplied by 1.5. $D(t)$ is exponential with base 1.5 and y-intercept 2.5. The equation is $D(t) = 2.5(1.5)^t$.

b. Complete a table for $S(t)$, assuming that the number of subscribers increases by 120% per year.

t (years)	$S(t)$ (thousands of subscribers)
0	1.71
1	$1.71(2.2) = 3.762$
2	$3.762(2.2) \approx 8.28$
3	$8.28(2.2) \approx 18.22$
4	$18.22(2.2) \approx 40.08$

For every increase in t of 1, $S(t)$ is multiplied by 2.2. $S(t)$ is exponential with base 2.2 and y-intercept 1.71. The equation is $S(t) = 1.71(2.2)^t$.

c.

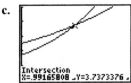

Intersection
X=.99165808   Y=3.7373376

It means that the number of stand-alone TiVo subscribers will pass the number of DIRECTV® subscribers in about 1 year, or 2008.

9. a. Complete a table for $f(t)$, assuming 8% interest compounded annually.

t (years)	$f(t)$ (dollars)
0	3000
1	$3000(1.08) = 3240$
2	$3240(1.08) = 3499.20$
3	$3499(1.08) \approx 3799.14$

As the value of t increases by 1, the value of $f(t)$ is multiplied by 1.08. $f(t)$ is exponential with base 1.08 and y-intercept 3000. The equation is $f(t) = 3000(1.08)^t$.

b. The base, b, of the model is 1.08. The base minus one is the interest rate.
$b - 1 = 1.08 - 1 = 0.08 = 8\%$

c. The coefficient, a, is the y-intercept. In this model, it is the initial amount invested, $3000.

d. $f(15) = 3000(1.08)^{15} \approx 9516.51$
In 15 years, the account's value will be $9516.51.

11. a. Complete the table for $f(t)$.

t (years)	$f(t)$ (dollars)
0	4000
6	$4000(2) = 4000(2)^1$
12	$4000(2)(2) = 4000(2)^2$
18	$4000(2)(2)(2) = 4000(2)^3$
t	$4000(2)^{t/6}$

As the value of t increases by 6, the value of $f(t)$ is multiplied by 2. $f(t)$ is exponential with base 2, has a y-intercept of 4000, and an exponent $\dfrac{t}{6}$ which allows t to remain in years even though the doubling occurs every 6 years. The equation is
$f(t) = 4000(2)^{t/6}$.

b. $f(20) = 4000(2)^{20/6} \approx 40{,}317.47$
In 20 years, the investment will be worth $40,317.47.

13. The first 3 years could be modeled using an exponential function with base 1.06 and y-intercept 5000. The equation is $f(t) = 5000(1.06)^t$.

After 3 years, there would be $f(3) = 5000(1.06)^3 = \$5955.08$ in the account. Then, all of this money is put into an account earning 8% interest (compounded annually) for 5 years. This can be modeled using an exponential function with base 1.08 and y-intercept 5955.08. The equation is $g(t) = 5955.08(1.08)^t$.

After 5 years, there will be $g(5) = 5955.08(1.08)^5 = \$8749.97$. This is the value of the investment after 8 years.

15. a. Complete a table for $g(t)$ based on the assumption that the sales are cut in half each year.

t (years)	$f(t)$ (sales)
0	984
1	492
2	246
3	123
4	61.5

When the value of t increases by 1, the value of $g(t)$ is multiplied by $\frac{1}{2}$. $g(t)$ is exponential with base $\frac{1}{2}$ and y-intercept 984. The equation is $g(t) = 984\left(\frac{1}{2}\right)^t$.

b. The coefficient, a, is the s-intercept. In this model, it is the number of new textbooks sold in 2011, which is 984.

c. $g(3) = 984\left(\frac{1}{2}\right)^3 = 123$

In this situation, it means that the college bookstore sold 123 new textbooks in 2014.

d. The half-life of new textbook sales is 1 year, as textbook sales are half of the previous year's sales.

17. a. As the value of t increases by 1600, the value of $f(t)$ is multiplied by $\frac{1}{2}$. $f(t)$ is exponential with base $\frac{1}{2}$, a y-intercept of 100, and an exponent $\frac{t}{1600}$ which allows t to remain in years even though the halving occurs every 1600 years. The equation is

$$f(t) = 100\left(\frac{1}{2}\right)^{t/1600}$$

b. $f(100) = 100\left(\frac{1}{2}\right)^{100/1600} = 95.76$

This means that after 100 years, 95.76% of the radium will remain.

c. $f(3200) = 100\left(\frac{1}{2}\right)^{3200/1600} = 25$

This means that after 3200 years, 25% of the radium will remain. This result can be found without using the equation. We know that after 1600 years, one half of the radium will remain. In an additional 1600 years (for a total of 3200 years), one half of the one half of the radium will remain. We know that $\frac{1}{2} \cdot \frac{1}{2} = \frac{1}{4}$ or 25% will remain.

19. a. As the value of t increases by 7.56, the value of $f(t)$ is multiplied by $\frac{1}{2}$. $f(t)$ is exponential with base $\frac{1}{2}$, a y-intercept of 100, and an exponent $\frac{t}{7.56}$ which allows t to remain in days even though the halving occurs every 7.56 days. The equation is

$$f(t) = 100\left(\frac{1}{2}\right)^{t/7.56}.$$

b. $f(3) = 100\left(\frac{1}{2}\right)^{3/7.56} = 75.95$

This means that after 3 days, 75.95% of the iodine-131 will remain.

c.

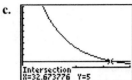

Intersection
X=32.673776 Y=5

A patient can safely spend a lot of time near a child after 33 days.

21. As the value of t increases by 6, the value of $f(t)$ is multiplied by $\frac{1}{2}$. $f(t)$ is exponential with base $\frac{1}{2}$, a y-intercept of 80, and an exponent $\frac{t}{6}$. The equation is $f(t) = 80\left(\frac{1}{2}\right)^{t/6}$.

$$f(14) = 80\left(\frac{1}{2}\right)^{14/6} = 15.87$$

After 14 hours, 15.87 milligrams of caffeine will remain in the bloodstream.

23. $f(t)$ is exponential with base b, a y-intercept of 12, and an exponent t. The equation is $f(t) = 12b^t$. When $t = 7$, $f(7) = 105$. Solve for b.

$$f(7) = 12b^7$$
$$105 = 12b^7$$
$$\frac{105}{12} = b^7$$
$$1.36 \approx b$$
$$f(9) = 12(1.36)^9 \approx 191$$

The number of wireless Internet users was 191 million in 2012.

25. $f(t)$ is exponential with base b, a y-intercept of 198, and an exponent t. The equation is $f(t) = 198b^t$. When $t = 39$, $f(39) = 715$. Solve for b.

$$f(39) = 198b^{39}$$
$$715 = 198b^{39}$$
$$\frac{715}{198} \approx b^{39}$$
$$1.03 \approx b$$
$$f(9) = 198(1.03)^9 \approx 258$$

The average annual spending on healthcare by 30-year-old adults who have private health insurance is $258.

27. a. Make a scattergram of the data.

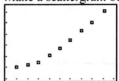

Use an exponential function to describe the data since the data points appear to bend upwards.

b. Use an exponential regression.

ExpReg
y=a*b^x
a=1.213372683
b=1.016064588

The equation is $f(t) = 1.21(1.0161)^t$.

c. $f(110) = 1.21(1.0161)^{110} \approx 7.011$

This means that the world population in 2010 was 7.011 billion people. I have performed interpolation. Answers may vary. Example: I performed interpolation because I used the model to estimate the population in a year that falls within the range of known years.

d. $f(118) = 1.21(1.0161)^{118} \approx 7.967$

This means that the world population in 2018 will be 7.967 billion people. I have performed extrapolation. Answers may vary. Example: I performed extrapolation because I used the model to predict the population in a year that falls outside the range of known years.

29. a. Make a scattergram of the data.

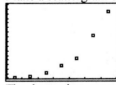

The data points appear to bend upwards, so model with an exponential function. Use an exponential regression.

ExpReg
y=a*b^x
a=100.8448529
b=1.410176259

The equation is $f(t) = 100.84(1.41)^t$.

Compare this model to the data.

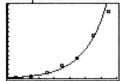

The model appears to fit the data well.

b. The rate of growth is the base, b, minus 1.
$b - 1 = 1.41 - 1 = 0.41$, or 41% growth per year.

c. $f(18) = 100.84(1.41)^{18} \approx 48{,}930$
According to the model, there will be 48,930 Starbucks stores in 2008. Because the actual number of stores in 2008 was only 15,256, the rate of growth of stores between 1991 and 2003 must decrease for the period 2003–2008.

d.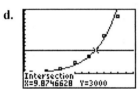

In $1990 + 9.87 \approx 2000$, there were about 3000 Starbucks stores.

31. a. Make a scattergram of the data.

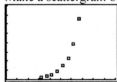

The data points appear to bend upwards, so model with an exponential function. Use an exponential regression.

The equation is $g(t) = 0.66(1.096)^t$.

b. $a = 0.66$
This is the g-intercept of the function. It corresponds to the cost of insurance for a newborn child. Model breakdown has occurred.

c. $b = 1.096$
This means that the insurance rate increases by a factor of 1.096 for every year of the faculty member's age.

d. $g(47) = 0.66(1.096)^{47} = 49.05$
It means that a 47-year-old faculty member should pay $49.05 per month in insurance rates. However, the actual rate, according to the table, is $46.00.

e. Men would pay higher monthly rates because there is a higher likelihood that the man will die at a younger age than a woman.

33. a.

Year	Population (millions)	Population Ratio
1790	3.9	
1800	5.3	1.36
1810	7.2	1.36
1820	9.6	1.33
1830	12.9	1.34
1840	17.1	1.33
1850	23.2	1.36
1860	31.4	1.35

b. The ratios stay approximately the same, with the average being about 1.35.

c. It is better to use an exponential model. The populations appear to be increasing by a constant multiplicative factor.

d. As the value of t (years) increases by 10, the value of $f(t)$ (population) is multiplied by approximately 1.35. Thus, the base is about $1.35^{1/10} \approx 1.03$, the y-intercept is 3.94, and the exponent is t. The equation is $f(t) = 3.94(1.03)^t$.

e.

Year	Population (millions)	Population Ratio
1860	31.4	
1870	39.8	1.27
1880	50.2	1.26
1890	62.9	1.25
1900	76.0	1.21

f. No, it is not likely that f gives reasonable population estimates after 1860. The population ratio changes from approximately 1.35 to 1.25. This indicates that model breakdown may occur in years after 1860.

g. $2012 - 1790 = 222$

$$f(222) = 3.94(1.03)^{222} \approx 2788.4$$

The model predicts that the population of the United States was approximately 2788.4 million in the year 2012. The actual population was 313.9 million.
The error in the estimate is
$2788.4 - 313.9 = 2474.5$ million or 2.4745 billion people.

35. a. Make a scattergram of the data.

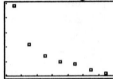

The data points appear to bend downwards, so model with an exponential function. Use an exponential regression.

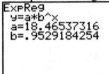

The equation is $f(t) = 18.47(0.95)^t$.

b. $f(117) = 18.47(0.95)^{117} \approx 0.05$

In 2017, the lightening fatality rate will be 0.05 death per million people per year.

c. 0.05 of 332 is about 17 lightening deaths.

d. There will be 10 times as many injuries as deaths from lightening, so there will be 170 injuries in 2017.

e. Make a scattergram of the data.

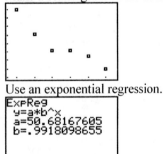

Use an exponential regression.

The equation is $g(t) = 50.68(0.992)^t$.

f. The base of f is $b = 0.95$, which means the lightning fatality rate is decreasing by 5% per year, but the base of g is 0.992, which means the percentage of Americans who live in rural areas is decreasing by only 0.8%, so migration of Americans from rural areas to urban ones can't be the only reason the lightning fatality rate is decreasing.

37. a. The graph passes through $p = 50$ when $t = 10$. So, the half-life of the element is 10 years.

b. $f(t)$ is exponential with base $\dfrac{1}{2}$, a

y-intercept of 100, and an exponent $\dfrac{t}{10}$.

The equation is $f(t) = 100\left(\dfrac{1}{2}\right)^{t/10}$.

$$f(40) = 100\left(\dfrac{1}{2}\right)^{40/10} = 6.25$$

After 40 years, 6.25% of the element will remain in the tank.

39.

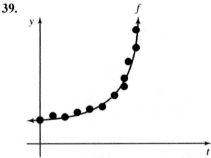

Decrease the base, b, of the model to slow the growth of the function so that it will pass through the middle of the data points.

41. Answers may vary. Example:
First make a scattergram of the data. If the points appear to fall in a straight line, use a linear model. If the points appear to bend upward or downward, use an exponential model.

43. Answers may vary. Example:
If $f(t) = ab^t$, where $a > 0$, models a quantity at time t, then the percent rate of change is constant. In particular, if $b > 1$, then the quantity grows exponentially at a rate of $b - 1$ percent (in decimal form) per unit of time. If $0 < b < 1$, then the quantity decays exponentially at a rate of $1 - b$ percent (in decimal form) per unit of time.

45. a. $C(t) = 800(1.03)^t$

b. $S(t) = 24t + 800$

c. $C(1) = 800(1.03)^1 = 824$
$C(2) = 800(1.03)^2 = 848.72$
$S(1) = 800(1 + 0.03(1)) = 824$
$S(2) = 800(1 + 0.03(2)) = 848$
$C(1)$ and $S(1)$ are equal because they are both $800(1.03)$. However,
$C(2) = 800(1.0609)$, whereas
$S(2) = 800(1.06)$.

d. $C(20) = 800(1.03)^{20} = 1444.89$
$S(20) = 800(1 + 0.03(20)) = 1280$
For $C(20)$, interest has compounded for 20 years, whereas for $S(20)$ simple interest has accumulated for 20 years. That is why $C(20)$ is much higher than $S(20)$.

47. a. Make a scattergram of the data.

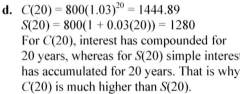

A linear function is better as the scattergram shows a relatively straight line.

b. Use a linear regression.

The equation is $f(t) = 0.16t - 0.44$.

c. $f(17) = 0.16(17) - 0.44 = 2.28$

The model predicts that the retail sales in China in 2017 will be $2.28 trillion.

d. Substitute 4.65 for $f(t)$ and solve for t.
$f(t) = 0.16t - 0.44$
$4.65 = 0.16t - 0.44$
$5.09 = 0.16t$
$31.81 \approx t$
Retail sales in China will reach $4.65 trillion in 2032.

49. Answers may vary. Example:
$f(x) = 2(3)^x$
$f(3) = 2(3)^3 = 54$

51. Answers may vary. Example:
$y = 3(2)^x$

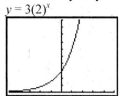

53. Answers may vary. Example:
The growth of a population of frogs is modeled by the exponential expression $85(1.02)^t$. In 8 years, the population of frogs will be $85(1.02)^8 = 99.6$.

Chapter 4 Review

1. $\dfrac{2^{-400}}{2^{-405}} = 2^{-400-(-405)} = 2^5 = 32$

2. $\dfrac{4b^{-3}c^{12}}{16b^{-4}c^3} = \dfrac{b^{-3-(-4)}c^{12-3}}{4} = \dfrac{bc^9}{4}$

3. $\left(2b^{-5}c^{-2}\right)^3\left(3b^4c^{-6}\right)^{-2} = 2^3b^{-15}c^{-6}3^{-2}b^{-8}c^{12}$
$= \dfrac{2^3b^{-15+(-8)}c^{-6+12}}{3^2}$
$= \dfrac{8b^{-23}c^6}{9}$
$= \dfrac{8c^6}{9b^{23}}$

4. $\dfrac{\left(20b^{-2}c^{-9}\right)\left(27b^5c^3\right)}{\left(18b^3c^{-1}\right)\left(30b^{-1}c^{-4}\right)} = \dfrac{20\cdot27b^{-2+5}c^{-9+3}}{18\cdot30b^{3+(-1)}c^{-1+(-4)}}$
$= \dfrac{540b^3c^{-6}}{540b^2c^{-5}}$
$= b^{3-2}c^{-6-(-5)}$
$= bc^{-1}$
$= \dfrac{b}{c}$

5. $32^{4/5} = \left(32^{1/5}\right)^4 = 2^4 = 16$

6. $16^{-3/4} = \dfrac{1}{16^{3/4}} = \dfrac{1}{\left(16^{1/4}\right)^3} = \dfrac{1}{2^3} = \dfrac{1}{8}$

7. $\dfrac{b^{-1/3}}{b^{4/3}} = b^{-\frac{1}{3}-\frac{4}{3}} = b^{-\frac{5}{3}} = \dfrac{1}{b^{5/3}}$

8. $\dfrac{\left(16b^8c^{-4}\right)^{1/4}}{\left(25b^{-6}c^4\right)^{3/2}} = \dfrac{16^{1/4}b^{8(1/4)}c^{-4(1/4)}}{25^{3/2}b^{-6(3/2)}c^{4(3/2)}}$

$= \dfrac{2b^2c^{-1}}{\left(25^{1/2}\right)^3 b^{-9}c^6}$

$= \dfrac{2b^{2-(-9)}c^{-1-6}}{5^3}$

$= \dfrac{2b^{11}c^{-7}}{125}$

$= \dfrac{2b^{11}}{125c^7}$

9. $\left(\dfrac{32b^2c^5}{2b^{-6}c^1}\right)^{1/4} = \left(16b^{2-(-6)}c^{5-1}\right)^{1/4}$

$= \left(16b^8c^4\right)^{1/4}$

$= 16^{1/4}b^{8(1/4)}c^{4(1/4)}$

$= 2b^2c$

10. $\left(8^{2/3}b^{-1/3}c^{3/4}\right)\left(64^{-1/3}b^{1/2}c^{-5/2}\right)$

$= 8^{2/3}64^{-1/3}b^{-1/3}b^{1/2}c^{3/4}c^{-5/2}$

$= \left(8^{1/3}\right)^2\left(64^{1/3}\right)^{-1}b^{-\frac{1}{3}+\frac{1}{2}}c^{\frac{3}{4}+\left(-\frac{5}{2}\right)}$

$= 2^2 4^{-1}b^{-\frac{2}{6}+\frac{3}{6}}c^{\frac{3}{4}+\left(-\frac{10}{4}\right)}$

$= 4\left(\dfrac{1}{4}\right)b^{1/6}c^{-7/4}$

$= \dfrac{b^{1/6}}{c^{7/4}}$

11. $b^{2n-1}b^{4n+3} = b^{(2n-1)+(4n+3)}$

$= b^{2n-1+4n+3}$

$= b^{6n+2}$

12. $\dfrac{b^{n/2}}{b^{n/3}} = b^{\frac{n}{2}-\frac{n}{3}} = b^{\frac{3n}{6}-\frac{2n}{6}} = b^{n/6}$

13. $3^{2x} = \left(3^2\right)^x = 9^x$

14. $f(-2) = 3(5)^{-2} = \dfrac{3}{5^2} = \dfrac{3}{25}$

15. $g(a+2) = 6^{a+2} = 6^a \cdot 6^2 = 36 \cdot 6^a$

16. $f\left(\dfrac{1}{2}\right) = 49^{1/2} = 7$

17. $g\left(-\dfrac{3}{4}\right) = 2(81)^{-3/4}$

$= \dfrac{2}{81^{3/4}}$

$= \dfrac{2}{\left(81^{1/4}\right)^3}$

$= \dfrac{2}{3^3}$

$= \dfrac{2}{27}$

18. $4.4487 \times 10^7 = 4.4487 \cdot 10,000,000$

$= 44,487,000$

19. $3.85 \times 10^{-5} = 3.85 \cdot \dfrac{1}{100,000}$

$= 0.0000385$

20. $54,000,000 = 5.4 \times 10^7$

21. $-0.00897 = -8.97 \times 10^{-3}$

22. $f(x) = 2(3)^x$

x	$f(x)$
-1	$\dfrac{2}{3}$
0	2
1	6
2	18

23. $h(x) = -3(2)^x$

x	$h(x)$
-1	-1.5
0	-3
1	-6
2	-12

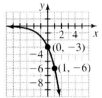

Domain: all real numbers
Range: $y < 0$

24. $g(x) = 12\left(\dfrac{1}{2}\right)^x$

x	$g(x)$
-1	24
0	12
1	6
2	3

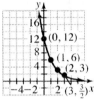

Domain: all real numbers
Range: $y > 0$

25. $3.9b^7 = 283.5$
$$b^7 \approx 72.69$$
$$b \approx 72.69^{1/7}$$
$$b \approx 1.84$$

26. $5b^4 - 13 = 67$
$$5b^4 = 80$$
$$b^4 = 16$$
$$b = \pm(16)^{1/4}$$
$$b = \pm 2$$

27. $\dfrac{1}{3}b^2 - \dfrac{1}{5} = \dfrac{2}{3}$
$$15\left(\dfrac{1}{3}b^2 - \dfrac{1}{5}\right) = 15\left(\dfrac{2}{3}\right)$$
$$5b^2 - 3 = 10$$
$$5b^2 = 13$$
$$b^2 = \dfrac{13}{5}$$
$$b = \pm\left(\dfrac{13}{5}\right)^{1/2}$$
$$b \approx \pm 1.61$$

28. $f(x)$ is linear. As x increases by 1, $f(x)$ decreases by 4. The linear equation is $f(x) = -4x + 34$.

$g(x)$ is exponential. As x increases by 1, $g(x)$ is multiplied by 3. The exponential equation is
$$g(x) = \dfrac{5}{3}(3)^x.$$

$h(x)$ does not appear to be linear or exponential.
$k(x)$ is exponential. As x increases by 1, $k(x)$ is multiplied by $\dfrac{1}{2}$. The exponential equation is
$$k(x) = 192\left(\dfrac{1}{2}\right)^x.$$

29. $f(4) = 18$

30. When $k(x) = 6$, $x = 5$.

31. $(0, 2)$ and $(5, 3)$
The y-intercept is $(0, 2)$, and the equation is of the form $y = 2b^x$. Substitute 5 for x and 3 for y to find b.
$$3 = 2b^5$$
$$b^5 = \dfrac{3}{2}$$
$$b = \left(\dfrac{3}{2}\right)^{1/5}$$
$$b \approx 1.08$$
The equation is $y = 2(1.08)^x$.

32. $(3, 30)$ and $(9, 7)$
We can form a system of equations since both points satisfy $y = a(b)^x$.
$$30 = ab^3$$
$$7 = ab^9$$
Combine the two equations.
$$\dfrac{7}{30} = \dfrac{ab^9}{ab^3}$$
$$\dfrac{7}{30} = b^6$$
$$b = \left(\dfrac{7}{30}\right)^{1/6}$$
$$b \approx 0.78$$
Use b and one of the points to solve for a. Substitute 3 for x and 30 for y.
$$30 = a(0.78)^3$$
$$a = \dfrac{30}{(0.78)^3}$$
$$a \approx 63.22$$
The equation of the curve is $y = 63.22(0.78)^x$.

33.

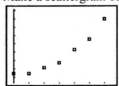

First, increase the value of the coefficient a to raise the y-intercept until it matches the data point at $t = 0$. Then, decrease the value of the base b to slow the increase of the function until it passes through the middle of the data points.

34. a. For every increase of t by 1, $f(t)$ is multiplied by 1.07. $f(t)$ is exponential with base 1.07. Since $2000 is the initial investment, $a = 2000$. The equation is $f(t) = 2000(1.07)^t$.

  b. $f(5) = 2000(1.07)^5 = 2805.10$
  The value of the account after 5 years will be $2805.10.

35. a. Since revenue doubles every year, $g(t)$ must be exponential with base 2 and exponent t. In the first year, total revenue was $17,000, so let $a = 17$. The equation is of the form $g(t) = 17(2)^t$.

  b. $g(8) = 17(2)^8 = 4352$
  In 2018, the corporation's total revenue will be $4,352,000.

36. a. For every increase in t of 5730 years, the value of $f(t)$ is multiplied by $\frac{1}{2}$. $f(t)$ is exponential with base $\frac{1}{2}$ and exponent $\frac{t}{5730}$. Since at time $t = 0$, 100% of the carbon-14 remains, $a = 100$. The equation is $f(t) = 100\left(\frac{1}{2}\right)^{t/5730}$.

  b. $f(100) = 100\left(\frac{1}{2}\right)^{100/5730} \approx 98.80$
  After 100 years, 98.8% of the carbon-14 remains in the tank.

37. We know that $a = 8.5$.
  Use $t = 30$ and $f(30) = 14.3$ to solve for b.
  $$14.3 = 8.5b^{30}$$
  $$\frac{14.3}{8.5} = b^{30}$$
  $$b \approx 1.017$$
  Solve for $t = 37$ (for 2017).
  $$f(37) = 8.5(1.017)^{37} \approx 15.9$$
  About 15.9 million homes will be unoccupied in 2017.

38. a. Make a scattergram of the data.

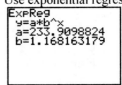

  Since the data seem to lie in an upward curve, use an exponential model.

  b. Use exponential regression.

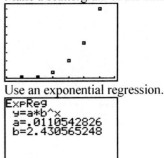

  The equation is $f(t) = 233.91(1.17)^t$.

  c. The coefficient a is 233.91. This means that the price of one ounce of gold was about $234 in 2000.

  d. The base b is 1.17. This means that the price of one ounce of gold is growing exponentially by 17% per year.

  e. $f(18) = 233.91(1.17)^{18} \approx 3948.16$
  According to the model, the price of one ounce of gold will be about $3948 in 2018.

39. a. Make a scattergram of the data.

  Use an exponential regression.

  The equation is $f(t) = 0.011(2.43)^t$.

b. The percentage rate of growth of the number of users of Zimride is about 143% per year.

c. $f(17) = 0.011(2.43)^{17} \approx 39{,}509$

There will be about 39,509 thousand (39.509 million) users of Zimride in 2017.

d.

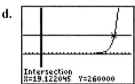

Intersection
X=19.122045 Y=260000

According to the model, all people at least 18 years old will be users of Zimride by 2019. Model breakdown has likely occurred.

Chapter 4 Test

1. $32^{2/5} = \left(32^{1/5}\right)^2 = 2^2 = 4$

2. $-8^{-4/3} = -\dfrac{1}{8^{4/3}} = -\dfrac{1}{\left(8^{1/3}\right)^4} = -\dfrac{1}{(2)^4} = -\dfrac{1}{16}$

3. $\left(2b^3 c^8\right)^3 = 2^3 b^{3\cdot3} c^{8\cdot3} = 8b^9 c^{24}$

4. $\left(\dfrac{4b^{-3}c}{25b^5 c^{-9}}\right)^0 = 1$

5. $\dfrac{b^{1/2}}{b^{1/3}} = b^{\frac{1}{2}-\frac{1}{3}} = b^{\frac{3}{6}-\frac{2}{6}} = b^{1/6}$

6. $\dfrac{25b^{-9}c^{-8}}{35b^{-10}c^{-3}} = \dfrac{5b^{-9-(-10)}c^{-8-(-3)}}{7}$

$= \dfrac{5bc^{-5}}{7}$

$= \dfrac{5b}{7c^5}$

7. $\left(\dfrac{6b\left(b^3 c^{-2}\right)}{3b^2 c^5}\right)^2 = \left(\dfrac{2bb^3 c^{-2}}{b^2 c^5}\right)^2$

$= \left(2b^{1+3-2}c^{-2-5}\right)^2$

$= \left(2b^2 c^{-7}\right)^2$

$= 2^2 b^{2\cdot2} c^{-7\cdot2}$

$= \dfrac{4b^4}{c^{14}}$

8. $\dfrac{\left(25b^8 c^{-6}\right)^{3/2}}{\left(7b^{-2}\right)\left(2c^3\right)^{-1}} = \dfrac{25^{3/2}2b^{8(3/2)}b^{-2}c^{-6(3/2)}c^3}{7}$

$= \dfrac{\left(25^{1/2}\right)^3 2b^{12}b^{-2}c^{-9}c^3}{7}$

$= \dfrac{5^3 \cdot 2b^{12+(-2)}c^{-9+3}}{7}$

$= \dfrac{125 \cdot 2b^{10}c^{-6}}{7}$

$= \dfrac{250b^{10}}{7c^6}$

9. $8^{x/3}2^{x+3} = 8(4)^x$

$\left(2^3\right)^{x/3}2^{x+3} = 2^3\left(2^2\right)^x$

$2^{3(x/3)}2^{x+3} = 2^3\left(2^{2x}\right)$

$2^x 2^{x+3} = 2^3\left(2^{2x}\right)$

$2^{x+x+3} = 2^{3+2x}$

$2^{2x+3} = 2^{2x+3}$

10. $f(-2) = 4^{-2} = \dfrac{1}{4^2} = \dfrac{1}{16}$

11. $f\left(-\dfrac{3}{2}\right) = 4^{-3/2} = \dfrac{1}{4^{3/2}} = \dfrac{1}{\left(4^{1/2}\right)^3} = \dfrac{1}{2^3} = \dfrac{1}{8}$

12. $f(x) = -5(2)^x$

x	$f(x)$
-1	-2.5
0	-5
1	-10
2	-20

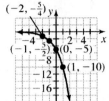

The domain is all real numbers. The range is all negative real numbers.

13. $f(x) = 18\left(\dfrac{1}{3}\right)^x$

x	$g(x)$
-1	54
0	18
1	6
2	2
3	$\dfrac{2}{3}$

The domain is all real numbers. The range is all positive real numbers.

14. Answers may vary. Example:

$$f(x) = 2\left(\dfrac{2}{3}\right)^x \qquad h(x) = 3(2)^x$$

$$g(x) = 4\left(\dfrac{9}{8}\right)^x \qquad k(x) = 3\left(\dfrac{1}{2}\right)^x$$

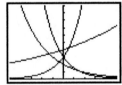

15. For every increase of 1 in t, the value of $f(t)$ is multiplied by $\dfrac{1}{2}$. $f(t)$ is exponential with base $\dfrac{1}{2}$ and y-intercept 160. $f(t) = 160\left(\dfrac{1}{2}\right)^x$.

16. $3b^6 + 5 = 84$
$$3b^6 = 79$$
$$b^6 = 26.33$$
$$b = \pm(26.33)^{1/6}$$
$$b = \pm 1.72$$

17. $(0, 70)$ and $(6, 20)$
The y-intercept is $(0, 70)$, so the equation is of the form $y = 70b^x$. Substitute 6 for x and 20 for y to find b.
$$20 = 70b^6$$
$$\dfrac{20}{70} = b^6$$
$$b = \left(\dfrac{20}{70}\right)^{1/6}$$
$$b \approx 0.812$$
The equation is $y = 70(0.812)^x$.

18. $(4, 9)$ and $(7, 50)$
We can form a system of equations since both points satisfy $y = a(b)^x$.
$$50 = ab^7$$
$$9 = ab^4$$
Combine the two equations.
$$\dfrac{50}{9} = \dfrac{ab^7}{ab^4}$$
$$\dfrac{50}{9} = b^3$$
$$b = \left(\dfrac{50}{9}\right)^{1/3}$$
$$b \approx 1.771$$
Use b and one of the points to solve for a. Substitute 4 for x and 9 for y.
$$9 = a(1.771)^4$$
$$a = \dfrac{9}{(1.771)^4}$$
$$a \approx 0.915$$
The equation of the curve is
$$y = 0.915(1.771)^x.$$

19. $f(0) = 6$

20. When $f(x) = 3$, $x = 1$.

21. From the graph, we see that the points $(0, 6)$ and $(1, 3)$ lie on the curve. The y-intercept is $(0, 6)$, so the equation is of the form $y = 6b^x$. Substitute 1 for x and 3 for y to find b.

$$3 = 6b^1$$

$$b = \frac{1}{2}$$

The equation is $y = 6\left(\dfrac{1}{2}\right)^x$.

22. a. Complete a table for $f(t)$.

t (weeks)	$f(t)$ (leaves)
0	400
1	1200
2	3600
3	10,800
4	32,400

For every increase of 1 in t, the value of $f(t)$ is multiplied by 3. $f(t)$ is exponential with base 3 and y-intercept 400. The equation is $f(t) = 400(3)^t$.

b. $f(6) = 400(3)^6 = 291{,}600$
There will be 291,600 leaves on the tree 6 weeks after March 1.

c. $f(52) = 400(3)^{52} = 2.58 \times 10^{27}$
The model predicts there will be 2.58×10^{27} leaves on the tree 1 year after March 1. This is unrealistic, model breakdown has occurred.

23. a. Make a scattergram of the data.

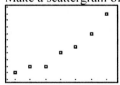

Use an exponential regression.

The equation is $f(t) = 0.27(1.16)^t$.

b. The base b is 1.16. This means that the number of fraud complaints by consumers is increasing by 16% per year.

c. $a = 0.27$ is the y-intercept. This means that there were 0.27 million (270 thousand) fraud complaints by consumers in 2000.

d. $f(18) = 0.27(1.16)^{18} \approx 3.9$
The model predicts there will be about 3.9 million complaints by consumers in 2018.

e. $f(20) = 0.27(1.16)^{20} \approx 5.25$

$$\frac{5.25}{341} \approx 0.015$$

About 1.5% of the U.S. population will file fraud complaints in 2020.

Chapter 5
Logarithmic Functions

1. $(f \circ g)(0) = f(g(0)) = f(4) = 2$

3. $(g \circ f)(0) = g(f(0)) = g(1) = 0$

5. $(g \circ g)(1) = g(g(1)) = g(0) = 4$

7.

x	$(f \circ g)(x)$
0	$f(g(0)) = f(4) = 2$
1	$f(g(1)) = f(0) = 1$
2	$f(g(2)) = f(3) = 4$
3	$f(g(3)) = f(2) = 0$
4	$f(g(4)) = f(1) = 3$

9. $(f \circ g)(8) = f(g(8)) = f(9) = 7$

11. $(g \circ f)(8) = g(f(8)) = g(6) = 5$

13. $(f \circ f)(6) = f(f(6)) = f(5) = 8$

15.

x	$(g \circ f)(x)$
5	$g(f(5)) = g(8) = 9$
6	$g(f(6)) = g(5) = 7$
7	$g(f(7)) = g(9) = 8$
8	$g(f(8)) = g(6) = 5$
9	$g(f(9)) = g(7) = 6$

17. a. $(f \circ g)(2) = f(g(2))$
$$= f(3(2) - 6)$$
$$= f(0)$$
$$= 4(0) + 1$$
$$= 1$$

b. $(g \circ f)(2) = g(f(2))$
$$= g(4(2) + 1)$$
$$= g(9)$$
$$= 3(9) - 6$$
$$= 21$$

19. a. $(f \circ g)(2) = f(g(2))$
$$= f\left(\frac{8}{2}\right)$$
$$= f(4)$$
$$= 6(4) - 8$$
$$= 16$$

b. $(g \circ f)(2) = g(f(2))$
$$= g(6(2) - 8)$$
$$= g(4)$$
$$= \frac{8}{4}$$
$$= 2$$

21. a. $(f \circ g)(2) = f(g(2))$
$$= f(4(2) - 5)$$
$$= f(3)$$
$$= 2(3)^3$$
$$= 54$$

b. $(g \circ f)(2) = g(f(2))$
$$= g\left(2(3)^2\right)$$
$$= g(18)$$
$$= 4(18) - 5$$
$$= 67$$

23. a. $(f \circ g)(2) = f(g(2))$
$$= f\left(16 - 3(2)^2\right)$$
$$= f(4)$$
$$= \frac{4 + 1}{5(4) - 7}$$
$$= \frac{5}{13}$$

b. $(g \circ f)(2) = g(f(2))$
$$= g\left(\frac{2 + 1}{5(2) - 7}\right)$$
$$= g(1)$$
$$= 16 - 3(1)^2$$
$$= 13$$

25. a. $(f \circ g)(x) = f(g(x)) = f(2x) = 2x + 4$

b. $(g \circ f)(2) = g(f(x))$
$$= g(x+4)$$
$$= 2(x+4)$$
$$= 2x+8$$

c. $(f \circ g)(3) = 2(3)+4 = 10$

d. $(g \circ f)(3) = 2(3)+8 = 14$

27. a. $(f \circ g)(x) = f(g(x))$
$$= f(2x-4)$$
$$= 3(2x-4)-7$$
$$= 6x-12-7$$
$$= 6x-19$$

b. $(g \circ f)(x) = g(f(x))$
$$= g(3x-7)$$
$$= 2(3x-7)-4$$
$$= 6x-14-4$$
$$= 6x-18$$

c. $(f \circ g)(3) = 6(3)-19 = -1$

d. $(g \circ f)(3) = 6(3)-18 = 0$

29. a. $(f \circ g)(x) = f(g(x)) = f(x+2) = 2^{x+2}$

b. $(g \circ f)(x) = g(f(x)) = g(2^x) = 2^x + 2$

c. $(f \circ g)(3) = 2^{3+2} = 32$

d. $(g \circ f)(3) = 2^3 + 2 = 10$

31. a. $(f \circ g)(x) = f(g(x)) = f(x-1) = (x-1)^4$

b. $(g \circ f)(x) = g(f(x)) = g(x^4) = x^4 - 1$

c. $(f \circ g)(3) = (3-1)^4 = 16$

d. $(g \circ f)(3) = 3^4 - 1 = 80$

33. $(f \circ g)(-2) = f(g(-2)) = f(3) = 1$

35. $(g \circ f)(5) = g(f(5)) = g(2) = 1$

37. $(f \circ f)(-5) = f(f(-5)) = f(-3) = -2$

39. $(f \circ g)(0) = f(g(0)) = f(2) = 0.5$
The y-intercept is $(0, 0.5)$.

41. $(f \circ g)(2) = f(g(2)) = f(3) = 8$

43. $(g \circ f)(2) = g(f(2)) = g(4) = 2$

45. $(f \circ f)(0) = f(f(0)) = f(1) = 2$

47. Answers may vary. Example:
To form 5^{x-9}, we can substitute $x-9$ for x in 5^x. This suggests that $g(x) = x-9$ and $f(x) = 5^x$.
Check: $f(g(x)) = f(x-9) = 5^{x-9}$

49. Answers may vary. Example:
To form $2^x + 6$, we can substitute 2^x for x in $x+6$. This suggests that $g(x) = 2^x$ and $f(x) = x+6$.
Check: $f(g(x)) = f(2^x) = 2^x + 6$

51. Answers may vary. Example:
To form $(5x-2)^2$, we can substitute $5x-2$ for x in x^2. This suggests that $g(x) = 5x-2$ and $f(x) = x^2$.
Check: $f(g(x)) = f(5x-2) = (5x-2)^2$

53. Answers may vary. Example:
To form $\dfrac{1}{3x-7}$, we can substitute $3x-7$ for x in $\dfrac{1}{x}$. This suggests that $g(x) = 3x-7$ and $f(x) = \dfrac{1}{x}$.
Check: $f(g(x)) = f(3x-7) = \dfrac{1}{3x-7}$

55. $f(19)$ is the U.S. population at 19 years since 2000, or in the year 2019. So $(g \circ f)(19) = 9.8$ means that Americans will eat 9.8 billion pounds of french fries in 2019.

57. $f(8.99)$ is the number (in thousands) of Americans who will purchase The Kills album *No Wow* at \$8.99. So $(g \circ f)(8.99) = 2.3$ means that the revenue will be \$2.3 million if the price of the album is \$8.99.

59. a. Since there are 3 feet in one yard,
$$f(y) = 3y.$$
Since there are 12 inches in 1 foot,
$$g(x) = 12x.$$

b. $(g \circ f)(y) = g(f(y))$
$$= g(3y)$$
$$= 12(3y)$$
$$= 36y$$

c. $(g \circ f)(5) = 36(5) = 180$
This means that there are 180 inches in 5 yards.

61. a. Since there are 60 seconds in 1 minute,
$$f(M) = 60M.$$
Since there are 60 minutes in 1 hour,
$$g(H) = 60H.$$

b. $(f \circ g)(H) = f(g(H))$
$$= f(60H)$$
$$= 60(60H)$$
$$= 3600H$$

c. $(f \circ g)(3) = 3600(3) = 10,800$
This means that there are 10,800 seconds in 3 hours.

63. a. Since it takes 1 gallon to drive 25 miles,
$$f(d) = \frac{d}{25}.$$
Since one gallon of gas costs \$4.25,
$$g(x) = 4.25x.$$

b. $(g \circ f)(d) = g(f(d))$
$$= g\left(\frac{d}{25}\right)$$
$$= 4.25\left(\frac{d}{25}\right)$$
$$= 0.17d$$

c. $(g \circ f)(300) = 0.17(300) = 51$
This means that for a 300-mile trip, the cost of gasoline is \$51.

65. a. Answers may vary. Example:
Since births occur either outside marriage or from married couples, the sum of the percentages of the two options must be 100. If p is the percentage of births outside marriage, then $100 - p$ is the percentage of births from married couples.

b. $h(t) = (g \circ f)(t)$
Answers may vary. Example:
$f(t)$ is the percentage of births outside marriage at t years since 1900. Since $g(p)$ uses the percentage of births outside marriage to calculate the percentage of births that were from married couples, $g(f(t))$ is the percentage of births from married couples at t years since 1900. So $h(t) = (g \circ f)(t)$.

c. $h(t) = (g \circ f)(t)$
$$= g(f(t))$$
$$= g(0.76t - 42.04)$$
$$= 100 - (0.76t - 42.04)$$
$$= -0.76t + 142.04$$

d. $h(117) = -0.76(117) + 142.04 = 53.12$
This means that about 53.1% of births will be from married couples in 2017.

e. $h(t) = -0.76t + 142.04$
$$52 = -0.76t + 142.04$$
$$-90.04 = -0.76t$$
$$118.47 \approx t$$
This means that 52% of births will be from married couples in 2018.

67. a. Answers may vary. Example:
To convert any unit from billions to millions, we multiply by 1000.

b. $h(t) = (M \circ D)(t)$
Answers may vary. Example:
$D(t)$ is the annual revenue (in billions of dollars) from downloaded music at t years since 2005. Since $M(b)$ converts billions

of dollars to millions of dollars, $M(D(t))$ is the annual revenue (in millions of dollars) from downloaded music at t years since 2005. So, $h(t) = (M \circ D)(t)$.

c.
$$h(t) = (M \circ D)(t)$$
$$= M(D(t))$$
$$= M(0.34t + 0.56)$$
$$= 1000(0.34t + 0.56)$$
$$= 340t + 560$$

d. $h(-1) = 340(-1) + 560 = 220$

This means that in 2004, the revenue from downloaded music was $220 million.

e.
$$h(t) = 340t + 560$$
$$0 = 340t + 560$$
$$-340t = 560$$
$$t \approx -1.65$$

The t-intercept is $(-1.65, 0)$.

This means there was no revenue from downloaded music in 2003.
The t-intercept of the graph of h is the same as the t-intercept of the graph of D. Answers may vary. Example:
The only difference between $h(t)$ and $D(t)$ is a factor of 1000, which does not affect the t-intercept.

69. a. $h(C) = (g \circ f)(C)$

Answers may vary. Example:
$g(F)$ is the number of chirps per minute a cricket makes when the temperature is F degrees Fahrenheit. Since $f(C)$ converts Celsius readings to equivalent Fahrenheit readings, $g(f(C))$ is the number of chirps per minute a cricket makes when the temperature is C degrees Celsius. So, $h(C) = (g \circ f)(C)$.

b.
$$h(C) = (g \circ f)(C)$$
$$= g(f(C))$$
$$= g(1.8C + 32)$$
$$= 4.3(1.8C + 32) - 172$$
$$= 7.74C + 137.6 - 172$$
$$= 7.74C - 34.4$$

c. $h(23) = 7.74(23) - 34.4 \approx 144$

This means that when the temperature is $23°C$, the crickets chirp at a rate of 144 times per minute.

e.
$$h(C) = 7.74C - 34.4$$
$$200 = 7.74C - 34.4$$
$$234.4 = 7.74C$$
$$30 \approx C$$

Crickets chirp at the rate of 200 times per minute when the temperature is about $30°C$.

71. The student found the product of $x + 8$ and $x + 5$, which is incorrect.
$$(f \circ g)(x) = f(x + 5) = (x + 5) + 8 = x + 13$$

73. The student substituted $4x - 2$ for x in $g(x) = -7x + 3$, which is incorrect.
$$(f \circ g)(x) = f(-7x + 3)$$
$$= 4(-7x + 3) - 2$$
$$= -28x + 12 - 2$$
$$= -28x + 10$$

75. a.
$$(f \circ g)(3) = f(4(3) - 10)$$
$$= f(2)$$
$$= 5(2) - 6$$
$$= 4$$

b.
$$(g \circ f)(3) = g(5(3) - 6)$$
$$= g(9)$$
$$= 4(9) - 10$$
$$= 26$$

c. No, the results from parts (a) and (b) are not equal.

77. a. $g(2) = 3(2) - 2 = 4$
$$(f \circ g)(2) = f(4) = 4(4) - 7 = 9$$

b.
$$(f \circ g)(x) = f(3x - 2)$$
$$= 4(3x - 2) - 7$$
$$= 12x - 8 - 7$$
$$= 12x - 15$$

c. $(f \circ g)(2) = 12(2) - 15 = 9$
The result is equal to the one in part (a).

79. a. $(f \circ f)(x) = f(2x) = 2(2x) = 4x$

b. $(f \circ (f \circ f))(x) = f(4x) = 2(4x) = 8x$

c. $(f \circ (f \circ (f \circ f)))(x) = f(8x)$
$$= 2(8x)$$
$$= 16x$$

d. $\underbrace{(f \circ (f \circ (f \circ ... \circ f)...))}_{n \text{ functions}}(x) = 2^n x$

81. Answers may vary. Example:
$f(x) = x - 4$, $g(x) = 2x + 3$
$(f \circ g)(x) = f(2x + 3) = 2x + 3 - 4 = 2x - 1$
$(g \circ f)(x) = g(x - 4)$
$$= 2(x - 4) + 3$$
$$= 2x - 8 + 3$$
$$= 2x - 5$$

83. a. Each input x of f has exactly one output $f(x)$, because f is a function, and each input $f(x)$ of g has exactly one output $g(f(x))$, because g is a function.

b. Domain: set A; range: set C

85. $y = 2(3)^x$

This is an exponential function.

87. $f(x) = 2 + 3x$

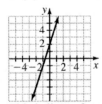

This is linear function.

Homework 5.2

1. $f^{-1}(7) = 4$, since $f(4) = 7$.

3. $f(4) = 6$

5. $f^{-1}(4) = 5$, since $f(5) = 4$.

7.

x	$f^{-1}(x)$
2	6
4	5
6	4
8	3
10	2

9. $f^{-1}(f(6)) = f^{-1}(2) = 6$

11. $g(2) = 6$

13. $g^{-1}(2) = 1$, since $g(1) = 2$.

15.

x	$g^{-1}(x)$
2	1
6	2
18	3
54	4
162	5
486	6

17. $g^{-1}(g(4)) = g^{-1}(54) = 4$

19.

x	$f^{-1}(x)$
4	6
10	5
16	4
22	3
28	2
34	1

21. Answers may vary. Example:
Begin by creating a table of values for $f(x)$, then build a table for $f^{-1}(x)$ from that information.

x	$f(x)$	x	$f^{-1}(x)$
0	3	3	0
1	6	6	1
2	12	12	2
3	24	24	3
4	48	48	4

23. $f(3) = 3(2)^3 = 3(8) = 24$

25. $f^{-1}(3) = 0$, since $f(0) = 3$.

27. $f(x) = 3(2)^x$

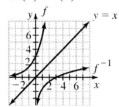

29. $f(x) = 3^x$

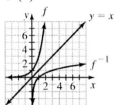

31. $f(x) = 2x$

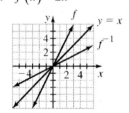

33. $f(x) = 3x - 2$

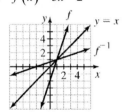

35. $f(x) = \dfrac{1}{2}x + 1$

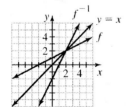

37. $f(x) = 4\left(\dfrac{1}{2}\right)^x$

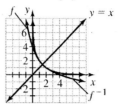

39. $f(x) = \left(\dfrac{1}{3}\right)^x$

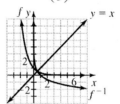

41. Since g sends 2 to 3, $g(2) = 3$.

43. Since g sends 1 to 2, g^{-1} sends 2 back to 1.
So, $g^{-1}(2) = 1$.

45. Since g sends 0 to 0, g^{-1} sends 0 back to 0.
So, $g^{-1}(0) = 0$.

47. Since f sends 2 to 1, $f(2) = 1$.

49. Since f sends 3 to 4, f^{-1} sends 4 back to 3.
So, $f^{-1}(4) = 3$.

51.

graph

53. a. $f(t) = 0.76t - 42.04$
Replace $f(t)$ with p.
$p = 0.76t - 42.04$
Solve for t.
$$p = 0.76t - 42.04$$
$$p + 42.04 = 0.76t$$
$$\frac{p + 42.04}{0.76} = t$$
An approximate equation is
$t = 1.32p + 55.32$.

Replace t with $f^{-1}(p)$.

$f^{-1}(p) = 1.32p + 55.32$

b. $f(100) = 0.76(100) - 42.04 = 33.96$

According to the model, in 2000, about 34% of births were outside of marriage.

c. $f^{-1}(100) = 1.32(100) + 55.32 = 187.32$

According to the inverse model, 100% of births will be outside marriage in 2087. Model breakdown has likely occurred.

d. The slope of $f^{-1}(p)$ is 1.32. This means that the rate of change of t with respect to p is 1.32. According to the model f^{-1}, the percentage of births outside marriage increases by 1 percentage point every 1.32 years

55. a.

The plotted data lie mostly in a straight line. Therefore, a linear model better fits the data.

b. Using the linear regression feature, $f(t) = 3.07t + 14.07$.

c. $f(t) = 3.07t + 14.07$

Replace $f(t)$ with p.
$p = 3.07t + 14.07$
Solve for t.
$$p = 3.07t + 14.07$$
$$p - 14.07 = 3.07t$$
$$\frac{p - 14.07}{3.07} = t$$
An approximate equation is $0.33p - 4.58 = t$.

Replace t with $f^{-1}(p)$.

$f^{-1}(p) = 0.33p - 4.58$

d. $f(t) = 3.07t + 14.07$
$69 = 3.07t + 14.07$
$54.93 = 3.07t$
$17.89 \approx t$
$2000 + 17.89 \approx 2018$

In 2018, 69% of households will have two or more working computers.

e. $f^{-1}(p) = 0.33p - 4.58$

$f^{-1}(69) = 0.33(69) - 4.58 = 18.19$

$2000 + 18.19 \approx 2018$

In 2018, 69% of households will have two or more working computers.

f. The results are the same, when rounding is accounted for.

57. a.

The plotted data lie mostly in a straight line. Therefore, a linear model better fits the data.

b. Using the linear regression feature, $f(a) = 2.17a + 581.49$.

c. $f(a) = 2.17a + 581.49$

Replace $f(a)$ with c.
$c = 2.17a + 581.49$
Solve for t.
$c - 581.49 = 2.17t$
$$\frac{c - 581.49}{2.17} = t$$
An approximate equation is $0.46c - 267.97 = t$.

Replace t with $f^{-1}(a)$.

$f^{-1}(c) = 0.46c - 267.97$

d. $f^{-1}(677) = 0.46(677) - 267.97 = 43.45$

According to the model, the age of adults whose average credit score is 677 points is about 43 years.

e. $f^{-1}(830) = 0.46(830) - 267.97 = 113.83$

According to the model, the age of adults whose average credit score is 830 points is about 114 years.

f. The slope of f^{-1} is 0.46 and it represents a credit score increase by 1 point for each age increase of 0.46 years.

59. $f(x) = x + 8$

Replace $f(x)$ with y.

$y = x + 8$

Solve for x.

$y - 8 = x$

Replace x with $f^{-1}(y)$.

$f^{-1}(y) = y - 8$

Write equation in terms of x.

$f^{-1}(x) = x - 8$

61. $f(x) = -4x$

Replace $f(x)$ with y.

$y = -4x$

Solve for x.

$\dfrac{y}{-4} = x$

Replace x with $f^{-1}(y)$.

$f^{-1}(y) = -\dfrac{y}{4}$

Write equation in terms of x.

$f^{-1}(x) = -\dfrac{x}{4} = -\dfrac{1}{4}x$

63. $f(x) = \dfrac{x}{7}$

Replace $f(x)$ with y.

$y = \dfrac{x}{7}$

Solve for x.

$7y = x$

Replace x with $f^{-1}(y)$.

$f^{-1}(y) = 7y$

Write equation in terms of x.

$f^{-1}(x) = 7x$

65. $f(x) = -6x - 2$

Replace $f(x)$ with y.

$y = -6x - 2$

Solve for x.

$y + 2 = -6x$

$-6x = y + 2$

$x = \dfrac{y + 2}{-6}$

$x = -\dfrac{1}{6}y - \dfrac{2}{6}$

Replace x with $f^{-1}(y)$.

$f^{-1}(y) = -\dfrac{1}{6}y - \dfrac{1}{3}$

Write equation in terms of x.

$f^{-1}(x) = -\dfrac{1}{6}x - \dfrac{1}{3}$

67. $f(x) = 0.4x - 7.9$

Replace $f(x)$ with y.

$y = 0.4x - 7.9$

Solve for x.

$y + 7.9 = 0.4x$

$\dfrac{y + 7.9}{0.4} = x$

$x = \dfrac{y}{0.4} + \dfrac{7.9}{0.4}$

Replace x with $f^{-1}(y)$.

$f^{-1}(y) = 2.5y + 19.75$

Write equation in terms of x.

$f^{-1}(x) = 2.5x + 19.75$

69. $f(x) = \dfrac{7}{3}x + 1$

Replace $f(x)$ with y.

$y = \dfrac{7}{3}x + 1$

Solve for x.

$y - 1 = \dfrac{7}{3}x$

$\dfrac{3}{7}(y - 1) = x$

$x = \dfrac{3y - 3}{7}$

Replace x with $f^{-1}(y)$.

$f^{-1}(y) = \dfrac{3}{7}y - \dfrac{3}{7}$

Write equation in terms of x.

$f^{-1}(x) = \dfrac{3}{7}x - \dfrac{3}{7}$

71. $f(x) = -\dfrac{5}{6}x - 3$

Replace $f(x)$ with y.

$y = -\dfrac{5}{6}x - 3$

Solve for x.

$y + 3 = -\dfrac{5}{6}x$

$$-\frac{6}{5}(y+3)=x$$

$$x=\frac{-6y-18}{5}$$

Replace x with $f^{-1}(y)$.

$$f^{-1}(y)=-\frac{6}{5}y-\frac{18}{5}$$

Write equation in terms of x.

$$f^{-1}(x)=-\frac{6}{5}x-\frac{18}{5}$$

73. $f(x)=\dfrac{6x-2}{5}$

Replace $f(x)$ with y.

$$y=\frac{6x-2}{5}$$

Solve for x.

$$5y=6x-2$$
$$5y+2=6x$$
$$\frac{5y+2}{6}=x$$
$$x=\frac{5y}{6}+\frac{2}{6}$$

Replace x with $f^{-1}(y)$.

$$f^{-1}(y)=\frac{5}{6}y+\frac{1}{3}$$

Write equation in terms of x.

$$f^{-1}(x)=\frac{5}{6}x+\frac{1}{3}$$

75. $f(x)=7-8(x+1)$

Replace $f(x)$ with y.

$y=7-8(x+1)$

Solve for x.

$$y=7-8x-8$$
$$y=-8x-1$$
$$y+1=-8x$$
$$\frac{y+1}{-8}=x$$
$$x=-\frac{1}{8}y-\frac{1}{8}$$

Replace x with $f^{-1}(y)$.

$$f^{-1}(y)=-\frac{1}{8}y-\frac{1}{8}$$

Write equation in terms of x.

$$f^{-1}(x)=-\frac{1}{8}x-\frac{1}{8}$$

77. $f(x)=x$

Replace $f(x)$ with y.

$y=x$

Solve for x.

$x=y$

Replace x with $f^{-1}(y)$.

$f^{-1}(y)=y$

Write equation in terms of x.

$f^{-1}(x)=x$

79. $f(x)=x^3$

Replace $f(x)$ with y.

$y=x^3$

Solve for x.

$\sqrt[3]{y}=x$

Replace x with $f^{-1}(y)$.

$f^{-1}(y)=\sqrt[3]{y}=y^{1/3}$

Write equation in terms of x.

$f^{-1}(x)=\sqrt[3]{x}$ or $f^{-1}(x)=x^{1/3}$

81. a. $\left(f^{-1}\circ f\right)(x)=f^{-1}(x+7)=x+7-7=x$

b. $\left(f\circ f^{-1}\right)(x)=f(x-7)=x-7+7=x$

83. a. $\left(f^{-1}\circ f\right)(x)=f^{-1}(2x-5)$

$$=\frac{1}{2}(2x-5)+\frac{5}{2}$$
$$=x-\frac{5}{2}+\frac{5}{2}$$
$$=x$$

b. $\left(f\circ f^{-1}\right)(x)=f\left(\frac{1}{2}x+\frac{5}{2}\right)$

$$=2\left(\frac{1}{2}x+\frac{5}{2}\right)-5$$
$$=x+5-5$$
$$=x$$

85. a. $\left(f^{-1}\circ f\right)(x)=f^{-1}\left(\frac{3}{4}x-2\right)$

$$=\frac{4}{3}\left(\frac{3}{4}x-2\right)+\frac{8}{3}$$
$$=x-\frac{8}{3}+\frac{8}{3}$$
$$=x$$

b. $\left(f \circ f^{-1}\right)(x) = f\left(\dfrac{4}{3}x + \dfrac{8}{3}\right)$

$\qquad = \dfrac{3}{4}\left(\dfrac{4}{3}x + \dfrac{8}{3}\right) - 2$

$\qquad = x + 2 - 2$

$\qquad = x$

87. a. $f(x) = 5x - 9$

Replace $f(x)$ with y.

$y = 5x - 9$

Solve for x.

$\qquad y = 5x - 9$

$\qquad y + 9 = 5x$

$\qquad \dfrac{y + 9}{5} = \dfrac{5x}{5}$

$\qquad \dfrac{1}{5}y + \dfrac{9}{5} = x$

Replace x with $f^{-1}(y)$.

$f^{-1}(y) = \dfrac{1}{5}y + \dfrac{9}{5}$

Write equation in terms of x.

$f^{-1}(x) = \dfrac{1}{5}x + \dfrac{9}{5}$

b. $f(x) = 5x - 9$

$f(4) = 5(4) - 9 = 11$

c. $f^{-1}(x) = \dfrac{1}{5}x + \dfrac{9}{5}$

$f^{-1}(4) = \dfrac{1}{5}(4) + \dfrac{9}{5} = \dfrac{13}{5}$

89. a. $f(x) = 3x - 5$

Replace $f(x)$ with y.

$y = 3x - 5$

Solve for x.

$\qquad \dfrac{y + 5}{3} = \dfrac{3x}{3}$

$\qquad \dfrac{1}{3}y + \dfrac{5}{3} = x$

Replace x with $f^{-1}(y)$.

$f^{-1}(y) = \dfrac{1}{3}y + \dfrac{5}{3}$

Write equation in terms of x.

$f^{-1}(x) = \dfrac{1}{3}x + \dfrac{5}{3}$

b. Answers may vary. Example:

x	$f^{-1}(x)$
-2	$\dfrac{1}{3}(-2) + \dfrac{5}{3} = 1$
0	$\dfrac{1}{3}(0) + \dfrac{5}{3} = \dfrac{5}{3}$
1	$\dfrac{1}{3}(1) + \dfrac{5}{3} = 2$
4	$\dfrac{1}{3}(4) + \dfrac{5}{3} = 3$
7	$\dfrac{1}{3}(7) + \dfrac{5}{3} = 4$

c.

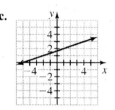

d. For each input-output pair, the output is $\dfrac{5}{3}$ more than $\dfrac{1}{3}$ times the input.

91. No. Answers may vary. Example:

$(f \circ g)(x) = f\left(\dfrac{x}{2}\right) = 2^{x/2}$

$(g \circ f)(x) = g\left(2^x\right) = \dfrac{2^x}{2} = 2^{x-1}$

Since $(f \circ g)(x) \neq (g \circ f)(x) \neq x$, $g(x)$ is not the inverse of $f(x)$.

93. Answers may vary. Example:

It makes sense that $g(x) = x - 5$ is the inverse function of $f(x) = x + 5$ because addition and subtraction are inverse operations of each other (or, 5 and -5 are additive inverses).

95. Answers may vary. Example:

It makes sense that if a function g is the inverse function of an invertible function f, then f is the inverse function of g, because if g undoes f, then undoing f should result in g.

97. No, $f(x) = 3$ is not one-to-one and it is not invertible.

Answers may vary. Example:

$f(1) = 3$ and $f(2) = 3$. So, the inverse of f would send 3 to both 1 and 2, making the inverse of f not a function. Since f^{-1} is not a function, f is not invertible.

99. a. $f(x) = mx + b$

Replace $f(x)$ with y.

$y = mx + b$

Solve for x.

$y - b = mx$

$\dfrac{y - b}{m} = x$

$x = \dfrac{y - b}{m}$

$x = \dfrac{1}{m}y - \dfrac{b}{m}$

Replace x with $f^{-1}(y)$.

$f^{-1}(y) = \dfrac{1}{m}y - \dfrac{b}{m}$

Write equation in terms of x.

$f^{-1}(x) = \dfrac{1}{m}x - \dfrac{b}{m}$

b. The function $f(x)$ given in part (a) is a linear function with nonzero slope (since $m \neq 0$). The inverse function is also a linear function as its slope is nonzero $\left(\dfrac{1}{m} \text{ is nonzero since } m \neq 0 \right)$.

101. a

The point of intersection is (4, 5).

b. $f(x) = 2x - 3$

Replace $f(x)$ with y.

$y = 2x - 3$

Solve for x.

$\dfrac{y + 3}{2} = \dfrac{2x}{2}$

$\dfrac{1}{2}y + \dfrac{3}{2} = x$

Replace x with $f^{-1}(y)$.

$f^{-1}(y) = \dfrac{1}{2}y + \dfrac{3}{2}$

Write equation in terms of x.

$f^{-1}(x) = \dfrac{1}{2}x + \dfrac{3}{2}$

c. $g(x) = \dfrac{1}{2}x + 3$

Replace $g(x)$ with y.

$y = \dfrac{1}{2}x + 3$

Solve for x.

$2(y - 3) = 2\left(\dfrac{1}{2}x \right)$

$2y - 6 = x$

Replace x with $g^{-1}(y)$.

$g^{-1}(y) = 2y - 6$

Write equation in terms of x.

$g^{-1}(x) = 2x - 6$

d.

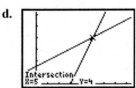

The point of intersection is (5, 4).

e. The x- and y-coordinates are exchanged. An inverse function exchanges the input and output values.

103. Multiply the first equation by 5 and the second equation by 4. Then add the equations.

$12x - 20y = 40$

$\underline{35x + 20y = 195}$

$47x = 235$

$x = 5$

Solve for y.

$3x - 5y = 10$

$3(5) - 5y = 10$

$15 - 5y = 10$

$\dfrac{-5y}{-5} = \dfrac{-5}{-5}$

$y = 1$

The solution is (5, 1). The two equations form a linear system in two variables.

105.

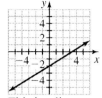

This is a linear equation in two variables.

Homework 5.3

1. $\log_9(81) = 2$, since $9^2 = 81$.

3. $\log_3(27) = 3$, since $3^3 = 27$.

5. $\log_4(256) = 4$, since $4^4 = 256$.

7. $\log_6(216) = 3$, since $6^3 = 216$.

9. $\log(100) = 2$, since $10^2 = 100$.

11. $\log_4\left(\dfrac{1}{4}\right) = -1$, since $4^{-1} = \dfrac{1}{4}$.

13. $\log_2\left(\dfrac{1}{8}\right) = -3$, since $2^{-3} = \dfrac{1}{8}$.

15. $\log\left(\dfrac{1}{10,000}\right) = -4$, since $10^{-4} = \dfrac{1}{10,000}$.

17. $\log_5(1) = 0$, since $5^0 = 1$.

19. $\log_9(9) = 1$, since $9^1 = 9$.

21. $\log_9(3) = \dfrac{1}{2}$, since $9^{1/2} = 3$.

23. $\log_8(2) = \dfrac{1}{3}$, since $8^{1/3} = 2$.

25. $\log_7\left(\sqrt{7}\right) = \dfrac{1}{2}$, since $7^{1/2} = \sqrt{7}$.

27. $\log_5\left(\sqrt[4]{5}\right) = \dfrac{1}{4}$, since $5^{1/4} = \sqrt[4]{5}$.

29. $\log_2(\log_2(16)) = \log_2(4)$ (since $2^4 = 16$)
                  $= 2$ (since $2^2 = 4$)

31. $\log_{10}(\log_{10}(10)) = \log_{10}(1)$ (since $10^1 = 10$)
                 $= 0$ (since $10^0 = 1$)

33. $\log_b(b) = 1$, since $b^1 = b$.

35. $\log_b\left(b^4\right) = 4$, since $b^4 = b^4$.

37. $\log_b\left(\dfrac{1}{b^5}\right) = -5$, since $b^{-5} = \dfrac{1}{b^5}$.

39. $\log_b\left(\sqrt{b}\right) = \dfrac{1}{2}$, since $b^{\frac{1}{2}} = \sqrt{b}$.

41. $\log_b(\log_b(b)) = \log_b(1)$ (since $b^1 = b$)
              $= 0$ (since $b^0 = 1$)

43. $f(x) = 3^x$
     $f^{-1}(x) = \log_3(x)$

45. $h(x) = 10^x$
     $h^{-1}(x) = \log(x)$

47. $f(x) = \log_5(x)$
     $f^{-1}(x) = 5^x$

49. $h(x) = \log(x)$
     $h^{-1}(x) = 10^x$

51. $f(2) = 2^2 = 4$

53. $f^{-1}(2) = \log_2(2) = 1$, since $2^1 = 2$.

55. $g(3) = \log_3(3) = 1$

57. $g^{-1}(3) = 3^3 = 27$

59. $f(1) = 3$

61. $f^{-1}(1) = \log_3(1) = 0$

63. $\log_2\left(2^6\right) = 6$

65. $5^{\log_5(8)} = 8$

67. $\log\left(10^7\right) = 7$

69. $10^{\log(3)} = 3$

71.

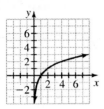

73.

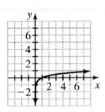

75.

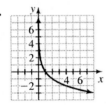

77. a.

x	$f(x) = \log_5(x)$
$\dfrac{1}{5}$	$\log_5\left(\dfrac{1}{5}\right) = -1$
1	$\log_5(1) = 0$
5	$\log_5(5) = 1$
25	$\log_5(25) = 2$
125	$\log_5(125) = 3$

b.

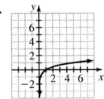

c. For each input-output pair, the output is the logarithm, base 5, of the input.

79. a. $x = 2$, since $5^2 = 25$.

b. $\log_5(25) = 2$, since $5^2 = 25$.

c. 5^2 and $\log_5(25)$ are inverses of each other.

81. a. $R = \log\left(\dfrac{1.6 \times 10^9 \, A_0}{A_0}\right)$

$= \log\left(1.6 \times 10^9\right)$

$= 9.2$

The Richter number for the Indian Ocean earthquake was 9.2.

b. $R = \log\left(\dfrac{6.3 \times 10^7 \, A_0}{A_0}\right)$

$= \log\left(6.3 \times 10^7\right)$

$= 7.8$

The Richter number for the Mexico City earthquake was 7.8.

c. $\dfrac{1.6 \times 10^9}{6.3 \times 10^7} = 25.40$

The ratio of the Indian Ocean earthquake's amplitude to the Mexico City earthquake's amplitude is 25.40.

83. $L = 10\log\left(\dfrac{1}{I_0}\right)$

Sound	Intensity (W/m^2)	Decibels (L)
Faintest sound heard by humans	10^{-12}	0
Whisper	10^{-10}	20
Inside a running car	10^{-8}	40
Conversation	10^{-6}	60
Noisy street corner	10^{-4}	80
Soft-rock concert	10^{-2}	100
Threshold of pain	1	120

85. (a)

87. (d)

89. Answers may vary. Example:

$a = \log_2(7) \Rightarrow 2^a = 7$

$b = \log_3(7) \Rightarrow 3^b = 7$

Because 2 is less than 3, 2 requires a larger exponent to get 7 than 3 does.

91. Answers may vary. Example:

Make a table of values for $f(x) = 5^x$, and then build a table for $f^{-1}(x)$.

x	$f(x)$		x	$f^{-1}(x)$
0	1		1	0
1	5		5	1
2	25		25	2
3	125		125	3
4	625		625	4

Another name for $f^{-1}(x)$ is $\log_5(x)$.

93. a.

x	$f(x)$ $= \log_2(x)$	$g(x)$ $= 2x$	$h(x)$ $= 2^x$
1	0	2	2
2	1	4	4
4	2	8	16
8	3	16	256
16	4	32	65,536

b. The outputs of h are growing the fastest, followed by g.

95. a. $\left(f^{-1} \circ f\right)(x) = f^{-1}\left(3^x + 5\right)$

$$= \log_3\left(3^x + 5 - 5\right)$$
$$= \log_3\left(3^x\right)$$
$$= x$$

b. $\left(f \circ f^{-1}\right)(x) = f\left(\log_3(x - 5)\right)$

$$= 3^{\log_3(x-5)} + 5$$
$$= x - 5 + 5$$
$$= x$$

97. a. $\left(f^{-1} \circ f\right)(x) = f^{-1}\left(5(2)^x - 6\right)$

$$= \log_2\left(\frac{5(2)^x - 6 + 6}{5}\right)$$
$$= \log_2\left(\frac{5(2)^x}{5}\right)$$
$$= \log_2\left((2)^x\right)$$
$$= x$$

b. $\left(f \circ f^{-1}\right)(x) = f\left(\log_2\left(\frac{x+6}{5}\right)\right)$

$$= 5(2)^{\log_2\left(\frac{x+6}{5}\right)} - 6$$
$$= 5\left(\frac{x+6}{5}\right) - 6$$
$$= x + 6 - 6$$
$$= x$$

99. Use the data points (3, 5) and (7, 89) and the exponential regression feature on a graphing calculator.

$$y = 0.58(2.05)^x$$

This is an exponential equation in two variables.

101. $4b^{2/3}c^{-5/4}\left(2b^{-1/5}c^{3/4}\right)$

$$= 8b^{10/15}b^{-3/15}c^{-5/4}c^{3/4}$$
$$= 8b^{7/15}c^{-2/4}$$
$$= \frac{8b^{7/15}}{c^{1/2}}$$

This is an expression in two variables that involves exponents.

Homework 5.4

1. $\log_3(243) = 5$

$$3^5 = 243$$

3. $\log(100) = 2$

$$10^2 = 100$$

5. $\log_b(a) = c$

$$b^c = a$$

7. $\log(m) = n$

$$10^n = m$$

9. $5^3 = 125$

$$\log_5(125) = 3$$

11. $10^3 = 1000$

$$\log(1000) = 3$$

13. $y^w = x$

$$\log_y(x) = w$$

15. $10^p = q$

$\log(q) = p$

17. $\log_4(x) = 2$

$4^2 = x$

$x = 16$

19. $\log(x) = -2$

$10^{-2} = x$

$x = \dfrac{1}{100}$

$x = 0.01$

21. $\log_4(x) = 0$

$4^0 = x$

$x = 1$

23. $\log_{27}(t) = \dfrac{4}{3}$

$27^{4/3} = x$

$x = 81$

25. $2\log_8(2x - 5) = 4$

$\log_8(2x - 5) = 2$

$8^2 = 2x - 5$

$\dfrac{64 + 5}{2} = \dfrac{2x}{2}$

$\dfrac{69}{2} = x$

27. $4\log_{81}(x) - 3 = -2$

$\dfrac{4\log_{81}(x)}{4} = \dfrac{1}{4}$

$\log_{81}(x) = \dfrac{1}{4}$

$81^{1/4} = x$

$3 = x$

29. $\log_2\left(\log_3(y)\right) = 3$

$2^3 = \log_3(y)$

$8 = \log_3(y)$

$3^8 = y$

$6561 = y$

31. $\log_6\left(x^3\right) = 2$

$6^2 = x^3$

$36 = x^3$

$36^{1/3} = x$

$3.3019 \approx x$

33. $\log_b(49) = 2$

$b^2 = 49$

$b = 49^{1/2}$

$b = 7$

35. $\log_b(8) = 3$

$b^3 = 8$

$b = 8^{1/3}$

$b = 2$

37. $\log_b(16) = 5$

$b^5 = 16$

$b = 16^{1/5}$

$b \approx 1.7411$

39. $4^x = 9$

$\log\left(4^x\right) = \log(9)$

$x\log(4) = \log(9)$

$x = \dfrac{\log(9)}{\log(4)}$

$x \approx 1.5850$

41. $5\left(4^x\right) = 80$

$4^x = 16$

$\log\left(4^x\right) = \log(16)$

$x\log(4) = \log(16)$

$x = \dfrac{\log(16)}{\log(4)}$

$x = 2$

43. $3.83(2.18)^t = 170.91$

$$2.18^t = \frac{170.91}{3.83}$$

$$\log(2.18^t) = \log\left(\frac{170.91}{3.83}\right)$$

$$t\log(2.18) = \log\left(\frac{170.91}{3.83}\right)$$

$$t = \frac{\log\left(\frac{170.91}{3.83}\right)}{\log(2.18)}$$

$$t \approx 4.8738$$

45. $8 + 5(2)^x = 79$

$$5(2^x) = 71$$

$$2^x = \frac{71}{5}$$

$$\log(2^x) = \log\left(\frac{71}{5}\right)$$

$$x\log(2) = \log\left(\frac{71}{5}\right)$$

$$x = \frac{\log\left(\frac{71}{5}\right)}{\log(2)}$$

$$x \approx 3.8278$$

47. $2^{4x+5} = 17$

$$\log\left(2^{4x+5}\right) = \log(17)$$

$$(4x+5)\log(2) = \log(17)$$

$$4x+5 = \frac{\log(17)}{\log(2)}$$

$$4x = \frac{\log(17)}{\log(2)} - 5$$

$$x = \frac{\left(\frac{\log(17)}{\log(2)} - 5\right)}{4}$$

$$x \approx -0.2281$$

49. $6(3)^x - 7 = 85 + 4(3)^x$

$$6(3)^x - 4(3)^x = 92$$

$$\frac{2(3)^x}{2} = \frac{92}{2}$$

$$(3)^x = 46$$

$$x\log(3) = \log(46)$$

$$x = \frac{\log(46)}{\log(3)}$$

$$x \approx 3.4850$$

51. $4^{3p} \cdot 4^{2p-1} = 100$

$$4^{5p-1} = 100$$

$$\log\left(4^{5p-1}\right) = \log(100)$$

$$(5p-1)\log(4) = 2$$

$$5p-1 = \frac{2}{\log(4)}$$

$$5p = \frac{2}{\log(4)} + 1$$

$$p = \frac{\left(\frac{2}{\log(4)} + 1\right)}{5}$$

$$p \approx 0.8644$$

53. $3^x = -8$

No real-number solution. 3 raised to any power will *always* be positive.

55. $\log_4(x) = 3$

$$4^3 = x$$

$$x = 64$$

57. $3(4)^t + 15 = 406$

$$\frac{3(4)^t}{3} = \frac{391}{3}$$

$$\log(4^t) = \log\left(\frac{391}{3}\right)$$

$$t = \frac{\log\left(\frac{391}{3}\right)}{\log(4)}$$

$$t \approx 3.5130$$

59. $\log_b(73) = 5$

$$b^5 = 73$$

$$b = 73^{\frac{1}{5}}$$

$$b \approx 2.3587$$

61. $3\log_{27}(y-1)=2$

$$\log_{27}(y-1)=\frac{2}{3}$$
$$27^{2/3}=y-1$$
$$9+1=y$$
$$10=y$$

63. $3(2)^{4x-2}=83$

$$\log\left(2^{4x-2}\right)=\log\left(\frac{83}{3}\right)$$
$$(4x-2)\log(2)=\log\left(\frac{83}{3}\right)$$
$$4x-2=\frac{\log\left(\dfrac{83}{3}\right)}{\log(2)}$$
$$\frac{4x}{4}=\frac{\dfrac{\log\left(\dfrac{83}{3}\right)}{\log(2)}+2}{4}$$
$$x\approx1.6975$$

65. $x=1$ because $y=2^x$ and $y=4\left(\dfrac{1}{2}\right)^x$ intersect at $x=1$.

67. $x=0$ and $x\approx3.7$ because $y=4\left(\dfrac{1}{2}\right)^x$ intersects $y=4-x$ in two points with the given x-values.

69. $x=2$ because $y=4\left(\dfrac{1}{2}\right)^x$ is 1 when $x=2$.

71. Graph $y=3^x$ and $y=5-x$. Then use the intersect feature on a graphing calculator. The x-coordinate, 1.2122, is the approximate solution of the equation $3^x=5-x$.

73. Graph $y=7\left(\dfrac{1}{2}\right)^x$ and $y=2x$. Then use the intersect feature on a graphing calculator. The x-coordinate, 1.3618, is the approximate solution of the equation $7\left(\dfrac{1}{2}\right)^x=2x$.

75. Graph $y=\log(x+1)$ and $y=3-\dfrac{2}{5}x$. Then use the intersect feature on a graphing calculator. The x-coordinate, 5.4723, is the approximate solution of the equation $\log(x+1)=3-\dfrac{2}{5}x$.

77. $x=2$
Functions representing each side of the equation have the same y-value when $x=2$.

79. $x=5$
Functions representing each side of the equation have the same y-value when $x=5$.

81. The solution is (3, 1.5).
Functions representing each equation in the system have the same x- and y-values at (3, 1.5).

83. $ab^x=c$
$$b^x=\frac{c}{a}$$
$$\log\left(b^x\right)=\log\left(\frac{c}{a}\right)$$
$$x\log(b)=\log\left(\frac{c}{a}\right)$$
$$x=\frac{\log\left(\dfrac{c}{a}\right)}{\log(b)}$$

85. $b^x+c=d$
$$b^x=d-c$$
$$\log\left(b^x\right)=\log(d-c)$$
$$x\log(b)=\log(d-c)$$
$$x=\frac{\log(d-c)}{\log(b)}$$

87. $ab^x-c=d$
$$ab^x=d+c$$
$$b^x=\frac{d+c}{a}$$
$$\log\left(b^x\right)=\log\left(\frac{d+c}{a}\right)$$
$$x\log(b)=\log\left(\frac{d+c}{a}\right)$$
$$x=\frac{\log\left(\dfrac{d+c}{a}\right)}{\log(b)}$$

89.

$$ab^{x+p} - c = d$$

$$ab^{x+p} = c + d$$

$$b^{x+p} = \frac{c+d}{a}$$

$$b^x \cdot b^p = \frac{c+d}{a}$$

$$\log\left(b^x \cdot b^p\right) = \log\left(\frac{c+d}{a}\right)$$

$$\log\left(b^x\right) + \log\left(b^p\right) = \log\left(\frac{c+d}{a}\right)$$

$$x\log\left(b\right) + p\log\left(b\right) = \log\left(\frac{c+d}{a}\right)$$

$$x\log\left(b\right) = \log\left(\frac{c+d}{a}\right) - p\log\left(b\right)$$

$$x = \frac{\log\left(\dfrac{c+d}{a}\right) - p\log\left(b\right)}{\log\left(b\right)}$$

91. Line 3 includes an error.

$$\log[3(8^x)] \neq x\log[3(8)]$$

The power property for logarithms would only work in this case if *both* the 3 and the 8 were raised to the *x* power. The first step the student should have made was to divide both sides of the equation by 3.

$$3\left(8^x\right) = 7$$

$$8^x = \frac{7}{3}$$

Then take the log of both sides and solve for *x*.

$$\log\left(8^x\right) = \log\left(\frac{7}{3}\right)$$

$$x\log\left(8\right) = \log\left(\frac{7}{3}\right)$$

$$x = \frac{\log\left(\dfrac{7}{3}\right)}{\log\left(8\right)}$$

$$x \approx 0.4075$$

93. $f(4) = 4^{(4)} = 256$

95.

$$4^x = 3$$

$$\log\left(4^x\right) = \log\left(3\right)$$

$$x\log\left(4\right) = \log\left(3\right)$$

$$x = \frac{\log\left(3\right)}{\log\left(4\right)}$$

$$x \approx 0.7925$$

97. $g(8) = \log_2(8) = 3$

99. $\log_2(a) = 5$

$$2^5 = a$$

$$a = 32$$

101. a. False. $\dfrac{\log_2(4)}{\log_2(16)} \neq \dfrac{4}{16}$ because we cannot divide out logarithms of different numbers.

$$\frac{\log_2(4)}{\log_2(16)} = \frac{2}{4} = \frac{1}{2} \neq \frac{4}{16}$$

b. False. We cannot divide out logarithms of different numbers.

$$\frac{\log_3(1)}{\log_3(27)} = \frac{0}{3} = 0 \neq \frac{1}{27}$$

c. False. We cannot divide out logarithms of different numbers.

$$\frac{\log(1000)}{\log(10,000)} = \frac{3}{4} \neq \frac{1000}{10,000}$$

d. False. We cannot divide out logarithms of different numbers.

103. a.

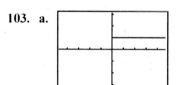

The part of the line $y = 1$ where $x > 0$.

b. We cannot take the logarithm of a negative number.

c. $f(x) = \log\left(x^3\right) - 3\log\left(x\right) + 1$

$$= 3\log\left(x\right) - 3\log\left(x\right) + 1$$

$$= 1$$

Answers may vary. Example:
This result is the same as the graph from part (a), with the domain of $x > 0$.

105. Answers may vary. Example:
For $x > 0$, $b > 0$, and $b \neq 1$,

$$\log_b(x^p) = p\log_b(x).$$

This does not imply $x^p = px$.

$$x^p \overset{?}{=} px$$

$$\log(x^p) \overset{?}{=} \log(px)$$

$$p\log(x) \overset{?}{=} \log(px) \quad \text{false}$$

107. $5(3p-7)-9p=-4p+23$

$15p-35-9p=-4p+23$

$10p=58$

$p=\dfrac{58}{10}$

$p=\dfrac{29}{5}$

109. $5b^6-88=56$

$\dfrac{5b^6}{5}=\dfrac{144}{5}$

$b=\left(\dfrac{144}{5}\right)^{\frac{1}{6}}$

$b\approx\pm1.7508$

111. $\dfrac{3}{8}r=\dfrac{5}{6}r-\dfrac{2}{3}$

$24\left(\dfrac{3}{8}r\right)=24\left(\dfrac{5}{6}r-\dfrac{2}{3}\right)$

$9r=20r-16$

$\dfrac{-11r}{-11}=\dfrac{-16}{-11}$

$r=\dfrac{16}{11}$

113. $\log_2(x)=-5$

$2^{-5}=x$

$\dfrac{1}{32}=x$

This is a logarithmic equation in one variable.

115.

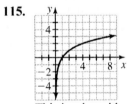

This is a logarithmic function.

Homework 5.5

1. a. $V=f(t)$

We know we can model the situation well by using an exponential model $f(t)=ab^t$. The V-intercept is (0, 2000), so $a=2000$ and $f(t)=2000b^t$. Since the interest rate is 5%, the base must be 1.05, so

$f(t)=2000(1.05)^t$.

b. The V-intercept is (0, 2000). The original ($t=0$) value in the account is $2000.

c. $f(t)=2000(1.05)^t$

$f(5)=2000(1.05)^5=2552.56$

In 5 years, the investment will be worth $2552.56.

d. $2000(1.05)^t=3000$

$1.05^t=1.5$

$\log(1.05^t)=\log(1.5)$

$t\log(1.05)=\log(1.5)$

$t=\dfrac{\log(1.5)}{\log(1.05)}$

$t\approx8.3104$

The balance will be $3000 after 8.31 years.

3. $f(t)=9300(1.06)^t$

$9300(1.06)^t=13{,}700$

$1.06^t=\dfrac{13{,}700}{9300}$

$\log(1.06^t)=\log\left(\dfrac{13{,}700}{9300}\right)$

$t\log(1.06)=\log\left(\dfrac{13{,}700}{9300}\right)$

$t=\dfrac{\log\left(\dfrac{13{,}700}{9300}\right)}{\log(1.06)}$

$t\approx6.65$

The balance will be $13,700 after 6.65 years.

5. $f(t)=6000(1.10)^t$

$6000(1.10)^t=12000$

$1.10^t=2$

$\log(1.10^t)=\log(2)$

$t\log(1.10)=\log(2)$

$t=\dfrac{\log(2)}{\log(1.10)}$

≈7.27

The balance will double after 7.27 years. The interest is compounded annually, so that interest on the previous year's interest earnings will also grow at 10% per year.

7. $f(t) = 13.9(1.19)^t$

$13.9(1.19)^t = 40$

$1.19^t = \dfrac{40}{13.9}$

$\log(1.19^t) = \log\left(\dfrac{40}{13.9}\right)$

$t\log(1.19) = \log\left(\dfrac{40}{13.9}\right)$

$t = \dfrac{\log\left(\dfrac{40}{13.9}\right)}{\log(1.19)}$

$t \approx 6.08$

The annual production of ethanol will reach 40 billion in $2011 + 6 = 2017$.

9. **a.** We know we can model the situation well by using an exponential model $f(t) = ab^t$. Let $p = f(t)$, the total number of people. The p-intercept is $(0, 30)$, and the base is 3.

$f(t) = 30(3)^t$

b. $f(t) = 30(3)^t$

$f(8) = 30(3)^8$

$= 196{,}830$

The total number of Americans who will have heard the rumor after 8 days is 196,830.

c. $30(3)^t = 315{,}000{,}000$

$(3)^t = 10{,}500{,}000$

$\log(3^t) = \log(10{,}500{,}000)$

$t\log(3) = \log(10{,}500{,}000)$

$t = \dfrac{\log(10{,}500{,}000)}{\log(3)}$

$t \approx 14.72$

The model predicts that all Americans will have heard the rumor after about 15 days.

11. **a.** Use an exponential decay model $T = f(d) = ab^d$. The T-intercept is $(0, 8)$ so $a = 8$. For every increase of 5 decibels d, the exposure time is halved so the base is $\dfrac{1}{2}$ and the exponent is $\dfrac{d}{5}$.

$f(d) = 8\left(\dfrac{1}{2}\right)^{d/5}$

b. $f(24) = 8\left(\dfrac{1}{2}\right)^{24/5} \approx 0.2872$

The model predicts that at 114 decibels, the bands could play for 0.29 hours (about 17.4 minutes!) without the fans experiencing hearing loss. The average rock concert lasts longer than 18 minutes, so the model predicts that these fans experience hearing loss.

c. $f^{-1}(3)$

$3 = 8\left(\dfrac{1}{2}\right)^{d/5}$

$\left(\dfrac{1}{2}\right)^{d/5} = \dfrac{3}{8}$

$\log\left(\left(\dfrac{1}{2}\right)^{d/5}\right) = \log\left(\dfrac{3}{8}\right)$

$\dfrac{d}{5}\log\left(\dfrac{1}{2}\right) = \log\left(\dfrac{3}{8}\right)$

$\dfrac{d}{5} = \dfrac{\log\left(\dfrac{3}{8}\right)}{\log\left(\dfrac{1}{2}\right)}$

$t = \dfrac{5\log\left(\dfrac{3}{8}\right)}{\log\left(\dfrac{1}{2}\right)}$

$t \approx 7.08$

To play for 3 hours, the rock bands should play at 97 decibels.

13. Use the exponential model $y = ab^t$ with 1975 as year 0. Substituting the data point $(0, 0.21)$ yields $a = 0.21$. Solve for b using \$8.53 million in 2012, when $t = 37$.

$y = ab^t$

$8.53 = 0.21(b)^{37}$

$\dfrac{8.53}{0.21} = (b)^{37}$

$\left(\dfrac{8.53}{0.21}\right)^{1/37} = b$

$1.11 \approx b$

The model is $y = 0.21(1.11)^t$.

Predict when the prize will be \$19 million.

$$19 = 0.21(1.11)^t$$

$$\frac{19}{0.21} = \frac{0.21(1.11)^t}{0.21}$$

$$\log\left(\frac{19}{0.21}\right) = \log(1.11)^t$$

$$\frac{\log\left(\frac{19}{0.21}\right)}{\log(1.11)} = \frac{t\log(1.11)}{\log(1.11)}$$

$$43.17 \approx t$$

The prize will be $19 million in
1975 + 43 = 2018.

15. Use the exponential model $y = ab^t$ with 1990
as year 0. This provides the data point
$(0, 471)$ so that $a = 471$. Solve for b using
36 million in 2010, when $t = 20$.

$$y = ab^t$$

$$36 = 471(b)^{20}$$

$$\frac{36}{471} = (b)^{20}$$

$$\left(\frac{36}{471}\right)^{1/20} = b$$

$$0.88 \approx b$$

The model is $y = 471(0.88)^t$.

Predict when the harvest will be 13 million
board feet.

$$13 = 471(0.88)^t$$

$$\frac{13}{471} = \frac{471(0.88)^t}{471}$$

$$\log\left(\frac{13}{471}\right) = \log(0.88)^t$$

$$\frac{\log\left(\frac{13}{471}\right)}{\log(0.88)} = \frac{t\log(0.88)}{\log(0.88)}$$

$$28.08 \approx t$$

The harvest will be 13 million board feet in
1990 + 28 = 2018.

17. **a.**

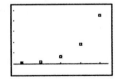

The scattergram shows that the points
"bend" and that an exponential function
will model the data better than a linear
function.

b. Answers may vary depending on the points
chosen. Example:
Use the points (4, 29) and (6, 350) to find
an equation of the form $y = ab^t$.

$$29 = ab^4$$

$$350 = ab^6$$

Divide and solve for b.

$$\frac{29}{350} = b^{-2}$$

$$\left(\frac{29}{350}\right)^{-1/2} = \left(b^{-2}\right)^{-1/2}$$

$$3.47 \approx b$$

Solve for a.

$$29 = a(3.47)^4$$

$$\frac{29}{(3.47)^4} = a$$

$$0.20 \approx a$$

The model is $f(t) = 0.20(3.47)^t$. (Your
equation may be slightly different if you
chose different points.) Use the graphing
calculator to check your results.

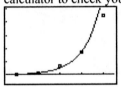

c. $b = 3.47$
The number of Twitter employees is
growing exponentially by 247% per year.

d. $f(9) = 0.20(3.47)^9 \approx 14{,}588$
In 2014, Twitter will employ about
14,588 people.

e.
$$f(t) = 0.20(3.47)^t$$

$$394{,}000 = 0.20(3.47)^t$$

$$\frac{394{,}000}{0.20} = \frac{0.20(3.47)^t}{0.20}$$

$$\log\left(\frac{394{,}000}{0.20}\right) = \log(3.47)^t$$

$$\frac{\log\left(\frac{394{,}000}{0.20}\right)}{\log(3.47)} = \frac{t\log(3.47)}{\log(3.47)}$$

$$11.65 \approx t$$

In 2005 + 12 = 2017, Twitter will employ
about 394 thousand people.

19. a.

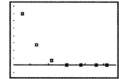

The scattergram shows that the points "bend" and that an exponential function will model the data better than a linear function.

b. Answers may vary depending on the points chosen. Example:
Use the points (8, 350) and (16, 33) to find an equation of the form $y = ab^t$.

$$350 = ab^8$$
$$33 = ab^{16}$$

Divide and solve for b.

$$\frac{350}{33} = b^{-8}$$
$$\left(\frac{350}{33}\right)^{-1/8} = \left(b^{-8}\right)^{-1/8}$$
$$0.74 \approx b$$

Solve for a.

$$350 = a(0.74)^8$$
$$\frac{350}{(0.74)^8} = a$$
$$3892 \approx a$$

The model is $f(t) = 3892(0.74)^t$. (Your equation may be slightly different if you chose different points.) Use the graphing calculator to check your results.

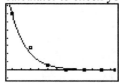

c. $f(t) = 3892(0.74)^t$

$$f(38) = 3892(0.74)^{38} \approx 0.04$$

In 2083, the model predicts that there will be about 0.04 thousand, or 40, polio cases.

d.
$$f(t) = 3892(0.74)^t$$
$$0.001 = 3892(0.74)^t$$
$$\frac{0.001}{3892} = (0.74)^t$$
$$\log\left(\frac{0.001}{3892}\right) = \log(0.74)^t$$
$$\frac{\log\left(\dfrac{0.001}{3892}\right)}{\log(0.74)} = \frac{t\log(0.74)}{\log(0.74)}$$
$$50.40 \approx t$$

There will be 1 case of polio in $1980 + 50 = 2030$.

e. Half the number of cases in 1980, or half of 3892 thousand, is 1946 thousand.

$$f(t) = 3892(0.74)^t$$
$$1946 = 3892(0.74)^t$$
$$\frac{1946}{3892} = \frac{3892(0.74)^t}{3892}$$
$$\frac{1}{2} = (0.74)^t$$
$$\log\left(\frac{1}{2}\right) = \log(0.74)^t$$
$$\frac{\log\left(\dfrac{1}{2}\right)}{\log(0.74)} = \frac{t\log(0.74)}{\log(0.74)}$$
$$2.30 \approx t$$

The approximate half-life of the number of polio cases is 2.30 years.

21. a.
$$f(t) = 1.21(1.0161)^t$$
$$9.3 = 1.21(1.0161)^t$$
$$1.0161^t = \frac{9.3}{1.21}$$
$$\log(1.0161^t) = \log\left(\frac{9.3}{1.21}\right)$$
$$t\log(1.0161) = \log\left(\frac{9.3}{1.21}\right)$$
$$t = \frac{\log\left(\dfrac{9.3}{1.21}\right)}{\log(1.0161)}$$
$$t \approx 127.69$$

The model predicts the world population will reach 9.3 billion in $1900 + 128 = 2028$.

b.

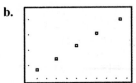

The plotted data lie mostly in a straight line. Therefore, a linear model better fits the data.

Using the linear regression feature on a graphing calculator, $g(t) = 0.080t - 1.92$.

c. $g(t) = 0.080t - 1.92$
$9.3 = 0.080t - 1.92$
$11.22 = 0.080t$
$$\frac{11.22}{0.080} = \frac{0.080t}{0.080}$$
$140.25 = t$

The year the population will reach 9.3 billion is $1900 + 140 = 2040$, which is after the model in part (a) of 2028.
Answers may vary. Example:
This is because the linear increase is more gradual than the exponential increase in part (a).

d.

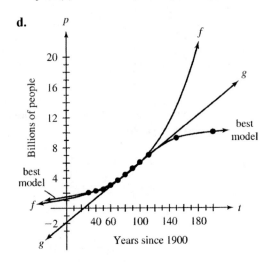

23. a. Answers may vary depending on the points chosen. Example:
Use the points $(67, 1.1)$ and $(88, 12.9)$ to find an equation of the form $y = ab^t$.

$1.1 = ab^{67}$
$12.9 = ab^{88}$

Divide and solve for b.

$$\frac{1.1}{12.9} = \frac{ab^{67}}{ab^{88}}$$

$$\frac{1.1}{12.9} = b^{-21}$$

$$\left(\frac{1.1}{12.9}\right)^{-1/21} = \left(b^{-21}\right)^{-1/21}$$

$$1.12 \approx b$$

Solve for a.

$$1.1 = a(1.12)^{67}$$

$$\frac{1.1}{(1.12)^{67}} = a$$

$$0.00055 \approx a$$

The model is $f(t) = 0.00055(1.12)^t$.

(Your equation may be slightly different if you chose different points.) Use the graphing calculator to check your results.

b.   The base b is 1.12 and it represents the percentage of seniors with severe memory impairment increases by 12% for each additional year of age. (Your equation may be slightly different if you chose different points.) Use the graphing calculator to check your results.

c. $f(t) = 0.00055(1.12)^t$

$f(70) = 0.00055(1.12)^{70} = 1.53$

Based on this model, about 1.53% of 70-year-old seniors have severe memory impairment. (Your equation may be slightly different if you chose different points.) Use the graphing calculator to check your results.

d. $$f(t) = 0.00055(1.12)^t$$

$$10 = 0.00055(1.12)^t$$

$$\frac{10}{0.00055} = (1.12)^t$$

$$\log\left(\frac{10}{0.00055}\right) = \log(1.12)^t$$

$$\frac{\log\left(\frac{10}{0.00055}\right)}{\log(1.12)} = \frac{t\log(1.12)}{\log(1.12)}$$

$$86.54 \approx t$$

This model predicts 10% of seniors will have severe memory impairment at age 87. (Your equation may be slightly different if you chose different points.) Use the graphing calculator to check your results.

e. Answers may vary. Example: No, memory speed and memory impairment are different phenomena that may correlate better with different mathematical relationships.

25. a. Using a graphing calculator and the exponential regression feature gives:

$$E(s) = 0.36(1.0036)^s$$

$$R(s) = 0.037(1.0049)^s$$

b. $E(1425) = 0.36(1.0036)^{1425} \approx 60.3\%$

$R(1425) = 0.037(1.0049)^{1425} \approx 39.2\%$

c. Half = 50%
Use $E(s)$ for early decision applicants.
$$E(s) = 50$$
$$0.36(1.0036)^s = 50$$
$$1.0036^s = \frac{50}{0.36}$$
$$\log\left(1.0036^s\right) = \log\left(\frac{50}{0.36}\right)$$
$$s\log\left(1.0036\right) = \log\left(\frac{50}{0.36}\right)$$
$$s = \frac{\log\left(\dfrac{50}{0.36}\right)}{\log\left(1.0036\right)}$$
$$s \approx 1373$$
Use $R(s)$ for regular decision applicants.
$$R(s) = 50$$
$$0.037(1.0049)^s = 50$$
$$1.0049^s = \frac{50}{0.037}$$
$$\log\left(1.0049^s\right) = \log\left(\frac{50}{0.037}\right)$$
$$s\log\left(1.0049\right) = \log\left(\frac{50}{0.037}\right)$$
$$s = \frac{\log\left(\dfrac{50}{0.037}\right)}{\log\left(1.0049\right)}$$
$$s \approx 1475$$

d. $1475 - 1373 = 102$ points

e.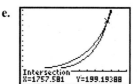

The intersection is (1757.58, 199.19). Students who score 1758 points on their SAT have the same chance (199%) of being selected from the early decision and regular decision systems. Model breakdown has occurred.
Challenge problem:
$$0.36(1.0036)^s = 0.037(1.0049)^s$$
$$\frac{(1.0036)^s}{(1.0049)^s} = \frac{0.037}{0.36}$$
$$\left(\frac{1.0036}{1.0049}\right)^s = \frac{0.037}{0.36}$$
$$\log\left(\frac{1.0036}{1.0049}\right)^s = \log\left(\frac{0.037}{0.36}\right)$$
$$\frac{s\log\left(\dfrac{1.0036}{1.0049}\right)}{\log\left(\dfrac{1.0036}{1.0049}\right)} = \frac{\log\left(\dfrac{0.037}{0.36}\right)}{\log\left(\dfrac{1.0036}{1.0049}\right)}$$
$$s \approx 1757.58$$

27. a. Use the exponential model $f(t) = ab^t$.
At $t = 0$, 100% of the gallium citrate-67 remains. Substitute (0, 100).
$$100 = ab^0$$
$$100 = a$$
To solve for b, substitute (3.25, 50).
$$50 = 100(b)^{3.25}$$
$$\frac{50}{100} = (b)^{3.25}$$
$$\left(\frac{1}{2}\right)^{1/3.25} = b$$
The equation is $f(t) = 100\left(\dfrac{1}{2}\right)^{t/3.25}$

b. $f(t) = 100\left(\dfrac{1}{2}\right)^{2/3.25} = 65.28$
After 2 days, 65.28% of the gallium citrate-67 remains.

c.
$$f(t) = 100\left(\frac{1}{2}\right)^{t/3.25}$$
$$0.39 = 100\left(\frac{1}{2}\right)^{t/3.25}$$
$$\log\left(\frac{0.39}{100}\right) = \log\left(\frac{1}{2}\right)^{t/3.25}$$
$$\frac{\log\left(\frac{0.39}{100}\right)}{\log\left(\frac{1}{2}\right)} = \frac{t}{3.25}\cdot\frac{\log\left(\frac{1}{2}\right)}{\log\left(\frac{1}{2}\right)}$$
$$\frac{\log\left(\frac{0.39}{100}\right)}{\log\left(\frac{1}{2}\right)} = \frac{t}{3.25}$$
$$3.25\left(\frac{\log\left(\frac{0.39}{100}\right)}{\log\left(\frac{1}{2}\right)}\right) = t$$
$$26.0 \approx t$$

After 26 days, 0.39% of the gallium citrate-67 remains.

29. Use an exponential decay model $f(t) = ab^t$. Every 5730 years the amount of carbon-14 is halved, so the base is $\frac{1}{2}$ and the exponent is $\frac{t}{5730}$. 24.46% of the carbon-14 remains, so $f(t) = 24.46$.

$$f(t) = 100\left(\frac{1}{2}\right)^{t/5730}$$
$$24.46 = 100\left(\frac{1}{2}\right)^{t/5730}$$
$$\log\left(\frac{24.46}{100}\right) = \log\left(\frac{1}{2}\right)^{t/5730}$$
$$\frac{\log\left(\frac{24.46}{100}\right)}{\log\left(\frac{1}{2}\right)} = \frac{t}{5730}\cdot\frac{\log\left(\frac{1}{2}\right)}{\log\left(\frac{1}{2}\right)}$$
$$\frac{\log\left(\frac{24.46}{100}\right)}{\log\left(\frac{1}{2}\right)} = \frac{t}{5730}$$

$$5730\left(\frac{\log\left(\frac{24.46}{100}\right)}{\log\left(\frac{1}{2}\right)}\right) = t$$
$$11640.5 \approx t$$

The ice sheet advanced about 11,641 years ago.

31. a. Since 50% of the wood's carbon-14 remains, one half-life has passed. The half-life of carbon-14 is 5730 years. The age of the wood is 5730 years.

b. Since 25% of the wood's carbon-14 remains, two half-lives have passed (half of 100% is 50% and then half of 50% is 25%). The age of the wood is $2 \cdot 5730$ or 11,460 years.

c. After three half-lives have passed 12.5% of the wood's carbon-14 remains. This corresponds to $3 \cdot 5730$ or 17,190 years passing. The wood has only 10% of its carbon-14 remaining, so guess a little longer than 17,190 years.

$$f(t) = 100\left(\frac{1}{2}\right)^{t/5730}$$
$$10 = 100\left(\frac{1}{2}\right)^{t/5730}$$
$$\left(\frac{1}{2}\right)^{t/5730} = 0.1$$
$$\log\left(\frac{1}{2}\right)^{t/5730} = \log(0.1)$$
$$\frac{t}{5730}\log\left(\frac{1}{2}\right) = -1$$
$$t = \frac{-5730}{\log\left(\frac{1}{2}\right)}$$
$$t \approx 19035$$

The age of the wood is 19,035 years.

33. Use an exponential decay model $f(t) = ab^t$. Let $P = f(t)$ be the percentage remaining of the element at t years. The P-intercept is (0, 100). Every 100 years the amount of the element is halved so the base is $\frac{1}{2}$ and the exponent is $\frac{t}{100}$.

$$f(t) = 100\left(\frac{1}{2}\right)^{t/100}$$

Find $f^{-1}(0.01)$.

$$0.01 = 100\left(\frac{1}{2}\right)^{t/100}$$

$$\left(\frac{1}{2}\right)^{t/100} = 0.0001$$

$$\log\left(\frac{1}{2}^{t/100}\right) = \log(0.0001)$$

$$\frac{t}{100}\log\left(\frac{1}{2}\right) = -4$$

$$\frac{t}{100} = \frac{-4}{\log\left(\frac{1}{2}\right)}$$

$$t = \frac{-400}{\log\left(\frac{1}{2}\right)}$$

$$t \approx 1328.77$$

The tank must remain intact for 1329 years.

35. 3% account: $A = P(1 + 0.03)^t = P(1.03)^t$

6% account: $A = P(1 + 0.06)^t = P(1.06)^t$

$$\frac{P(1.06)^t}{P(1.03)^t} = 2$$

$$\frac{(1.06)^t}{(1.03)^t} = 2$$

$$\log\left(\frac{(1.06)^t}{(1.03)^t}\right) = \log(2)$$

$$\log\left((1.06)^t\right) - \log\left((1.03)^t\right) = \log(2)$$

$$t\log(1.06) - t\log(1.03) = \log(2)$$

$$t\left(\log(1.06) - \log(1.03)\right) = \log(2)$$

$$t = \frac{\log(2)}{\log(1.06) - \log(1.03)}$$

$$t \approx 24.14$$

There will be twice as much money in the 6% account as in the 3% account after 24.14 years.

37. a. Use the linear regression feature of a graphing calculator.
The linear model is
$L(t) = -5.51t + 122.84$.

Use the exponential regression feature of a graphing calculator. The exponential model is $E(t) = 177.7(0.876)^t$.

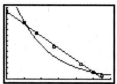

The linear model appears to fit the data better as shown in the graph above.

b.
$$45 = -5.51t + 122.84$$
$$45 - 122.84 = -5.51t$$
$$\frac{-77.84}{-5.51} = \frac{-5.51t}{-5.51}$$
$$14.13 \approx t$$

The bar of soap weighed 45 grams after 14 days of use.

c. The slope is -5.51. This means that the weight of the soap bar decreases by 5.51 grams each day.

d.
$$0 = -5.51t + 122.84$$
$$0 - 122.84 = -5.51t$$
$$\frac{-122.84}{-5.51} = \frac{-5.51t}{-5.51}$$
$$22.3 \approx t$$

There is no soap left after the 22nd day of use.
Answers may vary. Example:
Model breakdown has occurred, because in fact there was enough soap left on the 23rd day for the bar to break into two pieces.

39. $\dfrac{-25b^{3/8}}{40b^{2/5}} = \dfrac{-5b^{15/40}}{8b^{16/40}} = -\dfrac{5}{8b^{1/40}}$

This is an expression in one variable that involves exponents.

41. $4(6)^x - 31 = 180$

$$\frac{4(6)^x}{4} = \frac{211}{4}$$

$$\log(6)^x = \log\left(\frac{211}{4}\right)$$

$$\frac{x\log(6)}{\log(6)} = \frac{\log\left(\frac{211}{4}\right)}{\log(6)}$$

$$x \approx 2.2132$$

This is an exponential equation in one variable.

Homework 5.6

1. $\log_b(x) + \log_b(3x) = \log_b\left[(x)(3x)\right]$
$$= \log_b\left(3x^2\right)$$

3. $\log_b(8x) - \log_b(2) = \log_b\left(\dfrac{8x}{2}\right) = \log_b(4x)$

5. $4\log_b(t) + \log_b(5t) = \log_b(t)^4 + \log_b(5t)$
$$= \log_b\left[\left(t^4\right)(5t)\right]$$
$$= \log_b\left(5t^5\right)$$

7. $\log_b\left(3x^2\right) - 5\log_b(x) = \log_b\left(3x^2\right) - \log_b(x)^5$
$$= \log_b\left(\dfrac{3x^2}{x^5}\right)$$
$$= \log_b\left(\dfrac{3}{x^3}\right)$$

9. $2\log_b(3x) + 3\log_b\left(x^3\right)$
$$= \log_b(3x)^2 + \log_b\left(x^3\right)^3$$
$$= \log_b\left[\left(9x^2\right)\left(x^9\right)\right]$$
$$= \log_b\left(9x^{11}\right)$$

11. $3\log_b(2m) + 5\log_b\left(m^2\right) - \log_b(3m)$
$$= \log_b(2m)^3 + \log_b\left(m^2\right)^5 - \log_b(3m)$$
$$= \log_b\left[\dfrac{\left(8m^3\right)\left(m^{10}\right)}{3m}\right]$$
$$= \log_b\left(\dfrac{8m^{12}}{3}\right)$$

13. $\log_5(6x) + \log_5(x) = 2$
$$\log_5\left[(6x)(x)\right] = 2$$
$$5^2 = 6x^2$$
$$\dfrac{25}{6} = x^2$$
$$\left(\dfrac{25}{6}\right)^{1/2} = x$$
$$2.0412 \approx x$$

15. $\log_2(9x) - \log_2(3) = 5$
$$\log_2\left(\dfrac{9x}{3}\right) = 5$$
$$2^5 = \dfrac{9x}{3}$$
$$32 = 3x$$
$$\dfrac{32}{3} = x$$
$$10.6667 \approx x$$

17. $\log_7\left(w^2\right) + 2\log_7(3w) = 2$
$$\log_7\left(w^2\right) + \log_7(3w)^2 = 2$$
$$\log_7\left[\left(w^2\right)\left(9w^2\right)\right] = 2$$
$$7^2 = 9w^4$$
$$\dfrac{49}{9} = w^4$$
$$\left(\dfrac{49}{9}\right)^{1/4} = w$$
$$1.5275 \approx w$$

19. $\log\left(x^{13}\right) - 2\log\left(x^4\right) = 1$
$$\log\left(x^{13}\right) - \log\left(x^4\right)^2 = 1$$
$$\log\left(\dfrac{x^{13}}{x^8}\right) = 1$$
$$\log\left(x^5\right) = 1$$
$$10^1 = x^5$$
$$(10)^{1/5} = x$$
$$1.5849 \approx x$$

21. $3\log\left(x^2\right) + 4\log(2x) = 2$
$$\log\left(x^2\right)^3 + \log(2x)^4 = 2$$
$$\log\left[\left(x^6\right)\left(16x^4\right)\right] = 2$$
$$10^2 = 16x^{10}$$
$$\left(\dfrac{100}{16}\right)^{1/10} = x$$
$$1.2011 \approx x$$

23. $3\log_5\left(p^4\right) - 5\log_5\left(2p\right) = 3$

$\log_5\left(p^4\right)^3 - \log_5\left(2p\right)^5 = 3$

$\log_5\left(\dfrac{p^{12}}{32p^5}\right) = 3$

$5^3 = \dfrac{p^7}{32}$

$\left(4000\right)^{1/7} = p$

$3.2702 \approx p$

25. $\log_3\left(7\right) = \dfrac{\log(7)}{\log(3)} \approx 1.7712$

27. $\log_9\left(3.58\right) = \dfrac{\log(3.58)}{\log(9)} \approx 0.5804$

29. $\log_8\left(\dfrac{1}{70}\right) = \dfrac{\log\left(\dfrac{1}{70}\right)}{\log(8)} \approx -2.0431$

31. Graph $y = \log\left(x+5\right) + \log\left(x+2\right)$ and $y = 3 - x$. Then use "intersect."

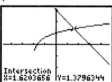

The x-coordinate, 1.6204, is the approximate solution of the equation $\log\left(x+5\right) + \log\left(x+2\right) = 3 - x$.

33. Given $\log_5\left(x+3\right) + \log_2\left(x+4\right) = -2x+9$, use the change of base formula to graph

$y = \dfrac{\log\left(x+3\right)}{\log 5} + \dfrac{\log\left(x+4\right)}{\log 2}$ and $y = -2x + 9$.

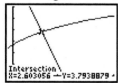

The x-coordinate, 2.6031, is the approximate solution of the equation $\log_5\left(x+3\right) - \log_2\left(x+4\right) = -2x+9$.

35. Given $\log_2\left(x+4\right) + \log_3\left(x+5\right) = 2^x + 1$, use the change of base formula to graph

$y = \dfrac{\log\left(x+4\right)}{\log 2} + \dfrac{\log\left(x+5\right)}{\log 3}$ and $y = 2^x + 1$.

The x-coordinates, -2.6876 and 1.6964, are the approximate solutions of the equation $\log_2\left(x+4\right) + \log_3\left(x+5\right) = 2^x + 1$.

37. $\dfrac{\log_2(x)}{\log_2(7)} = \log_7(x)$

39. $\dfrac{\log_b(r)}{\log_b(s)} = \log_s(r)$

41. $g\left(17\right) = \log_{12}\left(17\right) = \dfrac{\log\left(17\right)}{\log(12)} \approx 1.1402$

43. $g\left(8\right) = \log_{12}\left(8\right) = \dfrac{\log\left(8\right)}{\log(12)} \approx 0.8368$

45. All three students did the problem correctly.

47. $\log_b\left(b^2\right) = 2$

$\log_b\left(\dfrac{b^6}{b^4}\right) = \log_b(b^2) = 2$

$\log_b(b^6) = 6$

$\log_b(b^6) - \log_b(b^4) = \log_b\left(\dfrac{b^6}{b^4}\right) = \log_b(b^2) = 2$

$\dfrac{\log_b(b^6)}{\log_b(b^4)} = \dfrac{6}{4} = 1.5$

The expressions $\log_b(b^2)$, $\log_b(b^6) - \log_b(b^4)$, 2, and $\log_b\left(\dfrac{b^6}{b^4}\right)$ are all equal.

49. $\log_b(x) - \log_b(x) = \log_b\left(\dfrac{x}{x}\right) = \log_b(1) = 0$

51. a. $\log_2(x^3) + \log_2(x^5)$
$= \log_2[(x^3)(x^5)]$
$= \log_2(x^8)$

   b. $\log_2(x^3) + \log_2(x^5) = 7$
$\log_2[(x^3)(x^5)] = 7$
$\log_2(x^8) = 7$
$x^8 = 2^7$
$x = \left(2^7\right)^{1/8}$
$x \approx 1.8340$

   c. Answers may vary. Example:
   Simplifying an expression involves
   combining separate logarithms into one
   logarithm. Solving the equation involves
   simplifying and using the definition of a
   logarithm to modify the statement into an
   exponential equation that can then be
   solved.

   d. Answers may vary. Example:
   Simplifying an expression allows you to
   combine similar variables and reduce
   coefficients to their simplest forms,
   making the process of solution easier.

53. $\log_2\left(x^4\right) + \log_2\left(x^3\right)$
$= \log_2\left[\left(x^4\right)\left(x^3\right)\right]$
$= \log_2\left(x^7\right)$

55. $\log_2\left(x^4\right) + \log_2\left(x^3\right) = 4$
$\log_2\left[\left(x^4\right)\left(x^3\right)\right] = 4$
$\log_2\left(x^7\right) = 4$
$2^4 = x^7$
$\left(2^4\right)^{1/7} = x$
$1.4860 \approx x$

57. $2\log_9\left(x^3\right) - 3\log_9\left(2x\right) = 2$
$\log_9\left(x^3\right)^2 - \log_9\left(2x\right)^3 = 2$
$\log_9\left(\dfrac{x^6}{8x^3}\right) = 2$
$9^2 = \dfrac{x^3}{8}$
$9^2 = \dfrac{x^3}{8}$
$\left(9^2 \cdot 8\right)^{1/3} = x$
$8.6535 \approx x$

59. $2\log_9\left(x^3\right) - 3\log_9\left(2x\right)$
$= \log_9\left(x^3\right)^2 - \log_9\left(2x\right)^3$
$= \log_9\left(\dfrac{x^6}{8x^3}\right)$
$= \log_9\left(\dfrac{x^3}{8}\right)$

61. $\left(16b^{16}c^{-7}\right)^{1/4}\left(27b^{27}c^5\right)^{1/3}$
$= \left(16^{1/4}b^{16/4}c^{-7/4}\right)\left(27^{1/3}b^{27/3}c^{5/3}\right)$
$= \left(2b^4c^{-7/4}\right)\left(3b^9c^{5/3}\right)$
$= 6b^{13}c^{-\frac{21}{12}+\frac{20}{12}}$
$= 6b^{13}c^{-1/12}$
$= \dfrac{6b^{13}}{c^{1/12}}$

63. $3\log_b\left(2x^5\right) + 2\log_b\left(3x^4\right)$
$= \log_b\left(2x^5\right)^3 + \log_b\left(3x^4\right)^2$
$= \log_b\left[\left(8x^{15}\right)\left(9x^8\right)\right]$
$= \log_b\left(72x^{23}\right)$

65. Substitute $3x - 7$ for y in the second equation.
$y = -2x + 3$
$3x - 7 = -2x + 3$
$5x = 10$
$x = 2$
Solve for y when $x = 2$.
$y = 3x - 7 = 3(2) - 7 = -1$
The solution is $(2, -1)$.

67. Substitute $\log_2(x) + 2$ for y in the first equation.

$$y = \log_2\left(4x^2\right) - 3$$
$$\log_2(x) + 2 = \log_2\left(4x^2\right) - 3$$
$$\log_2(x) - \log_2\left(4x^2\right) = -5$$
$$\log_2\left(\frac{x}{4x^2}\right) = -5$$
$$2^{-5} = \frac{1}{4x}$$
$$\frac{1}{32} = \frac{1}{4x}$$
$$4x = 32$$
$$x = 8$$

Solve for y when $x = 8$.

$$y = \log_2(x) + 2$$
$$= \log_2(8) + 2$$
$$= \frac{\log(8)}{\log(2)} + 2$$
$$= 3 + 2 = 5$$

The solution is $(8, 5)$.

69. Substitute $\frac{2}{3}x - 2$ for y in the first equation.

$$2x - 3\left(\frac{2}{3}x - 2\right) = 6$$
$$2x - 2x + 6 = 6$$
$$6 = 6 \ \text{True}$$

The solution is all points on the line

$$y = \frac{2}{3}x - 2.$$
$$2x - 3y = 6$$
$$y = \frac{2}{3}x - 2$$

This is a dependent linear system of equations in two variables.

71.

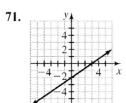

This is a linear equation in two variables.

Homework 5.7

1. $\ln(54.8) \approx 4.0037$

3. $\ln\left(\frac{1}{2}\right) \approx -0.6931$

5. $\ln(e^4) = 4$

7. $\ln(e) = \ln(e^1) = 1$

9. $\ln\left(\frac{1}{e}\right) = \ln(e^{-1}) = -1$

11. $\frac{1}{2}\ln(e^6) = \frac{1}{2}(6) = 3$

13. $e^{\ln 7} = 7$

15. $\ln(x) = 2$
$$e^2 = x$$
$$7.3891 \approx x$$

17. $\ln(p + 5) = 3$
$$e^3 = p + 5$$
$$e^3 - 5 = p$$
$$p \approx 15.0855$$

19. $7e^x = 44$
$$e^x = \frac{44}{7}$$
$$x = \ln\left(\frac{44}{7}\right)$$
$$x \approx 1.8383$$

21. $5\ln(3x) + 2 = 7$
$$\frac{5\ln(3x)}{5} = \frac{5}{5}$$
$$\ln(3x) = 1$$
$$3x = e^1$$
$$x = \frac{e}{3}$$
$$x \approx 0.9061$$

23. $4e^{3m-1} = 68$
$$e^{3m-1} = 17$$
$$3m - 1 = \ln(17)$$
$$3m = \ln(17) + 1$$
$$m = \frac{\ln(17) + 1}{3}$$
$$m \approx 1.2777$$

25. $e^{3x-5} \cdot e^{2x} = 135$

$$e^{5x-5} = 135$$
$$5x - 5 = \ln(135)$$
$$\frac{5x}{5} = \frac{\ln(135) + 5}{5}$$
$$x \approx 1.9811$$

27. $\qquad 3.1^x = 49.8$

$$\ln(3.1^x) = \ln(49.8)$$
$$x \ln(3.1) = \ln(49.8)$$
$$x = \frac{\ln(49.8)}{\ln(3.1)}$$
$$x \approx 3.4541$$

$$3(6^x) - 1 = 97$$
$$3(6^x) = 98$$
$$6^x = \frac{98}{3}$$

29. $\quad \ln(6^x) = \ln\left(\frac{98}{3}\right)$

$$x \ln(6) = \ln\left(\frac{98}{3}\right)$$
$$x = \frac{\ln\left(\frac{98}{3}\right)}{\ln(6)}$$
$$x \approx 1.9458$$

31. $5e^x - 20 = 2e^x + 67$

$$3e^x = 87$$
$$e^x = 29$$
$$x = \ln(29)$$
$$x \approx 3.3673$$

33. $\ln(4x) + \ln(3x^4) = \ln\left[(4x)(3x^4)\right] = \ln(12x^5)$

35. $\ln(25x^4) - \ln(5x^3) = \ln\left(\frac{25x^4}{5x^3}\right) = \ln(5x)$

37. $2\ln(w^4) + 3\ln(2w)$

$$= \ln(w^4)^2 + \ln(2w)^3$$
$$= \ln\left[(w^8)(8w^3)\right]$$
$$= \ln(8w^{11})$$

39. $3\ln(3x) - 2\ln(x^2)$

$$= \ln(3x)^3 - \ln(x^2)^2$$
$$= \ln(27x^3) - \ln(x^4)$$
$$= \ln\left(\frac{27x^3}{x^4}\right)$$
$$= \ln\left(\frac{27}{x}\right)$$

41. $3\ln(2k) + 4\ln(k^2) - \ln(k^7)$

$$= \ln(2k)^3 + \ln(k^2)^4 - \ln(k^7)$$
$$= \ln\left[\frac{(8k^3)(k^8)}{(k^7)}\right]$$
$$= \ln(8k^4)$$

43. $\quad \ln(3x) + \ln(x) = 4$

$$\ln[(3x)(x)] = 4$$
$$\ln(3x^2) = 4$$
$$3x^2 = e^4$$
$$x^2 = \frac{e^4}{3}$$
$$x = \left(\frac{e^4}{3}\right)^{1/2}$$
$$x \approx 4.2661$$

45. $\ln(4x^5) - 2\ln(x^2) = 5$

$$\ln(4x^5) - \ln(x^2)^2 = 5$$
$$\ln(4x^5) - \ln(x^4) = 5$$
$$\ln\left(\frac{4x^5}{x^4}\right) = 5$$
$$\ln(4x) = 5$$
$$4x = e^5$$
$$x = \frac{e^5}{4}$$
$$x \approx 37.1033$$

47. $2\ln(3x) + 2\ln\left(x^3\right) = 8$

$$\ln(3x)^2 + \ln\left(x^3\right)^2 = 8$$

$$\ln\left(9x^2\right) + \ln\left(x^6\right) = 8$$

$$\ln\left[\left(9x^2\right)\left(x^6\right)\right] = 8$$

$$\ln\left(9x^8\right) = 8$$

$$9x^8 = e^8$$

$$x^8 = \frac{e^8}{9}$$

$$x = \left(\frac{e^8}{9}\right)^{1/8}$$

$$x \approx 2.0654$$

49. $5\ln\left(2m\right) - 3\ln\left(m^4\right) = 7$

$$\ln\left(2m\right)^5 - \ln\left(m^4\right)^3 = 7$$

$$\ln\left(32m^5\right) - \ln\left(m^{12}\right) = 7$$

$$\ln\left(\frac{32m^5}{m^{12}}\right) = 7$$

$$\ln\left(\frac{32}{m^7}\right) = 7$$

$$\frac{32}{m^7} = e^7$$

$$\frac{1}{m^7} = \frac{e^7}{32}$$

$$m^{-7} = \frac{e^7}{32}$$

$$m = \left(\frac{e^7}{32}\right)^{-1/7}$$

$$m \approx 0.6036$$

51. Graph $y = e^x$ and $y = 5 - x$. Then use "intersect."

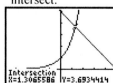

The x-coordinate, 1.3066, is the approximate solution of the equation $e^x = 5 - x$.

53. Graph $y = 3\ln(x + 2)$ and $y = -2x + 6$. Then use "intersect."

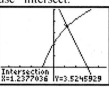

The x-coordinate, 1.2377 is the approximate solution of the equation $3\ln(x + 2) = -2x + 6$.

55. Graph $y = 3\ln(x + 3)$ and $y = 0.7x + 2$. Then use "intersect."

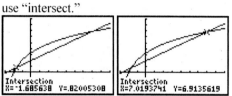

The x-coordinates, -1.6856 and 7.0194, are the approximate solutions of the equation $3\ln(x + 3) = 0.7x + 2$.

57. $f(x) = 4\ln(x)$

$$f\left(e^5\right) = 4\ln\left(e^5\right) = 4 \cdot 5 = 20$$

59. $f(x) = 4\ln(x)$

$$-8 = 4\ln\left(x\right)$$

$$\frac{-8}{4} = \frac{4\ln\left(x\right)}{4}$$

$$-2 = \ln(x)$$

$$x = e^{-2}$$

$$x \approx 0.1353$$

61. $ae^{bx} = c$

$$e^{bx} = \frac{c}{a}$$

$$bx = \ln\left(\frac{c}{a}\right)$$

$$x = \frac{\ln\left(\frac{c}{a}\right)}{b}$$

63. a.

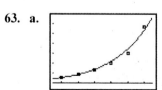

The function appears to be a reasonable model.

b. $f(0) = 29.89e^{0.45(0)} = 29.89$

There were about 30 laser incidents involving aircraft in 2000.

c.
$$f(t) = 29.89e^{0.45t}$$
$$60{,}000 = 29.89e^{0.45t}$$
$$\frac{60{,}000}{29.89} = e^{0.45t}$$
$$\ln\left(\frac{60{,}000}{29.89}\right) = 0.45t$$
$$\frac{\ln\left(\dfrac{60{,}000}{29.89}\right)}{0.45} = t$$
$$16.90 \approx t$$

There will be 60,000 laser incidents involving aircraft in 2017.

d. $f(18) = 29.89e^{0.45(18)} \approx 98{,}472$

67% of 98,472 is about 65,976. Cockpits will be illuminated about 65,976 times in 2018.

65. a. The time when the person bought it is $t = 0$.
$$70 + 137e^{-0.66(0)} = 70 + 137e^{0}$$
$$= 70 + 137$$
$$= 207$$

The temperature was 207°F when the coffee was purchased.

b. $180 = 70 + 137e^{-0.06t}$
$$110 = 137e^{-0.06t}$$
$$e^{-0.06t} = \frac{110}{137}$$
$$\ln\left(e^{-0.06t}\right) = \ln\left(\frac{110}{137}\right)$$
$$-0.06t \ln(e) = \ln\left(\frac{110}{137}\right)$$
$$-0.06t = \ln\left(\frac{110}{137}\right)$$
$$t = \frac{\ln\left(\dfrac{110}{137}\right)}{-0.06}$$
$$t \approx 3.6583$$

He will be able to drink the coffee in 3.66 minutes.

c.

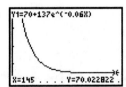

Looking at the graph of y we can see that as t gets larger, y approaches 70. So the temperature of the store is 70°F.

67. a. The poles are at $x = 10$ or $x = -10$
$$h(10) = 10\left(e^{0.03(10)} + e^{-0.03(10)}\right)$$
$$= 10\left(e^{0.3} + e^{-0.3}\right)$$
$$\approx 20.91$$
The poles are 20.91 feet high.

b. $h(6) = 10\left(e^{0.03(6)} + e^{-0.03(6)}\right)$
$$= 10\left(e^{0.18} + e^{-0.18}\right)$$
$$\approx 20.32$$
The cable is 20.32 feet high when it is 6 feet away from the center (or 4 feet to the left of the rightmost pole).

c. The shortest height happens in the center, where $x = 0$.
$$h(0) = 10\left(e^{0.03(0)} + e^{-0.03(0)}\right)$$
$$= 10(1 + 1)$$
$$= 10(2)$$
$$= 20$$
The minimum height of the cable is 20 feet.

69. Both students solve the equation correctly. The base of the logarithm used does not matter.

71. $3\ln(x) = 3\ln(x)$
$$\ln(x^7) - \ln(x^4) = \ln\left(\frac{x^7}{x^4}\right) = \ln(x^3) = 3\ln(x)$$
$$\frac{\ln(x^7)}{\ln(x^4)} = \log_{x^4}(x^7)$$
$$2\ln(x)\ln(x) = 2[\ln(x)]^2$$
$$\ln(x^3) = 3\ln(x)$$
$$\ln(3x) = \ln(3x)$$
$3\ln(x)$, $\ln(x^7) - \ln(x^4)$, and $\ln(x^3)$ are all equal.

73. $\ln(e) = 1$ because $e^1 = e$.

75. a. i.
$$3^x = 58$$
$$\ln\left(3^x\right) = \ln(58)$$
$$\frac{x\ln(3)}{\ln(3)} = \frac{\ln(58)}{\ln(3)}$$
$$x \approx 3.6960$$

ii.
$$3^x = 58$$
$$\log\left(3^x\right) = \log(58)$$
$$\frac{x\log(3)}{\log(3)} = \frac{\log(58)}{\log(3)}$$
$$x \approx 3.6960$$

iii. Both results are the same.

b. i. Answers may vary. Example:
$$5^x = 25$$
$$\ln\left(5^x\right) = \ln(25)$$
$$\frac{x\ln(5)}{\ln(5)} = \frac{\ln(25)}{\ln(5)}$$
$$x = 2$$

ii.
$$5^x = 25$$
$$\log\left(5^x\right) = \log(25)$$
$$\frac{x\log(5)}{\log(5)} = \frac{\log(25)}{\log(5)}$$
$$x = 2$$

iii. Both results are the same.

c. Answers may vary. Example:
We can take either the common logarithm or the natural logarithm to solve an exponential equation.

77. $\ln\left(x^8\right) - \ln\left(x^3\right) = \ln\left(\dfrac{x^8}{x^3}\right) = \ln\left(x^5\right)$

79.
$$\ln\left(x^8\right) - \ln\left(x^3\right) = 4$$
$$\ln\left(\frac{x^8}{x^3}\right) = 4$$
$$\ln\left(x^5\right) = 4$$
$$x^5 = e^4$$
$$x = \left(e^4\right)^{1/5}$$
$$x \approx 2.2255$$

81.
$$3e^x - 5 = 7$$
$$\frac{3e^x}{3} = \frac{12}{3}$$
$$e^x = 4$$
$$x = \ln(4)$$
$$x \approx 1.3863$$

83.
$$7 - 3(2t - 4) = 5t + 6$$
$$7 - 6t + 12 = 5t + 6$$
$$\frac{-11t}{-11} = \frac{-13}{-11}$$
$$t = \frac{13}{11}$$

85.
$$\frac{b^7}{b^3} = 16$$
$$b^4 = 16$$
$$b = (16)^{1/4}$$
$$b = \pm 2$$

87. Answers may vary. Example:
$$5x^2 - 2x + 1$$

89. Answers may vary. Example:
$$\log(2x - 1) = 5$$
$$10^{\log(2x-1)} = 10^5$$
$$2x - 1 = 10^5$$
$$2x = 10^5 + 1$$
$$x = \frac{10^5 + 1}{2}$$

91. Answers may vary. Example:
$$y = 3x + 2$$

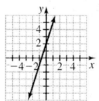

Chapter 5 Review

1. $f(2) = 3$

2. $f^{-1}(2) = 0$

3. $(f \circ g)(1) = f\left(g(1)\right) = f(2) = 3$

4. $(g \circ f)(1) = g(f(1)) = g(4) = 0$

5. $(g^{-1} \circ g)(3) = g^{-1}(g(3)) = g^{-1}(4) = 3$

6. $(f \circ g^{-1})(0) = f(g^{-1}(0)) = f(4) = 1$

7.

x	$g(x)$	$(f \circ g)(x)$
0	3	0
1	2	3
2	1	4
3	4	1
4	0	2

8. a. $(f \circ g)(x) = f(g(x))$
$= f(-4x + 7)$
$= 5(-4x + 7) - 2$
$= -20x + 35 - 2$
$= -20x + 33$

b. $(g \circ f)(x) = g(f(x))$
$= g(5x - 2)$
$= -4(5x - 2) + 7$
$= -20x + 8 + 7$
$= -20x + 15$

c. $(f \circ g)(3) = -20(3) + 33 = -27$

d. $(g \circ f)(3) = -20(3) + 15 = -45$

9. a. $(f \circ g)(x) = f(g(x))$
$= f(2x - 4)$
$= 4(2)^{2x-4}$

b. $(g \circ f)(x) = g(f(x))$
$= g(4(2)^x)$
$= 2(4(2)^x) - 4$
$= 8(2)^x - 4$

c. $(f \circ g)(3) = 4(2)^{2(3)-4} = 4(2)^2 = 16$

d. $(g \circ f)(3) = 8(2)^3 - 4 = 64 - 4 = 60$

10. a. $(f \circ g)(x) = f(g(x))$
$= f(x + 6)$
$= \log_3(x + 6)$

b. $(g \circ f)(x) = g(f(x))$
$= g(\log_3(x))$
$= \log_3(x) + 6$

c. $(f \circ g)(3) = \log_3(3 + 6) = \log_3(9) = 2$

d. $(g \circ f)(3) = \log_3(3) + 6 = 1 + 6 = 7$

11. Answers may vary. Example:
$g(x) = x - 5$ and $f(x) = e^x$
Check: $f(g(x)) = f(x - 5) = e^{x-5}$

12. a. $f(n) = 8n$ and $g(d) = 0.06d$

b. $g(f(n)) = f(8n) = 0.06(8n) = 0.48n$

c. $g(f(7)) = 0.48(7) = 3.36$
This means the sales tax on the purchase of 7 books is $3.36.

13.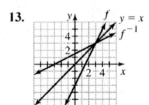

14.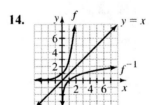

15. a.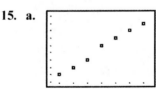

The plotted data lie mostly in a straight line. Therefore, a linear model better fits the data.

b. Using the linear regression feature, $f(t) = 1.21t + 2.86$.

c. $f(18) = 1.21(18) + 2.86 = 24.64$

This means there will be about 25 million background checks in 2018.

d. First find $f^{-1}(x)$.

Replace $f(t)$ with p and solve for t.
$$p = 1.21t + 2.86$$
$$\frac{p - 2.86}{1.21} = t$$
$$0.83p - 2.36 \approx t$$

An approximate equation is
$t = 0.83p - 2.36$.

Replace t with $f^{-1}(p)$.

$f^{-1}(p) = 0.83p - 2.36$

Now replace p with 18.

$f^{-1}(18) = 0.83(18) - 2.36 = 12.58$

In 2013, there were about 18 million background checks.

e. $C(n) = 15n$

f. $h(t) = (C \circ f)(t)$

Answers may vary. Example:
$C(t)$ is the total cost (in millions of dollars) of n million background checks. Since $f(t)$ is the number (in millions) of FBI background checks for firearm purchases in the year that is t years since 2000, $(C \circ f)(t)$ is the total cost (in millions of dollars) of FBI background checks for all firearm purchases in the year that is t years since 2000. So $h(t) = (C \circ f)(t)$.

g. $h(t) = (C \circ f)(t)$
$$= C(f(t))$$
$$= C(1.21t + 2.86)$$
$$= 15(1.21t + 2.86)$$
$$= 18.15t + 42.90$$

h. $h(17) = 18.15(17) + 42.90 = 351.45$

In 2017, the total cost of background checks will be about $351 million.

16. a. $f(x) = 3x$

Replace $f(x)$ with y.
$y = 3x$
Solve for x.
$$x = \frac{1}{3}y$$
Replace x with $f^{-1}(y)$.
$$f^{-1}(y) = \frac{1}{3}y$$
Write in terms of x.
$$f^{-1}(x) = \frac{1}{3}x$$

b. $(f^{-1} \circ f)(x) = f^{-1}(f(x))$
$$= f^{-1}(3x)$$
$$= \frac{1}{3}(3x)$$
$$= x$$

c. $(f \circ f^{-1})(x) = f(f^{-1}(x))$
$$= f\left(\frac{1}{3}x\right)$$
$$= 3\left(\frac{1}{3}x\right)$$
$$= x$$

17. a. $f(x) = \frac{5}{6}x - 2$

Replace $f(x)$ with y.
$$y = \frac{5}{6}x - 2$$
Solve for x.
$$6y = 5x - 12$$
$$6y + 12 = 5x$$
$$\frac{6y + 12}{5} = x$$
$$\frac{6}{5}y + \frac{12}{5} = x$$
Replace x with $f^{-1}(y)$.
$$f^{-1}(y) = \frac{6}{5}y + \frac{12}{5}$$
Write in terms of x.
$$f^{-1}(x) = \frac{6}{5}x + \frac{12}{5}$$

b. $\left(f^{-1} \circ f\right)(x) = f^{-1}\left(f(x)\right)$

$\qquad = f^{-1}\left(\dfrac{5}{6}x - 2\right)$

$\qquad = \dfrac{6}{5}\left(\dfrac{5}{6}x - 2\right) + \dfrac{12}{5}$

$\qquad = x - \dfrac{12}{5} + \dfrac{12}{5}$

$\qquad = x$

c. $\left(f \circ f^{-1}\right)(x) = f\left(f^{-1}(x)\right)$

$\qquad = f\left(\dfrac{6}{5}x + \dfrac{12}{5}\right)$

$\qquad = \dfrac{5}{6}\left(\dfrac{6}{5}x + \dfrac{12}{5}\right) - 2$

$\qquad = x + 2 - 2$

$\qquad = x$

18. $\log_5(25) = 2$, since $5^2 = 25$.

19. $\log(100,000) = 5$, since $10^5 = 100,000$.

20. $\log_3\left(\dfrac{1}{9}\right) = -2$, since $3^{-2} = \dfrac{1}{9}$.

21. $\ln\left(\dfrac{1}{e^3}\right) = \ln(e^{-3}) = -3$.

22. $\log_4\left(\sqrt[3]{4}\right) = \dfrac{1}{3}$, since $4^{\frac{1}{3}} = \sqrt[3]{4}$.

23. $\log_3(7) = \dfrac{\log(7)}{\log(3)} \approx 1.7712$

24. $\ln(5) \approx 1.6094$

25. $\log_b(b^7) = 7$, since $b^7 = b^7$.

26. $h(x) = 3^x$

$\quad h^{-1}(x) = \log_3(x)$

27. $h(x) = \log(x)$

$\quad h^{-1}(x) = 10^x$

28.

29. $\log_d(k) = x$

30. $y^r = w$

31. $6(2)^x = 30$

$\qquad 2^x = 5$

$\qquad \log(2^x) = \log(5)$

$\qquad x\log(2) = \log(5)$

$\qquad x = \dfrac{\log(5)}{\log(2)}$

$\qquad x \approx 2.3219$

32. $\log_3(x) = -4$

$\qquad x = 3^{-4}$

$\qquad x = \dfrac{1}{81}$

33. $4.3(9.8)^x - 3.3 = 8.2$

$\qquad 4.3(9.8)^x = 11.5$

$\qquad 9.8^x = \dfrac{11.5}{4.3}$

$\qquad \log(9.8^x) = \log\left(\dfrac{11.5}{4.3}\right)$

$\qquad x\log(9.8) = \log\left(\dfrac{11.5}{4.3}\right)$

$\qquad x = \dfrac{\log\left(\dfrac{11.5}{4.3}\right)}{\log(9.8)}$

$\qquad x \approx 0.4310$

34. $\log_b(83) = 6$

$\qquad b^6 = 83$

$\qquad b = 83^{\frac{1}{6}}$

$\qquad b \approx 2.0886$

35. $5\log_{32}(m) - 3 = -1$

$\qquad \log_{32}(m)^5 = 2$

$\qquad m^5 = 32^2$

$\qquad m = \left(32^2\right)^{1/5}$

$\qquad m = 4$

36.
$$5(4)^{3r-7} = 40$$
$$4^{3r-7} = 8$$
$$\log(4^{3r-7}) = \log(8)$$
$$(3r-7)\log(4) = \log(8)$$
$$3r - 7 = \frac{\log(8)}{\log(4)}$$
$$3r = \frac{\log(8)}{\log(4)} + 7$$
$$r = \frac{\frac{\log(8)}{\log(4)} + 7}{3}$$
$$r \approx 2.8333$$

37.
$$2^{4t} \cdot 2^{3t-5} = 94$$
$$2^{7t-5} = 94$$
$$\log\left(2^{7t-5}\right) = \log(94)$$
$$\left(7t - 5\right)\log(2) = \log(94)$$
$$7t - 5 = \frac{\log(94)}{\log(2)}$$
$$7t = \frac{\log(94)}{\log(2)} + 5$$
$$t = \frac{\frac{\log(94)}{\log(2)} + 5}{7}$$
$$t \approx 1.6507$$

38. $x = 4$ because $y = \log_2(x)$ and $y = -\frac{3}{4}x + 5$ intersect at $x = 4$.

39. $x = 0$ because $y = 2^x - 3$ is -2 when $x = 0$.

40. The solution to the system $y = \log_2(x)$ and $y = -\frac{3}{4}x + 5$ is the point of intersection on the graph, $(4, 2)$.

41. $f(x) = 3^x$
$f(4) = 3^4 = 81$

42. $f(x) = 3^x$, so $f^{-1}(x) = \log_3(x)$.
$$f^{-1}(25) = \log_3(25) = \frac{\log(25)}{\log(3)} \approx 2.9299$$

43. $f(x) = 3^x$
$$6 = 3^x$$
$$\log(6) = \log\left(3^x\right)$$
$$\log(6) = x\log(3)$$
$$\frac{\log(6)}{\log(3)} = x$$
$$1.6309 \approx x$$

44. $f(x) = 3^x$, so $f^{-1}(x) = \log_3(x)$.
Substitute using $f^{-1}(x) = 6$.
$$6 = \log_3(x)$$
$$3^6 = x$$
$$729 = x$$

45. a. Use an exponential model $f(t) = ab^t$. Let $V = f(t)$, where V is the value of the account. The V-intercept is (0, 8000), so $a = 8000$. Since the interest rate is 5%, the base is $b = 1.05$. $f(t) = 8000(1.05)^t$.

b. $f(9) = 8000(1.05)^9 \approx 12410.63$
The balance in the account after 9 years is $12,410.63.

c. The balance will have doubled when it is $16,000. Find $f^{-1}(16,000)$
$$16,000 = 8000(1.05)^t$$
$$1.05^t = 2$$
$$\log(1.05^t) = \log(2)$$
$$t\log(1.05) = \log(2)$$
$$t = \frac{\log(2)}{\log(1.05)}$$
$$t \approx 14.2067$$
The balance will be doubled in 14.2 years.

46. a. Use an exponential model $f(t) = ab^t$. Let $n = f(t)$ be the number of leaves. The n-intercept is (0, 30), so $a = 30$. The number of leaves quadruples every week so the base $b = 4$.
$$f(t) = 30(4)^t$$

b. $f(5) = 30(4)^5 = 30720$
There are 30,720 leaves on the tree 5 weeks after April 1.

c.
$$f^{-1}(10000)$$
$$100000 = 30(4)^t$$
$$4^t = \frac{100000}{30}$$
$$\log(4^t) = \log\left(\frac{10000}{3}\right)$$
$$t\log(4) = \log\left(\frac{10000}{3}\right)$$
$$t = \frac{\log\left(\frac{10000}{3}\right)}{\log(4)}$$
$$t \approx 5.85$$

There are 100,000 leaves on the tree 6 weeks after April 1.

47. a. Exponential regression on a graphing calculator yields the equation
$$f(t) = 0.12(1.083)^t .$$

b. The x-intercept is 0.12. The national health spending in 1970 was 0.12 trillion (120 billion) dollars.

c. The percentage rate of growth is 8.3% per year.

d. $f(49) = 0.12(1.083)^{49} \approx 5.97$

The national health spending in 2019 ($t = 49$) will be about 6.0 trillion dollars.

e.
$$f(t) = 0.12(1.083)^t$$
$$5 = 0.12(1.083)^t$$
$$\frac{5}{0.12} = \frac{0.12(1.083)^t}{0.12}$$
$$\log\left(\frac{5}{0.12}\right) = \log(1.083)^t$$
$$\log\left(\frac{5}{0.12}\right) = t\log(1.083)$$
$$\frac{\log\left(\frac{5}{0.12}\right)}{\log(1.083)} = t$$
$$46.78 \approx t$$

National health spending will be 5 trillion dollars in $1970 + 47 = 2017$.

48. a. Exponential regression on a graphing calculator yields the equation
$$f(n) = 9.33(1.31)^n .$$

b. The base is 1.31. As each cassette is added to the bag, the length increases by 31%.

c. The coefficient is 9.33. The initial length of the rubber band is 9.33 inches.

d. $f(8) = 9.33(1.31)^8 \approx 80.92$

The rubber band is stretched to 80.92 inches with 8 cassettes. Answers may vary. Example: There are two scenarios which might cause model breakdown. The rubber band reaches a point where it can stretch no farther, or the rubber band breaks.

e.
$$f^{-1}(139)$$
$$139 = 9.33(1.31)^n$$
$$1.31^n = \frac{139}{9.33}$$
$$\log(1.31^n) = \log\left(\frac{139}{9.33}\right)$$
$$n\log(1.31) = \log\left(\frac{139}{9.33}\right)$$
$$n = \frac{\log\left(\frac{139}{9.33}\right)}{\log(1.31)}$$
$$n \approx 10.00$$

It would take 10 cassettes to stretch the rubber band to 139 inches. If model breakdown occurs with 8 cassettes, then it definitely occurs with 10. Either the rubber band has stopped stretching with the addition of the last two cassettes, or the rubber band has broken.

49. Use an exponential decay model $f(t) = ab^t$. Every 5.3 years the amount of cobalt-60 is halved, so the base is $\frac{1}{2}$ and the exponent is $\frac{t}{5.3}$. 15% of the cobalt-60 remains, so $f(t) = 15$:

$$f(t) = 100\left(\frac{1}{2}\right)^{\frac{t}{5.3}}$$

$$15 = 100\left(\frac{1}{2}\right)^{\frac{t}{5.3}}$$

$$\left(\frac{15}{100}\right) = \left(\frac{1}{2}\right)^{\frac{t}{5.3}}$$

$$\log\left(\frac{15}{100}\right) = \left(\frac{t}{5.3}\right)\log\left(\frac{1}{2}\right)$$

$$(5.3)\frac{\log\left(\frac{15}{100}\right)}{\log\left(\frac{1}{2}\right)} = \frac{t}{5.3}(5.3)$$

$$14.51 \approx t$$

About 15% of the cobalt-60 remains after 14.5 years.

50. $\log_b(p) + \log_b(6p) - \log_b(2p)$

$$= \log_b[p(6p)] - \log_b(2p)$$

$$= \log_b(6p^2) - \log_b(2p)$$

$$= \log_b\left(\frac{6p^2}{2p}\right)$$

$$= \log_b(3p)$$

51. $3\log_b(2x) + 2\log_b(3x)$

$$= \log_b(2x)^3 + \log_b(3x)^2$$

$$= \log_b[(2x)^3(3x)^2]$$

$$= \log_b[72x^5]$$

52. $4\log_b\left(x^2\right) - 2\log_b\left(x^5\right)$

$$= \log_b\left(x^2\right)^4 - \log_b\left(x^5\right)^2$$

$$= \log_b\left(\frac{x^8}{x^{10}}\right)$$

$$= \log_b\left(\frac{1}{x^2}\right)$$

53. $\dfrac{\log_b(w)}{\log_b(y)} = \log_y(w)$

54. $\log_b(b^5) - \log_b(b^2) = 5 - 2 = 3$

$$3 = 3$$

$$\frac{\log_b(b^5)}{\log_b(b^2)} = \frac{5}{2}$$

$$\log_b(b^3) = 3$$

$$\log_b(b^5) = 5$$

$$\log_b\left(\frac{b^5}{b^2}\right) = \log_b\left(b^3\right) = 3$$

The expressions

$$\log_b(b^5) - \log_b(b^2) = 5 - 2 = 3, 3,$$

$$\log_b(b^3) = 3, \text{ and } \log_b\left(\frac{b^5}{b^2}\right) = \log_b\left(b^3\right) = 3$$

are all equal.

55. $2\log_9(3w) + 3\log_9\left(w^2\right) = 5$

$$\log_9(3w)^2 + \log_9\left(w^2\right)^3 = 5$$

$$\log_9\left[\left(9w^2\right)\left(w^6\right)\right] = 5$$

$$\log_9\left(9w^8\right) = 5$$

$$9w^8 = 9^5$$

$$w^8 = \frac{9^5}{9}$$

$$w = \left(9^4\right)^{1/8}$$

$$w = \pm 3$$

Substitute each answer into the original equation. Because the equation is defined only for $w = 3$, the only solution is 3.

56. $5\log_6(2x) - 3\log_6(4x) = 2$

$$\log_6(2x)^5 - \log_6(4x)^3 = 2$$

$$\log_6\left(\frac{32x^5}{64x^3}\right) = 2$$

$$\log_6\left(\frac{x^2}{2}\right) = 2$$

$$\frac{x^2}{2} = 6^2$$

$$x^2 = 2 \cdot 36$$

$$x = (72)^{1/2}$$

$$x \approx 8.4853$$

57. $3\ln(4x) + 2\ln(2x)$

$$= \ln(4x)^3 + \ln(2x)^2$$

$$= \ln\left[\left(64x^3\right)\left(4x^2\right)\right]$$

$$= \ln\left(256x^5\right)$$

58. $\ln\left(2m^7\right) - 4\ln\left(m^3\right) + 3\ln\left(m^2\right)$

$= \ln\left(2m^7\right) - \ln\left(m^3\right)^4 + \ln\left(m^2\right)^3$

$= \ln\left[\dfrac{\left(2m^7\right)\left(m^6\right)}{\left(m^{12}\right)}\right]$

$= \ln\left(2m\right)$

59. $4e^x = 75$

$e^x = \dfrac{75}{4}$

$x = \ln\left(\dfrac{75}{4}\right) \approx 2.9312$

60. $-3\ln\left(p\right) + 7 = 1$

$-3\ln\left(p\right) = -6$

$\ln(p) = 2$

$e^2 = p$

$7.3891 \approx p$

61. $3\ln\left(t^5\right) - 5\ln\left(2t\right) = 7$

$\ln\left(t^5\right)^3 - \ln\left(2t\right)^5 = 7$

$\ln\left(\dfrac{t^{15}}{32t^5}\right) = 7$

$\dfrac{t^{10}}{32} = e^7$

$t^{10} = 32e^7$

$t = \left(32e^7\right)^{1/10}$

$t \approx 2.8479$

Chapter 5 Test

1. a. $(f \circ g)(x) = f\left(g(x)\right)$

$= f\left(2x - 5\right)$

$= 3\left(2x - 5\right) + 4$

$= 6x - 15 + 4$

$= 6x - 11$

b. $(g \circ f)(x) = g\left(f(x)\right)$

$= g\left(3x + 4\right)$

$= 2\left(3x + 4\right) - 5$

$= 6x + 8 - 5$

$= 6x + 3$

c. $(f \circ g)(2) = 6(2) - 11 = 1$

d. $(g \circ f)(2) = 6(2) + 3 = 15$

2. a. $(f \circ g)(x) = f\left(g(x)\right) = f(x - 4) = 3^{x-4}$

b. $(g \circ f)(x) = g\left(f(x)\right) = g\left(3^x\right) = 3^x - 4$

c. $(f \circ g)(2) = 3^{2-4} = 3^{-2} = \dfrac{1}{9}$

d. $(g \circ f)(2) = 3^2 - 4 = 5$

3. $(f \circ g)(1) = f\left(g(1)\right) = f(4) = 2$

4. $(g \circ f)(1) = g\left(f(1)\right) = g(0) = 5$

5. $g^{-1}(3) = 2$

6.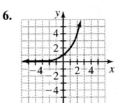

7.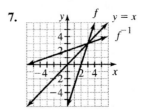

8. a. Create a scattergram to determine the type of regression to use.

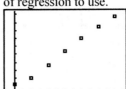

The plotted data lie mostly in a straight line. Therefore, a linear model fits the data. Using the linear regression feature on a graphing calculator, $f(t) = 3.75t + 44.21$.

b. $f(11) = 3.75(11) + 44.21 = 85.46$

An adult ticket cost about $85.46 in 2011.

c.   Replace $f(t)$ with p and solve for t.

$$p = 3.75t + 44.21$$
$$p - 44.21 = 3.75t$$
$$\frac{p - 44.21}{3.75} = \frac{3.75t}{3.75}$$
$$0.27p - 11.79 \approx t$$

An approximate equation is
$t = 0.27p - 11.79$.

Replace t with $f^{-1}(p)$.

$$f^{-1}(p) = 0.27p - 11.79$$

d. $f^{-1}(110) = 0.27(110) - 11.79 = 17.91$

An adult one-day ticket will be \$110 in
$2000 + 18 = 2018$.

e. $S(d) = 0.065d$

f. $h(t) = (S \circ f)(t)$

Answers may vary. Example:

$S(t)$ is the sales tax (in dollars) on a ticket

worth d dollars. Since $f(t)$ is the price (in

dollars) of an adult one-day ticket at t years

since 2000, $(S \circ f)(t)$ is the sales tax (in

dollars) on an adult one-day ticket at t

years since 2000. So $h(t) = (S \circ f)(t)$.

g. $h(t) = (S \circ f)(t)$
$$= S(3.75t + 44.21)$$
$$= 0.065(3.75t + 44.21)$$
$$\approx 0.24t + 2.87$$

An approximate equation is
$h(t) = 0.24t + 2.87$.

h. $h(17) = 0.24(17) + 2.87 = 6.95$

This means that in 2017, the sales tax on
an adult one-day ticket will be \$6.95.

9. $g(x) = \dfrac{2x - 9}{5}$

Replace $g(x)$ with y.

$$y = \frac{2x - 9}{5}$$

Solve for x:

$$5y = 2x - 9$$
$$5y + 9 = 2x$$
$$\frac{5y + 9}{2} = x$$
$$\frac{5}{2}y + \frac{9}{2} = x$$

Replace x with $g^{-1}(y)$.

$$g^{-1}(y) = \frac{5}{2}y + \frac{9}{2}$$

Write in terms of x.

$$g^{-1}(x) = \frac{5}{2}x + \frac{9}{2}$$

10. $\log_2(16) = 4$, since $2^4 = 16$.

11. $\log_4\left(\dfrac{1}{64}\right) = -3$, since $4^{-3} = \dfrac{1}{64}$.

12. $\log_7(10) = \dfrac{\log(10)}{\log(7)} \approx 1.1833$

13. $\log(0.1) = -1$, since $10^{-1} = 0.1$.

14. $\log_b(\sqrt{b}) = \dfrac{1}{2}$, since $b^{\frac{1}{2}} = \sqrt{b}$.

15. $\ln\left(\dfrac{1}{e^2}\right) = -2$, since $e^{-2} = \dfrac{1}{e^2}$.

16. $h(x) = 4^x$

$$h^{-1}(x) = \log_4(x)$$

17. $f(x) = \log_5(x)$

$$f^{-1}(x) = 5^x$$

18. $\log_s(w) = k$

19. $c^d = a$

20. $\log_b(50) = 4$
$$b^4 = 50$$
$$b = (50)^{\frac{1}{4}}$$
$$b \approx 2.6591$$

21. $6(2)^x - 9 = 23$

$$6(2)^x = 32$$

$$2^x = \frac{32}{6}$$

$$\log(2^x) = \log\left(\frac{32}{6}\right)$$

$$x\log(2) = \log\left(\frac{32}{6}\right)$$

$$x = \frac{\log\left(\frac{32}{6}\right)}{\log(2)}$$

$$x \approx 2.4150$$

22. $\log_4(7p + 5) = -\frac{3}{2}$

$$7p + 5 = 4^{-3/2}$$

$$7p = 4^{-3/2} - 5$$

$$p = \frac{4^{-3/2} - 5}{7}$$

$$= -0.6964$$

23. Graph $y = 4^x - 8$ and $y = -\frac{1}{2}x + 3$. Then use "intersect."

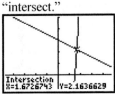

The x-coordinate, 1.67, is the approximate solution of the equation

$$4^x - 8 = -\frac{1}{2}x + 3.$$

24. a. Create a scattergram to determine the type of model to use.

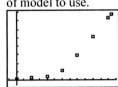

The scattergram shows that the points "bend" and that an exponential function will model the data.
Answers may vary depending on the points chosen. Example:

Use the exponential model $y = ab^t$ with 1950 as year 0. This provides the data point $(0, 600)$, so that $a = 600$.
Solve for b using $(60, 36,640)$.

$$y = ab^t$$

$$36,640 = 600(b)^{60}$$

$$\frac{36,640}{600} = (b)^{60}$$

$$\left(\frac{36,640}{600}\right)^{1/60} = b$$

$$1.071 \approx b$$

The model is $f(t) = 600(1.071)^t$. (Your equation may be slightly different if you chose different points.)

b. $(0, 600)$. The tuition in 1950 was about $600. (Your answer may be slightly different if you chose different points.)

c. The tuition rate is growing at a rate of about 7.1% per year. (Your answer may be slightly different if you chose different points.)

d. $f(67) = 600(1.071)^{67} \approx 59,435$

According to this model, the tuition in 2017 is predicted to be $59,435. (Your answer may be slightly different if you chose different points.)

e.
$$f(t) = 600(1.071)^t$$

$$80,000 = 600(1.071)^t$$

$$\frac{80,000}{600} = \frac{600(1.071)^t}{600}$$

$$\log\left(\frac{80,000}{600}\right) = \log(1.071)^t$$

$$\frac{\log\left(\frac{80,000}{600}\right)}{\log(1.071)} = \frac{t\log(1.071)}{\log(1.071)}$$

$$71.33 \approx t$$

According to this model, the tuition is predicted to be $80,000 in $1950 + 71 = 2021$. (Your answer may be slightly different if you chose different points.)

25. Use an exponential decay model $f(t) = ab^t$. Every 5730 years the amount of carbon-14 is halved so the base is $\frac{1}{2}$ and the exponent is

$\frac{t}{5730}$. 78.04% of the carbon-14 remains, so $f(t) = 78.04$:

$$f(t) = 100\left(\frac{1}{2}\right)^{t/5730}$$

$$78.04 = 100\left(\frac{1}{2}\right)^{t/5730}$$

$$\left(\frac{78.04}{100}\right) = \left(\frac{1}{2}\right)^{t/5730}$$

$$\log\left(\frac{78.04}{100}\right) = \left(\frac{t}{5730}\right)\log\left(\frac{1}{2}\right)$$

$$(5730)\frac{\log\left(\dfrac{78.04}{100}\right)}{\log\left(\dfrac{1}{2}\right)} = \frac{t}{5730}(5760)$$

$$2049.70 \approx t$$

The mummy is about 2050 years old.

26. $\log_b(x^3) + \log_b(5x)$

$= \log_b[x^3(5x)]$

$= \log_b(5x^4)$

27. $3\log_b\left(4p^2\right) - 2\log_b\left(8p^5\right) + \log_b\left(2p^4\right)$

$= \log_b\left(4p^2\right)^3 - \log_b\left(8p^5\right)^2 + \log_b\left(2p^4\right)$

$= \log_b\left[\dfrac{\left(64p^6\right)\left(2p^4\right)}{64p^{10}}\right]$

$= \log_b(2)$

28. $\log_3(x) + \log_3(2x) = 5$

$\log_3[x(2x)] = 5$

$\log_3(2x^2) = 5$

$2x^2 = 3^5$

$x^2 = \dfrac{243}{2}$

$x = \left(\dfrac{243}{2}\right)^{1/2}$

$x \approx 11.0227$

29. $2\log_4\left(x^4\right) - 3\log_4(3x) = 3$

$\log_4\left(x^4\right)^2 - \log_4(3x)^3 = 3$

$\log_4\left(x^8\right) - \log_4\left(27x^3\right) = 3$

$\log_4\left(\dfrac{x^8}{27x^3}\right) = 3$

$\log_4\left(\dfrac{x^5}{27}\right) = 3$

$\dfrac{x^5}{27} = 4^3$

$x^5 = 27 \cdot 4^3$

$x = \left(27 \cdot 4^3\right)^{1/5}$

$x \approx 4.4413$

30. $2\ln(5w) + 3\ln\left(w^6\right) = \ln(5w)^2 + \ln\left(w^6\right)^3$

$= \ln\left[\left(25w^2\right)\left(w^{18}\right)\right]$

$= \ln\left(25w^{20}\right)$

31. $2e^{3x-1} = 54$

$e^{3x-1} = 27$

$\ln\left(e^{3x-1}\right) = \ln(27)$

$(3x-1)\ln e = \ln(27)$

$3x - 1 = \ln(27)$

$3x = \ln(27) + 1$

$x = \dfrac{\ln(27) + 1}{3}$

$x \approx 1.4319$

32. $7\ln(x-2) - 1 = 4$

$7\ln(x-2) = 5$

$\ln(x-2) = \dfrac{5}{7}$

$x - 2 = e^{5/7}$

$x = e^{5/7} + 2$

$x \approx 4.0427$

Cumulative Review of Chapters 1–5

1. $2(4)^{5x-1} = 17$

$$(4)^{5x-1} = \frac{17}{2}$$

$$\log(4^{5x-1}) = \log\left(\frac{17}{2}\right)$$

$$(5x-1)\log(4) = \log\left(\frac{17}{2}\right)$$

$$5x - 1 = \frac{\log\left(\frac{17}{2}\right)}{\log(4)}$$

$$5x = \frac{\log\left(\frac{17}{2}\right)}{\log(4)} + 1$$

$$x = \frac{\frac{\log\left(\frac{17}{2}\right)}{\log(4)} + 1}{5}$$

$$x \approx 0.5087$$

2. $\log_3(x-5) = 4$

$$x - 5 = 3^4$$
$$x = 81 + 5$$
$$x = 86$$

3. $3b^7 - 18 = 7$

$$3b^7 = 25$$
$$b^7 = \frac{25}{3}$$
$$b = \left(\frac{25}{3}\right)^{1/7}$$
$$b \approx 1.3538$$

4. $8 + 2e^x = 15$

$$2e^x = 7$$
$$e^x = \frac{7}{2}$$
$$\ln(e^x) = \ln\left(\frac{7}{2}\right)$$
$$x = \ln\left(\frac{7}{2}\right)$$
$$x \approx 1.2528$$

5. $4\log_5(3x^2) + 3\log_5(6x^4) = 3$

$$\log_5(3x^2)^4 + \log_5(6x^4)^3 = 3$$
$$\log_5(81x^8) + \log_5(216x^{12}) = 3$$
$$\log_5[(81x^8)(216x^{12})] = 3$$
$$\log_5(17496x^{20}) = 3$$
$$17496x^{20} = 5^3$$
$$x^{20} = \frac{125}{17496}$$
$$x = \pm\left(\frac{125}{17496}\right)^{1/20}$$
$$x \approx \pm 0.7811$$

6. $7 - 3(4w - 2) = 2(3w + 5) - 4(2w + 1)$

$$7 - 12w + 6 = 6w + 10 - 8w - 4$$
$$13 - 12w = -2w + 6$$
$$7 = 10w$$
$$\frac{7}{10} = w$$

7. $x = 1$ because $y = 3^x$ and $y = 9\left(\frac{1}{3}\right)^x$ intersect at $x = 1$.

8. $x = 2$ because $y = 9\left(\frac{1}{3}\right)^x$ and $y = x - 1$ intersect at $x = 2$.

9. Substitute $x = 2y - 5$ into $4x - 5y = -14$.

$$4(2y - 5) - 5y = -14$$
$$8y - 20 - 5y = -14$$
$$3y = 6$$
$$y = 2$$

Find x.

$$x = 2(2) - 5 = -1$$

The solution is $(-1, 2)$.

10. Simplify $3(2 - 4x) = -10 - 2y$.

$$3(2 - 4x) = -10 - 2y$$
$$6 - 12x = -10 - 2y$$
$$-12x + 2y = -16$$

Multiply the second equation by 6 and add.

$$-12x + 2y = -16$$
$$\underline{12x - 18y = -48}$$
$$-16y = -64$$
$$y = 4$$

Find x.
$$2x - 3(4) = -8$$
$$2x - 12 + 12 = -8 + 12$$
$$2x = 4$$
$$x = 2$$
The solution is $(2, 4)$.

11. $8x - 3 \geq -3(4x - 5)$
$$8x - 3 \geq -12x + 15$$
$$20x \geq 18$$
$$x \geq \frac{18}{20}$$
$$x \geq \frac{9}{10}$$
$$\left[\frac{9}{10}, \infty\right)$$

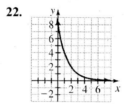

12. $(4b^{-3}c^2)^3(5b^{-7}c^{-1})^2$
$$= (64b^{-9}c^6)(25b^{-14}c^{-2})$$
$$= (64 \cdot 25)(b^{-9} \cdot b^{-14})(c^6 \cdot c^{-2})$$
$$= 1600b^{-23}c^4$$
$$= \frac{1600c^4}{b^{23}}$$

13. $\dfrac{8b^{1/3}c^{-1/2}}{6b^{-1/2}c^{3/4}} = \dfrac{4b^{\frac{1}{3}-\left(-\frac{1}{2}\right)}c^{-\frac{1}{2}-\frac{3}{4}}}{3}$
$$= \frac{4b^{\frac{2}{6}+\frac{3}{6}}c^{-\frac{2}{4}-\frac{3}{4}}}{3}$$
$$= \frac{4b^{5/6}c^{-5/4}}{3}$$
$$= \frac{4b^{5/6}}{3c^{5/4}}$$

14. $4\log_b\left(x^7\right) - 2\log_b\left(7x\right)$
$$= \log_b\left(x^7\right)^4 - \log_b\left(7x\right)^2$$
$$= \log_b\left(x^{28}\right) - \log_b\left(49x^2\right)$$
$$= \log_b\left(\frac{x^{28}}{49x^2}\right)$$
$$= \log_b\left(\frac{x^{26}}{49}\right)$$

15. $3\ln\left(p^6\right) + 4\ln\left(p^2\right) = \ln\left(p^6\right)^3 + \ln\left(p^2\right)^4$
$$= \ln\left(p^{18}\right) + \ln\left(p^8\right)$$
$$= \ln\left[\left(p^{18}\right)\left(p^8\right)\right]$$
$$= \ln\left(p^{26}\right)$$

16. $f(x) = 5(3)^x$

17. $g(x) = 3x + 25$

18. For each increase of 1 in the x-values, the y-values decrease by 7. The slope is -7.

19. $k(5) = 40$

20. $(g \circ k)(8) = g(k(8)) = g(5) = 40$

21. $f^{-1}(5) = 0$, since $f(0) = 5$.

22.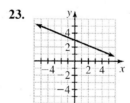

23.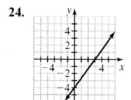

24.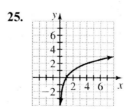

25.

26. First find the slope.

$$m = \frac{-3-7}{5+4} = -\frac{10}{9}$$

So $y = -\frac{10}{9}x + b$. Substitute a point, say

$(-4, 7)$, to solve for b.

$$7 = -\frac{10}{9}(-4) + b$$

$$7 = \frac{40}{9} + b$$

$$7 - \frac{40}{9} = b$$

$$\frac{63}{9} - \frac{40}{9} = b$$

$$b = \frac{23}{9}$$

So the equation is $y = -\frac{10}{9}x + \frac{23}{9}$.

27. Both points satisfy the equation $y = ab^x$. This produces a system of equations.

$$13 = ab^7$$

$$85 = ab^3$$

Combine the equations.

$$\frac{13}{85} = \frac{ab^7}{ab^3}$$

$$\frac{13}{85} = b^4$$

$$b = \left(\frac{13}{85}\right)^{1/4}$$

$$b \approx 0.6254$$

Substitute this value of the base into the equation $y = a(0.6254)^x$. Substitute $(3, 85)$ to find a.

$$85 = a(0.6254)^3$$

$$a = \frac{85}{(0.6254)^3}$$

$$a \approx 347.49$$

So the equation is $y = 347.49(0.63)^x$.

28. a. $(f \circ g)(x) = f(g(x)) = f(2^x) = 2^x - 3$

$(f \circ g)(5) = 2^5 - 3 = 32 - 3 = 29$

b. $(g \circ f)(x) = g(f(x)) = g(x-3) = 2^{x-3}$

29. $f(-4) = 2(3)^{-4} = \frac{2}{81}$

30.

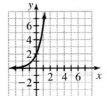

31.

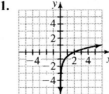

32. $f^{-1}(35)$

$$2(3)^x = 35$$

$$3^x = \frac{35}{2}$$

$$\log(3^x) = \log\left(\frac{35}{2}\right)$$

$$x\log(3) = \log\left(\frac{35}{2}\right)$$

$$x = \frac{\log\left(\frac{35}{2}\right)}{\log(3)}$$

$$x \approx 2.6053$$

33. $\log_3\left(\frac{1}{81}\right) = -4$, since $3^{-4} = \frac{1}{81}$.

34. $\log_b\left(\sqrt[7]{b}\right) = \frac{1}{7}$, since $b^{\frac{1}{7}} = \sqrt[7]{b}$.

35. $\log_8(73) = \frac{\log(73)}{\log(8)} \approx 2.0633$

36. $g(-1) = 3$

37. $(f \circ g)(3) = f(g(3)) = f(1) = 4$

38. The *y*-intercept is (0, 2). So the equation is
$y = 2b^x$. Substitute (−1, 1) to find *b*.

$$1 = 2b^{-1}$$
$$b^{-1} = \frac{1}{2}$$
$$b = \left(\frac{1}{2}\right)^{-1}$$
$$b = 2$$

So the equation is $f(x) = 2(2)^x$ or
$f(x) = (2)^{x+1}$.

39.

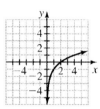

40. $f^{-1}(2)$

$$2 = 2(2)^x$$
$$2^x = 1$$
$$\log(2^x) = \log(1)$$
$$x\log(2) = 0$$
$$x = \frac{0}{\log(2)}$$
$$x = 0$$

41. $f(x) = \frac{2}{7}x - 3$

Replace $f(x)$ with *y*.

$$y = \frac{2}{7}x - 3$$

Solve for *x*.

$$y + 3 = \frac{2}{7}x$$
$$\frac{7}{2}(y + 3) = x$$
$$x = \frac{7}{2}y + \frac{21}{2}$$

Replace *x* with $f^{-1}(y)$.

$$f^{-1}(y) = \frac{7}{2}y + \frac{21}{2}$$

Replace *y* with *x*.

$$f^{-1}(x) = \frac{7}{2}x + \frac{21}{2}$$

42. $g(x) = 8^x$

Replace $g(x)$ with *y*.

$$y = 8^x$$

Solve for *x*.

$$\log(y) = \log(8^x)$$
$$\log(y) = x\log(8)$$
$$x = \frac{\log(y)}{\log(8)} = \log_8(y)$$

Replace *x* with $g^{-1}(y)$.

$$g^{-1}(y) = \log_8(y)$$

Replace *y* with *x*.

$$g^{-1}(x) = \log_8(x)$$

43. Answers may vary. Example:

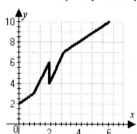

x	*y*
0	2
1	3
2	4
2	5
2	6
3	7

The relation is not a function because the input
$x = 2$ has three different outputs: $y = 4$, $y = 5$,
and $y = 6$.

44. a. The *y*-intercept for both functions is (0, 2).

b. For *f*, as the value of *x* increases by 1, the
value of $f(x)$ increases by 3.

For *g*, as the value of *x* increases by 1, the
value of $g(x)$ is multiplied by 3.

c. The function *g* will have a greater output
value of *y*.
Answers may vary. Example:
Raising 3 to a large power (as happens in *g*
for large *x* values) will yield a larger
number than multiplying that number by 3
(as happens in *f*).

d.

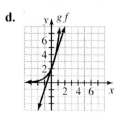

45. a. First find the slope.

$$m = \frac{8-3}{7-4} = \frac{5}{3}$$

So $y = \frac{5}{3}x + b$. Substitute (4. 3) and solve for b.

$$3 = \frac{5}{3}(4) + b$$

$$3 = \frac{20}{3} + b$$

$$3 - \frac{20}{3} = b$$

$$\frac{9}{3} - \frac{20}{3} = b$$

$$b = -\frac{11}{3}$$

So the equation is $y = \frac{5}{3}x - \frac{11}{3}$.

b. Both points satisfy the equation $y = ab^x$, which produces a system of equations.

$$8 = ab^7$$

$$3 = ab^4$$

Combine the equations.

$$\frac{8}{3} = \frac{ab^7}{ab^4}$$

$$\frac{8}{3} = b^3$$

$$b = \left(\frac{8}{3}\right)^{1/3}$$

$$b \approx 1.3867$$

Substitute this value of the base into the equation $y = a(1.3867)^x$. Substitute (4, 3) to find a.

$$3 = a(1.3867)^4$$

$$\frac{3}{(1.3867)^4} = a$$

$$0.81 \approx a$$

So the equation is $y = 0.81(1.39)^x$.

c.

46. a. $f(2) = 3(2) = 6$

$g(2) = 3^2 = 9$

b. Find f^{-1}.

$f(x) = 3x$

Replace $f(x)$ with y.

$y = 3x$

Solve for x.

$$x = \frac{1}{3}y$$

Replace x with $f^{-1}(y)$.

$$f^{-1}(y) = \frac{1}{3}y$$

Replace y with x.

$$f^{-1}(x) = \frac{1}{3}x$$

Find g^{-1}.

$g(x) = 3^x$

Replace $g(x)$ with y.

$$y = 3^x$$

Solve for x.

$$\log(y) = \log(3^x)$$

$$\log(y) = x \log(3)$$

$$x = \frac{\log(y)}{\log(3)}$$

$$x = \log_3(y)$$

Replace x with $g^{-1}(y)$.

$$g^{-1}(y) = \log_3(y)$$

Replace y with x.

$$g^{-1}(x) = \log_3(x)$$

c. $f^{-1}(81) = \frac{1}{3}(81) = 27$

$g^{-1}(81) = \log_3(81) = 4$, since $3^4 = 81$.

47. a. $U(x) = 0.69x + 19.95$

$B(x) = 0.45x + 29.95$

b. The slope of U is 0.69, so U-Haul charges $0.69 per mile. The slope of B is 0.45, so Budget charges $0.45 per mile.

c. $0.69x + 19.95 = 0.45x + 29.95$

$$0.24x = 10$$

$$x = \frac{10}{0.24}$$

$$x \approx 41.67 \text{ miles}$$

d.
$$U(x) < B(x)$$
$$0.69x + 19.95 < 0.45x + 29.95$$
$$0.24x < 10$$
$$x < 41.67$$
U-Haul costs less than Budget for mileage less than 41.67 miles.

48. Write the system.
$$x + y = 15000$$
$$43x + 60y = 721500$$
Use substitution.
$$43(15000 - y) + 60y = 721500$$
$$645000 - 43y + 60y = 721500$$
$$17y = 76500$$
$$y = 4500$$
Find x.
$$x + 4500 = 15000$$
$$x = 10500$$
10,500 tickets should be sold at $43 and 4500 tickets should be sold at $60.

49. a. $f(t) = 9.0(1.14)^t$, where $t = 0$ in 2010.

b. The r-intercept is the value when $t = 0$, which is 9.0. This means that in 2010, the revenue was $9.0 billion.

c. The base is 1.14. This means that the revenue is growing exponentially by 14% per year.

d.
$$f(t) = 9.0(1.14)^t$$
$$25 = 9.0(1.14)^t$$
$$\frac{25}{9.0} = \frac{9.0(1.14)^t}{9.0}$$
$$\log\left(\frac{25}{9.0}\right) = \log(1.14)^t$$
$$\frac{\log\left(\frac{25}{9.0}\right)}{\log(1.14)} = \frac{t\log(1.14)}{\log(1.14)}$$
$$7.80 \approx t$$
The model predicts that the annual revenue will be $25 billion in 2010 + 8 = 2018.

50. a. Create a scattergram to determine the type of regression to use.

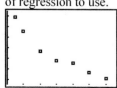

The scattergram shows that the points "bend" and that an exponential function will model the data.
Using exponential regression, the points fit the curve $f(t) = 76.58(0.97)^t$.

b. The n-intercept is 76.58. There were 76,580 tuberculosis cases in 1950.

c. Rate of decay $= 1.00 - 0.97 = 0.03$
The number of tuberculosis cases decreases by 3% per year.

d. $f(9) = 76.58(0.97)^9 \approx 58.22$
There were 58,220 tuberculosis cases in 1959.

e. To find $f^{-1}(9)$, substitute 9 for f and solve for t.
$$9 = 76.58(0.97)^t$$
$$\frac{9}{76.58} = \frac{76.58(0.97)^t}{76.58}$$
$$\log\left(\frac{9}{76.58}\right) = \log(0.97)^t$$
$$\frac{\log\left(\frac{9}{76.58}\right)}{\log(0.97)} = \frac{t\log(0.97)}{\log(0.97)}$$
$$70.29 \approx t$$
There will be 9 thousand tuberculosis cases in 1950 + 70 = 2020.

f. In 1950, $t = 0$.
$$f(0) = 76.58(0.97)^0 = 76.58$$
So, there were 76,580 tuberculosis cases in 1950.
Find when 38,290 cases (which is exactly half 76,580) are predicted.
$$f(t) = 76.58(0.97)^t$$
$$38.29 = 76.58(0.97)^t$$
$$\frac{38.29}{76.58} = \frac{76.58(0.97)^t}{76.58}$$
$$\log\left(\frac{1}{2}\right) = \log(0.97)^t$$
$$\frac{\log\left(\frac{1}{2}\right)}{\log(0.97)} = \frac{t\log(0.97)}{\log(0.97)}$$
$$22.76 \approx t$$
The approximate half-life of the number of cases is 22.76 years.

51. a. Create a scattergram to determine the type of regression to use.

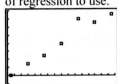

The plotted data lie mostly in a straight line. Therefore, a linear model fits the data. Using the linear regression feature on a graphing calculator, $f(t) = 0.53t - 0.17$.

b. The slope is 0.53. The slope means the number of women who place in the top 100 is increasing by 0.53 woman per year, on average.

c.
$$0 = 0.53t - 0.17$$
$$\frac{0.17}{0.53} = \frac{0.53t}{0.53}$$
$$0.32 \approx t$$
The t-intercept is $(0.32, 0)$. This means that no women placed in the top 100 in 1980.

d. $f(32) = 0.53(32) - 0.17 = 16.79$

The model estimates that 17 women would have placed in the top 100 if the marathon had not been canceled.

e. Replace $f(t)$ with n and solve for t.
$$n = 0.53t - 0.17$$
$$n + 0.17 = 0.53t$$
$$\frac{n + 0.17}{0.53} = \frac{0.53t}{0.53}$$
$$1.89n + 0.32 \approx t$$
Replace t with $f^{-1}(n)$.
An approximate equation is
$$f^{-1}(n) = 1.89n + 0.32.$$

f. $f^{-1}(20) = 1.89(20) + 0.32 = 38.12$

This means that in 2018, 20 women will place in the top 100.

g.
$$h(t) = (g \circ f)(t)$$
$$= g(f(t))$$
$$= g(0.53t - 0.17)$$
$$= 100 - (0.53t - 0.17)$$
$$= -0.53t + 100.17$$

h. $h(37) = -0.53(37) + 100.17 = 80.56$

This means that in 2017, 81 men will place in the top 100.

Chapter 6
Polynomial Functions

Homework 6.1

1. $5x^2 - 6x + 2$ is a quadratic (2nd-degree) polynomial of one variable.

3. $-2x^3 - 4x^2 + 5x - 1$ is a cubic (3rd-degree) polynomial of one variable.

5. $6p^4q^3 + 3p^2q^4 - 2q^5$ is a 7th-degree polynomial of two variables.

7. $6x^2 - 3x - 2x^2 + 4x = 6x^2 - 2x^2 - 3x + 4x$
$$= (6-2)x^2 + (-3+4)x$$
$$= 4x^2 + x$$

9. $-5x^3 - 4x + 2x^2 - 7x^3 + 5 - x$
$$= -5x^3 - 7x^3 + 2x^2 - 4x - x + 5$$
$$= (-5-7)x^3 + 2x^2 + (-4-1)x + 5$$
$$= -12x^3 + 2x^2 - 5x + 5$$

11. $4a^4b^2 - 7ab^3 - 9a^4b^2 + 2ab^3$
$$= 4a^4b^2 - 9a^4b^2 - 7ab^3 + 2ab^3$$
$$= (4-9)a^4b^2 + (-7+2)ab^3$$
$$= -5a^4b^2 - 5ab^3$$

13. $2x^4 - 4x^3y + 2x^2y^2 + x^3y - 2x^2y^2 + xy^3$
$$= 2x^4 - 4x^3y + x^3y + 2x^2y^2 - 2x^2y^2 + xy^3$$
$$= 2x^4 + (-4+1)x^3y + (2-2)x^2y^2 + xy^3$$
$$= 2x^4 - 3x^3y + xy^3$$

15. $\left(3x^2 - 5x - 2\right) + \left(6x^2 + 2x - 7\right)$
$$= 3x^2 + 6x^2 - 5x + 2x - 2 - 7$$
$$= 9x^2 - 3x - 9$$

17. $\left(-2x^3 + 4x - 3\right) + \left(5x^3 - 6x^2 + 2\right)$
$$= -2x^3 + 5x^3 - 6x^2 + 4x - 3 + 2$$
$$= 3x^3 - 6x^2 + 4x - 1$$

19. $\left(8a^2 - 7ab + 2b^2\right) + \left(3a^2 + 4ab - 7b^2\right)$
$$= 8a^2 + 3a^2 - 7ab + 4ab + 2b^2 - 7b^2$$
$$= 11a^2 - 3ab - 5b^2$$

21. $\left(2m^4p + m^3p^2 - 7m^2p^3\right)$
$$+ \left(m^3p^2 + 7m^2p^3 - 8mp^3\right)$$
$$= 2m^4p + m^3p^2 + m^3p^2 - 7m^2p^3$$
$$+ 7m^2p^3 - 8mp^3$$
$$= 2m^4p + 2m^3p^2 - 8mp^3$$

23. $\left(2x^2 + 4x - 7\right) - \left(9x^2 - 5x + 4\right)$
$$= 2x^2 - 9x^2 + 4x + 5x - 7 - 4$$
$$= -7x^2 + 9x - 11$$

25. $\left(6x^3 - 3x^2 + 4\right) - \left(-7x^3 + x - 1\right)$
$$= 6x^3 + 7x^3 - 3x^2 - x + 4 + 1$$
$$= 13x^3 - 3x^2 - x + 5$$

27. $\left(8m^2 + 3mp - 5p^2\right) - \left(-2m^2 - 7mp - 4p^2\right)$
$$= 8m^2 + 2m^2 + 3mp + 7mp - 5p^2 + 4p^2$$
$$= 10m^2 + 10mp - p^2$$

29. $\left(a^3b - 5a^2b^2 + ab^3\right) - \left(5a^2b^2 - 7ab^3 + b^3\right)$
$$= a^3b - 5a^2b^2 - 5a^2b^2 + ab^3 + 7ab^3 - b^3$$
$$= a^3b - 10a^2b^2 + 8ab^3 - b^3$$

31. $f(3) = -2(3)^2 - 5(3) + 3 = -30$

33. $g(-4) = 3(-4)^2 - 8(-4) - 1 = 79$

35. $f(0) = -2(0)^2 - 5(0) + 3 = 3$

37. $h(3) = 2(3)^3 - 4(3) = 42$

39. $h(-2) = 2(-2)^3 - 4(-2) = -8$

41. $f(-1) = 3$

43. $f(1) = -1$

45. $a = -1$ or $a = 3$

47. $a = 1$

49. $f(0) = 19$

51. $f(4) = 3$

53. $x = 0$ or $x = 6$

55. $x = 3$

57. a. $x = 1$ or $x = 5$

    b. $f(x)$ does not have an inverse function because $f(2) = f(4) = 3$. There would be two values of $f^{-1}(x)$ for one value of x, and such a relation is not a function.

59.
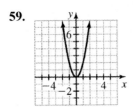

61.
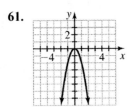

63.
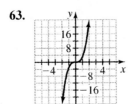

65.
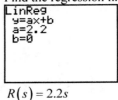

67. $(f + g)(x) = \left(4x^2 - 2x + 8\right) + \left(7x^2 + 5x - 1\right)$
$$= 4x^2 + 7x^2 - 2x + 5x + 8 - 1$$
$$= 11x^2 + 3x + 7$$
$$(f + g)(3) = 11(3)^2 + 3(3) + 7 = 115$$

69. $(f - h)(x) = \left(4x^2 - 2x + 8\right) - \left(-3x^2 - 4x - 9\right)$
$$= 4x^2 + 3x^2 - 2x + 4x + 8 + 9$$
$$= 7x^2 + 2x + 17$$
$$(f - h)(4) = 7(4)^2 + 2(4) + 17 = 137$$

71. $(f + g)(x) = \left(2x^3 - 4x + 1\right) + \left(-3x^2 + 5x - 3\right)$
$$= 2x^3 - 3x^2 - 4x + 5x + 1 - 3$$
$$= 2x^3 - 3x^2 + x - 2$$
$$(f + g)(2) = 2(2)^3 - 3(2)^2 + 2 - 2 = 4$$

73. $(f - h)(x) = \left(2x^3 - 4x + 1\right) - \left(x^3 - 3x^2 + 2x\right)$
$$= 2x^3 - x^3 + 3x^2 - 4x - 2x + 1$$
$$= x^3 + 3x^2 - 6x + 1$$
$$(f - h)(-1) = (-1)^3 + 3(-1)^2 - 6(-1) + 1 = 9$$

75. a. $(M + S)(t)$
$$= (-0.28t + 36.67) + (0.82t + 8.96)$$
$$= -0.28t + 0.82t + 36.67 + 8.96$$
$$= 0.54t + 45.63$$

    b. For the expression $M(t) + S(t)$, we have $M(t) =$ gallons of milk per person $+ S(t) =$ gallons of soft drinks per person. The units of the expression are gallons per person.

    c. $(M + S)(53) = 0.54(53) + 45.63 \approx 74.3$
    In $1950 + 53 = 2003$, 74.3 gallons of milk and soft drinks (combined) were consumed per person.

    d. $(M - S)(t)$
$$= (-0.28t + 36.67) - (0.82t + 8.96)$$
$$= -0.28t - 0.82t + 36.67 - 8.96$$
$$= -1.1t + 27.71$$

    e. $(M - S)(53) = -1.1(53) + 27.71 \approx -30.6$
    In 2003, 30.6 fewer gallons of milk than soft drinks were consumed per person.

77. a. Find the regression line for the data.

```
LinReg
y=ax+b
a=2.2
b=0
```

$R(s) = 2.2s$

    b. $(R + B)(t) = (2.2s) + \left(0.063s^2\right)$
$$= 0.063s^2 + 2.2s$$

c. For the expression $R(s) + B(s)$, we have $R(s)$ = reaction distance in feet + $B(s)$ = braking distance in feet. The expression is the total distance traveled in feet.

d. $(R+B)(26) = 0.063(26)^2 + 2.2(26)$
$$= 42.588 + 57.2$$
$$\approx 99.79$$
If you are driving at 26 miles per hour, it will take about 100 feet to stop.

e. $(R+B)(38) = 0.063(38)^2 + 2.2(38)$
$$= 90.972 + 83.6$$
$$\approx 174.57$$
Yes; she would need about 175 feet to stop.

79. The student did not apply the negative sign to $4x$ and 3.
$$\left(6x^2 + 8x + 5\right) - \left(2x^2 + 4x + 3\right)$$
$$= 6x^2 + 8x + 5 - 2x^2 - 4x - 3$$
$$= 4x^2 + 4x + 2$$

81. a. $(f-g)(x) = (3x+7) - (5x+2)$
$$= 3x - 5x + 7 - 2$$
$$= -2x + 5$$
$(g-f)(x) = (5x+2) - (3x+7)$
$$= 5x - 3x + 2 - 7$$
$$= 2x - 5$$

b. $(f-g)(2) = -2(2) + 5 = 1$
$(g-f)(2) = 2(2) - 5 = -1$
$(f-g)(2) = -(g-f)(2)$

c. $(f-g)(4) = -2(4) + 5 = -3$
$(g-f)(4) = 2(4) - 5 = 3$
$(f-g)(4) = -(g-f)(4)$

d. $(f-g)(7) = -2(7) + 5 = -9$
$(g-f)(7) = 2(7) - 5 = 9$
$(f-g)(7) = -(g-f)(7)$

e. In each case, $(f-g)(x) = -(g-f)(x)$. This makes sense because $f - g = (-1)(g-f)$.

83. Answers may vary. Example: Rearrange the equation.
$$A - B = -10$$
$$A = B - 10$$
This expression translates to "A is 10 less than B" in words.

85. Answers may vary. Example: To add two polynomials, combine like terms. To subtract polynomials, first distribute -1 to the second polynomial and then combine like terms.

87. b

89. c

91. a

93. $(f \circ g)(x) = f(g(x))$
$$= f(4x+5)$$
$$= 2(4x+5) - 3$$
$$= 8x + 10 - 3$$
$$= 8x + 7$$
$(f \circ g)(3) = 8(3) + 7 = 31$

95. $(f-g)(x) = f(x) - g(x)$
$$= (2x-3) - (4x+5)$$
$$= 2x - 3 - 4x - 5$$
$$= -2x - 8$$
$(f-g)(2) = -2(2) - 8 = -12$

97. $\left(\dfrac{25b^5 c^{-7}}{4b^{-3}c}\right)^{1/2} = \dfrac{25^{\frac{1}{2}} b^{5 \cdot \frac{1}{2}} c^{-7 \cdot \frac{1}{2}}}{4^{\frac{1}{2}} b^{-3 \cdot \frac{1}{2}} c^{\frac{1}{2}}}$

$$= \frac{5b^{5/2} c^{-7/2}}{2b^{-3/2} c^{1/2}}$$

$$= \frac{5b^{\frac{5}{2} - \left(-\frac{3}{2}\right)} c^{-\frac{7}{2} - \frac{1}{2}}}{2}$$

$$= \frac{5b^4 c^{-4}}{2}$$

$$= \frac{5b^4}{2c^4}$$

This is an expression in two variables involving exponents (or an exponential expression in two variables).

99. $f(x) = -2(2)^x$

This is an exponential function in one variable.

Homework 6.2

1. $3x^2\left(6x^4\right) = 18x^6$

3. $2a^3b^5\left(-4a^2b^3\right) = -8a^5b^8$

5. $-6x(5x-2) = -30x^2 + 12x$

7. $5ab^2\left(4a^2 - 7ab + 3b^2\right)$
$$= 20a^3b^2 - 35a^2b^3 + 15ab^4$$

9. $(x+3)(x+6) = x^2 + 3x + 6x + 18$
$$= x^2 + 9x + 18$$

11. $(3m-2)(5m+4) = 15m^2 + 12m - 10m - 8$
$$= 15m^2 + 2m - 8$$

13. $(8x-3)(4x-1) = 32x^2 - 8x - 12x + 3$
$$= 32x^2 - 20x + 3$$

15. $(1.7x - 2.4)(2.3x + 1.2)$
$$= 3.91x^2 + 2.04x - 5.52x - 2.88$$
$$= 3.91x^2 - 3.48x - 2.88$$

17. $(2a+5b)(3a-7b) = 6a^2 - 14ab + 15ab - 35b^2$
$$= 6a^2 + ab - 35b^2$$

19. $(4x-9y)(5x-2y)$
$$= 20x^2 - 8xy - 45xy + 18y^2$$
$$= 20x^2 - 53xy + 18y^2$$

21. $\left(2a^2 - 5b^2\right)\left(7a^2 + 3b^2\right)$
$$= 14a^4 + 6a^2b^2 - 35a^2b^2 - 15b^4$$
$$= 14a^4 - 29a^2b^2 - 15b^4$$

23. $3x^2\left(2x-5\right)\left(4x+1\right)$
$$= 3x^2\left(8x^2 + 2x - 20x - 5\right)$$
$$= 3x^2\left(8x^2 - 18x - 5\right)$$
$$= 24x^4 - 54x^3 - 15x^2$$

25. $5x\left(x^2 + 3\right)(x-4)$
$$= 5x\left(x^3 - 4x^2 + 3x - 12\right)$$
$$= 5x^4 - 20x^3 + 15x^2 - 60x$$

27. $(3x+2)\left(4x^2 + 5x - 3\right)$
$$= 12x^3 + 15x^2 - 9x + 8x^2 + 10x - 6$$
$$= 12x^3 + 23x^2 + x - 6$$

29. $(a+b)\left(a^2 - ab + b^2\right)$
$$= a^3 - a^2b + ab^2 + a^2b - ab^2 + b^3$$
$$= a^3 + b^3$$

31. $(4x-3y)\left(2x^2 - xy + 5y^2\right)$
$$= 8x^3 - 4x^2y + 20xy^2 - 6x^2y + 3xy^2 - 15y^3$$
$$= 8x^3 - 10x^2y + 23xy^2 - 15y^3$$

33. $\left(x^2 + 2x - 3\right)\left(x^2 - x + 2\right)$
$$= x^4 - x^3 + 2x^2 + 2x^3 - 2x^2 + 4x - 3x^2$$
$$\quad + 3x - 6$$
$$= x^4 + x^3 - 3x^2 + 7x - 6$$

35. $\left(2x^2 + xy - 3y^2\right)\left(x^2 - 2xy + y^2\right)$
$$= 2x^4 - 4x^3y + 2x^2y^2 + x^3y - 2x^2y^2 + xy^3$$
$$\quad - 3x^2y^2 + 6xy^3 - 3y^4$$
$$= 2x^4 - 3x^3y - 3x^2y^2 + 7xy^3 - 3y^4$$

37. $(x+5)^2 = (x+5)(x+5)$
$$= x^2 + 5x + 5x + 25$$
$$= x^2 + 10x + 25$$

39. $(x-8)^2 = (x-8)(x-8)$
$$= x^2 - 8x - 8x + 64$$
$$= x^2 - 16x + 64$$

41. $(3x+5)^2 = (3x+5)(3x+5)$
$= 9x^2 + 15x + 15x + 25$
$= 9x^2 + 30x + 25$

43. $(2.6x - 3.2)^2 = (2.6x - 3.2)(2.6x - 3.2)$
$= 6.76x^2 - 8.32x - 8.32x + 10.24$
$= 6.76x^2 - 16.64x + 10.24$

45. $(4a+3b)^2 = (4a+3b)(4a+3b)$
$= 16a^2 + 12ab + 12ab + 9b^2$
$= 16a^2 + 24ab + 9b^2$

47. $(2x^2 - 6y^2)^2 = (2x^2 - 6y^2)(2x^2 - 6y^2)$
$= 4x^4 - 12x^2y^2 - 12x^2y^2 + 36y^4$
$= 4x^4 - 24x^2y^2 + 36y^4$

49. $-2x(2x+5)^2 = -2x(2x+5)(2x+5)$
$= -2x(4x^2 + 10x + 10x + 25)$
$= -2x(4x^2 + 20x + 25)$
$-8x^3 - 40x^2 - 50x$

51. $(x-4)(x+4) = x^2 + 4x - 4x - 16 = x^2 - 16$

53. $(3x+6)(3x-6) = 9x^2 - 18x + 18x - 36$
$= 9x^2 - 36$

55. $(2r - 8t)(2r + 8t) = 4r^2 + 16rt - 16rt - 64t^2$
$= 4r^2 - 64t^2$

57. $(3rt - 9w)(3rt + 9w)$
$= 9r^2t^2 - 27rtw + 27rtw - 81w^2$
$= 9r^2t^2 - 81w^2$

59. $(8a^2 + 3b^2)(8a^2 - 3b^2)$
$= 64a^4 - 24a^2b^2 + 24a^2b^2 - 9b^4$
$= 64a^4 - 9b^4$

61. $(x-2)(x+2)(x^2+4)$
$= (x^2 + 2x - 2x - 4)(x^2 + 4)$
$= (x^2 - 4)(x^2 + 4)$
$= x^4 + 4x^2 - 4x^2 - 16$
$= x^4 - 16$

63. $(3a+2b)(3a-2b)(9a^2+4b^2)$
$= (9a^2 - 6ab + 6ab - 4b^2)(9a^2 + 4b^2)$
$= (9a^2 - 4b^2)(9a^2 + 4b^2)$
$= 81a^4 + 36a^2b^2 - 36a^2b^2 - 16b^4$
$= 81a^4 - 16b^4$

65. $f(5b) = (5b)^2 - 3(5b) = 25b^2 - 15b$

67. $f(c+4) = (c+4)^2 - 3(c+4)$
$= c^2 + 8c + 16 - 3c - 12$
$= c^2 + 5c + 4$

69. $f(b-3) = (b-3)^2 - 3(b-3)$
$= b^2 - 6b + 9 - 3b + 9$
$= b^2 - 9b + 18$

71. $f(a+2) - f(a)$
$= ((a+2)^2 - 3(a+2)) - ((a)^2 - 3(a))$
$= a^2 + 4a + 4 - 3a - 6 - a^2 + 3a$
$= 4a - 2$

73. $f(a+h) - f(a)$
$= ((a+h)^2 - 3(a+h)) - ((a)^2 - 3(a))$
$= a^2 + 2ah + h^2 - 3a - 3h - a^2 + 3a$
$= 2ah + h^2 - 3h$

75. $f(x) = (x+6)^2$
$= (x+6)(x+6)$
$= x^2 + 6x + 6x + 36$
$= x^2 + 12x + 36$

77. $f(x) = 2(x+3)^2 + 1$
$= 2(x+3)(x+3) + 1$
$= 2(x^2 + 3x + 3x + 9) + 1$
$= 2(x^2 + 6x + 9) + 1$
$= 2x^2 + 12x + 18 + 1$
$= 2x^2 + 12x + 19$

79. $f(x) = -3(x-5)^2 - 1$
$= -3(x-5)(x-5) - 1$
$= -3(x^2 - 5x - 5x + 25) - 1$
$= -3(x^2 - 10x + 25) - 1$
$= -3x^2 + 30x - 75 - 1$
$= -3x^2 + 30x - 76$

81. $(f \cdot g)(x) = (2x-3)(3x+2)$
$= 6x^2 + 4x - 9x - 6$
$= 6x^2 - 5x - 6$
$(f \cdot g)(3) = 6(3)^2 - 5(3) - 6 = 33$

83. $(f \cdot h)(x) = (2x-3)(2x^2 - 4x + 3)$
$= 4x^3 - 8x^2 + 6x - 6x^2 + 12x - 9$
$= 4x^3 - 14x^2 + 18x - 9$
$(f \cdot h)(2) = 4(2)^3 - 14(2)^2 + 18(2) - 9 = 3$

85. $(f \cdot f)(x) = (2x-3)(2x-3)$
$= 4x^2 - 6x - 6x + 9$
$= 4x^2 - 12x + 9$
$(f \cdot f)(4) = 4(4)^2 - 12(4) + 9 = 25$

87. $(f \cdot g)(x) = (4x+1)(5x+3)$
$= 20x^2 + 12x + 5x + 3$
$= 20x^2 + 17x + 3$
$(f \cdot g)(-1) = 20(-1)^2 + 17(-1) + 3 = 6$

89. $(f \cdot h)(x) = (4x+1)(3x^2 - x - 2)$
$= 12x^3 - 4x^2 - 8x + 3x^2 - x - 2$
$= 12x^3 - x^2 - 9x - 2$
$(f \cdot h)(-2) = 12(-2)^3 - (-2)^2 - 9(-2) - 2$
$= -84$

91. $(h \cdot h)(x) = (3x^2 - x - 2)(3x^2 - x - 2)$
$= 9x^4 - 3x^3 - 6x^2 - 3x^3 + x^2 + 2x$
$\quad - 6x^2 + 2x + 4$
$= 9x^4 - 6x^3 - 11x^2 + 4x + 4$
$(h \cdot h)(1) = 9(1)^4 - 6(1)^3 - 11(1)^2 + 4(1) + 4$
$= 0$

93. a. $V(t) = 62t + 508$

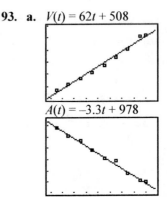

$A(t) = -3.3t + 978$

Both models fit the data well.

b. $(V \cdot A)(t) = (62t + 508)(-3.3t + 978)$
$= -204.6t^2 + 58,959.6t + 496,824$

c. For the expression $V(t) \cdot A(t)$, we have $V(t)$ = average value of farmland in dollars per acre $\times A(t)$ = amount of farmland in millions of acres. The units of the expression are total value of farmland in millions of dollars.

d. $(V \cdot A)(27)$
$= -204.6(27)^2 + 58,959.6(27) + 496,824$
$= 1,939,579.8$

The total value of U.S. farmland in $1990 + 27 = 2017$ will be $\$1,939,579.8$ million, or about $\$1.94$ trillion.

e.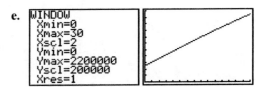

The function is increasing. This means that between 1990 and 2017, the total value of U.S. farmland will increase every year. Answers may vary. Example: The average value of farmland is increasing from 1990 to 2017 at a faster rate than the amount of farmland is decreasing.

95. a. $B(t) = -0.197t^2 + 6.15t + 3$

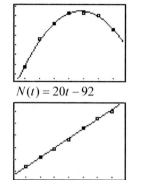

$N(t) = 20t - 92$

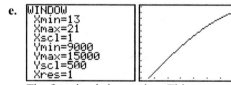

Both models fit the data well.

b. $(B \cdot N)(t)$

$= \left(-0.197t^2 + 6.15t + 3\right)(20t - 92)$

$= -3.94t^3 + 18.124t^2 + 123t^2 - 565.8t$
$\quad + 60t - 276$

$= -3.94t^3 + 141.124t^2 - 505.8t - 276$

c. For the expression $B(t) \cdot N(t)$, we have $B(t)$ = average bill in dollars per month $\times$ $N(t)$ = number of subscribers in millions of people. The units of the expression are millions of dollars per month.

d. $(B \cdot N)(24)$

$= -3.94(24)^3 + 141.124(24)^2$
$\quad -505.8(24) - 276$

$= 14,405.664$

$\approx 14,406$

In $1990 + 24 = 2014$, total monthly cell phone revenue will be about $14,406 million, or $14.4 billion.

e.

```
WINDOW
 Xmin=13
 Xmax=21
 Xscl=1
 Ymin=9000
 Ymax=15000
 Yscl=500
 Xres=1
```

The function is increasing. This means that between 2004 and 2010, the total monthly revenue from cell phones increased. Answers may vary. Example: The number of subscribers increased, so the total monthly cell phone revenue continued to increase.

97. The student did not complete the square.

$(x + 8)^2 = (x + 8)(x + 8)$

$\qquad = x^2 + 8x + 8x + 64$

$\qquad = x^2 + 16x + 64$

99. a. $(x + 2)^2$

X	Y₁
-3	1
-2	0
-1	1
0	4
1	9
2	16
3	25

X= -3

$x^2 + 2^2$

X	Y₁
-3	13
-2	8
-1	5
0	4
1	5
2	8
3	13

X= -3

b. $(x + 2)^2 = (x + 2)(x + 2)$

$\qquad = x^2 + 2x + 2x + 4$

$\qquad = x^2 + 4x + 4$

c. $(x + 2)^2$

X	Y₁
-3	1
-2	0
-1	1
0	4
1	9
2	16
3	25

X= -3

$x^2 + 4x + 4$

X	Y₁
-3	1
-2	0
-1	1
0	4
1	9
2	16
3	25

X= -3

101. The student performed subtraction instead of multiplication.

$7x(-2x) = -14x^2$

103. $(2x - 5)(3x + 4) = 6x^2 + 8x - 15x - 20$

$\qquad\qquad\qquad = 6x^2 - 7x - 20$

$3x(2x - 2) - x - 20 = 6x^2 - 6x - x - 20$

$\qquad\qquad\qquad = 6x^2 - 7x - 20$

$(3x + 4)(2x - 5) = 6x^2 - 15x + 8x - 20$

$\qquad\qquad\qquad = 6x^2 - 7x - 20$

$(3x - 4)(2x + 5) = 6x^2 + 15x - 8x - 20$

$\qquad\qquad\qquad = 6x^2 + 7x - 20$

105. $(A-B)^2 = (A-B)(A-B)$
$$= A^2 - AB - AB + B^2$$
$$= A^2 - 2AB + B^2$$

107. $(g-h)(x) = (4x-5) - (2x^2 - 3x + 1)$
$$= 4x - 5 - 2x^2 + 3x - 1$$
$$= -2x^2 + 7x - 6$$
$(g-h)(-2) = -2(-2)^2 + 7(-2) - 6$
$$= -8 - 14 - 6$$
$$= -28$$

109. $(h \cdot k)(x) = (2x^2 - 3x + 1)(x+3)$
$$= 2x^3 - 3x^2 + x + 6x^2 - 9x + 3$$
$$= 2x^3 + 3x^2 - 8x + 3$$
$(h \cdot k)(-1) = 2(-1)^3 + 3(-1)^2 - 8(-1) + 3$
$$= -2 + 3 + 8 + 3$$
$$= 12$$

111. $(f \circ g)(x) = f(g(x))$
$$= f(4x-5)$$
$$= (4x-5)^2$$
$$= 16x^2 - 40x + 25$$
$(f \circ g)(3) = 16(3)^2 - 40(3) + 25$
$$= 144 - 120 + 25$$
$$= 49$$

113. $(h \circ k)(x) = h(k(x))$
$$= h(x+3)$$
$$= 2(x+3)^2 - 3(x+3) + 1$$
$$= 2x^2 + 12x + 18 - 3x - 9 + 1$$
$$= 2x^2 + 9x + 10$$
$(h \circ k)(2) = 2(2)^2 + 9(2) + 10$
$$= 8 + 18 + 10$$
$$= 36$$

115. $f(x) = (2x-5)(3x-1)$
$$= 6x^2 - 2x - 15x + 5$$
$$= 6x^2 - 17x + 5$$
This is a quadratic function.

117. $f(x) = x^2 - (x+1)^2$
$$= x^2 - (x+1)(x+1)$$
$$= x^2 - (x^2 + 2x + 1)$$
$$= -2x - 1$$
This is a linear function.

119. $\log_b(x+5) + \log_b(x-3) = \log_b(x+5)(x-3)$
$$= \log_b(x^2 + 2x - 15)$$

121. $2\log_b(w-3) + \log_b(w+3)$
$$= \log_b(w-3)^2(w+3)$$
$$= \log_b(w-3)(w-3)(w+3)$$
$$= \log_b(w^2 - 6w + 9)(w+3)$$
$$= \log_b(w^3 - 3w^2 - 9w + 27)$$

123. $4x(3x+5)(2x-3)$
$$= 4x(6x^2 - 9x + 10x - 15)$$
$$= 4x(6x^2 + x - 15)$$
$$= 24x^3 + 4x^2 - 60x$$
This is a cubic polynomial in one variable.

125. $f(x) = -3(x-4)^2 + 5$
$$= -3(x-4)(x-4) + 5$$
$$= -3(x^2 - 4x - 4x + 16) + 5$$
$$= -3(x^2 - 8x + 16) + 5$$
$$= -3x^2 + 24x - 48 + 5$$
$$= -3x^2 + 24x - 43$$
This is a quadratic function.

Homework 6.3

1. $\dfrac{3x^5 + 7x^3}{x} = \dfrac{3x^5}{x} + \dfrac{7x^3}{x} = 3x^4 + 7x^2$
Check:

Plot1 Plot2 Plot3		X	Y1	Y2
\Y1◘(3X^5+7X^3)/		1	10	10
X		2	76	76
\Y2◘3X^4+7X²		3	306	306
\Y3=		4	880	880
\Y4=		5	2050	2050
\Y5=		6	4140	4140
\Y6=		7	7546	7546
		X=1		

3. $\dfrac{6x^3 + 12x^2}{3x} = \dfrac{6x^3}{3x} + \dfrac{12x^2}{3x} = 2x^2 + 4x$

Check:

5. $\dfrac{20p^9 - 5p^6 + 15p^3}{5p^3} = \dfrac{20p^9}{5p^3} - \dfrac{5p^6}{5p^3} + \dfrac{15p^3}{5p^3}$

$\qquad\qquad = 4p^6 - p^3 + 3$

Check:

7. $\dfrac{16x^5 - 3x^4 - 24x^3}{-8x^2} = \dfrac{16x^5}{-8x^2} + \dfrac{-3x^4}{-8x^2} + \dfrac{-24x^3}{-8x^2}$

$\qquad\qquad = -2x^3 + \dfrac{3}{8}x^2 + 3x$

Check:

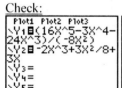

9. $\dfrac{4x^5 - 8x^4 + 7x^2}{2x^4} = \dfrac{4x^5}{2x^4} - \dfrac{8x^4}{2x^4} + \dfrac{7x^2}{2x^4}$

$\qquad\qquad = 2x - 4 + \dfrac{7}{2x^2}$

Check:

$\left(2x^4\right) \cdot \left(2x - 4 + \dfrac{7}{2x^2}\right) = 4x^5 - 8x^4 + 7x^2$

11. $\dfrac{-8k^4 - 6k^3 + 12k}{-4k^2} = \dfrac{-8k^4}{-4k^2} - \dfrac{6k^3}{-4k^2} + \dfrac{12k}{-4k^2}$

$\qquad\qquad = 2k^2 + \dfrac{3}{2}k - \dfrac{3}{k}$

Check:

$\left(-4k^2\right) \cdot \left(2k^2 + \dfrac{3}{2}k - \dfrac{3}{k}\right) = -8k^4 - 6k^3 + 12k$

13. $\dfrac{x^4y + x^3y^2 - xy^4}{xy} = \dfrac{x^4y}{xy} + \dfrac{x^3y^2}{xy} - \dfrac{xy^4}{xy}$

$\qquad\qquad = x^3 + x^2y - y^3$

Check:

$\left(xy\right) \cdot \left(x^3 + x^2y - y^3\right) = x^4y + x^3y^2 - xy^4$

15. $\dfrac{10m^4r^2 + 3m^2r^3 + 4mr^4}{2mr^2}$

$\quad = \dfrac{10m^4r^2}{2mr^2} + \dfrac{3m^2r^3}{2mr^2} + \dfrac{4mr^4}{2mr^2}$

$\quad = 5m^3 + \dfrac{3}{2}mr + 2r^2$

Check:

$\left(2mr^2\right) \cdot \left(5m^3 + \dfrac{3}{2}mr + 2r^2\right)$

$\quad = 10m^4r^2 + 3m^2r^3 + 4mr^4$

17. $\dfrac{12x^6y^4 + 18x^5y^2 - 8x^4y}{-4x^3y^2}$

$\quad = \dfrac{12x^6y^4}{-4x^3y^2} + \dfrac{18x^5y^2}{-4x^3y^2} - \dfrac{8x^4y}{-4x^3y^2}$

$\quad = -3x^3y^2 - \dfrac{9}{2}x^2 + \dfrac{2x}{y}$

Check:

$\left(-4x^3y^2\right) \cdot \left(-3x^3y^2 - \dfrac{9}{2}x^2 + \dfrac{2x}{y}\right)$

$\quad = 12x^6y^4 + 18x^5y^2 - 8x^4y$

19.

$$
\begin{array}{r}
3x + 2 \\
x + 4 \overline{)\,3x^2 + 14x + 8} \\
\underline{-3x^2 - 12x} \\
2x + 8 \\
\underline{-2x - 8} \\
0
\end{array}
$$

$\dfrac{3x^2 + 14x + 8}{x + 4} = 3x + 2$

Check:

$\left(x + 4\right) \cdot \left(3x + 2\right) = 3x^2 + 2x + 12x + 8$

$\qquad\qquad\qquad = 3x^2 + 14x + 8$

21.

$$x-5\overline{\smash{\big)}4x^2-22x+10}\quad\text{quotient: }4x-2$$

$$\underline{-4x^2+20x}$$
$$-2x+10$$
$$\underline{2x-10}$$
$$0$$

$$\frac{4x^2-22x+10}{x-5}=4x-2$$

Check:
$$(x-5)\cdot(4x-2)=4x^2-2x-20x+10$$
$$=4x^2-22x+10$$

23.

$$3p+2\overline{\smash{\big)}6p^2+19p+12}\quad\text{quotient: }2p+5$$

$$\underline{-6p^2-4p}$$
$$15p+12$$
$$\underline{-15p-10}$$
$$2$$

$$\frac{6p^2+19p+12}{3p+2}=2p+5+\frac{2}{3p+2}$$

Check:
$$(3p+2)\cdot(2p+5)+2$$
$$=6p^2+15p+4p+10+2$$
$$=6p^2+19p+12$$

25.

$$2x+5\overline{\smash{\big)}4x^2+2x-23}\quad\text{quotient: }2x-4$$

$$\underline{-4x^2-10x}$$
$$-8x-23$$
$$\underline{8x+20}$$
$$-3$$

$$\frac{4x^2+2x-23}{2x+5}=2x-4-\frac{3}{2x+5}$$

Check:
$$(2x+5)\cdot(2x-4)-3=4x^2-8x+10x-20-3$$
$$=4x^2+2x-23$$

27.

$$5x-1\overline{\smash{\big)}10x^2-7x+3}\quad\text{quotient: }2x-1$$

$$\underline{-10x^2+2x}$$
$$-5x+3$$
$$\underline{5x-1}$$
$$2$$

$$\frac{10x^2-7x+3}{5x-1}=2x-1+\frac{2}{5x-1}$$

Check:
$$(5x-1)\cdot(2x-1)+2=10x^2-5x-2x+1+2$$
$$=10x^2-7x+3$$

29. $2m-9+20m^2=20m^2+2m-9$

$$4m-2\overline{\smash{\big)}20m^2+2m-9}\quad\text{quotient: }5m+3$$

$$\underline{-20m^2+10m}$$
$$12m-9$$
$$\underline{-3m+6}$$
$$-3$$

$$\frac{2m-9+20m^2}{4m-2}=5m+3-\frac{3}{4m-2}$$

Check:
$$(4m-2)\cdot(5m+3)-3$$
$$=20m^2+12m-10m-6-3$$
$$=20m^2+2m-9$$

31.

$$2x+3\overline{\smash{\big)}4x^3+12x^2+15x+13}\quad\text{quotient: }2x^2+3x+3$$

$$\underline{-4x^3-6x^2}$$
$$6x^2+15x$$
$$\underline{-6x^2-9x}$$
$$6x+13$$
$$\underline{-6x-9}$$
$$4$$

$$\frac{4x^3+12x^2+15x+13}{2x+3}=2x^2+3x+3+\frac{4}{2x+3}$$

Check:

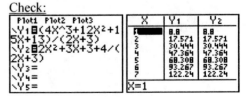

33. $17p + 12p^3 - 14 - 22p^2$
$= 12p^3 - 22p^2 + 17p - 14$

$$\begin{array}{r} 4p^2 - 2p + 3 \\ 3p - 4 \overline{\smash{\big)}\ 12p^3 - 22p^2 + 17p - 14} \\ \underline{-12p^3 + 16p^2} \\ -6p^2 + 17p \\ \underline{6p^2 - 8p} \\ 9p - 14 \\ \underline{-9p + 12} \\ -2 \end{array}$$

$$\frac{17p + 12p^3 - 14 - 22p^2}{3p - 4}$$
$$= 4p^2 - 2p + 3 - \frac{2}{3p - 4}$$

Check:

	X	Y₁	Y₂
Plot1 Plot2 Plot3	1	7	7
\Y₁ᗎ(17X+12X^3-1	2	14	14
4-22X²)/(3X-4)	3	32.6	32.6
\Y₂ᗎ4X²-2X+3-2/(	4	58.75	58.75
3X-4)	5	92.818	92.818
\Y₃=	6	134.86	134.86
\Y₄=	7	184.88	184.88
\Y₅=	X=1		

35. $x^2 + 7 = x^2 + 0x + 7$

$$\begin{array}{r} x - 3 \\ x + 3 \overline{\smash{\big)}\ x^2 + 0x + 7} \\ \underline{-x^2 - 3x} \\ -3x + 7 \\ \underline{3x + 9} \\ 16 \end{array}$$

$$\frac{x^2 + 7}{x + 3} = x - 3 + \frac{16}{x + 3}$$

Check:

	X	Y₁	Y₂
Plot1 Plot2 Plot3	1	2	2
\Y₁ᗎ(X²+7)/(X+3)	2	2.2	2.2
	3	2.6667	2.6667
\Y₂ᗎX-3+16/(X+3)	4	3.2857	3.2857
	5	4	4
\Y₃=	6	4.7778	4.7778
\Y₄=	7	5.6	5.6
\Y₅=	X=1		

37. $8x^3 - 125 = 8x^3 + 0x^2 + 0x - 125$

$$\begin{array}{r} 4x^2 + 10x + 25 \\ 2x - 5 \overline{\smash{\big)}\ 8x^3 + 0x^2 + 0x - 125} \\ \underline{-8x^3 + 20x^2} \\ 20x^2 + 0x \\ \underline{-20x^2 + 50x} \\ 50x - 125 \\ \underline{-50x + 125} \\ 0 \end{array}$$

$$\frac{8x^3 - 125}{2x - 5} = 4x^2 + 10x + 25$$

Check:
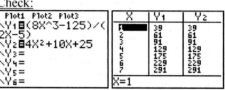

	X	Y₁	Y₂
Plot1 Plot2 Plot3	1	39	39
\Y₁ᗎ(8X^3-125)/(	2	61	61
2X-5)	3	91	91
\Y₂ᗎ4X²+10X+25	4	129	129
\Y₃=	5	175	175
\Y₄=	6	229	229
\Y₅=		291	291
\Y₆=	X=1		

39. $3x^3 - 10x^2 + 7 = 3x^3 - 10x^2 + 0x + 7$

$$\begin{array}{r} x^2 - 3x - 1 \\ 3x - 1 \overline{\smash{\big)}\ 3x^3 - 10x^2 + 0x + 7} \\ \underline{-3x^3 + x^2} \\ -9x^2 + 0x \\ \underline{9x^2 - 3x} \\ -3x + 7 \\ \underline{3x - 1} \\ 6 \end{array}$$

$$\frac{3x^3 - 10x^2 + 7}{3x - 1} = x^2 - 3x - 1 + \frac{6}{3x - 1}$$

Check:
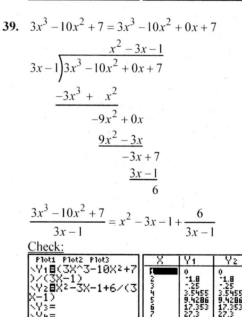

	X	Y₁	Y₂
Plot1 Plot2 Plot3	1	0	0
\Y₁ᗎ(3X^3-10X²+7	2	-1.8	-1.8
)/(3X-1)	3	-.25	-.25
\Y₂ᗎX²-3X-1+6/(3	4	3.5455	3.5455
X-1)	5	9.4286	9.4286
\Y₃=	6	17.353	17.353
\Y₄=		27.3	27.3
\Y₅=	X=1		

41.

$$\require{enclose}
\begin{array}{r}
3x-4 \\
x^2+3 \enclose{longdiv}{3x^3-4x^2+9x-16}
\end{array}$$

$$\begin{array}{r}
\underline{-3x^3+0x^2-9x} \\
-4x^2+0x-16 \\
\underline{-4x^2+0x+12} \\
-4
\end{array}$$

$$\frac{3x^3-4x^2+9x-16}{x^2+3}=3x-4-\frac{4}{x^2+3}$$

Check:

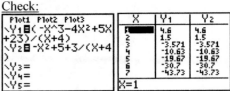

43.

$$\begin{array}{r}
2y+3 \\
3y^2-2 \enclose{longdiv}{6y^3+9y^2-4y-7}
\end{array}$$

$$\begin{array}{r}
\underline{-6y^3+0y^2+4y} \\
9y^2+0y-7 \\
\underline{-9y^2+0y+6} \\
-1
\end{array}$$

$$\frac{6y^3+9y^2-4y-7}{3y^2-2}=2y+3-\frac{1}{3y^2-2}$$

Check:

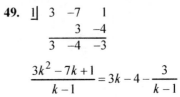

45.

$$\begin{array}{r|rrrr}
2 & 3 & -10 & 10 & -4 \\
  &   & 6 & -8 & 4 \\
\hline
  & 3 & -4 & 2 & 0
\end{array}$$

$$\frac{3x^3-10x^2+10x-4}{x-2}=3x^2-4x+2$$

Check:

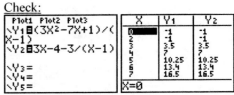

47.

$$\begin{array}{r|rrrr}
-4 & -1 & -4 & 5 & 23 \\
   &   & 4 & 0 & -20 \\
\hline
   & -1 & 0 & 5 & 3
\end{array}$$

$$\frac{-x^3-4x^2+5x+23}{x+4}=-x^2+5+\frac{3}{x+4}$$

Check:

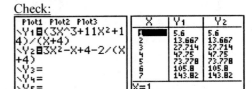

49.

$$\begin{array}{r|rrr}
1 & 3 & -7 & 1 \\
  &   & 3 & -4 \\
\hline
  & 3 & -4 & -3
\end{array}$$

$$\frac{3k^2-7k+1}{k-1}=3k-4-\frac{3}{k-1}$$

Check:

51. $-11x^2-17+8x+2x^3=2x^3-11x^2+8x-17$

$$\begin{array}{r|rrrr}
5 & 2 & -11 & 8 & -17 \\
  &   & 10 & -5 & 15 \\
\hline
  & 2 & -1 & 3 & -2
\end{array}$$

$$\frac{-11x^2-17+8x+2x^3}{x-5}=2x^2-x+3-\frac{2}{x-5}$$

Check:

53. $3x^3+11x^2+14=3x^3+11x^2+0x+14$

$$\begin{array}{r|rrrr}
-4 & 3 & 11 & 0 & 14 \\
   &   & -12 & 4 & -16 \\
\hline
   & 3 & -1 & 4 & -2
\end{array}$$

$$\frac{3x^3+11x^2+14}{x+4}=3x^2-x+4-\frac{2}{x+4}$$

Check:

55. The student did not divide the second term in the numerator, $7x$, by the denominator.

$$\frac{6x^2 + 7x}{2x^2} = \frac{6x^2}{2x^2} + \frac{7x}{2x^2} = 3 + \frac{7}{2x}$$

57. The student did not change the signs before adding.

$$
\begin{array}{r}
2x + 4 \\
x + 3 \overline{\smash{)}\, 2x^2 + 10x + 12} \\
\underline{-2x^2 - 6x} \\
4x + 12 \\
\underline{-4x - 12} \\
0
\end{array}
$$

$$\frac{2x^2 + 10x + 12}{x + 3} = 2x + 4$$

59. Answers may vary. Example:
The student performed the synthetic division correctly, but did not correctly use the results to find the quotient polynomial and the remainder.

$$\frac{4x^3 - 10x^2 - 9x + 15}{x - 3} = 4x^2 + 2x - 3 + \frac{6}{x - 3}$$

61. Answers may vary. Example:

$$\frac{2x^4 - 3x^3 + 7x^2}{x^2} = \frac{2x^4}{x^2} - \frac{3x^3}{x^2} + \frac{7x^2}{x^2}$$
$$= 2x^2 - 3x + 7$$

63. Multiply the binomial by the quotient and add the remainder.

$$(x - 3)(2x - 5) + 4 = 2x^2 - 5x - 6x + 15 + 4$$
$$= 2x^2 - 11x + 19$$

65. Answers may vary.

67. Answers may vary.

69. Answers may vary.

71. $\left(6x^2 - x - 2\right)\left(3x - 2\right)$

$$= 18x^3 - 12x^2 - 3x^2 + 2x - 6x + 4$$
$$= 18x^3 - 15x^2 - 4x + 4$$

Check:

73. $\left(6x^2 - x - 2\right) - \left(3x - 2\right) = 6x^2 - x - 2 - 3x + 2$

$$= 6x^2 - 4x$$

Check:

75.
$$
\begin{array}{r}
2x + 1 \\
3x - 2 \overline{\smash{)}\, 6x^2 - x - 2} \\
\underline{-6x^2 + 4x} \\
3x - 2 \\
\underline{-3x + 2} \\
0
\end{array}
$$

$$\frac{6x^2 - x - 2}{3x - 2} = 2x + 1$$

Check:

77. $\left(3x - 2\right)^2 = 9x^2 - 12x + 4$

Check:

79. Solve the second equation for y:

$$4x - y = 7$$
$$-y = -4x + 7$$
$$y = 4x - 7$$

Substitute the result for y in the first equation:

$$3x - 2y = 9$$
$$3x - 2(4x - 7) = 9$$
$$3x - 8x + 14 = 9$$
$$-5x = -5$$
$$x = 1$$

Use the value of x to find y:

$$y = 4x - 7 = 4(1) - 7 = -3$$

The solution is $(1, -3)$.

This is a system of linear equations in two variables.

81. $4\log_2(3x) + \log_2\left(5x^3\right) = 8$

$$\log_2\left[(3x)^4 \cdot \left(5x^3\right)\right] = 8$$

$$\log_2\left(405x^7\right) = 8$$

$$2^{\log_2\left(405x^7\right)} = 2^8$$

$$405x^7 = 256$$

$$x^7 = \frac{256}{405}$$

$$x = \left(\frac{256}{405}\right)^{1/7}$$

$$x \approx 0.9366$$

This is a logarithmic equation in one variable.

Homework 6.4

1. $x^2 + 11x + 28 = (x+4)(x+7)$

3. $x^2 - 8x + 12 = (x-6)(x-2)$

5. $r^2 - 4r - 32 = (r-8)(r+4)$

7. $x^2 + 5x - 14 = (x+7)(x-2)$

9. This expression is prime since there are no two integers with a product of –12 and a sum of –7.

11. $x^2 + 10x + 25 = (x+5)(x+5) = (x+5)^2$

13. $t^2 - 18t + 81 = (t-9)(t-9) = (t-9)^2$

15. $4x - 5 + x^2 = x^2 + 4x - 5 = (x+5)(x-1)$

17. $a^2 + 12ab + 20b^2 = (a+10b)(a+2b)$

19. $w^2 - 5wy + 4y^2 = (w-4y)(w-y)$

21. $p^2 + 3pq - 28q^2 = (p+7q)(p-4q)$

23. This expression is prime since there are no two integers with a product of –16 and a sum of 4.

25. $p^2 - 6pq - 16q^2 = (p-8q)(p+2q)$

27. $3x + 21 = 3(x+7)$

29. $16x^2 - 12x = 4x(4x-3)$

31. $9y^5 + 18y^3 = 9y^3\left(y^2 + 2\right)$

33. $3ab - 12a^2b = 3ab(1 - 4a)$

35. $18a^4b^2 + 12a^2b^3 = 6a^2b^2\left(3a^2 + 2b\right)$

37. $-14x^5y + 63x^2y^2 = 7x^2y\left(-2x^3 + 9y\right)$
$$= -7x^2y\left(2x^3 - 9y\right)$$

39. $2x^2 + 12x + 18 = 2\left(x^2 + 6x + 9\right)$
$$= 2(x+3)(x+3)$$
$$= 2(x+3)^2$$

41. $3x^2 - 3x - 18 = 3\left(x^2 - x - 6\right) = 3(x-3)(x+2)$

43. $15k - 50 + 5k^2 = 5\left(k^2 + 3k - 10\right)$
$$= 5(k+5)(k-2)$$

45. $-4x^2 + 24x - 36 = -4\left(x^2 - 6x + 9\right)$
$$= -4(x-3)(x-3)$$
$$= -4(x-3)^2$$

47. $-x^2 + 11x - 10 = -1\left(x^2 - 11x + 10\right)$
$$= -(x-10)(x-1)$$

49. $3w^2 - 27w - 60 = 3\left(w^2 - 9w - 20\right)$

51. $4x^3 - 24x^2 + 32x = 4x\left(x^2 - 6x + 8\right)$
$$= 4x(x-4)(x-2)$$

53. $a^4 - 21a^3 + 20a^2 = a^2\left(a^2 - 21a + 20\right)$
$$= a^2(a-20)(a-1)$$

55. $5x^2y + 45xy^2 + 40y^3 = 5y\left(x^2 + 9xy + 8y^2\right)$
$$= 5y(x+8y)(x+y)$$

57. $4x^4y - 12x^3y^2 - 40x^2y^3$
$$= 4x^2y\left(x^2 - 3xy - 10y^2\right)$$
$$= 4x^2y(x-5y)(x+2y)$$

59. $-2x^3 y^2 + 16x^2 y^3 - 32xy^4$
$$= -2xy^2 \left(x^2 - 8xy + 16y^2 \right)$$
$$= -2xy^2 \left(x - 4y \right)\left(x - 4y \right)$$
$$= -2xy^2 \left(x - 4y \right)^2$$

61. The student did not complete the factoring.
$$2x^2 + 16x + 30 = 2\left(x^2 + 8x + 15 \right)$$
$$= 2\left(x + 5 \right)\left(x + 3 \right)$$

63. The student did not complete the factoring.
$$12x^3 + 18x^2 = 6x^2 \left(2x + 3 \right)$$

65. $\left(x - 3 \right)\left(x + 6 \right) = x^2 + 6x - 3x - 18$
$$= x^2 + 3x - 18$$
$$\left(x + 6 \right)\left(x - 3 \right) = x^2 - 3x + 6x - 18$$
$$= x^2 + 3x - 18$$

67. a. $x^2 - 5x + 4 = \left(x - 4 \right)\left(x - 1 \right)$

b.

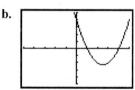

The x-intercepts are $(1, 0)$ and $(4, 0)$.

c. The factors in part (a) are equal to the x-intercept of f. This makes sense when $f(x) = 0$

69. Answers may vary. Examples:
$$x^2 + 5x + 4 = \left(x + 4 \right)\left(x + 1 \right)$$
$$x^2 + 8x + 16 = \left(x + 4 \right)\left(x + 4 \right) = \left(x + 4 \right)^2$$
$$x^2 + 2x - 8 = \left(x + 4 \right)\left(x - 2 \right)$$

71. To factor $x^2 + kx + 28$, find all the values of p and q such that $pq = 28$ and $p + q = k$.

Product = 28	Sum = k
$1(28) = 28$	$1 + 28 = 29$
$2(14) = 28$	$2 + 14 = 16$
$4(7) = 28$	$4 + 7 = 11$
$-1(-28) = 28$	$-1 + (-28) = -29$
$-2(-14) = 28$	$-2 + (-14) = -16$
$-4(-7) = 28$	$-4 + (-7) = -11$

73. Answers may vary. Example:
Factoring is the reverse process of multiplying polynomials. In factoring, the polynomial is separated into binomials which, when multiplied, again, yield the polynomial.

75. $\left(x + 5 \right)\left(x - 3 \right) = x^2 - 3x + 5x - 15$
$$= x^2 + 2x - 15$$

77. $k^2 - 7k - 30 = \left(k - 10 \right)\left(k + 3 \right)$

79. $\left(7x - 5 \right)\left(7x + 5 \right) = 49x^2 - 35x + 35x - 25$
$$= 49x^2 - 25$$

81. $81r^2 - 49 = \left(9r + 7 \right)\left(9r - 7 \right)$

83. $x^2 + 3x - 28 = \left(x + 7 \right)\left(x - 4 \right)$
This is a quadratic expression in one variable.

85. $\left(3w - 4 \right)\left(2w^2 + 3w - 5 \right) = 6w^3 + w^2 - 27w + 20$
This is a cubic expression in one variable.

Homework 6.5

1. $x^3 + 3x^2 + 4x + 12 = \left(x^2 + 4 \right)\left(x + 3 \right)$

3. $5x^3 - 20x^2 + 3x - 12 = \left(5x^2 + 3 \right)\left(x - 4 \right)$

5. $6m^3 - 15m^2 + 2m - 5 = \left(3m^2 + 1 \right)\left(2m - 5 \right)$

7. $10x^3 + 25x^2 - 2x - 5 = \left(5x^2 - 1 \right)\left(2x + 5 \right)$

9. $ax - 3ay - 2bx + 6by = \left(x - 3y \right)\left(a - 2b \right)$

11. $5a^2 x + 2a^2 y - 5bx - 2by = \left(a^2 - b \right)\left(5x + 2y \right)$

13. $3x^2 + 11x + 10 = \left(3x + 5 \right)\left(x + 2 \right)$

15. $2x^2 - x - 15 = \left(2x + 5 \right)\left(x - 3 \right)$

17. $5p^2 - 21p + 4 = \left(5p - 1 \right)\left(p - 4 \right)$

19. $4x^2 + 16x + 15 = \left(2x + 5 \right)\left(2x + 3 \right)$

21. This expression is prime.

23. $1 + 9w^2 - 6w = (1 - 3w)(1 - 3w)$
$= (1 - 3w)^2$
$= (3w - 1)^2$

25. $15x^2 + x - 6 = (5x - 3)(3x + 2)$

27. $6x^2 - 17x + 12 = (3x - 4)(2x - 3)$

29. $16y^2 - 29y - 6 = (16y + 3)(y - 2)$

31. $10a^2 + 21ab + 9b^2 = (5a + 3b)(2a + 3b)$

33. $20x^2 + 17xy - 3y^2 = (20x - 3y)(x + y)$

35. This expression is prime.

37. $4r^2 - 20ry + 25y^2 = (2r - 5y)(2r - 5y)$
$= (2r - 5y)^2$

39. $6x^2 + 26x - 20 = 2(3x^2 + 13x - 10)$
$= 2(3x - 2)(x + 5)$

41. $-12x^2 + 3x + 9 = 3(-4x^2 + x + 3)$
$= 3(4x + 3)(-x + 1)$

43. $12x - 32x^2 + 16x^3 = 4x(3 - 8x + 4x^2)$
$= 4x(2x - 3)(2x - 1)$

45. $30x^4 + 4x^3 - 2x^2 = 2x^2(15x^2 + 2x - 1)$
$= 2x^2(5x - 1)(3x + 1)$

47. $36t^3 + 48t^2w + 16tw^2 = 4t(9t^2 + 12tw + 4w^2)$
$= 4t(3t + 2w)(3t + 2w)$

49. $20a^3b^2 + 30a^2b^3 - 140ab^4$
$= 10ab^2(2a^2 + 3ab - 14b^2)$
$= 10ab^2(2a + 7b)(a - 2b)$

51. $x^2 - 6x - 40 = (x + 4)(x - 10)$

53. $3w^3 - 6w^2 + 5w - 10 = (3w^2 + 5)(w - 2)$

55. $3x^4 - 21x^3y - 54x^2y^2 = 3x^2(x^2 - 7xy - 18y^2)$
$= 3x^2(x + 2y)(x - 9y)$

57. This expression is prime.

59. $6x^2 - 19x + 10 = (3x - 2)(2x - 5)$

61. $x^2 + xy - 30y^2 = (x + 6y)(x - 5y)$

63. $-6r^3 + 24r^2 - 24r = -6r(r^2 - 4r + 4)$
$= -6r(r - 2)(r - 2)$
$= -6r(r - 2)^2$

65. $-10 + 12x^2 + 2x$
$= 2(6x^2 + x - 5)$
$= 2(6x - 5)(x + 1)$

67. $a^2x - 3a^2y - 2bx + 6by = (a^2 - 2b)(x - 3y)$

69. This expression is prime.

71. $10p^3t^2 + 22p^2t^3 - 24pt^4$
$= 2pt^2(5p^2 + 11pt - 12t^2)$
$= 2pt^2(5p - 4t)(p + 3t)$

73. The student did not complete the factoring.
$x^3 + 5x^2 - 3x - 15 = x^2(x + 5) - 3(x + 5)$
$= (x^2 - 3)(x + 5)$

75. The student did not first factor out the 3.
$3x^2 - 9x - 30 = 3(x^2 - 3x - 10)$
$= 3(x - 5)(x + 2)$

77. $2(x - 2)(x - 6) = 2x^2 - 16x + 24$
$2(x^2 - 8x + 12) = 2x^2 - 16x + 24$
$(x - 2)(2x - 12) = 2x^2 - 16x + 24$
$2(x - 4)^2 - 8 = 2x^2 - 16x + 24$
$(2x - 4)(x - 6) = 2x^2 - 16x + 24$

79. Answers may vary. Example:
You will have a simpler expression to factor if you first factor out the GCF.

$$8x^2 + 40x + 48 = 8\left(x^2 + 5x + 6\right)$$
$$= 8(x+2)(x+3)$$

81. $12x^3 - 27x = 3x\left(4x^2 - 9\right)$
$$= 3x(2x-3)(2x+3)$$

83. $-2(3p+5)(4p-3)$
$$= -2\left(12p^2 - 9p + 20p - 15\right)$$
$$= -2\left(12p^2 + 11p - 15\right)$$
$$= -24p^2 - 22p + 30$$

85. $2x^3 - 5x^2 - 18x + 45 = \left(x^2 - 9\right)(2x-5)$
$$= (x+3)(x-3)(2x-5)$$

87. $(3k+4)\left(2k^2 - k + 3\right)$
$$= 6k^3 - 3k^2 + 9k + 8k^2 - 4k + 12$$
$$= 6k^3 + 5k^2 + 5k + 12$$

89. $-2x(3x-5)^2 = -2x(3x-5)(3x-5)$
$$= -2x\left(9x^2 - 30x + 25\right)$$
$$= -18x^3 + 60x^2 - 50x$$
This is a cubic polynomial in one variable.

91. $8x^3 - 40x^2 + 50x = 2x\left(4x^2 - 20x + 25\right)$
$$= 2x(2x-5)(2x-5)$$
$$= 2x(2x-5)^2$$
This is a cubic polynomial in one variable.

Homework 6.6

1. $x^2 - 25 = (x+5)(x-5)$

3. $a^2 - 36 = (a+6)(a-6)$

5. $4x^2 - 49 = (2x+7)(2x-7)$

7. This expression is prime.

9. $16p^2 - 25t^2 = (4p+5t)(4p-5t)$

11. $75x^2 - 12 = 3\left(25x^2 - 4\right) = 3(5x+2)(5x-2)$

13. $18a^3b - 32ab^3 = 2ab\left(9a^2 - 16b^2\right)$
$$= 2ab(3a+4b)(3a-4b)$$

15. $16x^4 - 81 = \left(4x^2 + 9\right)\left(4x^2 - 9\right)$
$$= \left(4x^2 + 9\right)(2x+3)(2x-3)$$

17. $t^4 - w^4 = \left(t^2 + w^2\right)\left(t^2 - w^2\right)$
$$= \left(t^2 + w^2\right)(t+w)(t-w)$$

19. $x^3 + 27 = (x+3)\left(x^2 - 3x + 9\right)$

21. $x^3 - 8 = (x-2)\left(x^2 + 2x + 4\right)$

23. $m^3 + 1 = (m+1)\left(m^2 - m + 1\right)$

25. $8x^3 + 27 = (2x+3)\left(4x^2 - 6x + 9\right)$

27. $125x^3 - 8 = (5x-2)\left(25x^2 + 10x + 4\right)$

29. $27p^3 + 8t^3 = (3p+2t)\left(9p^2 - 6pt + 4t^2\right)$

31. $27x^3 - 64y^3 = (3x-4y)\left(9x^2 + 12xy + 16y^2\right)$

33. $5x^3 + 40 = 5\left(x^3 + 8\right) = 5(x+2)\left(x^2 - 2x + 4\right)$

35. $2x^4 - 54xy^3 = 2x\left(x^3 - 27y^3\right)$
$$= 2x(x-3y)\left(x^2 + 3xy + 9y^2\right)$$

37. $k^6 - 1 = \left(k^3 - 1\right)\left(k^3 + 1\right)$
$$= (k-1)\left(k^2 + k + 1\right)(k+1)\left(k^2 - k + 1\right)$$

39. $64x^6 - y^6$
$$= \left(8x^3 - y^3\right)\left(8x^3 + y^3\right)$$
$$= (2x-y)\left(4x^2 + 2xy + y^2\right)(2x+y)\left(4x^2 - 2xy + y^2\right)$$

41. $a^2 - 3ab - 28b^2 = (a-7b)(a+4b)$

43. $2x^4 - 16xy^3 = 2x\left(x^3 - 8y^3\right)$
$$= 2x(x - 2y)\left(x^2 + 2xy + 4y^2\right)$$

45. $-7x - 18 + x^2 = (x - 9)(x + 2)$

47. $4x^3y - 8x^2y^2 - 96xy^3$
$$= 4xy\left(x^2 - 2xy - 24y^2\right)$$
$$= 4xy(x - 6y)(x + 4y)$$

49. $-k^2 + 12k - 36 = -1\left(k^2 - 12k + 36\right)$
$$= -1(k - 6)(k - 6)$$
$$= -(k - 6)^2$$

51. This expression is prime.

53. $x^3 - 2x^2 - 9x + 18 = \left(x^2 - 9\right)(x - 2)$
$$= (x + 3)(x - 3)(x - 2)$$

55. $6x^4 - 33x^3 + 45x^2 = 3x^2\left(2x^2 - 11x + 15\right)$
$$= 3x^2(2x - 5)(x - 3)$$

57. $32m^2 - 98t^2 = 2\left(16m^2 - 49t^2\right)$
$$= 2(4m + 7t)(4m - 7t)$$

59. $8x^2 + 10x - 3 = (2x + 3)(4x - 1)$

61. $12x^2y - 26xy^2 - 10y^3$
$$= 2y\left(6x^2 - 13xy - 5y^2\right)$$
$$= 2y(3x + y)(2x - 5y)$$

63. This expression is prime.

65. $125x^3 + 27 = (5x + 3)\left(25x^2 - 15x + 9\right)$

67. $p^2 + 18p + 81 = (p + 9)(p + 9) = (p + 9)^2$

69. $20x^3 - 8x^2 - 5x + 2 = \left(4x^2 - 1\right)(5x - 2)$
$$= (2x + 1)(2x - 1)(5x - 2)$$

71. $49x^2 + 14x + 1 = (7x + 1)(7x + 1) = (7x + 1)^2$

73. This expression is prime.

75. $2w^3y + 250y^4$
$$= 2y\left(w^3 + 125y^3\right)$$
$$= 2y(w + 5y)\left(w^2 - 5wy + 25y^2\right)$$

77. $-3x^3 + 3x^2 + 90x = -3x\left(x^2 - x - 30\right)$
$$= -3x(x + 5)(x - 6)$$

79. $27x^3 - 75x = 3x\left(9x^2 - 25\right)$
$$= 3x(3x + 5)(3x - 5)$$

81. $81p^4 - 16q^4$
$$= \left(9p^2 + 4q^2\right)\left(9p^2 - 4q^2\right)$$
$$= \left(9p^2 + 4q^2\right)(3p + 2q)(3p - 2q)$$

83. The student factored the AB portion of
$$A^3 - B^3 = (A - B)\left(A^2 + AB + B^2\right)$$
incorrectly, as $A = x$ and $B = 2$.
$$x^3 - 8 = (x - 2)\left(x^2 + 2x + 4\right)$$

85. The student incorrectly factors $x^2 + 25$, which is a prime expression.
$$4x^2 + 100 = 4\left(x^2 + 25\right)$$

87. a. $x^2 - 4 = (x + 2)(x - 2)$

b.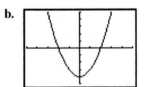

The x-intercepts are $(-2, 0)$ and $(2, 0)$.

c. The factors in part (a) match the x-intercepts of f. This makes sense because the quadratic function equals 0 when $x = 2$ or $x = -2$.

89. $(A - B)\left(A^2 + AB + B^2\right)$
$$= A^3 + A^2B + AB^2 - A^2B - AB^2 - B^3$$
$$= A^3 - B^3$$

91. Answers may vary. Example:
To factor a difference of two squares, find the product of the sum of the terms and the difference of the terms. To factor the sum of two cubes, find the product of the sum of the terms and the sum of the squares of the terms minus the difference of the product of the terms. To factor the difference of two cubes, find the product of the difference of the terms and the sum of the squares of the terms plus the product of the terms.

93. $(3x-7)(3x+7) = 9x^2 - 49$

95. $36p^2 - 49 = (6p+7)(6p-7)$

97. $(t-5)(t^2 + 5t + 25) = t^3 - 125$

99. $27p^3 + 1 = (3p+1)(9p^2 - 3p + 1)$

101. $2(3x - 2y) = 3x + 4$
$6x - 4y = 3x + 4$
$-4y = -3x + 4$
$y = \dfrac{3}{4}x - 1$

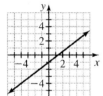

This is a linear equation in two variables.

103. $3x - 5y = 21$
$2x + 7y = -17$
$x - 12y = 38$
$\quad x = 38 + 12y$
$3(38 + 12y) - 5y = 21$
$114 + 36y - 5y = 21$
$\quad\quad\quad 31y = -93$
$\quad\quad\quad\quad y = -3$
$x - 12(-3) = 38$
$\quad\quad x = 2$
The solution is (2, –3).
This is a system of linear equations in two variables.

Homework 6.7

1. $(x+4)(w-7) = 0$
$x = -4$ or $x = 7$

3. $w^2 + w - 12 = 0$
$(w+4)(w-3) = 0$
$w = -4$ or $w = 3$

5. $x^2 - 8x + 15 = 0$
$(x-5)(x-3) = 0$
$x = 5$ or $x = 3$

7. $14x + 49 + x^2 = 0$
$(x+7)(x+7) = 0$
$x = -7$

9. $-24 - 2t + t^2 = 0$
$(t-6)(t+4) = 0$
$t = 6$ or $t = -4$

11. $25x^2 - 49 = 0$
$(5x+7)(5x-7) = 0$
$x = -\dfrac{7}{5}$ or $x = \dfrac{7}{5}$

13. $6m^2 - 11m + 3 = 0$
$(3m-1)(2m-3) = 0$
$m = \dfrac{1}{3}$ or $m = \dfrac{3}{2}$

15. $3x^2 + 3x - 90 = 0$
$3\left(x^2 + x - 30\right) = 0$
$3(x+6)(x-5) = 0$
$x = -6$ or $x = 5$

17. $8x^3 - 12x^2 - 20x = 0$
$4x\left(2x^2 - 3x - 5\right) = 0$
$4x(2x-5)(x+1) = 0$
$x = 0, \; x = \dfrac{5}{2}, \text{ or } x = -1$

19. $x^2 = 5x + 14$
$x^2 - 5x - 14 = 0$
$(x-7)(x+2) = 0$
$x = 7$ or $x = -2$

21.
$$4x^2 - 8x = 32$$
$$4x^2 - 8x - 32 = 0$$
$$4\left(x^2 - 2x - 8\right) = 0$$
$$4(x-4)(x+2) = 0$$
$$x = 4 \text{ or } x = -2$$

23.
$$12t - 36 = t^2$$
$$t^2 - 12t + 36 = 0$$
$$(t-6)(t-6) = 0$$
$$t = 6$$

25.
$$16x^2 = 25$$
$$16x^2 - 25 = 0$$
$$(4x+5)(4x-5) = 0$$
$$x = -\frac{5}{4} \text{ or } x = \frac{5}{4}$$

27.
$$6x^3 - 24x = 0$$
$$6x\left(x^2 - 4\right) = 0$$
$$6x(x-2)(x+2) = 0$$
$$x = 0, \ x = 2, \text{ or } x = -2$$

29.
$$3r^2 = 6r$$
$$3r^2 - 6r = 0$$
$$3r(r-2) = 0$$
$$r = 0 \text{ or } r = 2$$

31.
$$9x = -2x^2 + 5$$
$$2x^2 + 9x - 5 = 0$$
$$(2x-1)(x+5) = 0$$
$$x = \frac{1}{2} \text{ or } x = -5$$

33.
$$2x^3 = 6x^2 + 36x$$
$$2x^3 - 6x^2 - 36x = 0$$
$$2x\left(x^2 - 3x - 18\right) = 0$$
$$2x(x-6)(x+3) = 0$$
$$x = 0, \ x = 6, \text{ or } x = -3$$

35.
$$18y^3 + 3y^2 = 6y$$
$$18y^3 + 3y^2 - 6y = 0$$
$$3y\left(6y^2 + y - 2\right) = 0$$
$$3y(3y+2)(2y-1) = 0$$
$$y = 0, \ y = -\frac{2}{3}, \text{ or } y = \frac{1}{2}$$

37.
$$20x = -4x^2 - 25$$
$$4x^2 + 20x + 25 = 0$$
$$(2x+5)(2x+5) = 0$$
$$x = -\frac{5}{2}$$

39.
$$\frac{1}{4}x^2 - \frac{1}{2}x = 6$$
$$\frac{1}{4}x^2 - \frac{1}{2}x - 6 = 0$$
$$\frac{1}{4}\left(x^2 - 2x - 24\right) = 0$$
$$\frac{1}{4}(x-6)(x+4) = 0$$
$$x = 6 \text{ or } x = -4$$

41.
$$\frac{a^2}{2} - \frac{a}{6} = \frac{1}{3}$$
$$\frac{a^2}{2} - \frac{a}{6} - \frac{1}{3} = 0$$
$$\frac{1}{6}\left(3a^2 - a - 2\right) = 0$$
$$\frac{1}{6}(3a+2)(a-1) = 0$$
$$a = -\frac{2}{3} \text{ or } a = 1$$

43.
$$x^2 - \frac{1}{25} = 0$$
$$\left(x - \frac{1}{5}\right)\left(x + \frac{1}{5}\right) = 0$$
$$x = \frac{1}{5} \text{ or } x = -\frac{1}{5}$$

45.
$$(x+2)(x+5) = 40$$
$$x^2 + 7x + 10 - 40 = 0$$
$$x^2 + 7x - 30 = 0$$
$$(x-3)(x+10) = 0$$
$$x = 3 \text{ or } x = -10$$

47. $4r^3 - 2r^2 - 36r + 18 = 0$

$\left(2r^2 - 18\right)\left(2r - 1\right) = 0$

$2\left(r^2 - 9\right)\left(2r - 1\right) = 0$

$2\left(r + 3\right)\left(r - 3\right)\left(2r - 1\right) = 0$

$r = -3,\ r = 3,\ \text{or}\ r = \dfrac{1}{2}$

49. $9x^3 - 12 = 4x - 27x^2$

$9x^3 + 27x^2 - 4x - 12 = 0$

$\left(9x^2 - 4\right)\left(x + 3\right) = 0$

$\left(3x + 2\right)\left(3x - 2\right)\left(x + 3\right) = 0$

$x = -\dfrac{2}{3},\ x = \dfrac{2}{3},\ \text{or}\ x = -3$

51. $2x\left(x + 1\right) = 5x\left(x - 7\right)$

$5x^2 - 35x - 2x^2 - 2x = 0$

$3x^2 - 37x = 0$

$x\left(3x - 37\right) = 0$

$x = 0\ \text{or}\ x = \dfrac{37}{3}$

53. $4p\left(p - 1\right) - 24 = 3p\left(p - 2\right)$

$4p^2 - 4p - 24 - 3p^2 + 6p = 0$

$p^2 + 2p - 24 = 0$

$\left(p + 6\right)\left(p - 4\right) = 0$

$p = -6\ \text{or}\ p = 4$

55. $\left(x^2 + 5x + 6\right)\left(x^2 - 5x - 24\right) = 0$

$\left(x + 3\right)\left(x + 2\right)\left(x - 8\right)\left(x + 3\right) = 0$

$x = -3,\ x = -2,\ \text{or}\ x = 8$

57.   Substitute 0 for $f(x)$ and solve for x:

$x^2 - 9x + 20 = 0$

$\left(x - 5\right)\left(x - 4\right) = 0$

$x = 5\ \text{or}\ x = 4$

The x-intercepts are $\left(5, 0\right)$ and $\left(4, 0\right)$.

59.   Substitute 0 for $f(x)$ and solve for x:

$36x^2 - 25 = 0$

$\left(6x + 5\right)\left(6x - 5\right) = 0$

$x = \dfrac{5}{6}\ \text{or}\ x = -\dfrac{5}{6}$

The x-intercepts are $\left(-\dfrac{5}{6}, 0\right)$ and $\left(\dfrac{5}{6}, 0\right)$.

61.   Substitute 0 for $f(x)$ and solve for x:

$24x^3 - 14x^2 - 20x = 0$

$2x\left(12x^2 - 7x - 10\right) = 0$

$2x\left(4x - 5\right)\left(3x + 2\right) = 0$

$x = 0,\ x = \dfrac{5}{4},\ \text{or}\ x = -\dfrac{2}{3}$

The x-intercepts are $\left(0, 0\right)$, $\left(\dfrac{5}{4}, 0\right)$, and

$\left(-\dfrac{2}{3}, 0\right)$.

63.   Substitute 0 for $f(x)$ and solve for x:

$x^3 + 2x^2 - x - 2 = 0$

$\left(x^2 - 1\right)\left(x + 2\right) = 0$

$\left(x - 1\right)\left(x + 1\right)\left(x + 2\right) = 0$

$x = 1,\ x = -1,\ \text{or}\ x = -2$

The x-intercepts are $\left(1, 0\right)$, $\left(-1, 0\right)$, and $\left(-2, 0\right)$.

65. $f(3) = \left(3\right)^2 - \left(3\right) - 6 = 0$

67. $x^2 - x - 6 = 14$

$x^2 - x - 20 = 0$

$\left(x - 5\right)\left(x + 4\right) = 0$

$x = 5\ \text{or}\ x = -4$

69. $x = -5\ \text{or}\ x = 3$

71. $x = -1$

73. $x = -1\ \text{or}\ x = 2$

75. $x = -1,\ x = 1,\ \text{or}\ x = 3$

77.

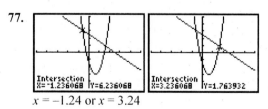

$x = -1.24\ \text{or}\ x = 3.24$

79.

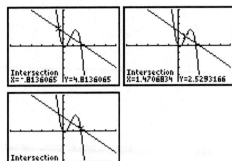

$x = -0.81$, $x = 1.47$, or $x = 3.34$

81. $x = -2$ or $x = 4$

83. There is no solution.

85. a.

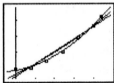

The quadratic model $Q(t)$ describes the situation best.

b.

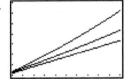

The quadratic model $Q(t)$ predicts the largest participation between 2010 and 2020.

c. For $L(t)$, the n-intercept is $(0, 1301)$. For $E(t)$, the n-intercept is $(0, 1325)$. For $Q(t)$, the n-intercept is $(0, 1376)$. The quadratic model $Q(t)$ describes the situation best.

d.
$$t^2 + 8t + 1376 = 1396$$
$$t^2 + 8t - 20 = 0$$
$$(t + 10)(t - 2) = 0$$
$$t = -10 \text{ or } t = 2$$

Assuming that the competition took place before 1990, there would have been 1396 participants in $1990 - 10 = 1980$, and in $1990 + 2 = 1992$.

87. a.

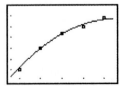

The model fits the data well.

b.
$$f(7) = -\frac{1}{3}(7)^2 + \frac{22}{3}(7) = 35$$

This means that 7 years after being rated B2, 35% of companies defaulted on their bonds.

c.
$$7 = -\frac{1}{3}t^2 + \frac{22}{3}t$$
$$\frac{1}{3}t^2 - \frac{22}{3}t + 7 = 0$$
$$\frac{1}{3}\left(t^2 - 22t + 21\right) = 0$$
$$\frac{1}{3}(t - 21)(t - 1) = 0$$
$$t = 1 \text{ or } t = 21$$

This means that 1 year and 21 years after being rated B2, 7% of companies default on their bonds. Model breakdown has occurred for the estimate of 21 years.

d.
$$0 = -\frac{1}{3}t^2 + \frac{22}{3}t$$
$$0 = -\frac{1}{3}t(t - 22)$$
$$t = 0 \text{ or } t = 22$$

This means that at the time of the B2 rating and 22 years after receiving a B2 rating, no companies defaulted on their bonds. Model breakdown has occurred for the estimate of 22 years.

89. a.

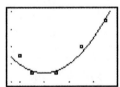

The model fits the data well.

b. The p-intercept is $(0, 28)$, which means that in 2000, 28% of Americans thought that labor unions would become stronger.

c. $f(17) = \frac{1}{4}(17)^2 - 3(17) + 28 = 49.25$

In 2017, 49% of Americans will think that labor unions will become stronger.

d. $\dfrac{1}{4}t^2 - 3t + 28 = 44$

$\dfrac{1}{4}t^2 - 3t - 16 = 0$

$t^2 - 12t - 64 = 0$

$(t + 4)(t - 16) = 0$

$t = -4$ or $t = 16$

Assuming that the model holds for years before 2000, 44% of Americans thought that labor unions would become stronger in $2000 - 4 = 1996$, and in $2000 + 16 = 2016$.

e. The function $f(t)$ is a line with a negative slope. The percentage of private-sector workers who are in a union decreased from 2007 to 2011.

91. a.

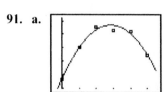

The model fits the data well.

b. $f(7) = -98(7)^2 + 561(7) + 11,536 = 10,661$

The model shows that the total attendance (in thousands) at Broadway shows in the year 2012 was 10,661 thousand people, or 10.661 million people.

c. $-98t^2 + 561t + 11,536 = 11,536$

$-98t^2 + 561t = 0$

$t(-98t + 561) = 0$

$t = 0$ or $t = \dfrac{561}{98} \approx 6$

The total attendance was 11,536 thousand (11.536 million) in 2005 and 2011.

d. No; the model predicts attendance will continue to decrease.

93. $A = l \times w = 60$

$l = 2w + 2$

$(2w + 2)w = 60$

$2w^2 + 2w - 60 = 0$

$2(w^2 + w - 30) = 0$

$2(w + 6)(w - 5) = 0$

$w = -6$ or $w = 5$

$l = 2(5) + 2 = 12$

The rug is 12 feet long by 5 feet wide.

95. $l = w + 4$

$2l \times 2w = 48$

$2(w + 4)(2w) - 48 = 0$

$2(2w^2 + 8w) - 48 = 0$

$4w^2 + 16w - 48 = 0$

$4(w^2 + 4w - 12) = 0$

$4(w + 6)(w - 2) = 0$

$w = -6$ or $w = 2$

$l = 2 + 4 = 6$

The original rectangle is 6 centimeters long and 2 centimeters wide.

97. $A = 6 \times 10 = 60$

$(6 + 2x)(10 + 2x) - 60 = 80$

$4x^2 + 32x - 80 = 0$

$4(x^2 + 8x - 20) = 0$

$4(x + 10)(x - 2) = 0$

$x = -10$ or $x = 2$

The border is 2 feet wide.

99. $A = (14 \times 10) + 52 = 140 + 52 = 192$

$(14 + 2x)(10 + 2x) = A$

$(14 + 2x)(10 + 2x) = 192$

$140 + 48x + 4x^2 - 192 = 0$

$4x^2 + 48x - 52 = 0$

$4(x^2 + 12x - 13) = 0$

$4(x + 13)(x - 1) = 0$

$x = -13$ or $x = 1$

The frame has a border width of 1 inch. The actual width of the frame is 12 inches.

101. You cannot factor out solutions, as this eliminates possible answers.

$x^2 = x$

$x^2 - x = 0$

$x(x - 1) = 0$

$x = 0$ or $x = 1$

103. The student overlooked one of the factors in the third step.

$x^3 + 4x^2 - 9x - 36 = 0$

$x^2(x + 4) - 9(x + 4) = 0$

$(x^2 - 9)(x + 4) = 0$

$(x + 3)(x - 3)(x + 4) = 0$

$x = -3$, $x = 3$, or $x = -4$

105. Answers may vary. Example:

The function $f(x) = x^2 + 4x - 5$ has the

x-intercepts $(-5, 0)$ and $(1, 0)$.

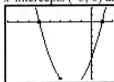

107. Answers may vary. Example:

The cubic function $f(x) = x^3 + 2x^2 - 8x$ has

solutions -4, 0, and 2.

109. Answers may vary. Example:

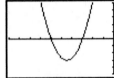

The quadratic function $h(x) = \dfrac{4}{3}x^2 + 12x + 24$

is a possible match for the graph.

111. Answers may vary. Example:

A quadratic equation in one variable can only

cross the x-axis up to two times, so it can have

a maximum of two solutions.

113. $x^2 + 5x + 6 = (x+3)(x+2)$

115. $x^2 + 5x + 6 = 0$

$(x+3)(x+2) = 0$

$x = -3$ or $x = -2$

117. $3p^3 + 8p^2 + 4p = 0$

$p(3p^2 + 8p + 4) = 0$

$p(3p+2)(p+2) = 0$

$p = 0,\ p = -\dfrac{2}{3},\ \text{or}\ p = -2$

119. $3p^3 + 8p^2 + 4p = p(3p^2 + 8p + 4)$

$\qquad\qquad\qquad = p(3p+2)(p+2)$

121. $3x^2(x-4) = 12x(x-3)$

$3x^3 - 12x^2 = 12x^2 - 36x$

$3x^3 - 24x^2 + 36x = 0$

$3x(x^2 - 8x + 12) = 0$

$3x(x-6)(x-2) = 0$

$x = 0,\ x = 6,\ \text{or}\ x = 2$

123. $3b^8 + 39 = 217$

$3b^8 = 178$

$b^8 = \dfrac{178}{3}$

$b \approx \pm 1.6660$

125. $5(2)^t - 24 = 97$

$5(2)^t = 121$

$2^t = 24.2$

$t \approx 4.5969$

127. $\log(x+3) + \log(x+6) = 1$

$\log\big[(x+3)(x+6)\big] = 1$

$x^2 + 9x + 18 = 10$

$x^2 + 9x + 8 = 0$

$(x+8)(x+1) = 0$

$x = -1$ or $x = -8$

By substitution, $x \ne -8$. So, $x = -1$.

129. Answers may vary. Example:

$x^2 + 8x + 16 = (x+4)(x+4)$

131. Answers may vary. Example:

$x^2 + 7x + 6 = 0$

$(x+6)(x+1) = 0$

$x = -6$ or $x = -1$

133. Answers may vary. Example:

$f(x) = 2^x - 3$

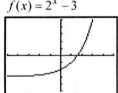

Chapter 6 Review

1. $\left(-7x^3 + 5x^2 - 9\right) + \left(2x^3 - 8x^2 + 3x\right)$

   $= -5x^3 - 3x^2 + 3x - 9$

2. $\left(5a^3b - 2a^2b^2 + 9ab^3\right) - \left(8a^3b + 4a^2b^2 - ab^3\right)$

   $= -3a^3b - 6a^2b^2 + 10ab^3$

3. $f(-2) = 3(-2)^2 - 5(-2) + 2 = 24$

4. $f(2) = 9$

5. $x = 2$ or $x = 4$

6. $(f+g)(x)$

   $= \left(3x^3 - 7x^2 - 4x + 2\right) + \left(-2x^3 + 5x^2 - 3x + 1\right)$

   $= x^3 - 2x^2 - 7x + 3$

   $(f+g)(2) = (2)^3 - 2(2)^2 - 7(2) + 3 = -11$

7. $(f-g)(x)$

   $= \left(3x^3 - 7x^2 - 4x + 2\right) - \left(-2x^3 + 5x^2 - 3x + 1\right)$

   $= 5x^3 - 12x^2 - x + 1$

   $(f-g)(-3) = 5(-3)^3 - 12(-3)^2 - (-3) + 1$

   $\qquad\qquad = -239$

8. $(x-7)(x+7) = x^2 - 49$

9. $8a^2b\left(-5a^3b^5\right) = -40a^5b^6$

10. $(4p + 9t)(2p - 5t) = 8p^2 - 2pt - 45t^2$

11. $(4x - 3)\left(5x^2 - 2x + 4\right) = 20x^3 - 23x^2 + 22x - 12$

12. $(3x + 7y)^2 = 9x^2 + 42xy + 49y^2$

13. $\left(6p^2 - 9t^3\right)\left(6p^2 + 9t^3\right) = 36p^4 - 81t^6$

14. $-3rt^3\left(2r^2 - 5rt + 3t^2\right) = -6r^3t^3 + 15r^2t^4 - 9rt^5$

15. $-4x(3x - 2)^2 = -36x^3 + 48x^2 - 16x$

16. $\left(3m^2 - mp + 2p^2\right)\left(2m^2 + 3mp - 4p^2\right)$

    $= 6m^4 + 7m^3p - 11m^2p^2 + 10mp^3 - 8p^4$

17. $f(a-4) = (a-4)^2 - 2(a-4) = a^2 - 10a + 24$

18. $f(a+3) - f(a)$

    $= (a+3)^2 - 2(a+3) - \left(a^2 - 2a\right)$

    $= 6a + 3$

19. $f(x) = -2(x-4)^2 + 3 = -2x^2 + 16x - 29$

20. $(f \cdot g)(x) = (3x - 7)\left(2x^2 - 4x + 3\right)$

    $= 6x^3 - 26x^2 + 37x - 21$

    $(f \cdot g)(3) = 6(3)^3 - 26(3)^2 + 37(3) - 21 = 18$

21. $\dfrac{6x^5 - 2x^3 + 9x}{-3x^2} = \dfrac{6x^5}{-3x^2} + \dfrac{-2x^3}{-3x^2} + \dfrac{9x}{-3x^2}$

    $\qquad\qquad = -2x^3 + \dfrac{2}{3}x - \dfrac{3}{x}$

22. $\dfrac{20w^4p + 15w^3p^2 - 35w^2p^3}{5w^2p}$

    $= \dfrac{20w^4p}{5w^2p} + \dfrac{15w^3p^2}{5w^2p} + \dfrac{-35w^2p^3}{5w^2p}$

    $= 4w^2 + 3wp - 7p^2$

23. $2x+1\overline{)8x^2 - 6x - 8}$ quotient $4x - 5$

    $\qquad\underline{-8x^2 - 4x}$

    $\qquad\qquad -10x - 8$

    $\qquad\qquad \underline{10x + 5}$

    $\qquad\qquad\qquad -3$

    $\dfrac{8x^2 - 6x - 8}{2x + 1} = 4x - 5 - \dfrac{3}{2x + 1}$

24.
$$
\begin{array}{r}
2x^2 - 4x + 3 \\
3x - 2 \overline{)6x^3 - 16x^2 + 17x - 2} \\
\underline{-3x^3 + 4x^2} \\
-12x^2 + 17x \\
\underline{12x^2 - 8x} \\
9x - 2 \\
\underline{-9x + 6} \\
4
\end{array}
$$

$$\frac{6x^3 - 16x^2 + 17x - 2}{3x - 2} = 2x^2 - 4x + 3 + \frac{4}{3x - 2}$$

25.
$$
\begin{array}{r}
16b^2 + 12b + 9 \\
4b - 3 \overline{)64b^3 + 0b^2 + 0b - 27} \\
\underline{-64b^3 + 48b^2} \\
48b^2 + 0b \\
\underline{-48b^2 + 36b} \\
36b - 27 \\
\underline{-36b + 27} \\
0
\end{array}
$$

$$\frac{64b^3 - 27}{4b - 3} = 16b^2 + 12b + 9$$

26.
$$
\begin{array}{r|rrrr}
3 & 2 & -10 & 15 & -4 \\
  &   & 6   & -12 & 9 \\
\hline
  & 2 & -4 & 3   & 5
\end{array}
$$

$$\frac{2x^3 - 10x^2 + 15x - 4}{x - 3} = 2x^2 - 4x + 3 + \frac{5}{x - 3}$$

27. $5y^2 - 8 + 3y^3 - 6y = 3y^3 + 5y^2 - 6y - 8$

$$
\begin{array}{r|rrrr}
-2 & 3 & 5 & -6 & -8 \\
   &   & -6 & 2 & 8 \\
\hline
   & 3 & -1 & -4 & 0
\end{array}
$$

$$\frac{5y^2 - 8 + 3y^3 - 6y}{y + 2} = 3y^2 - y - 4$$

28. $x^3 + x^2 + 16 = x^3 + x^2 + 0x + 16$

$$
\begin{array}{r|rrrr}
-3 & 1 & 1 & 0 & 16 \\
   &   & -3 & 6 & -18 \\
\hline
   & 1 & -2 & 6 & -2
\end{array}
$$

$$\frac{x^3 + x^2 + 16}{x + 3} = x^2 - 2x + 6 - \frac{2}{x + 3}$$

29. $x^2 - 25 = (x + 5)(x - 5)$

30. $x^2 - 12x + 36 = (x - 6)(x - 6) = (x - 6)^2$

31. $a^2 + 5ab - 36b^2 = (a + 9b)(a - 4b)$

32. $16a^5b^3 - 20a^3b^2 = 4a^3b^2(4a^2b - 5)$

33. This expression is prime.

34. $3w^2 - 5wy - 8y^2 = (3w - 8y)(w + y)$

35. $81t^4 - 16w^4 = (9t^2 + 4w^2)(3t + 2w)(3t - 2w)$

36. $6x^4 + 20x^3 - 16x^2 = 2x^2(3x - 2)(x + 4)$

37. This expression is prime.

38. $x^2 - 3x - 54 = (x + 6)(x - 9)$

39. $2y^3 - 54 = 2(y - 3)(y^2 + 3y + 9)$

40. $5r^2t + 30rt^2 + 45t^3 = 5t(r + 3t)(r + 3t)$
$$= 5t(r + 3t)^2$$

41. $2ax - 10ay - 3bx + 15by = (2a - 3b)(x - 5y)$

42. $\quad x^2 - 2x - 24 = 0$
$(x - 6)(x + 4) = 0$
$x = 6$ or $x = -4$

43. $\qquad 64t^2 = 9$
$64t^2 - 9 = 0$
$(8t + 3)(8t - 3) = 0$
$t = -\dfrac{3}{8}$ or $t = \dfrac{3}{8}$

44. $\qquad 3x(x + 10) = 6x^3$
$6x^3 - 3x^2 - 30x = 0$
$3x(2x^2 - x - 10) = 0$
$3x(2x - 5)(x + 2) = 0$
$x = 0,\ x = \dfrac{5}{2}$ or $x = -2$

45.
$$x^3 - 4x = 12 - 3x^2$$
$$x^3 + 3x^2 - 4x - 12 = 0$$
$$(x-2)(x+2)(x+3) = 0$$
$$x = 2, \ x = -2 \text{ or } x = -3$$

46.
$$\frac{m^2}{2} - \frac{7m}{6} + \frac{1}{3} = 0$$
$$3m^2 - 7m + 2 = 0$$
$$(3m-1)(m-2) = 0$$
$$m = \frac{1}{3} \text{ or } m = 2$$

47.
$$32x^2 = 24x$$
$$32x^2 - 24x = 0$$
$$8x(4x-3) = 0$$
$$x = 0 \text{ or } x = \frac{3}{4}$$

48.
$$4p(5p-6) = (2p+3)(2p-3)$$
$$20p^2 - 24p = 4p^2 - 9$$
$$16p^2 - 24p + 9 = 0$$
$$(4p-3)(4p-3) = 0$$
$$p = \frac{3}{4}$$

49. $f(x) = 3x^3 + 3x^2 - 18x$
$$= 3x(x^2 - x - 6)$$
$$= 3x(x+3)(x-2)$$
The x-intercepts are $(0,0)$, $(-3,0)$, and $(2,0)$.

50. $x = -5$ and $x = 1$

51. $x = -3$ and $x = 1$

52. a. $(B+A)(t)$
$$= \left(-2.42t^2 + 80.8t - 76\right)$$
$$+ \left(-2.29t^2 + 76.8t - 227\right)$$
$$= -2.42t^2 + 80.8t - 76 - 2.29t^2 + 76.8t$$
$$- 227$$
$$= -4.71t^2 + 157.6t - 303$$

b. For the expression $B(t) + A(t)$, we have $B(t)$ = number of bank tellers in thousands + $A(t)$ = number of ATMs in thousands. The units of the expression are in thousands of bank tellers and ATMs.

c. $(B+A)(22)$
$$= -4.71(22)^2 + 157.6(22) - 303$$
$$= 884.56$$
It means that in $1990 + 22 = 2012$, there were a total of about 885 thousand bank tellers and ATMs.

d. $(B-A)(t)$
$$= \left(-2.42t^2 + 80.8t - 76\right)$$
$$- \left(-2.29t^2 + 76.8t - 227\right)$$
$$= -2.42t^2 + 80.8t - 76 + 2.29t^2 - 76.8t$$
$$+ 227$$
$$= -0.13t^2 + 4t + 151$$

e. $(B-A)(22) = -0.13(22)^2 + 4(22) + 151$
$$= 176.08$$
It means that, in 2012, there were about 176 thousand more bank tellers than ATMs.

53. a.

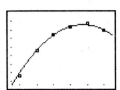

The model fits the data well.

b. $f(7) = -2(7)^2 + 19(7) + 203 = 238$
In 2012, there will be 238 million prescriptions of synthetic narcotics, morphine/opium derivatives, and codeine.

c. $-2t^2 + 19t + 203 = 193$
$$-2t^2 + 19t + 10 = 0$$
$$2t^2 - 19t - 10 = 0$$
$$(2t+1)(t-10) = 0$$
$$t = -\frac{1}{2} \text{ or } t = 10$$
In $2005 + 10 = 2015$, the total number of prescriptions of synthetic narcotics, morphine/opium derivatives, and codeine will be 193 million prescriptions.

54. $l = w + 8$

$2l \times 2w = 192$

$2(w+8)(2w) = 192$

$4w^2 + 32w - 192 = 0$

$4(w^2 + 8w - 48) = 0$

$4(w+12)(w-4) = 0$

$w = -12$ or $w = 4$

$l = 4 + 8 = 12$

The original rectangle is 12 meters long and 4 meters wide.

Chapter 6 Test

1. $\left(4a^3b - 9a^2b^2 - 2ab^3\right)$

$\qquad + \left(-5a^3b + 4a^2b^2 + 3ab^3\right)$

$\qquad = -a^3b - 5a^2b^2 + ab^3$

2. $f - g = \left(4x^2 + 5x - 9\right) - \left(6x^2 - 3x + 7\right)$

$\qquad = -2x^2 + 8x - 16$

$(f-g)(-2) = -2(-2)^2 + 8(-2) - 16 = -40$

3. $f(-3) = -3$

4. There is no solution.

5. $x = 1$

6. $x = -3$ and $x = 5$

7. $-2xy^2\left(7x^2 - 3xy + 6y^2\right)$

$\qquad = -14x^3y^2 + 6x^2y^3 - 12xy^4$

8. $(4x - 7y)(3x + 5y) = 12x^2 - xy - 35y^2$

9. $(2w - 5t)\left(3w^2 - wt + 4t^2\right)$

$\qquad = 6w^3 - 17w^2t + 13wt^2 - 20t^3$

10. $3x(2x+3)^2 = 12x^3 + 36x^2 + 27x$

11. $\left(3x^2 + x - 5\right)\left(2x^2 + 4x - 1\right)$

$\qquad = 6x^4 + 14x^3 - 9x^2 - 21x + 5$

12. $\left(4x^2 + 9y^2\right)\left(4x^2 - 9y^2\right) = 16x^4 - 81y^4$

13. $f(a-5) = (a-5)^2 - 3(a-5) = a^2 - 13a + 40$

14. $f(x) = -3(x+4)^2 - 7 = -3x^2 - 24x - 55$

15. $f \cdot g = \left(2x^2 - 5x + 4\right)(3x - 2)$

$\qquad = 6x^3 - 19x^2 + 22x - 8$

$(f \cdot g)(3) = 6(3)^3 - 19(3)^2 + 22(3) - 8 = 49$

16. $\dfrac{8a^4b - 12a^3b^2 - 20a^2b^3}{4ab^2}$

$\qquad = \dfrac{8a^4b}{4ab^2} + \dfrac{-12a^3b^2}{4ab^2} + \dfrac{-20a^2b^3}{4ab^2}$

$\qquad = \dfrac{2a^3}{b} - 3a^2 - 5ab$

17. $14x - 18 + 8x^2 = 8x^2 + 14x - 18$

$$
\begin{array}{r}
2x + 5 \\
4x - 3 \overline{)\ 8x^2 + 14x - 18} \\
\underline{-8x^2 + 6x} \\
20x - 18 \\
\underline{-20x + 15} \\
-3
\end{array}
$$

$\dfrac{14x - 18 + 8x^2}{4x - 3} = 2x + 5 - \dfrac{3}{4x - 3}$

18.

$$
\begin{array}{r|rrrr}
-3 & 4 & 10 & -5 & 1 \\
 & & -12 & 6 & -3 \\
\hline
 & 4 & -2 & 1 & -2
\end{array}
$$

$\dfrac{4x^3 + 10x^2 - 5x + 1}{x + 3} = 4x^2 - 2x + 1 - \dfrac{2}{x + 3}$

19. $x^2 - 10x - 24 = (x - 12)(x + 2)$

20. $18x + 2x^3 - 12x^2 = 2x(x-3)(x-3)$

$\qquad\qquad\qquad\quad = 2x(x-3)^2$

21. $-16x^2 - 26x + 12 = -2(8x - 3)(x + 2)$

22. $9m^2 - 64t^2 = (3m + 8t)(3m - 8t)$

23. $16a^4b - 36a^3b^2 + 18a^2b^3$

$\qquad = 2a^2b(4a - 3b)(2a - 3b)$

24. $54m^3 + 128p^3$

$= 2(3m + 4p)\left(9m^2 - 12mp + 16p^2\right)$

25. $\qquad 25x^2 = 16$

$25x^2 - 16 = 0$

$(5x - 4)(5x + 4) = 0$

$x = \dfrac{4}{5}$ or $x = -\dfrac{4}{5}$

26. $5w^3 - 15w^2 - 50w = 0$

$5w\left(w^2 - 3w - 10\right) = 0$

$5w(w - 5)(w + 2) = 0$

$w = 0,\ w = 5$ or $w = -2$

27. $(2x - 7)(x - 3) = 10$

$2x^2 - 13x + 11 = 0$

$(2x - 11)(x - 1) = 0$

$x = \dfrac{11}{2}$ or $x = 1$

28. $\qquad 2t^3 + 3t^2 = 18t + 27$

$2t^3 + 3t^2 - 18t - 27 = 0$

$(2t + 3)\left(t^2 - 9\right) = 0$

$(2t + 3)(t + 3)(t - 3) = 0$

$t = -\dfrac{3}{2},\ t = -3$ or $t = 3$

29. $3x(2x - 5) + 4x = 2(x - 3)$

$6x^2 - 13x + 6 = 0$

$(3x - 2)(2x - 3) = 0$

$x = \dfrac{2}{3}$ or $x = \dfrac{3}{2}$

30. $f(x) = 10x^2 - 19x + 6 = (5x - 2)(2x - 3)$

The x-intercepts are $\left(\dfrac{2}{5}, 0\right)$ and $\left(\dfrac{3}{2}, 0\right)$.

31.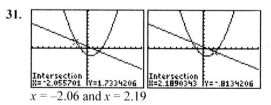

$x = -2.06$ and $x = 2.19$

32. a. $R(t) = -7.8t + 564$

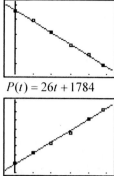

$P(t) = 26t + 1784$

The models fit the data well.

b. $(R \cdot P)(t) = (-7.8t + 564)(26t + 1784)$

$= -202.8t^2 - 13,915.2t + 14,664t$

$\qquad + 1,006,176$

$= -202.8t^2 + 748.8t + 1,006,176$

c. For the expression $R(t) \cdot P(t)$, we have $R(t)$ = number of deaths per 100,000 people $\times$ $P(t)$ = population in hundred-thousands. The units of the expression are total number of deaths in the U.S.

d. $(R \cdot P)(58)$

$= -202.8(58)^2 + 748.8(58) + 1,006,176$

$= 367,387.2$

This means that in $1960 + 58 = 2018$, there will be about 367,387 deaths from heart disease.

e.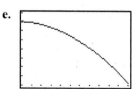

The function $R \cdot P$ is decreasing for values of t between 5 and 60. This means that the number of people dying from heart disease decreased and will continue to decrease between the years 1965 and 2020. The number of deaths from heart disease is decreasing at a faster rate than the population is growing.

33. a.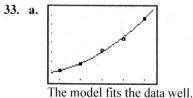

The model fits the data well.

b. $f(17) = (17)^2 - 13(17) + 70 = 138$

In $2000 + 17 = 2017$, the opium cultivation will be 138 thousand hectares.

c. $t^2 - 13t + 70 = 100$

$t^2 - 13t - 30 = 0$

$(t + 2)(t - 15) = 0$

$t = -2$ or $t = 15$

The model estimates that the opium cultivation was 100 thousand hectares in $2000 - 2 = 1998$ and the opium cultivation will be 100 thousand hectares in $2000 + 15 = 2015$.

34. $(11 + 2x)(15 + 2x) - 120 = 165$

$4x^2 + 52x - 120 = 0$

$4(x + 15)(x - 2) = 0$

$x = -15$ or $x = 2$

The width of the border is 2 inches. The actual width of the frame is 15 inches.

Chapter 7
Quadratic Functions

Homework 7.1

1.

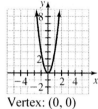

Vertex: (0, 0)

3.

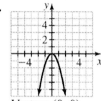

Vertex: (0, 0)

5.

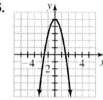

Vertex: (0, 5)

7.

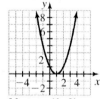

Vertex: (1, 0)

9.

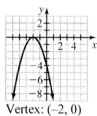

Vertex: (–2, 0)

11.

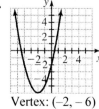

Vertex: (–2, – 6)

13.

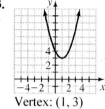

Vertex: (1, 3)

15.

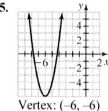

Vertex: (–6, –6)

17.

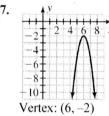

Vertex: (6, –2)

19.

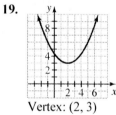

Vertex: (2, 3)

21.

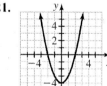

The domain is the set of all real numbers.
Since (0, –4) is the minimum point, the range
is the set of numbers where $y \geq -4$.

23.

The domain is the set of all real numbers.
Since $(0, -3)$ is the maximum point, the range
is the set of numbers where $y \le -3$.

25.

The domain is the set of all real numbers.
Since $(-4, 0)$ is the minimum point, the range
is the set of numbers where $y \ge 0$.

27.

The domain is the set of all real numbers.
Since $(-6, 2)$ is the minimum point, the range
is the set of numbers where $y \ge 2$.

29.

The domain is the set of all real numbers.
Since $(1, -4)$ is the minimum point, the range
is the set of numbers where $y \ge -4$.

31.

The domain is the set of all real numbers.
Since $(5, 2)$ is the maximum point, the range is
the set of numbers where $y \le 2$.

33. a. First make a scattergram of the data.

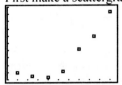

Imagine a parabola that comes close to (or
contains) the data points, and select a point
to be the vertex. We use the point
$(5.7, 265)$ as the vertex.

$$f(t) = a(t - 5.7)^2 + 265$$

Now use the point $(17, 551)$ to solve for a.

$$551 = a(17 - 5.7)^2 + 265$$
$$551 = a(11.3)^2 + 265$$
$$551 = 127.69a + 265$$
$$286 = 127.69a$$
$$2.24 \approx a$$

So the equation of f in vertex form is:

$$f(t) = 2.24(t - 5.7)^2 + 265$$

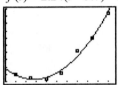

The graph of f is fairly close to the data
points.

b. The vertex of the model is $(5.7, 265)$. The
model predicts that in 1996, the U.S.
Department of Defense spent the least,
$265 billion.

c. To find the U.S. Department of Defense
spending in 2008, let $t = 18$ and solve for f.

$$f(18) = 2.24(18 - 5.7)^2 + 265$$
$$= 2.24(12.3)^2 + 265$$
$$= 2.24(151.29) + 265$$
$$= 338.8896 + 265$$
$$\approx 604$$

This model predicts that in 2008 the U.S.
Department of Defense will spend about
$604 billion.

d. To find the U.S. Department of Defense
spending in 2011, let $t = 21$ and solve for f.

$$f(21) = 2.24(21 - 5.7)^2 + 265$$
$$= 2.24(15.3)^2 + 265$$
$$= 2.24(234.09) + 265$$
$$= 524.3616 + 265$$
$$\approx 789$$

The ratio of federal spending on defense to federal spending on education is

$$\frac{789}{68.3} \approx 11.6.$$

35. a. First make a scattergram of the data.

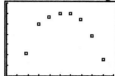

Imagine a parabola that comes close to (or contains) the data points, and select a point to be the vertex. We use the point $(52.3, 31)$ as the vertex.

$$f(t) = a(t - 52.3)^2 + 31$$

Now use the point $(90, 8)$ to solve for a.

$$f(t) = a(t - 52.3)^2 + 31$$
$$8 = a(90 - 52.3)^2 + 31$$
$$8 = a(37.7)^2 + 31$$
$$8 = a(1421.29) + 31$$
$$-23 = 1421.29a$$
$$-0.016 \approx a$$

So the equation of f in vertex form is:

$$f(t) = -0.016(t - 52.3)^2 + 31$$

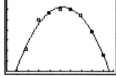

The graph of f is fairly close to the data points.

b. The vertex of the model is $(52.3, 31)$. This means that the largest percentage of Americans who are obese, 31%, occurs at age 52 years.

c. Let $t = 73$ and solve for f.

$$f(73) = -0.016(73 - 52.3)^2 + 31$$
$$= -0.016(20.7)^2 + 31$$
$$= -0.016(428.49) + 31$$
$$= -6.85584 + 31$$
$$\approx 24$$

Approximately 24% of Americans are obese at age 73.

d. Using the "zero" option on the graphing calculator, the t-intercepts are found to be $(8.28, 0)$ and $(96.32, 0)$. These indicate that no 8-year-old Americans and no 96-year-old Americans are obese. Model breakdown has occurred.

37. a.

x	y
−2	15
−1	5
0	−1
1	−3
2	−1

c. For each input-output pair, the output variable is 3 less than twice the square of the difference of the input variable and 1.

39. a.

b.
$$3 = (x - 3)^2 + 2$$
$$1 = (x - 3)^2$$
$$\pm 1 = x - 3$$
$$3 \pm 1 = x$$
$$x = 4 \text{ or } x = 2$$

c.
$$2 = (x - 3)^2 + 2$$
$$0 = (x - 3)^2$$
$$0 = x - 3$$
$$3 = x$$

d.
$$1 = (x - 3)^2 + 2$$
$$-1 = (x - 3)^2$$

The next step is to take the square root of both sides. However, this will require taking the square root of -1, which is not a valid step. There is no value of x such that $f(x) = 1$.

41. Answers may vary. Example:

$$y = a(x + 3)^2 + 4$$

where a is any negative number.

43. a. Because the parabola is face up, and the vertex is in the 3rd quadrant: $a > 0$, $h < 0$, $k < 0$.

b. Because the parabola is face down, and the vertex is in the 2nd quadrant: $a < 0$, $h < 0$, $k > 0$.

c. Because the parabola is face up, and the vertex is on the positive x-axis: $a > 0$, $h > 0$, $k = 0$.

d. Because the parabola is face down, and the vertex is on the negative y-axis: $a < 0$, $h = 0$, $k < 0$.

45. Answers may vary. Example:

$$y = a(x + 5)^2 + 3$$

where $a = -3, -2, -1, -½, ½, 1, 2, 3$

47. The graph shows that the vertex is (5, –6), so

$$f(x) = a(x - 5)^2 - 6.$$

To solve for a, substitute the point (1, 4) into the equation for f:

$$4 = a((1) - 5)^2 - 6$$
$$4 = a(-4)^2 - 6$$
$$4 = 16a - 6$$
$$10 = 16a$$
$$\frac{5}{8} = a$$

So the equation is:

$$f(x) = \frac{5}{8}(x - 5)^2 - 6$$

49. The value of a for the function f is the opposite of the value of a for the function g since g has a maximum point and f has a minimum point and we can assume that the graphs of f and g have the same "shape". Since the vertex (h, k) of g is $(-7, 3.71)$ and $a = -2.1$, an equation for g is:

$$g(x) = -2.1(x + 7)^2 + 3.71$$

51. It is possible. Example: $y = x^2 + 2$

53. It is possible. Example: $y = x^2 - 2$

55. Both equations have the same vertex (2, 5). From the graph notice this is the only point that lies on both graphs.

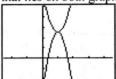

57. a.

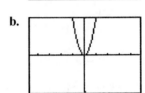

b.

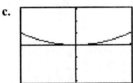

Answers may vary. Example:
The graph in part (a) is wider than the graph in part (b).

c.

Answers may vary. Example:
The graph in part (c) is much wider than the graphs in parts (a) and (b).

d. Answers may vary.

59. No, the student is not correct. Moving $y = x^2$ to the left by 4 units would result in $y = (x + 4)^2$. The equation $y = (x - 4)^2$ would move $y = x^2$ to the right by 4 units.

61. Adjust the WINDOW settings. Make your Xmin and Xmax much larger.

63. a.

b.

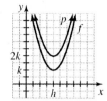

c.

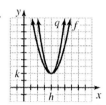

d.

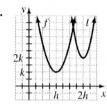

65. Answers may vary. Example:
To gather some idea of what the parabola will look like, remember that if $a > 0$, the parabola will open upward, whereas if $a < 0$, it will open downward. If $k > 0$, the vertex is always above the x-axis; if $k < 0$, the vertex is always below the x-axis; if $k = 0$, the vertex is on the x-axis. Likewise, if $h > 0$, the vertex is always to the right of the y-axis; if $h < 0$, the vertex is always to the left of the y-axis; if $h = 0$, the vertex is on the y-axis.
To draw the parabola, start at the vertex (h, k). Next make a table of the function, starting, for instance, with $x = h - 5$, up through $x = h + 5$. Plot these points on the graph. Connect the points in a parabola.

67. a.

Translate the graph of f 4 units to the right to get the graph of g. Translate the graph of f 4 units to the left to get the graph of h.

b.

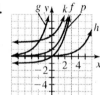

Translate the graph of f 4 units down to get the graph of g. Translate the graph of f 4 units up to get the graph of h.

c. Yes, the translations of x^3 match the translations of x^2. In each case add a constant to f to move the graph up or down; add a constant to x to move the graph left or right. The up and down movement is consistent with the sign, while the right and left movement is opposite its sign.

d.

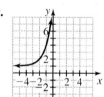

69.

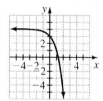

71.

73. $\log_4\left(3x^2\right) + 2\log_4\left(2x^4\right) = 6$

$\log_4\left(3x^2\right) + \log_4\left(2x^4\right)^2 = 6$

$\log_4\left(3x^2\right) + \log_4\left(4x^8\right) = 6$

$\log_4\left(3x^2 \cdot 4x^8\right) = 6$

$\log_4\left(12x^{10}\right) = 6$

$4^6 = 12x^{10}$

$4096 = 12x^{10}$

$341.33 = x^{10}$

$\pm 1.7919 = x$

The equation is logarithmic in one variable.

75. $\log_4\left(3x^2\right) + 2\log_4\left(2x^4\right)$

$= \log_4\left(3x^2\right) + \log_4\left(2x^4\right)^2$

$= \log_4\left(3x^2\right) + \log_4\left(4x^8\right)$

$= \log_4\left(3x^2 \cdot 4x^8\right)$

$= \log_4\left(12x^{10}\right)$

The expression is logarithmic in one variable.

77. $\log_4\left(3x^2\right) - 2\log_4\left(2x^4\right)$

$= \log_4\left(3x^2\right) - \log_4\left(2x^4\right)^2$

$= \log_4\left(3x^2\right) - \log_4\left(4x^8\right)$

$= \log_4\left(\dfrac{3x^2}{4x^8}\right)$

$= \log_4\left(\dfrac{3}{4x^6}\right)$

The expression is logarithmic in one variable.

Homework 7.2

1. Since $\dfrac{0+10}{2} = 5$, the x-coordinate of the vertex must be 5.

3. Since the points have the same y-coordinate, they are symmetric on the parabola.
Since $\dfrac{0+6}{2} = 3$, the x-coordinate of the vertex must be 3.

5. Since the points have the same y-coordinate, they are symmetric on the parabola.
Since $\dfrac{0+(-7)}{2} = -3.5$, the x-coordinate of the vertex must be -3.5.

7. Since the points have the same y-coordinate, they are symmetric on the parabola.
Since $\dfrac{0+7.29}{2} \approx 3.65$, the x-coordinate of the vertex must be 3.65.

9. A symmetric point to the y-intercept has a value of x that is 2 units to the right of $x = 2$ (value of x at the vertex). The value of y is the same as that of the y-intercept, so another point on the parabola is $(4, 9)$.

11. First, find the y-intercept by substituting 0 for x in the function:
$y = 0^2 - 6(0) + 7 = 7$
The y-intercept is $(0, 7)$. Next find the symmetric point to $(0, 7)$. Substitute 7 for y in the function and solve for x:
$7 = x^2 - 6x + 7$
$0 = x^2 - 6x$
$0 = x(x - 6)$
$x = 0$ or $x - 6 = 0$
$x = 0$ or $x = 6$
Therefore, the symmetric points are $(0, 7)$ and $(6, 7)$. Since $\dfrac{0+6}{2} = 3$, the x-coordinate of the vertex is 3. To find the y-coordinate of the vertex, substitute 3 for x and solve for y:
$y = 3^2 - 6(3) + 7 = -2$
So the vertex is $(3, -2)$.

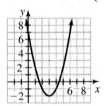

13. First, find the y-intercept by substituting 0 for x in the function:
$y = 0^2 + 8(0) + 9 = 9$
The y-intercept is $(0, 9)$. Next find the symmetric point of $(0, 9)$. Substitute 9 for y in the function and solve for x:
$9 = x^2 + 8x + 9$
$0 = x^2 + 8x$
$0 = x(x + 8)$
$x = 0$ or $x + 8 = 0$
$x = 0$ or $x = -8$
Therefore, the symmetric points are $(0, 9)$ and $(-8, 9)$. Since $\dfrac{0+(-8)}{2} = -4$, the x-coordinate of the vertex is -4. To find the y-coordinate of the vertex, substitute -4 for x and solve for y:
$y = (-4)^2 + 8(-4) + 9 = -7$
So the vertex is $(-4, -7)$.

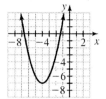

15. First, find the y-intercept by substituting 0 for x in the function:

$$y = -(0)^2 + 8(0) - 10 = -10$$

The y-intercept is $(0, -10)$. Next find the symmetric point of $(0, -10)$. Substitute -10 for y in the function and solve for x:

$$-10 = -x^2 + 8x - 10$$
$$0 = -x^2 + 8x$$
$$0 = -x(x - 8)$$
$$-x = 0 \quad \text{or} \quad x - 8 = 0$$
$$x = 0 \quad \text{or} \quad x = 8$$

Therefore, the symmetric points are $(0, -10)$ and $(8, -10)$. Since $\dfrac{0 + 8}{2} = 4$, the x-coordinate of the vertex is 4. To find the y-coordinate of the vertex, substitute 4 for x and solve for y:

$$y = -(4)^2 + 8(4) - 10 = 6$$

So the vertex is $(4, 6)$.

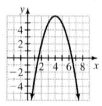

17. First, find the y-intercept by substituting 0 for x in the function:

$$y = 3(0)^2 + 6(0) - 4 = -4$$

The y-intercept is $(0, -4)$. Next find the symmetric point of $(0, -4)$. Substitute -4 for y in the function and solve for x:

$$-4 = 3x^2 + 6x - 4$$
$$0 = 3x^2 + 6x$$
$$0 = 3x(x + 2)$$
$$3x = 0 \quad \text{or} \quad x + 2 = 0$$
$$x = 0 \quad \text{or} \quad x = -2$$

Therefore, the symmetric points are $(0, -4)$ and $(-2, -4)$. Since $\dfrac{0 + (-2)}{2} = -1$, the

x-coordinate of the vertex is -1. To find the y-coordinate of the vertex, substitute -1 for x and solve for y:

$$y = 3(-1)^2 + 6(-1) - 4 = -7$$

So the vertex is $(-1, -7)$.

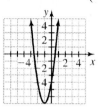

19. First, find the y-intercept by substituting 0 for x in the function:

$$y = -3(0)^2 + 12(0) - 5 = -5$$

The y-intercept is $(0, -5)$. Next find the symmetric point of $(0, -5)$. Substitute -5 for y in the function and solve for x:

$$-5 = -3x^2 + 12x - 5$$
$$0 = -3x^2 + 12x$$
$$0 = -3x(x - 4)$$
$$-3x = 0 \quad \text{or} \quad x - 4 = 0$$
$$x = 0 \quad \text{or} \quad x = 4$$

Therefore, the symmetric points are $(0, -5)$ and $(4, -5)$. Since $\dfrac{0 + 4}{2} = 2$, the x-coordinate of the vertex is 2. To find the y-coordinate of the vertex, substitute 2 for x and solve for y:

$$y = -3(2)^2 + 12(2) - 5 = 7$$

So the vertex is $(2, 7)$.

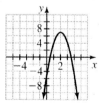

21. First, find the y-intercept by substituting 0 for x in the function:

$$y = -4(0)^2 - 9(0) - 5 = -5$$

The y-intercept is $(0, -5)$. Next find the symmetric point of $(0, -5)$. Substitute -5 for y in the function and solve for x:

$$-5 = -4x^2 - 9x - 5$$
$$0 = -4x^2 - 9x$$
$$0 = -x(4x + 9)$$

$-x = 0$ or $4x + 9 = 0$

$x = 0$ or $4x = -9$

$x = 0$ or $x = -\dfrac{9}{4} = -2.25$

Therefore, the symmetric points are $(0, -5)$

and $(-2.25, -5)$. Since $\dfrac{0 + (-2.25)}{2} \approx -1.13$,

the x-coordinate of the vertex is -1.13. To find
the y-coordinate of the vertex, substitute -1.13
for x and solve for y:

$y = -4(-1.13)^2 - 9(-1.13) - 5 \approx 0.06$

So the vertex is $(-1.13, 0.06)$.

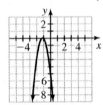

23. First, find the y-intercept by substituting 0 for
x in the function:

$y = 2(0)^2 - 7(0) + 7 = 7$

The y-intercept is $(0, 7)$. Next find the

symmetric point of $(0, 7)$. Substitute 7 for y in

the function and solve for x:

$7 = 2x^2 - 7x + 7$

$0 = 2x^2 - 7x$

$0 = x(2x - 7)$

$x = 0$ or $2x - 7 = 0$

$x = 0$ or $2x = 7$

$x = 0$ or $x = \dfrac{7}{2} = 3.5$

Therefore, the symmetric points are $(0, 7)$ and

$(3.5, 7)$. Since $\dfrac{0 + 3.5}{2} = 1.75$, the

x-coordinate of the vertex is 1.75. To find the
y-coordinate of the vertex, substitute 1.75 for x
and solve for y:

$y = 2(1.75)^2 - 7(1.75) + 7 \approx 0.88$

So the vertex is $(1.75, 0.88)$.

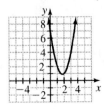

25. First, change the equation to standard form:

$4x^2 - y + 6 = 8x$

$y = 4x^2 - 8x + 6$

Next, find the y-intercept by substituting 0 for
x in the function:

$y = 4(0)^2 - 8(0) + 6 = 6$

The y-intercept is $(0, 6)$. Next find the

symmetric point of $(0, 6)$. Substitute 6 for y in

the function and solve for x:

$6 = 4x^2 - 8x + 6$

$0 = 4x^2 - 8x$

$0 = 4x(x - 2)$

$4x = 0$ or $x - 2 = 0$

$x = 0$ or $x = 2$

Therefore, the symmetric points are $(0, 6)$ and

$(2, 6)$. Since $\dfrac{0 + 2}{2} = 1$, the x-coordinate of the

vertex is 1. To find the y-coordinate of the
vertex, substitute 1 for x and solve for y:

$y = 4(1)^2 - 8(1) + 6 = 2$

So the vertex is $(1, 2)$.

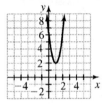

27. First, find the y-intercept by substituting 0 for
x in the function:

$y = 2.8(0)^2 - 8.7(0) + 4 = 4$

The y-intercept is $(0, 4)$. Next find the

symmetric point of $(0, 4)$. Substitute the 4 for

y in the function and solve for x:

$4 = 2.8x^2 - 8.7x + 4$

$0 = 2.8x^2 - 8.7x$

$0 = x(2.8x - 8.7)$

$x = 0$ or $2.8x - 8.7 = 0$

$x = 0$ or $2.8x = 8.7$

$x = 0$ or $x = \dfrac{8.7}{2.8} \approx 3.11$

Therefore, the symmetric points are $(0, 4)$ and

$(3.11, 4)$. Since $\dfrac{0 + 3.11}{2} \approx 1.56$, the

x-coordinate of the vertex is 1.56.

To find the y-coordinate of the vertex, substitute 1.56 for x and solve for y:

$$y = 2.8(1.56)^2 - 8.7(1.56) + 4 \approx -2.76$$

So the vertex is $(1.56, -2.76)$.

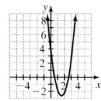

29. First, find the y-intercept by substituting 0 for x in the function:

$$y = 3.9(0)^2 + 6.9(0) - 3.4 = -3.4$$

The y-intercept is $(0, -3.4)$. Next find the symmetric point of $(0, -3.4)$. Substitute -3.4 for y in the function and solve for x:

$$-3.4 = 3.9x^2 + 6.9x - 3.4$$
$$0 = 3.9x^2 + 6.9x$$
$$0 = x(3.9x + 6.9)$$
$$x = 0 \text{ or } 3.9x + 6.9 = 0$$
$$x = 0 \text{ or } \quad 3.9x = -6.9$$
$$x = 0 \text{ or } \quad x = -\frac{6.9}{3.9} \approx -1.77$$

Therefore, the symmetric points are $(0, -3.4)$ and $(-1.77, -3.4)$. Since $\dfrac{0 + (-1.77)}{2} \approx -0.88$, the x-coordinate of the vertex is -0.88. To find the y-coordinate of the vertex, substitute -0.88 for x and solve for y:

$$y = 3.9(-0.88)^2 + 6.9(-0.88) - 3.4 \approx -6.45$$

So the vertex is $(-0.88, -6.45)$.

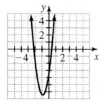

31. First, change the equation to standard form:

$$3.6y - 2.63x = 8.3x^2 - 7.1$$
$$3.6y = 8.3x^2 + 2.63x - 7.1$$
$$y = \frac{8.3x^2 + 2.63x - 7.1}{3.6}$$
$$y \approx 2.31x^2 + 7.31x - 1.97$$

Next, find the y-intercept by substituting 0 for x in the function:

$$y = 2.31(0)^2 + 7.31(0) - 1.97 = -1.97$$

The y-intercept is $(0, -1.97)$. Next find the symmetric point of $(0, -1.97)$. Substitute -1.97 for y in the function and solve for x:

$$-1.97 = 2.31x^2 + 7.31x - 1.97$$
$$0 = 2.31x^2 + 7.31x$$
$$0 = x(2.31x + 7.31)$$
$$x = 0 \text{ or } 2.31x + 7.31 = 0$$
$$x = 0 \text{ or } \quad 2.31x = -7.31$$
$$x = 0 \text{ or } \quad x = -\frac{7.31}{2.31} \approx -3.16$$

Therefore, the symmetric points are $(0, -1.97)$ and $(-3.16, -1.97)$. Since $\dfrac{0 + (-3.16)}{2} = -1.58$, the x-coordinate of the vertex is -1.58. To find the y-coordinate of the vertex, substitute -1.58 for x and solve for y:

$$y = 2.31(-1.58)^2 + 7.31(-1.58) - 1.97 \approx -7.75$$

So the vertex is $(-1.58, -7.75)$.

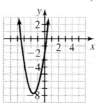

33. Since the x-intercepts are symmetric points and $\dfrac{2 + 6}{2} = 4$, the x-coordinate of the vertex is 4.

35. Since the x-intercepts are symmetric points and $\dfrac{-9 + 4}{2} = -\dfrac{5}{2}$, the x-coordinate of the vertex is $-\dfrac{5}{2}$.

37. To find the x-intercepts, let $y = 0$ and solve for x:

$$0 = 5x^2 - 10x$$
$$0 = 5x(x - 2)$$
$$5x = 0 \text{ or } x - 2 = 0$$
$$x = 0 \text{ or } \quad x = 2$$

The x-intercepts are $(0, 0)$ and $(2, 0)$. The y-intercept is, therefore, $(0, 0)$. Since the x-intercepts are symmetric points and $\dfrac{0 + 2}{2} = 1$, the x-coordinate of the vertex is 1.

Substitute 1 for x in the function and solve for y:

$$y = 5(1)^2 - 10(1) = -5$$

So, the vertex is $(1, -5)$.

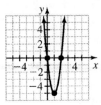

39. To find the x-intercepts, let $y = 0$ and solve for x:

$$0 = -2x^2 + 6x$$
$$0 = 2x(-x + 3)$$
$$2x = 0 \quad \text{or} \quad -x + 3 = 0$$
$$x = 0 \quad \text{or} \quad x = 3$$

The x-intercepts are $(0, 0)$ and $(3, 0)$. The y-intercept is, therefore, $(0, 0)$. Since the x-intercepts are symmetric points and $\dfrac{0 + 3}{2} = 1.5$, the x-coordinate of the vertex is 1.5. Substitute 1.5 for x in the function and solve for y:

$$y = -2(1.5)^2 + 6(1.5) = 4.5$$

So, the vertex is $(1.5, 4.5)$.

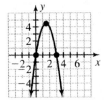

41. To find the x-intercepts, let $y = 0$ and solve for x:

$$0 = x^2 - 10x + 24$$
$$0 = (x - 6)(x - 4)$$
$$x - 6 = 0 \quad \text{or} \quad x - 4 = 0$$
$$x = 6 \quad \text{or} \quad x = 4$$

The x-intercepts are $(6, 0)$ and $(4, 0)$. To find the y-intercept, let $x = 0$ and solve for y:

$$y = (0)^2 - 10(0) + 24 = 24$$. The y-intercept is $(0, 24)$. Since the x-intercepts are symmetric points and $\dfrac{6 + 4}{2} = 5$, the x-coordinate of the vertex is 5. Substitute 5 for x in the function and solve for y:

$$y = (5)^2 - 10(5) + 24 = -1$$

So, the vertex is $(5, -1)$.

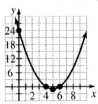

43. To find the x-intercepts, let $y = 0$ and solve for x:

$$0 = x^2 - 8x + 7$$
$$0 = (x - 7)(x - 1)$$
$$x - 7 = 0 \quad \text{or} \quad x - 1 = 0$$
$$x = 7 \quad \text{or} \quad x = 1$$

The x-intercepts are $(7, 0)$ and $(1, 0)$. To find the y-intercept, let $x = 0$ and solve for y:

$$y = (0)^2 - 8(0) + 7 = 7$$. The y-intercept is $(0, 7)$. Since the x-intercepts are symmetric points and $\dfrac{7 + 1}{2} = 4$, the x-coordinate of the vertex is 4. Substitute 4 for x in the function and solve for y:

$$y = (4)^2 - 8(4) + 7 = -9$$

So, the vertex is $(4, -9)$.

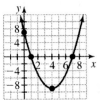

45. To find the x-intercepts, let $y = 0$ and solve for x:

$$0 = x^2 - 9$$
$$0 = (x - 3)(x + 3)$$
$$x - 3 = 0 \quad \text{or} \quad x + 3 = 0$$
$$x = 3 \quad \text{or} \quad x = -3$$

The x-intercepts are $(3, 0)$ and $(-3, 0)$. To find the y-intercept, let $x = 0$ and solve for y:

$$y = (0)^2 - 9 = -9$$. The y-intercept is $(0, -9)$. Since the x-intercepts are symmetric points and $\dfrac{3 + (-3)}{2} = 0$, the x-coordinate of the vertex is 0. So the vertex is $(0, -9)$.

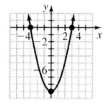

47. a. When the batter hits the ball, $t = 0$.
$$h(0) = -16(0)^2 + 140(0) + 3 = 0 + 0 + 3 = 3$$

b. The maximum height corresponds to the vertex of the graph.
$$t = -\frac{140}{2(-16)} = \frac{-140}{-32} = 4.375$$

$$\begin{aligned} h(4.375) &= -16(4.375)^2 + 140(4.375) + 3 \\ &= -16(19.14) + 612.5 + 3 \\ &= -306.25 + 615.5 \\ &= 309.25 \end{aligned}$$

So after 4.375 seconds the ball is at its maximum height of 309.25 feet.

c.

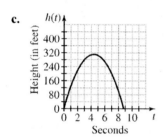

49. a.

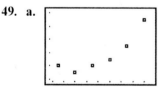

The function is a quadratic function, as the data appear to fit a quadratic function best.

b.

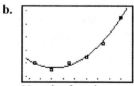

Yes, the function approximates the data very closely.

c. For 2017, $t = 27$.
$$\begin{aligned} f(27) &= 0.058(27)^2 - 0.98(27) + 15.43 \\ &= 31.252 \end{aligned}$$
The model predicts that about 31% of households will have outstanding student debt in 2017.

d. The percentage of households with outstanding student debt was the least at the $f(t)$ value of the vertex.
$$t = \frac{-b}{2a}$$
$$t = -\frac{-0.98}{2(0.058)} \approx 8.45$$

$$\begin{aligned} f(8.45) \\ &= 0.058(8.45)^2 - 0.98(8.45) + 15.43 \\ &= 11.29 \end{aligned}$$
The model estimates that the minimum percentage of households with outstanding student debt was about 11%, and that this happened in 1998.

e. For 2009, $t = 19$.
$$\begin{aligned} f(19) &= 0.058(19)^2 - 0.98(19) + 15.43 \\ &= 17.748 \end{aligned}$$
The number of households that had outstanding student debt in 2009 was about 17.748% of 117 million households, or $0.17748 \cdot 117 \approx 20.8$ million households .

51. a.

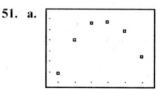

The function is a quadratic function, as the data appear to fit a quadratic function best.

b.

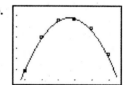

Yes, the function approximates the data very closely.

c. $\begin{aligned} f(18) &= -0.035(18)^2 + 3.25(18) - 26.34 \\ &= 20.82 \end{aligned}$
The average annual expenditure of 18-year-old Americans is $20.8 thousand.

d. The highest average annual expenditure corresponds with the vertex of the graph.
$$t = -\frac{3.25}{2(-0.035)} = \frac{-3.25}{-0.07} \approx 46.4$$

$$\begin{aligned} f(46.4) \\ &= -0.035(46.4)^2 + 3.25(46.4) - 26.34 \\ &\approx 49.1 \end{aligned}$$

The highest average annual expenditure is of 46-year-old Americans at $49.1 thousand.

53. **a.**

The quadratic function models the data the best.

b.

The linear function predicts the lowest student-to-faculty ratios.

c. $15.9 = 0.0119t^2 - 0.5t + 15.9$

$0 = 0.0119t^2 - 0.5t$

$0 = t(0.0119t - 0.5)$

$t = 0$ or $0.0119t - 0.5 = 0$

$t = 0$ or $0.0119t = 0.5$

$t = 0$ or $t \approx 42$

According to the model, during the year 2022, the student-to-faculty ratio will be 15.9.

d. The lowest student-to-faculty ratio corresponds with the vertex of the graph.

$t = -\dfrac{-0.5}{2(0.0119)} \approx 21$

$f(21) = 0.0119(21)^2 - 0.5(21) + 15.9$
≈ 10.65

The student-to-faculty ratio was at its lowest in 2001 at 10.65.

e. For 2011, $t = 31$.

$f(31) = 0.0119(31)^2 - 0.5(31) + 15.9$
≈ 11.84

The student-to-faculty ratio in 2011 was about 11.84, so the number of faculty was

$\dfrac{1769}{11.84} \approx 149.4$, or 150 faculty.

55. Let w be the width of the fenced in area, l be the length of the fenced in area, and A be the fenced in area. The perimeter and area of the fencing and fenced in area is given by:

$80 = 2w + 2l$

$A = wl$

Solve the first equation for l.

$80 = 2w + 2l$

$80 - 2w = 2l$

$40 - w = l$

Substitute this for l in the area equation:

$A = w(40 - w)$

$= 40w - w^2$

Find the maximum point of the parabola by finding the vertex.

$w = -\dfrac{40}{2(-1)} = \dfrac{-40}{-2} = 20$

So the width is 20 feet.

$A = 40(20) - (20)^2 = 800 - 400 = 400$

The maximum area is 400 square feet.

$80 = 2(20) + 2l$

$80 = 40 + 2l$

$40 = 2l$

$20 = l$

The length is 20 feet.

57. Let w be the width of the fenced in area, l be the length of the fenced in area, and A be the fenced in area. The perimeter and area of the fencing and fenced in area is given by:

$400 = 2w + l$

$A = wl$

Solve the first equation for l.

$400 = 2w + l$

$400 - 2w = l$

Substitute this for l in the area equation:

$A = w(400 - 2w)$

$= 400w - 2w^2$

Find the maximum point of the parabola by finding the vertex.

$w = -\dfrac{400}{2(-2)} = \dfrac{-400}{-4} = 100$

So the width is 100 feet.

$A = 400(100) - 2(100)^2$

$= 40,000 - 20,000$

$= 20,000$

The maximum area is 20,000 square feet.

$400 = 2(100) + l$

$400 = 200 + l$

$200 = l$

The length is 200 feet.

59. a.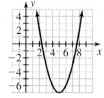

b. Answers may vary. Example:

x	y
3	−3
4	−6
5	−7
6	−6
7	−3

c. For each input-output pair, the output variable is 18 more than the difference between the square of the input and 10 times the input.

61. $f(-5) = -1$

63. When $x = -3, f(x) = 3.$

65. When $x = -2$ or $-4, f(x) = 2.$

67. The maximum value of f is 3.

69. a. $y = (0)^2 + 4(0) - 12 = -12$
Solve for x when $y = -12.$
$-12 = x^2 + 4x - 12$
$0 = x^2 + 4x$
$0 = x(x + 4)$
$x = 0$ or $x + 4 = 0$
$x = 0$ or $x = -4$
Take the average of the y-intercept and its symmetric point:
$\dfrac{0 + (-4)}{2} = \dfrac{-4}{2} = -2$
The x-coordinate of the vertex is −2.

b. $0 = x^2 + 4x - 12$
$0 = (x + 6)(x - 2)$
$x + 6 = 0$ or $x - 2 = 0$
$x = -6$ or $x = 2$
Taking the average of the x-intercepts:
$\dfrac{(-6) + 2}{2} = \dfrac{-4}{2} = -2$
The x-coordinate of the vertex is −2.

c. Yes, both methods produce the same result.

d. Averaging the x-coordinates of the y-intercept and its symmetric point is easier, because the function $g(x)$ is very difficult to factor.

e. Averaging the x-coordinates of the y-intercept and its symmetric point is the only method that can be used, because the function does not have any x-intercepts.

f. Answers may vary.

71. (3, 2) is the vertex for both f and k. The vertex of g is approximately (2.7, 1.8). The vertex of h is approximately (3.3, 1.7).

73. Answers may vary. Example:
If $a < 0$, then the graph of the function is a parabola that opens downward. Therefore the vertex, (h, k), is the maximum point, and the maximum value of f is k.

75. Answers may vary. See the box "Using Symmetric Points to Graph a Quadratic Function in Standard Form" on page 349 of the text and Examples 5 and 6 on p. 352 of the text.

77.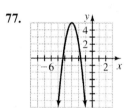

79.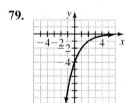

81. a. $f(x) = a(x - h)^2 + k$
$= a\left(x^2 - 2hx + h^2\right) + k$
$= ax^2 - 2ahx + ah^2 + k$
$= ax^2 - 2ahx + c$

b. $x = -\dfrac{b}{2(a)} = -\dfrac{-2ah}{2(a)} = \dfrac{2ah}{2a} = h$
Substitute h for x.
$f(h) = a(h - h)^2 + k = a(0)^2 + k = k$
The vertex of the parabola is (h, k).

83. $\dfrac{2b^{-2}c^4\left(3b^{-5}c^{-1}\right)^2}{8b^{-6}c^{-4}} = \dfrac{2b^{-2}c^4\left(9b^{-10}c^{-2}\right)}{8b^{-6}c^{-4}}$

$= \dfrac{18b^{-12}c^2}{8b^{-6}c^{-4}}$

$= \dfrac{9c^6}{4b^6}$

This is an expression in two variables involving exponents.

85.

This is an exponential function in one variable.

87. $-2(3)^x = -200$

$3^x = 100$

$\log_3 100 = x$

$4.19 = x$

This is an exponential equation in one variable.

Homework 7.3

1. $\sqrt{169} = \sqrt{13\cdot13} = 13$

3. $\sqrt{12} = \sqrt{4\cdot3} = \sqrt{4}\cdot\sqrt{3} = 2\sqrt{3}$

5. $\sqrt{\dfrac{4}{9}} = \dfrac{\sqrt{4}}{\sqrt{9}} = \dfrac{2}{3}$

7. $\sqrt{\dfrac{6}{49}} = \dfrac{\sqrt{6}}{\sqrt{49}} = \dfrac{\sqrt{6}}{7}$

9. $\dfrac{5}{\sqrt{2}} = \dfrac{5}{\sqrt{2}}\cdot\dfrac{\sqrt{2}}{\sqrt{2}} = \dfrac{5\sqrt{2}}{\sqrt{4}} = \dfrac{5\sqrt{2}}{2}$

11. $\dfrac{3}{\sqrt{32}} = \dfrac{3}{\sqrt{16\cdot2}} = \dfrac{3}{4\sqrt{2}}\cdot\dfrac{\sqrt{2}}{\sqrt{2}} = \dfrac{3\sqrt{2}}{4\sqrt{4}} = \dfrac{3\sqrt{2}}{4\cdot2} = \dfrac{3\sqrt{2}}{8}$

13. $\sqrt{\dfrac{3}{2}} = \dfrac{\sqrt{3}}{\sqrt{2}}\cdot\dfrac{\sqrt{2}}{\sqrt{2}} = \dfrac{\sqrt{6}}{\sqrt{4}} = \dfrac{\sqrt{6}}{2}$

15. $\sqrt{\dfrac{11}{20}} = \dfrac{\sqrt{11}}{\sqrt{20}} = \dfrac{\sqrt{11}}{\sqrt{4\cdot5}} = \dfrac{\sqrt{11}}{2\sqrt{5}}\cdot\dfrac{\sqrt{5}}{\sqrt{5}} = \dfrac{\sqrt{55}}{2\sqrt{25}} = \dfrac{\sqrt{55}}{10}$

17. $x^2 = 25$

$\sqrt{x^2} = \sqrt{25}$

$x = \pm5$

19. $x^2 - 3 = 0$

$x^2 = 3$

$x = \pm\sqrt{3}$

21. $t^2 = 32$

$\sqrt{t^2} = \pm\sqrt{32}$

$t = \pm\sqrt{16\cdot2} = \pm4\sqrt{2}$

23. $5x^2 = 3$

$5x^2 = 3$

$x^2 = \dfrac{3}{5}$

$\sqrt{x^2} = \sqrt{\dfrac{3}{5}}$

$x = \pm\dfrac{\sqrt{3}}{\sqrt{5}} = \pm\dfrac{\sqrt{3}}{\sqrt{5}}\cdot\dfrac{\sqrt{5}}{\sqrt{5}} = \pm\dfrac{\sqrt{15}}{5}$

25. $3p^2 - 11 = 3$

$3p^2 = 14$

$p^2 = \dfrac{14}{3}$

$\sqrt{p^2} = \pm\sqrt{\dfrac{14}{3}}$

$p = \pm\dfrac{\sqrt{14}}{\sqrt{3}} = \pm\dfrac{\sqrt{14}}{\sqrt{3}}\cdot\dfrac{\sqrt{3}}{\sqrt{3}} = \pm\dfrac{\sqrt{42}}{3}$

27. $(x+4)^2 = 7$

$\sqrt{(x+4)^2} = \pm\sqrt{7}$

$x + 4 = \pm\sqrt{7}$

$x = -4 \pm\sqrt{7}$

29. $(x-5)^2 = 27$

$\sqrt{(x-5)^2} = \pm\sqrt{27}$

$x - 5 = \pm3\sqrt{3}$

$x = 5 \pm3\sqrt{3}$

31. $(8y+3)^2 = 36$

$\sqrt{(8y+3)^2} = \pm\sqrt{36}$

$8y+3 = \pm 6$

$8y = -3 \pm 6$

$y = \dfrac{-3 \pm 6}{8}$

$y = \dfrac{-9}{8}$ or $y = \dfrac{3}{8}$

33. $(9x-5)^2 = 0$

$\sqrt{(9x-5)^2} = \sqrt{0}$

$9x-5 = 0$

$9x = 5$

$x = \dfrac{5}{9}$

35. $\left(x+\dfrac{3}{4}\right)^2 = \dfrac{41}{16}$

$\sqrt{\left(x+\dfrac{3}{4}\right)^2} = \pm\sqrt{\dfrac{41}{16}}$

$x+\dfrac{3}{4} = \pm\dfrac{\sqrt{41}}{\sqrt{16}}$

$x+\dfrac{3}{4} = \pm\dfrac{\sqrt{41}}{4}$

$x = \dfrac{-3 \pm \sqrt{41}}{4}$

37. $\left(w-\dfrac{7}{3}\right)^2 = \dfrac{5}{9}$

$\sqrt{\left(w-\dfrac{7}{3}\right)^2} = \pm\sqrt{\dfrac{5}{9}}$

$w-\dfrac{7}{3} = \pm\dfrac{\sqrt{5}}{\sqrt{9}}$

$w-\dfrac{7}{3} = \pm\dfrac{\sqrt{5}}{3}$

$w = \dfrac{7 \pm \sqrt{5}}{3}$

39. $5(x-6)^2 + 3 = 33$

$5(x-6)^2 = 30$

$(x-6)^2 = 6$

$\sqrt{(x-6)^2} = \pm\sqrt{6}$

$x-6 = \pm\sqrt{6}$

$x = 6 \pm \sqrt{6}$

41. $-3(x+1)^2 + 2 = -5$

$-3(x+1)^2 = -7$

$(x+1)^2 = \dfrac{7}{3}$

$\sqrt{(x+1)^2} = \pm\sqrt{\dfrac{7}{3}}$

$x+1 = \pm\dfrac{\sqrt{7}}{\sqrt{3}}$

$x+1 = \pm\dfrac{\sqrt{7}}{\sqrt{3}} \cdot \dfrac{\sqrt{3}}{\sqrt{3}}$

$x = -1 \pm \dfrac{\sqrt{21}}{3} = \dfrac{-3 \pm \sqrt{21}}{3}$

43. Solve for x when $f(x) = 0$.

$0 = x^2 - 17$

$x^2 = 17$

$x = \pm\sqrt{17}$

The x-intercepts are $\left(\sqrt{17}, 0\right)$ and $\left(-\sqrt{17}, 0\right)$.

45. Solve for x when $f(x) = 0$.

$2(x-3)^2 - 7 = 0$

$2(x-3)^2 = 7$

$(x-3)^2 = \dfrac{7}{2}$

$x-3 = \pm\sqrt{\dfrac{7}{2}}$

$x-3 = \pm\dfrac{\sqrt{7}}{\sqrt{2}}$

$x-3 = \pm\dfrac{\sqrt{7}}{\sqrt{2}} \cdot \dfrac{\sqrt{2}}{\sqrt{2}}$

$x-3 = \pm\dfrac{\sqrt{14}}{2}$

$x = 3 \pm \dfrac{\sqrt{14}}{2} = \dfrac{6 \pm \sqrt{14}}{2}$

The x-intercepts are $\left(\dfrac{6-\sqrt{14}}{2}, 0\right)$ and $\left(\dfrac{6+\sqrt{14}}{2}, 0\right)$.

47. Solve for x when $f(x) = 0$.

$0 = -4(x-2)^2 - 16$

$16 = -4(x-2)^2$

$-4 = (x-2)^2$

The next step would be to take the square root of both sides, resulting in an imaginary number. Since the *x*-intercept must be a real number, there are no *x*-intercepts.

49. $\sqrt{-36} = i\sqrt{36} = 6i$

51. $-\sqrt{-45} = -i\sqrt{45} = -i\sqrt{9 \cdot 5} = -3i\sqrt{5}$

53. $\sqrt{-\dfrac{5}{49}} = i\sqrt{\dfrac{5}{49}} = i\dfrac{\sqrt{5}}{\sqrt{49}} = i\dfrac{\sqrt{5}}{7}$

55. $\sqrt{-\dfrac{13}{5}} = i\sqrt{\dfrac{13}{5}} = i\dfrac{\sqrt{13}}{\sqrt{5}} = i\dfrac{\sqrt{13}}{\sqrt{5}} \cdot \dfrac{\sqrt{5}}{\sqrt{5}} = i\dfrac{\sqrt{65}}{5}$

57.
$$x^2 = -49$$
$$\sqrt{x^2} = \pm\sqrt{-49}$$
$$x = \pm i\sqrt{49} = \pm 7i$$

59.
$$x^2 = -18$$
$$\sqrt{x^2} = \pm\sqrt{-18}$$
$$x = \pm i\sqrt{18} = \pm i\sqrt{9 \cdot 2} = \pm 3i\sqrt{2}$$

61.
$$7x^2 + 26 = 5$$
$$7x^2 = -21$$
$$x^2 = -3$$
$$\sqrt{x^2} = \pm\sqrt{-3}$$
$$x = \pm i\sqrt{3}$$

63.
$$(m+4)^2 = -8$$
$$\sqrt{(m+4)^2} = \pm\sqrt{-8}$$
$$m + 4 = \pm i\sqrt{8}$$
$$m + 4 = \pm 2i\sqrt{2}$$
$$m = -4 \pm 2i\sqrt{2}$$

65.
$$\left(x - \dfrac{5}{4}\right)^2 = -\dfrac{3}{16}$$
$$\sqrt{\left(x - \dfrac{5}{4}\right)^2} = \pm\sqrt{-\dfrac{3}{16}}$$
$$x - \dfrac{5}{4} = \pm\dfrac{\sqrt{-3}}{\sqrt{16}}$$
$$x - \dfrac{5}{4} = \pm\dfrac{i\sqrt{3}}{4}$$
$$x = \dfrac{5 \pm i\sqrt{3}}{4}$$

67.
$$-2(y+3)^2 + 1 = 9$$
$$-2(y+3)^2 = 8$$
$$(y+3)^2 = -4$$
$$\sqrt{(y+3)^2} = \pm\sqrt{-4}$$
$$y + 3 = \pm i\sqrt{4}$$
$$y + 3 = \pm 2i$$
$$y = -3 \pm 2i$$

69. a.   Draw a scattergram of the data.

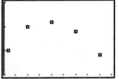

Imagine a parabola that comes close to (or contains) the data points, and select a point to be the vertex. We use the point $(19.9, 36.2)$ as the vertex.

$$f(t) = a(t - 19.9)^2 + 36.2$$

Now use the point $(40, 29)$ to solve for *a*.

$$29 = a(40 - 19.9)^2 + 36.2$$
$$29 = a(20.1)^2 + 36.2$$
$$29 = a(404.01) + 36.2$$
$$-7.2 = 404.01a$$
$$-0.018 \approx a$$
$$f(t) = -0.018(t - 19.9)^2 + 36.2$$

Check the model.

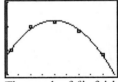

The graph of *f* is fairly close to the data points.

b. $f(49) = 2.24(49 - 5.7)^2 + 265 \approx 20.96$

In 2019, about 21% of Americans will say they are "very happy."

c. $-0.018(t-19.9)^2 + 36.2 = 23$

$$-0.018(t-19.9)^2 = -13.2$$

$$(t-19.9)^2 = \frac{13.2}{0.018}$$

$$t - 19.9 = \pm\sqrt{\frac{13.2}{0.018}}$$

$$t = 19.9 \pm \sqrt{\frac{13.2}{0.018}}$$

$$t \approx 19.9 \pm 27.08$$

$$t \approx -7.18 \text{ or } t \approx 46.98$$

In 1963, about 23% of Americans said they were "very happy," and in 2017, about 23% of Americans will say they are "very happy."

d. The vertex is $(19.9, 36.2)$. This means the largest percentage of Americans who said they were "very happy," about 36%, occurred in 1990.

71. a.
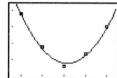

The model fits the data well.

b. The vertex is $(9.15, 4.45)$. This means that the greatest revenue from U.S. adult mattresses, about $4.5 billion, occurred in 2009.

c. $f(17) = 0.16(17 - 9.15)^2 + 4.45 \approx 14.3$

The revenue in 2017 will be $14.3 billion.

d. $0.16(t - 9.15)^2 + 4.45 = 10$

$$0.16(t - 9.15)^2 = 5.55$$

$$(t - 9.15)^2 = \frac{5.55}{0.16}$$

$$t - 9.15 = \pm\sqrt{\frac{5.55}{0.16}}$$

$$t = 9.15 \pm \sqrt{\frac{5.55}{0.16}}$$

$$t \approx 9.15 \pm 5.89$$

$$t \approx 3.26 \text{ or } t \approx 15.04$$

The annual revenue will be $10 billion in 2015.

73. The graphs of the two equations meet when $x \approx 1.4$ or 4.2.

75. The graph of the equation meets the line $y = 2$ when $x = 2$ or 4.

77. The graphs of the equations meet approximately at the points $(1.1, -3.5)$ and $(4.7, -1.7)$.

79. $a^2 + b^2 = c^2$

$$a^2 = c^2 - b^2$$

$$a = \pm\sqrt{c^2 - b^2}$$

81. $\dfrac{w^2}{r} + p = b$

$$\frac{w^2}{r} = b - p$$

$$w^2 = r(b - p)$$

$$w^2 = rb - rp$$

$$w = \pm\sqrt{rb - rp}$$

83. $(x + b)^2 = k$

$$x + b = \pm\sqrt{k}$$

$$x = -b \pm \sqrt{k}$$

85. $(py + a)^2 + b = c$

$$(py + a)^2 = c - b$$

$$py + a = \pm\sqrt{c - b}$$

$$py = -a \pm \sqrt{c - b}$$

$$y = \frac{-a \pm \sqrt{c - b}}{p}$$

87. No, the student did not solve it correctly. There is still an x on the right-hand side. The student should have factored the left hand side first.

$$x^2 - 10x + 25 = 0$$

$$(x - 5)(x - 5) = 0$$

$$(x - 5)^2 = 0$$

$$x - 5 = 0$$

$$x = 5$$

89. a. The vertex of the graph is $(3, 5)$.

b. Since $2 > 0$, the graph of f opens upward.

c. **i.** The equation has two solutions. Since the vertex is $(3, 5)$ and the graph opens upward, there must be two values of x that produce a y-value of 8.

ii. The equation has one solution ($x = 3$) because $y = 5$ at the vertex of the graph.

iii. The equation has no solutions, because the line $y = 1$ is below the vertex of the equation and the equation opens upward.

91. a.
$$25x^2 - 49 = 0$$
$$x^2 - \frac{49}{25} = 0$$
$$x^2 - \left(\frac{7}{5}\right)^2 = 0$$
$$\left(x + \frac{7}{5}\right)\left(x - \frac{7}{5}\right) = 0$$
$$x + \frac{7}{5} = 0 \quad \text{or} \quad x - \frac{7}{5} = 0$$
$$x = -\frac{7}{5} \quad \text{or} \quad x = \frac{7}{5}$$

b.
$$25x^2 - 49 = 0$$
$$25x^2 = 49$$
$$x^2 = \frac{49}{25}$$
$$\sqrt{x^2} = \pm\sqrt{\frac{49}{25}}$$
$$x = \pm\frac{\sqrt{49}}{\sqrt{25}} = \pm\frac{7}{5}$$

c. Both solutions are the same.

d. Answers may vary.

93. a. Yes it can be solved with the square root property.
$$(x + 4)^2 = 5$$
$$\sqrt{(x+4)^2} = \pm\sqrt{5}$$
$$x + 4 = \pm\sqrt{5}$$
$$x = -4 \pm \sqrt{5}$$

b. No, it cannot be solved by factoring.

c. No, not all equations that can be solved with the square root property can be solved by factoring. Not all equations involving quadratic terms will necessarily be factorable.

95. Answers may vary.

97.
$$x^6 = 142$$
$$\sqrt[6]{x^6} = \pm\sqrt[6]{142}$$
$$x = \pm 2.2841$$

99.
$$(t - 4)^3 = 88$$
$$\sqrt[3]{(t-4)^3} = \sqrt[3]{88}$$
$$t - 4 = 4.4480$$
$$t = 8.4480$$

101.
$$3(x + 1)^5 - 4 = 44$$
$$3(x + 1)^5 = 48$$
$$(x + 1)^5 = 16$$
$$\sqrt[5]{(x+1)^5} = \sqrt[5]{16}$$
$$x + 1 = 1.7411$$
$$x = 0.7411$$

103.
$$2w^3 + 3w^2 - 18w - 27 = w^2(2w + 3) - 9(2w + 3)$$
$$= (w^2 - 9)(2w + 3)$$
$$= (w + 3)(w - 3)(2w + 3)$$
This is a cubic polynomial in one variable.

105. From Exercise 103 we know that:
$$2w^3 + 3w^2 - 18w - 27 = (w - 3)(w + 3)(2w + 3)$$
So:
$$0 = 2w^3 + 3w^2 - 18w - 27$$
$$= (w - 3)(w + 3)(2w + 3)$$
$$w - 3 = 0 \quad \text{or} \quad w + 3 = 0 \quad \text{or} \quad 2w + 3 = 0$$
$$w = 3 \quad \text{or} \quad w = -3 \quad \text{or} \quad 2w = -3$$
$$w = 3 \quad \text{or} \quad w = -3 \quad \text{or} \quad w = -\frac{3}{2}$$
This is a cubic equation in one variable.

107. $\left(5w^2 - 2\right)(4w + 3) = 20w^3 + 15w^2 - 8w - 6$

This is a cubic polynomial in one variable.

Homework 7.4

1. $x^2 + 12x + c$
$$\left(\frac{12}{2}\right)^2 = 6^2 = 36 = c$$

This expression is $x^2 + 12x + 36$ and its factored form is $(x + 6)^2$.

3. $x^2 - 14x + c$

$$\left(\frac{-14}{2}\right)^2 = (-7)^2 = 49 = c$$

This expression is $x^2 - 14x + 49$ and its

factored form is $(x-7)^2$.

5. $x^2 + 7x + c$

$$\left(\frac{7}{2}\right)^2 = \frac{7^2}{2^2} = \frac{49}{4} = c$$

This expression is $x^2 + 7x + \frac{49}{4}$ and its

factored form is $\left(x + \frac{7}{2}\right)^2$.

7. $x^2 - 3 + c$

$$\left(\frac{-3}{2}\right)^2 = \frac{(-3)^2}{2^2} = \frac{9}{4} = c$$

This expression is $x^2 - 3x + \frac{9}{4}$ and its factored

form is $\left(x - \frac{3}{2}\right)^2$.

9. $x^2 + \frac{1}{2}x + c$

$$\left(\frac{1}{2} \cdot \frac{1}{2}\right)^2 = \left(\frac{1}{4}\right)^2 = \frac{1^2}{4^2} = \frac{1}{16} = c$$

This expression is $x^2 + \frac{1}{2}x + \frac{1}{16}$ and its

factored form is $\left(x + \frac{1}{4}\right)^2$.

11. $x^2 - \frac{4}{5}x + c$

$$\left(-\frac{4}{5} \cdot \frac{1}{2}\right)^2 = \left(\frac{-4}{10}\right)^2 = \frac{(-4)^2}{(10)^2} = \frac{16}{100} = \frac{4}{25} = c$$

This expression is $x^2 - \frac{4}{5}x + \frac{4}{25}$ and its

factored form is $\left(x - \frac{2}{5}\right)^2$.

13. Since $\left(\frac{6}{2}\right)^2 = 3^2 = 9$, add 9 to both sides of

the equation.
$$x^2 + 6x = 1$$
$$x^2 + 6x + 9 = 1 + 9$$
$$(x+3)^2 = 10$$
$$x + 3 = \pm\sqrt{10}$$
$$x = -3 \pm \sqrt{10}$$

15. Since $\left(\frac{2}{2}\right)^2 = 1^2 = 1$, add 1 to both sides of the

equation.
$$p^2 - 2p = 19$$
$$p^2 - 2p + 1 = 19 + 1$$
$$(p-1)^2 = 20$$
$$p - 1 = \pm\sqrt{20}$$
$$p = 1 \pm 2\sqrt{5}$$

17. Since $\left(\frac{4}{2}\right)^2 = 2^2 = 4$, add 4 to both sides of

the equation.
$$x^2 + 4x - 24 = 0$$
$$x^2 + 4x = 24$$
$$x^2 + 4x + 4 = 28$$
$$(x+2)^2 = 28$$
$$(x+2)^2 = 28$$
$$x + 2 = \pm\sqrt{28}$$
$$x = -2 \pm 2\sqrt{7}$$

19. Since $\left(\frac{-7}{2}\right)^2 = \frac{(-7)^2}{2^2} = \frac{49}{4}$, add $\frac{49}{4}$ to both

sides of the equation.
$$x^2 - 7x = 3$$
$$x^2 - 7x + \frac{49}{4} = 3 + \frac{49}{4}$$
$$\left(x - \frac{7}{2}\right)^2 = \frac{12}{4} + \frac{49}{4}$$
$$\left(x - \frac{7}{2}\right)^2 = \frac{61}{4}$$
$$x - \frac{7}{2} = \pm\sqrt{\frac{61}{4}}$$
$$x - \frac{7}{2} = \pm\frac{\sqrt{61}}{2}$$
$$x = \frac{7}{2} \pm \frac{\sqrt{61}}{2} = \frac{7 \pm \sqrt{61}}{2}$$

21. Since $\left(\dfrac{5}{2}\right)^2 = \dfrac{5^2}{2^2} = \dfrac{25}{4}$, add $\dfrac{25}{4}$ to both sides of the equation.

$$t^2 + 5t = 4$$
$$t^2 + 5t + \frac{25}{4} = 4 + \frac{25}{4}$$
$$\left(t + \frac{5}{2}\right)^2 = \frac{41}{4}$$
$$t + \frac{5}{2} = \pm\sqrt{\frac{41}{4}}$$
$$t + \frac{5}{2} = \pm\frac{\sqrt{41}}{\sqrt{4}}$$
$$t + \frac{5}{2} = \pm\frac{\sqrt{41}}{2}$$
$$t = \frac{-5 \pm \sqrt{41}}{2}$$

23. Since $\left(\dfrac{-5}{2} \cdot \dfrac{1}{2}\right)^2 = \left(\dfrac{-5}{4}\right)^2 = \dfrac{(-5)^2}{4^2} = \dfrac{25}{16}$, add

$\dfrac{25}{16}$ to both sides of the equation.

$$x^2 - \frac{5}{2}x = \frac{1}{2}$$
$$x^2 - \frac{5}{2}x + \frac{25}{16} = \frac{1}{2} + \frac{25}{16}$$
$$\left(x - \frac{5}{4}\right)^2 = \frac{8}{16} + \frac{25}{16}$$
$$\left(x - \frac{5}{4}\right)^2 = \frac{33}{16}$$
$$x - \frac{5}{4} = \pm\sqrt{\frac{33}{16}}$$
$$x - \frac{5}{4} = \pm\frac{\sqrt{33}}{4}$$
$$x = \frac{5}{4} \pm \frac{\sqrt{33}}{4} = \frac{5 \pm \sqrt{33}}{4}$$

25. First write the equation with an x^2 coefficient of 1.

$$2x^2 + 8x = 3$$
$$x^2 + 4x = \frac{3}{2}$$

Since $\left(\dfrac{4}{2}\right)^2 = 2^2 = 4$, add 4 to both sides of the equation.

$$x^2 + 4x = \frac{3}{2}$$
$$x^2 + 4x + 4 = \frac{3}{2} + 4$$
$$(x + 2)^2 = \frac{11}{2}$$
$$x + 2 = \pm\sqrt{\frac{11}{2}}$$
$$x + 2 = \pm\frac{\sqrt{11}}{\sqrt{2}}$$
$$x + 2 = \pm\frac{\sqrt{11}}{\sqrt{2}} \cdot \frac{\sqrt{2}}{\sqrt{2}}$$
$$x + 2 = \pm\frac{\sqrt{22}}{2}$$
$$x = \frac{-4 \pm \sqrt{22}}{2}$$

27. First write the equation with an x^2 coefficient of 1.

$$2r^2 - r - 7 = 0$$
$$r^2 - \frac{1}{2}r - \frac{7}{2} = 0$$

Since $\left(-\dfrac{1}{2} \cdot \dfrac{1}{2}\right)^2 = \left(-\dfrac{1}{4}\right)^2 = \dfrac{1}{16}$, add $\dfrac{1}{16}$ to both sides of the equation.

$$r^2 - \frac{1}{2}r - \frac{7}{2} = 0$$
$$r^2 - \frac{1}{2}r = \frac{7}{2}$$
$$r^2 - \frac{1}{2}r + \frac{1}{16} = \frac{7}{2} + \frac{1}{16}$$
$$\left(r - \frac{1}{4}\right)^2 = \frac{57}{16}$$
$$r - \frac{1}{4} = \pm\sqrt{\frac{57}{16}}$$
$$r - \frac{1}{4} = \pm\frac{\sqrt{57}}{\sqrt{16}}$$
$$r - \frac{1}{4} = \pm\frac{\sqrt{57}}{4}$$
$$r = \frac{1 \pm \sqrt{57}}{4}$$

29. First write the equation with an x^2 coefficient of 1.

$3x^2 + 4x - 5 = 0$

$x^2 + \dfrac{4}{3}x - \dfrac{5}{3} = 0$

Since $\left(\dfrac{4}{3}\cdot\dfrac{1}{2}\right)^2 = \left(\dfrac{2}{3}\right)^2 = \dfrac{2^2}{3^2} = \dfrac{4}{9}$, add $\dfrac{4}{9}$ to both sides of the equation.

$x^2 + \dfrac{4}{3}x - \dfrac{5}{3} = 0$

$x^2 + \dfrac{4}{3}x = \dfrac{5}{3}$

$x^2 + \dfrac{4}{3}x + \dfrac{4}{9} = \dfrac{5}{3} + \dfrac{4}{9}$

$\left(x + \dfrac{2}{3}\right)^2 = \dfrac{19}{9}$

$x + \dfrac{2}{3} = \pm\sqrt{\dfrac{19}{9}}$

$x + \dfrac{2}{3} = \pm\dfrac{\sqrt{19}}{\sqrt{9}}$

$x + \dfrac{2}{3} = \pm\dfrac{\sqrt{19}}{3}$

$x = \dfrac{-2\pm\sqrt{19}}{3}$

31. First write the equation with an x^2 coefficient of 1.

$6x^2 - 8x = -1$

$x^2 - \dfrac{4}{3}x = -\dfrac{1}{6}$

Since $\left(-\dfrac{4}{3}\cdot\dfrac{1}{2}\right)^2 = \left(\dfrac{2}{3}\right)^2 = \dfrac{2^2}{3^2} = \dfrac{4}{9}$, add $\dfrac{4}{9}$ to both sides of the equation.

$x^2 - \dfrac{4}{3}x = -\dfrac{1}{6}$

$x^2 - \dfrac{4}{3}x + \dfrac{4}{9} = -\dfrac{1}{6} + \dfrac{4}{9}$

$\left(x - \dfrac{2}{3}\right)^2 = \dfrac{5}{18}$

$x - \dfrac{2}{3} = \pm\sqrt{\dfrac{5}{18}}$

$x - \dfrac{2}{3} = \pm\dfrac{\sqrt{5}}{\sqrt{18}}$

$x - \dfrac{2}{3} = \pm\dfrac{\sqrt{5}}{3\sqrt{2}}$

$x - \dfrac{2}{3} = \pm\dfrac{\sqrt{5}}{3\sqrt{2}}\cdot\dfrac{\sqrt{2}}{\sqrt{2}}$

$x - \dfrac{2}{3} = \pm\dfrac{\sqrt{10}}{6}$

$x = \dfrac{4\pm\sqrt{10}}{6}$

33. First write the equation with an x^2 coefficient of 1.

$8w^2 + 4w - 3 = 0$

$w^2 + \dfrac{1}{2}w - \dfrac{3}{8} = 0$

Since $\left(\dfrac{1}{2}\cdot\dfrac{1}{2}\right)^2 = \left(\dfrac{1}{4}\right)^2 = \dfrac{1}{16}$ add $\dfrac{1}{16}$ to both sides of the equation.

$w^2 + \dfrac{1}{2}w - \dfrac{3}{8} = 0$

$w^2 + \dfrac{1}{2}w = \dfrac{3}{8}$

$w^2 + \dfrac{1}{2}w + \dfrac{1}{16} = \dfrac{3}{8} + \dfrac{1}{16}$

$\left(w + \dfrac{1}{4}\right)^2 = \dfrac{7}{16}$

$w + \dfrac{1}{4} = \pm\sqrt{\dfrac{7}{16}}$

$w + \dfrac{1}{4} = \pm\dfrac{\sqrt{7}}{\sqrt{16}}$

$w + \dfrac{1}{4} = \pm\dfrac{\sqrt{7}}{4}$

$w = \dfrac{-1\pm\sqrt{7}}{4}$

35. Since $\left(\dfrac{2}{2}\right)^2 = 1^2 = 1$, add 1 to both sides of the equation.

$x^2 + 2x = -7$

$x^2 + 2x + 1 = -7 + 1$

$(x+1)^2 = -6$

$x + 1 = \pm\sqrt{-6}$

$x = -1 \pm i\sqrt{6}$

37. Since $\left(\dfrac{6}{2}\right)^2 = 3^2 = 9$, add 9 to both sides of

the equation.

$$x^2 - 6x + 17 = 0$$
$$x^2 - 6x = -17$$
$$x^2 - 6x + 9 = -17 + 9$$
$$(x - 3)^2 = -8$$
$$x - 3 = \pm\sqrt{-8}$$
$$x = 3 \pm 2i\sqrt{2}$$

39. Since $\left(\dfrac{3}{2}\right)^2 = \dfrac{3^2}{2^2} = \dfrac{9}{4}$, add $\dfrac{9}{4}$ to both sides of

the equation.

$$k^2 + 3k + 4 = 0$$
$$k^2 + 3k = -4$$
$$k^2 + 3k + \dfrac{9}{4} = -4 + \dfrac{9}{4}$$
$$\left(k + \dfrac{3}{2}\right)^2 = \dfrac{-7}{4}$$
$$k + \dfrac{3}{2} = \pm\sqrt{\dfrac{-7}{4}}$$
$$k + \dfrac{3}{2} = \pm\dfrac{\sqrt{-7}}{\sqrt{4}}$$
$$k + \dfrac{3}{2} = \pm\dfrac{i\sqrt{7}}{2}$$
$$k = \dfrac{-3 \pm i\sqrt{7}}{2}$$

41. Since $\left(\dfrac{2}{3} \cdot \dfrac{1}{2}\right)^2 = \left(\dfrac{1}{3}\right)^2 = \dfrac{1^2}{3^2} = \dfrac{1}{9}$, add $\dfrac{1}{9}$ to

both sides of the equation.

$$x^2 + \dfrac{2}{3}x + \dfrac{7}{3} = 0$$
$$x^2 + \dfrac{2}{3}x = -\dfrac{7}{3}$$
$$x^2 + \dfrac{2}{3}x + \dfrac{1}{9} = -\dfrac{7}{3} + \dfrac{1}{9}$$
$$\left(x + \dfrac{1}{3}\right)^2 = -\dfrac{20}{9}$$
$$x + \dfrac{1}{3} = \pm\sqrt{\dfrac{-20}{9}}$$
$$x + \dfrac{1}{3} = \pm\dfrac{\sqrt{-20}}{\sqrt{9}}$$
$$x + \dfrac{1}{3} = \pm\dfrac{2i\sqrt{5}}{3}$$
$$x = \dfrac{-1 \pm 2i\sqrt{5}}{3}$$

43. First write the equation with an x^2 coefficient of 1.

$$4r^2 - 3r = -5$$
$$r^2 - \dfrac{3}{4}r = -\dfrac{5}{4}$$

Since $\left(-\dfrac{3}{4} \cdot \dfrac{1}{2}\right)^2 = \left(-\dfrac{3}{8}\right)^2 = \dfrac{(-3)^2}{8^2} = \dfrac{9}{64}$, add

$\dfrac{9}{64}$ to both sides of the equation.

$$r^2 - \dfrac{3}{4}r = -\dfrac{5}{4}$$
$$r^2 - \dfrac{3}{4}r + \dfrac{9}{64} = -\dfrac{5}{4} + \dfrac{9}{64}$$
$$\left(r - \dfrac{3}{8}\right)^2 = -\dfrac{71}{64}$$
$$r - \dfrac{3}{8} = \pm\sqrt{-\dfrac{71}{64}}$$
$$r - \dfrac{3}{8} = \pm\dfrac{\sqrt{-71}}{\sqrt{64}}$$
$$r - \dfrac{3}{8} = \pm\dfrac{i\sqrt{71}}{8}$$
$$r = \dfrac{3 \pm i\sqrt{71}}{8}$$

45. First write the equation with an x^2 coefficient of 1.

$$4p^2 + 6p + 3 = 0$$
$$p^2 + \dfrac{3}{2}p + \dfrac{3}{4} = 0$$

Since $\left(\dfrac{3}{2} \cdot \dfrac{1}{2}\right)^2 = \left(\dfrac{3}{4}\right)^2 = \dfrac{3^2}{4^2} = \dfrac{9}{16}$, add $\dfrac{9}{16}$ to

both sides of the equation.

$$p^2 + \dfrac{3}{2}p + \dfrac{3}{4} = 0$$
$$p^2 + \dfrac{3}{2}p = -\dfrac{3}{4}$$
$$p^2 + \dfrac{3}{2}p + \dfrac{9}{16} = -\dfrac{3}{4} + \dfrac{9}{16}$$
$$\left(p + \dfrac{3}{4}\right)^2 = -\dfrac{3}{16}$$
$$p + \dfrac{3}{4} = \pm\sqrt{-\dfrac{3}{16}}$$
$$p + \dfrac{3}{4} = \pm\dfrac{\sqrt{-3}}{\sqrt{16}}$$
$$p + \dfrac{3}{4} = \pm\dfrac{i\sqrt{3}}{4}$$
$$p = \dfrac{-3 \pm i\sqrt{3}}{4}$$

47. Solve for x when $f(x) = 0$.

$$0 = x^2 - 8x + 3$$
$$x^2 - 8x = -3$$
$$x^2 - 8x + 16 = -3 + 16$$
$$(x - 4)^2 = 13$$
$$x - 4 = \pm\sqrt{13}$$
$$x = 4 \pm \sqrt{13}$$

The x-intercepts are $\left(4 - \sqrt{13}, 0\right)$ and $\left(4 + \sqrt{13}, 0\right)$.

49. Solve for x when $f(x) = 0$.

$$0 = 2x^2 - 5x - 4$$
$$0 = x^2 - \frac{5}{2}x - 2$$
$$x^2 - \frac{5}{2}x = 2$$
$$x^2 - \frac{5}{2}x + \frac{25}{16} = 2 + \frac{25}{16}$$
$$\left(x - \frac{5}{4}\right)^2 = \frac{57}{16}$$
$$x - \frac{5}{4} = \pm\sqrt{\frac{57}{16}}$$
$$x - \frac{5}{4} = \pm\frac{\sqrt{57}}{\sqrt{16}}$$
$$x - \frac{5}{4} = \pm\frac{\sqrt{57}}{4}$$
$$x = \frac{5 \pm \sqrt{57}}{4}$$

The x-intercepts are
$\left(\dfrac{5 - \sqrt{57}}{4}, 0\right)$ and $\left(\dfrac{5 + \sqrt{57}}{4}, 0\right)$.

51. Solve for x when $f(x) = 0$.

$$0 = x^2 + 10x + 25$$
$$(x + 5)^2 = 0$$
$$x + 5 = 0$$
$$x = -5$$

The x-intercept is the vertex, $(-5, 0)$.

53. No, the student did not solve the equation correctly. The student should have first divided both sides by 4 and then completed the square and extracted the roots. The correct solution is:

$$4x^2 + 6x = 1$$
$$x^2 + \frac{3}{2}x = \frac{1}{4}$$
$$x^2 + \frac{3}{2}x + \frac{9}{16} = \frac{1}{4} + \frac{9}{16}$$
$$\left(x + \frac{3}{4}\right)^2 = \frac{4}{16} + \frac{9}{16}$$
$$\left(x + \frac{3}{4}\right)^2 = \frac{13}{16}$$
$$x + \frac{3}{4} = \pm\sqrt{\frac{13}{16}}$$
$$x + \frac{3}{4} = \pm\frac{\sqrt{13}}{\sqrt{16}}$$
$$x + \frac{3}{4} = \pm\frac{\sqrt{13}}{4}$$
$$x = -\frac{3}{4} \pm \frac{\sqrt{13}}{4} = \frac{-3 \pm \sqrt{13}}{4}$$

55. a.

$$3 = x^2 + 6x + 13$$
$$-10 = x^2 + 6x$$
$$-10 + 9 = x^2 + 6x + 9$$
$$-1 = (x + 3)^2$$

The next step is to take the square root of both sides. However, this will produce imaginary numbers. There is no real number such that $f(x) = 3$.

b.

$$4 = x^2 + 6x + 13$$
$$-9 = x^2 + 6x$$
$$-9 + 9 = x^2 + 6x + 9$$
$$0 = (x + 3)^2$$
$$0 = x + 3$$
$$-3 = x$$

When $x = -3$, $f(x) = 4$.

c.

$$6 = x^2 + 6x + 13$$
$$-7 = x^2 + 6x$$
$$-7 + 9 = x^2 + 6x + 9$$
$$2 = (x + 3)^2$$
$$\pm\sqrt{2} = x + 3$$
$$-3 \pm \sqrt{2} = x$$

When $x = -3 \pm \sqrt{2}$, $f(x) = 6$.

57. When $x = 1$ or 3, both sides of the equation are the same.

59. When $x = 2$ or 3, the left side of the equation is equal to -2.

61. The two equations have the points $(0, -0.5)$ and $(5, -8)$ in common.

63. a. $(x+k)^2 = x^2 + 2kx + k^2$

b. $\left(\dfrac{2k}{2}\right)^2 = k^2$

65. Answers may vary.

67. Answers may vary.

69. $w^2 - 10w + 25 = (w-5)^2$

71. $x^2 + \dfrac{5}{3}x + \dfrac{25}{36} = \left(x + \dfrac{5}{6}\right)^2$

73.

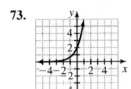

This is an exponential function.

75.
$$65 = 2(3)^x$$
$$32.5 = 3^x$$
$$\log_3 32.5 = x$$
$$3.17 = x$$
This is an exponential function.

77. $\log_3(3x+2) = 4$
$$3^4 = 3x + 2$$
$$81 = 3x + 2$$
$$79 = 3x$$
$$26.33 = x$$
This is a logarithmic equation in one variable.

Homework 7.5

1. $a = 2,\ b = 5,\ c = -2$
$$x = \dfrac{-5 \pm \sqrt{5^2 - 4(2)(-2)}}{2(2)}$$
$$x = \dfrac{-5 \pm \sqrt{41}}{4}$$

3. $a = 3,\ b = -6,\ c = 1$
$$x = \dfrac{-(-6) \pm \sqrt{(-6)^2 - 4(3)(1)}}{2(3)}$$
$$x = \dfrac{6 \pm \sqrt{24}}{6}$$
$$x = \dfrac{6 \pm 2\sqrt{6}}{6}$$
$$x = \dfrac{3 \pm \sqrt{6}}{3}$$

5.
$$t^2 = 4t + 3$$
$$t^2 - 4t - 3 = 0$$
$$a = 1,\ b = -4,\ c = -3$$
$$x = \dfrac{-(-4) \pm \sqrt{(-4)^2 - 4(1)(-3)}}{2(1)}$$
$$x = \dfrac{4 \pm \sqrt{28}}{2}$$
$$x = \dfrac{4 \pm 2\sqrt{7}}{2}$$
$$x = 2 \pm \sqrt{7}$$

7.
$$-2x^2 + 5x = 3$$
$$-2x^2 + 5x - 3 = 0$$
$$a = -2,\ b = 5,\ c = -3$$
$$x = \dfrac{-(5) \pm \sqrt{(5)^2 - 4(-2)(-3)}}{2(-2)}$$
$$x = \dfrac{-5 \pm \sqrt{1}}{-4}$$
$$x = \dfrac{-5 \pm 1}{-4}$$
$$x = 1 \text{ or } x = \dfrac{3}{2}$$

9. $a = 3,\ b = 0,\ c = -17$
$$x = \dfrac{0 \pm \sqrt{(0)^2 - 4(3)(-17)}}{2(3)}$$
$$x = \pm\dfrac{\sqrt{204}}{6}$$
$$x = \pm\dfrac{2\sqrt{51}}{6}$$
$$x = \pm\dfrac{\sqrt{51}}{3}$$

11.
$$2y^2 = -5y$$
$$2y^2 + 5y = 0$$
$$a = 2, \ b = 5, \ c = 0$$
$$x = \frac{-(5) \pm \sqrt{(5)^2 - 4(2)(0)}}{2(2)}$$
$$x = \frac{-5 \pm \sqrt{25}}{4}$$
$$x = \frac{-5 \pm 5}{4}$$
$$x = 0 \ \text{ or } \ x = -\frac{5}{2}$$

13.
$$\frac{2}{3}x^2 - \frac{5}{6}x = \frac{1}{3}$$
$$\frac{2}{3}x^2 - \frac{5}{6}x - \frac{1}{3} = 0$$
$$a = \frac{2}{3}, \ b = -\frac{5}{6}, \ c = -\frac{1}{3}$$
$$x = \frac{-\left(-\frac{5}{6}\right) \pm \sqrt{\left(-\frac{5}{6}\right)^2 - 4\left(\frac{2}{3}\right)\left(-\frac{1}{3}\right)}}{2\left(\frac{2}{3}\right)}$$
$$x = \frac{\frac{5}{6} \pm \sqrt{\frac{25}{36} + \frac{8}{9}}}{\frac{4}{3}}$$
$$x = \frac{\frac{5}{6} \pm \sqrt{\frac{57}{36}}}{\frac{4}{3}}$$
$$x = \frac{\frac{5}{6} \pm \frac{\sqrt{57}}{\sqrt{36}}}{\frac{4}{3}}$$
$$x = \frac{\frac{5}{6} \pm \frac{\sqrt{57}}{6}}{\frac{4}{3}}$$
$$x = \frac{5 \pm \sqrt{57}}{8}$$

15.
$$(3x + 2)(x - 1) = 1$$
$$3x^2 - x - 2 = 1$$
$$3x^2 - x - 3 = 0$$
$$a = 3, \ b = -1, \ c = -3$$
$$x = \frac{-(-1) \pm \sqrt{(-1)^2 - 4(3)(-3)}}{2(3)} = \frac{1 \pm \sqrt{37}}{6}$$

17.
$$2x^2 = 5x + 4$$
$$2x^2 - 5x - 4 = 0$$
$$a = 2, \ b = -5, \ c = -4$$
$$x = \frac{-(-5) \pm \sqrt{(-5)^2 - 4(2)(-4)}}{2(2)} = \frac{5 \pm \sqrt{57}}{4}$$
$$x \approx \frac{5 + 7.55}{4} \ \text{ or } \ x \approx \frac{5 - 7.55}{4}$$
$$x \approx 3.14 \ \text{ or } \ x \approx -0.64$$

19.
$$2.85p^2 - 7.12p = 4.49$$
$$2.85p^2 - 7.12p - 4.49 = 0$$
$$a = 2.85, b = -7.12, c = -4.49$$
$$p = \frac{-(-7.12) \pm \sqrt{(-7.12)^2 - 4(2.85)(-4.49)}}{2(2.85)}$$
$$p = \frac{7.12 \pm \sqrt{101.88}}{5.7}$$
$$p \approx \frac{7.12 + 10.09}{5.7} \ \text{ or } \ p \approx \frac{7.12 - 10.09}{5.7}$$
$$p \approx 3.02 \ \text{ or } \ p \approx -0.52$$

21.
$$-5.4x(x + 9.8) + 4.1 = 3.2 - 6.9x$$
$$-5.4x^2 - 52.92x + 4.1 - 3.2 + 6.9x = 0$$
$$-5.4x^2 - 46.02x + 0.9 = 0$$
$$a = -5.4, \ b = -46.02, \ c = 0.9$$
$$x = \frac{-(-46.02) \pm \sqrt{(-46.02)^2 - 4(-5.4)(0.9)}}{2(-5.4)}$$
$$x = \frac{46.02 \pm \sqrt{2137.2804}}{-10.8}$$
$$x \approx \frac{46.02 + 46.23}{-10.8} \ \text{ or } \ x \approx \frac{46.02 - 46.23}{-10.8}$$
$$x \approx -8.54 \ \text{ or } \ x \approx 0.02$$

23.
$$0 = 2x^2 - x - 7$$
$$x = \frac{-(-1) \pm \sqrt{(-1)^2 - 4(2)(-7)}}{2(2)} = \frac{1 \pm \sqrt{57}}{4}$$

The x-intercepts are $\left(\frac{1 - \sqrt{57}}{4}, 0\right)$ and
$\left(\frac{1 + \sqrt{57}}{4}, 0\right)$.

25. $0 = 3x^2 + 2x + 5$

$$x = \frac{-(2) \pm \sqrt{(2)^2 - 4(3)(5)}}{2(3)} = \frac{-2 \pm \sqrt{-56}}{6}$$

The only way to take the square root of a negative number is to use imaginary numbers. Therefore, there are no *x*-intercepts.

27. $0 = x^2 + 2x - 5$

$$x = \frac{-(2) \pm \sqrt{(2)^2 - 4(1)(-5)}}{2(1)}$$

$$x = \frac{-2 \pm \sqrt{24}}{2}$$

$$x = \frac{-2 \pm 2\sqrt{6}}{2}$$

$$x = -1 \pm \sqrt{6}$$

The *x*-intercepts are $(-1 - \sqrt{6}, 0)$ and $(-1 + \sqrt{6}, 0)$.

29. $x^2 - 3x + 8 = 0$

$a = 1,\ b = -3,\ c = 8$

$$x = \frac{-(-3) \pm \sqrt{(-3)^2 - 4(1)(8)}}{2(1)}$$

$$x = \frac{3 \pm \sqrt{-23}}{2}$$

$$x = \frac{3 \pm i\sqrt{23}}{2}$$

31. $-w^2 + 2w = 5$

$-w^2 + 2w - 5 = 0$

$a = -1,\ b = 2,\ c = -5$

$$w = \frac{-(2) \pm \sqrt{(2)^2 - 4(-1)(-5)}}{2(-1)}$$

$$w = \frac{-2 \pm \sqrt{-16}}{-2}$$

$$w = \frac{-2 \pm 4i}{-2}$$

$$w = 1 \pm 2i$$

33. $\frac{1}{4}x^2 = 2x - \frac{9}{2}$

$\frac{1}{4}x^2 - 2x + \frac{9}{2} = 0$

$a = \frac{1}{4},\ b = -2,\ c = \frac{9}{2}$

$$x = \frac{-(-2) \pm \sqrt{(-2)^2 - 4\left(\frac{1}{4}\right)\left(\frac{9}{2}\right)}}{2\left(\frac{1}{4}\right)}$$

$$x = \frac{2 \pm \sqrt{-\frac{1}{2}}}{\frac{1}{2}}$$

$$x = 4 \pm 2i\sqrt{\frac{1}{2}}$$

$$x = 4 \pm 2i\frac{\sqrt{1}}{\sqrt{2}}$$

$$x = 4 \pm 2i\frac{\sqrt{1}}{\sqrt{2}} \cdot \frac{\sqrt{2}}{\sqrt{2}}$$

$$x = 4 \pm 2i\frac{\sqrt{2}}{2}$$

$$x = 4 \pm i\sqrt{2}$$

35. $3x(3x - 2) = -2$

$9x^2 - 6x = -2$

$9x^2 - 6x + 2 = 0$

$a = 9,\ b = -6,\ c = 2$

$$x = \frac{-(-6) \pm \sqrt{(-6)^2 - 4(9)(2)}}{2(9)}$$

$$x = \frac{6 \pm \sqrt{-36}}{18}$$

$$x = \frac{6 \pm 6i}{18}$$

$$x = \frac{1 \pm i}{3}$$

37. $3k^2 = 4k - 5$

$3k^2 - 4k + 5 = 0$

$a = 3,\ b = -4,\ c = 5$

$$k = \frac{-(-4) \pm \sqrt{(-4)^2 - 4(3)(5)}}{2(3)}$$

$$k = \frac{4 \pm \sqrt{-44}}{6}$$

$$k = \frac{4 \pm 2i\sqrt{11}}{6}$$

$$k = \frac{2 \pm i\sqrt{11}}{3}$$

39. $4x^2 - 80 = 0$

$$4x^2 = 80$$
$$x^2 = 20$$
$$x = \pm\sqrt{20} = \pm 2\sqrt{5}$$

41. $5(w+3)^2 + 2 = 8$

$$5(w+3)^2 = 6$$
$$(w+3)^2 = \frac{6}{5}$$
$$w+3 = \pm\sqrt{\frac{6}{5}}$$
$$w+3 = \pm\frac{\sqrt{6}}{\sqrt{5}}$$
$$w+3 = \pm\frac{\sqrt{6}}{\sqrt{5}} \cdot \frac{\sqrt{5}}{\sqrt{5}}$$
$$w+3 = \pm\frac{\sqrt{30}}{5}$$
$$w = \frac{-15 \pm \sqrt{30}}{5}$$

43. $m^2 = -12m - 36$

$$m^2 + 12m + 36 = 0$$
$$(m+6)^2 = 0$$
$$m+6 = 0$$
$$m = -6$$

45. $-24x^2 + 18x = -60$

$$-24x^2 + 18x + 60 = 0$$
$$a = -24, \ b = 18, \ c = 60$$
$$x = \frac{-(18) \pm \sqrt{(18)^2 - 4(-24)(60)}}{2(-24)}$$
$$x = \frac{-18 \pm \sqrt{6084}}{-48}$$
$$x = \frac{-18 \pm 78}{-48}$$
$$x = -\frac{5}{4} \ \text{ or } \ x = 2$$

47. $\frac{1}{3}x^2 - \frac{3}{2}x = \frac{1}{6}$

$$2x^2 - 9x = 1$$
$$2x^2 - 9x - 1 = 0$$
$$a = 2, \ b = -9, \ c = -1$$
$$x = \frac{-(-9) \pm \sqrt{(-9)^2 - 4(2)(-1)}}{2(2)} = \frac{9 \pm \sqrt{89}}{4}$$

49. $(x-5)(x+2) = 3(x-1) + 2$

$$x^2 - 3x - 10 = 3x - 3 + 2$$
$$x^2 - 6x = 9$$
$$x^2 - 6x + 9 = 9 + 9$$
$$(x-3)^2 = 18$$
$$x - 3 = \pm\sqrt{18}$$
$$x - 3 = \pm 3\sqrt{2}$$
$$x = 3 \pm 3\sqrt{2}$$

51. $25r^2 = 49$

$$r^2 = \frac{49}{25}$$
$$r = \pm\sqrt{\frac{49}{25}} = \pm\frac{\sqrt{49}}{\sqrt{25}} = \pm\frac{7}{5}$$

53. $(x-1)^2 + (x+2)^2 = 6$

$$x^2 - 2x + 1 + x^2 + 4x + 4 = 6$$
$$2x^2 + 2x + 5 = 6$$
$$2x^2 + 2x - 1 = 0$$
$$a = 2, \ b = 2, \ c = -1$$
$$x = \frac{-(2) \pm \sqrt{(2)^2 - 4(2)(-1)}}{2(2)}$$
$$x = \frac{-2 \pm \sqrt{12}}{4}$$
$$x = \frac{-2 \pm 2\sqrt{3}}{4}$$
$$x = \frac{-1 \pm \sqrt{3}}{2}$$

55. $4x^2 = -25$

$$x^2 = -\frac{25}{4}$$
$$x = \pm\sqrt{-\frac{25}{4}} = \pm\frac{\sqrt{-25}}{\sqrt{4}} = \pm\frac{5i}{2}$$

57. $-2t^2 + 5t = 6$

$$-2t^2 + 5t - 6 = 0$$
$$a = -2, \ b = 5, \ c = -6$$
$$t = \frac{-(5) \pm \sqrt{(5)^2 - 4(-2)(-6)}}{2(-2)}$$
$$t = \frac{-5 \pm \sqrt{-23}}{-4}$$
$$t = \frac{5 \pm i\sqrt{23}}{4}$$

59. $(x-6)^2 + 5 = -43$

$(x-6)^2 = -48$

$x - 6 = \pm\sqrt{-48}$

$x - 6 = \pm 4i\sqrt{3}$

$x = 6 \pm 4i\sqrt{3}$

61. $(y-2)(y-5) = -4$

$y^2 - 7y + 10 = -4$

$y^2 - 7y = -14$

$y^2 - 7y + \dfrac{49}{4} = -14 + \dfrac{49}{4}$

$\left(y - \dfrac{7}{2}\right)^2 = \dfrac{-7}{4}$

$y - \dfrac{7}{2} = \pm\sqrt{\dfrac{-7}{4}}$

$y - \dfrac{7}{2} = \pm\dfrac{\sqrt{-7}}{\sqrt{4}}$

$y - \dfrac{7}{2} = \pm\dfrac{i\sqrt{7}}{2}$

$y = \dfrac{7 \pm i\sqrt{7}}{2}$

63. Since $(4)^2 - 4(3)(-5) = 76 > 0$, there are 2 real solutions.

65. Since $(-5)^2 - 4(2)(7) = -31 < 0$, there are 2 imaginary solutions.

67. Since $(-12)^2 - 4(4)(9) = 0$, there is 1 real solution.

69. a. Substitute 3 for $f(x)$:

$3 = x^2 - 4x + 8$

$x^2 - 4x + 5 = 0$

$a = 1, b = -4, c = 5$

$b^2 - 4ac = (-4)^2 - 4(1)(5) = 16 - 20 = -4 < 0$

So, there are no real number solutions, which means there are no points on f at $y = 3$.

b. Substitute 4 for $f(x)$:

$4 = x^2 - 4x + 8$

$x^2 - 4x + 4 = 0$

$a = 1, b = -4, c = 4$

$b^2 - 4ac = (-4)^2 - 4(1)(4) = 16 - 16 = 0$

So, there is one solution of the equation, which means there is one point on f at $y = 4$.

c. Substitute 5 for $f(x)$:

$5 = x^2 - 4x + 8$

$x^2 - 4x + 3 = 0$

$a = 1, b = -4, c = 3$

$b^2 - 4ac = (-4)^2 - 4(1)(3) = 16 - 12 = 4 > 0$

So, there are two real number solutions, which means there are two points on f at $y = 5$.

d.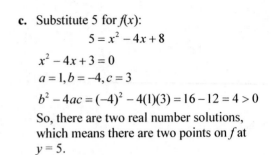

Answers may vary.

71. Solve for x when $f(x) = 2$:

$2 = x^2 - 6x + 7$

$x^2 - 6x + 5 = 0$

$(x-5)(x-1) = 0$

$x - 5 = 0$ or $x - 1 = 0$

$x = 5$ or $x = 1$

Therefore, two points at height 2 are $(1, 2)$ and $(5, 2)$. Since these points are symmetric and the average of the x-coordinates at these points is $\dfrac{1+5}{2} = 3$ the x-coordinate of the vertex is 3.

Substitute 3 for x in $f(x)$ to find the y-coordinate of the vertex:

$f(3) = 3^2 - 6(3) + 7 = -2$

So the vertex is $(3, -2)$.

73. a.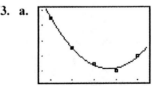

Yes, the model fits the data very well.

b. To predict the revenue in 2014, let $t = 9$ and solve for f.

$f(9) = 0.929(9)^2 - 8.73(9) + 50.8 = 47.479$

The model predicts that the revenue in 2014 will be \$47 billion.

c. To find the year when the annual revenue will be \$55 billion, substitute 55 for f and solve for t.

$$0.929t^2 - 8.73t + 50.8 = 55$$
$$0.929t^2 - 8.73t - 4.2 = 0$$
$$a = 0.929, b = -8.73, c = -4.2$$

$$t = \frac{-(-8.73) \pm \sqrt{(-8.73)^2 - 4(0.929)(-4.2)}}{2(0.929)}$$

$$t = \frac{8.73 \pm \sqrt{91.8201}}{1.858}$$

$$t \approx \frac{8.73 \pm 9.58}{1.858}$$

$$t \approx -0.457 \text{ or } t \approx 9.85$$

$$2005 + 9.85 \approx 2015$$

The annual revenue will be \$55 billion in 2015.

d. The revenue from boats and accessories was the least in 2010. The July new-home sales rate was the least in 2010. The consumer confidence index was the least in 2009. There seems to be a 1-year lag between when consumers begin to feel better about the economy and when they begin to spend more.

75. a.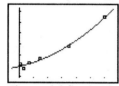

The model fits the data very well.

b. To find the percentage of women police officers in Glen Ellyn, Illinois, let $n = 27.04$ and solve for f.

$$f(27.04)$$
$$= 0.00006(27.04)^2 + 0.012(27.04) + 7.88$$
$$\approx 0.044 + 0.324 + 7.88$$
$$\approx 8.25$$

The model predicts that Glen Ellyn would have about 8.25% of police officers who are women.

c. $$10 = 0.00006n^2 + 0.012n + 7.88$$
$$0 = 0.00006n^2 + 0.012n - 2.12$$
$$a = 0.00006, b = 0.012, c = -2.12$$

$$n = \frac{-(0.012) \pm \sqrt{(0.012)^2 - 4(0.00006)(-2.12)}}{2(0.00006)}$$

$$n \approx \frac{-0.012 \pm 0.02555}{0.00012}$$

$$n \approx -312.92 \text{ or } n \approx 112.92$$

The model predicts that 10% of police officers will be women in cities with populations of about 112.9 thousand people. The population size of -312.9 thousand does not make sense, so model breakdown has occurred for that value.

77. a. $$h(3) = -16(3)^2 + 52(3) + 4$$
$$= -16(9) + 156 + 4$$
$$= -144 + 160$$
$$= 16$$

After 3 seconds the stone's height is 16 feet.

b. $$30 = -16t^2 + 52t + 4$$
$$26 = -16t^2 + 52t$$
$$-\frac{13}{8} = t^2 - \frac{13}{4}t$$
$$-\frac{13}{8} + \frac{169}{64} = t^2 - \frac{13}{4}t + \frac{169}{64}$$
$$\frac{65}{64} = \left(t - \frac{13}{8}\right)^2$$
$$\pm\sqrt{\frac{65}{64}} = t - \frac{13}{8}$$
$$\pm\frac{\sqrt{65}}{\sqrt{64}} = t - \frac{13}{8}$$
$$\pm\frac{\sqrt{65}}{8} = t - \frac{13}{8}$$
$$\frac{13 \pm \sqrt{65}}{8} = t$$
$$t \approx 2.63 \text{ or } t \approx 0.62$$

The stone is at a height of 30 feet at 0.62 second and 2.63 seconds.

c. $0 = -16t^2 + 52t + 4$

$a = -16,\ b = 52,\ c = 4$

$$t = \frac{-(52) \pm \sqrt{(52)^2 - 4(-16)(4)}}{2(-16)}$$

$$t = \frac{-52 \pm \sqrt{2960}}{-32}$$

$t \approx 3.33 \ \text{ or } \ t \approx -0.08$

Since we are looking for the time the stone hit the ground after it was thrown, we should not use –0.08 seconds.
The stone will hit the ground 3.33 seconds after being thrown.

79. The graphs of the two equations meet when $x = -2$ or 1.3.

81. The graph of the equation is equal to 4 when $x = \pm 2.4$.

83. The graphs of the two equations meet at the points $(-3.4, -3.4)$ and $(2.0, -1.2)$.

85. The student did not solve the equation correctly because they did not change the form into $ax^2 + bx + c = 0$ first.

$$2x^2 + 5x = 1$$

$2x^2 + 5x - 1 = 0$

So, $a = 2,\ b = 5,\ c = -1$

$$x = \frac{-5 \pm \sqrt{(5)^2 - 4(2)(-1)}}{2(2)} = \frac{-5 \pm \sqrt{33}}{4}$$

87. The student solved the equation correctly but did not simplify the result.

$$x = \frac{4 \pm \sqrt{56}}{10} = \frac{4 \pm 2\sqrt{14}}{10} = \frac{2 \pm \sqrt{14}}{5}$$

89. a. $mx + b = 0$

$mx + b - b = 0 - b$

$mx = -b$

$$\frac{mx}{m} = -\frac{b}{m}$$

$$x = -\frac{b}{m}$$

b. $7x + 21 = 0$

So, $m = 7$ and $b = 21$. Using the formula from part (a):

$$x = -\frac{21}{7} = -3$$

Solve for x:

$$7x + 21 = 0$$

$7x + 21 - 21 = 0 - 21$

$7x = -21$

$$\frac{7x}{7} = -\frac{21}{7}$$

$x = -3$

91. Factoring:

$$x^2 - x - 20 = 0$$

$(x - 5)(x + 4) = 0$

$x - 5 = 0 \ \text{ or } \ x + 4 = 0$

$x = 5 \ \text{ or } \qquad x = -4$

Completing the square:

$$x^2 - x = 20$$

$$x^2 - x + \frac{1}{4} = 20 + \frac{1}{4}$$

$$\left(x - \frac{1}{2}\right)^2 = \frac{81}{4}$$

$$x - \frac{1}{2} = \pm\sqrt{\frac{81}{4}}$$

$$x - \frac{1}{2} = \pm\frac{9}{2}$$

$x = \frac{1}{2} + \frac{9}{2} \ \text{ or } \ x = \frac{1}{2} - \frac{9}{2}$

$x = \frac{10}{2} \qquad \text{ or } \quad x = -\frac{8}{2}$

$x = 5 \qquad \text{ or } \quad x = -4$

Using the quadratic formula:

$x^2 - x - 20 = 0$

$$x = \frac{-(-1) \pm \sqrt{(-1)^2 - 4(1)(-20)}}{2(1)}$$

$$x = \frac{1 \pm \sqrt{81}}{2}$$

$$x = \frac{1 \pm 9}{2}$$

$x = \frac{1 + 9}{2} \ \text{ or } \ x = \frac{1 - 9}{2}$

$x = 5 \qquad \text{ or } \ x = -4$

93. Answers may vary.

95. $(x+2)(x-5) = x^2 - 5x + 2x - 10$
$$= x^2 - 3x - 10$$

97. $(x+2)(x-5) = 3$
$$x^2 - 5x + 2x - 10 = 3$$
$$x^2 - 3x = 13$$
$$x^2 - 3x + \frac{9}{4} = 13 + \frac{9}{4}$$
$$\left(x - \frac{3}{2}\right)^2 = \frac{61}{4}$$
$$x - \frac{3}{2} = \pm\sqrt{\frac{61}{4}}$$
$$x - \frac{3}{2} = \pm\frac{\sqrt{61}}{\sqrt{4}}$$
$$x - \frac{3}{2} = \pm\frac{\sqrt{61}}{2}$$
$$x = \frac{3 \pm \sqrt{61}}{2}$$

99. $-4(x-2)^2 + 3 = -1$
$$-4(x-2)^2 = -4$$
$$(x-2)^2 = 1$$
$$x - 2 = \pm\sqrt{1}$$
$$x - 2 = \pm 1$$
$$x = 2 \pm 1$$
$$x = 2 - 1 \quad \text{or} \quad x = 2 + 1$$
$$x = 1 \qquad \text{or} \quad x = 3$$

101. $-4(x-2)^2 + 3 = -4(x^2 - 4x + 4) + 3$
$$= -4x^2 + 16x - 16 + 3$$
$$= -4x^2 + 16x - 13$$

103. $4b^5 - 12 = 173$
$$4b^5 = 185$$
$$b^5 = 46.25$$
$$b \approx 2.1529$$

105. $4x - (7x - 5) = 3x + 1$
$$4x - 7x + 5 = 3x + 1$$
$$-3x + 5 = 3x + 1$$
$$4 = 6x$$
$$x = \frac{2}{3}$$

107. $7(3)^t + 8 = 271$
$$7(3)^t = 263$$
$$(3)^t = \frac{263}{7}$$
$$\log_3 \frac{263}{7} = t$$
$$\frac{\log \frac{263}{7}}{\log 3} \approx t$$
$$3.3007 \approx t$$

109. $8x^2 - 18x + 9 = (4x - 3)(2x - 3)$

This is a quadratic polynomial in one variable.

111. $f(-2) = 8(-2)^2 - 18(-2) + 9$
$$= 8(4) + 36 + 9$$
$$= 32 + 45$$
$$= 77$$

This is a quadratic function.

113. $0 = 8x^2 - 18x + 9$
$$0 = (4x - 3)(2x - 3)$$
$$4x - 3 = 0 \quad \text{or} \quad 2x - 3 = 0$$
$$4x = 3 \quad \text{or} \qquad 2x = 3$$
$$x = \frac{3}{4} \quad \text{or} \qquad x = \frac{3}{2}$$

This is a quadratic equation in one variable.

Homework 7.6

1. Add the first and third equations:
$$x + y + z = 0$$
$$\underline{x + 2y - z = -7}$$
$$2x + 3y \quad\;\; = -7 \quad (4)$$
Add the second and the third equations:
$$x - y + z = 6$$
$$\underline{x + 2y - z = -7}$$
$$2x + y \quad\quad = -1 \quad (5)$$
Subtract equation (5) from equation (4):
$$2x + 3y = -7$$
$$\underline{-2x - y = 1}$$
$$2y = -6$$
$$y = -3$$
Solve for x:
$$2x + 3(-3) = -7$$
$$2x - 9 = -7$$
$$2x = 2$$
$$x = 1$$

Solve for z:
$$(1) + (-3) + z = 0$$
$$1 - 3 + z = 0$$
$$-2 + z = 0$$
$$z = 2$$
The solution of the system is $(1, -3, 2)$.

3. Multiply the first equation by 2 and add it to the second equation.
$$2x + 2y - 2z = -2$$
$$\underline{2x - 2y + 3z = 8}$$
$$4x \qquad + z = 6 \quad (4)$$
Add equations (1) and (3):
$$x + y - z = -1$$
$$\underline{2x - y + 2z = 9}$$
$$3x \qquad + z = 8 \quad (5)$$
Subtract equations (4) and (5):
$$4x + z = 6$$
$$\underline{-3x - z = -8}$$
$$x \qquad = -2$$
Solve for z:
$$4(-2) + z = 6$$
$$-8 + z = 6$$
$$z = 14$$
Solve for y:
$$(-2) + y - (14) = -1$$
$$-2 + y - 14 = -1$$
$$y - 16 = -1$$
$$y = 15$$
The solution of the system is $(-2, 15, 14)$.

5. Add the first and third equations:
$$3x - y + 2z = 0$$
$$\underline{x + y + 6z = 0}$$
$$4x \qquad + 8z = 0 \quad (4)$$
Add 3 times the first equation to the second equation:
$$9x - 3y + 6z = 0$$
$$\underline{2x + 3y + 8z = 8}$$
$$11x \qquad + 14z = 8 \quad (5)$$
Subtract 11 times the fourth equation from 4 times the fifth equation:
$$44x + 56z = 32$$
$$\underline{-44x - 88z = 0}$$
$$-32z = 32$$
$$z = -1$$
Solve for x:
$$4x + 8(-1) = 0$$
$$4x - 8 = 0$$
$$4x = 8$$
$$x = 2$$

Solve for y:
$$(2) + y + 6(-1) = 0$$
$$2 + y - 6 = 0$$
$$y - 4 = 0$$
$$y = 4$$
The solution of the system is $(2, 4, -1)$.

7. Add the first and second equations:
$$2x + y + z = 3$$
$$\underline{2x - y - z = 9}$$
$$4x \qquad = 12$$
$$x \qquad = 3$$
Add the first and third equations:
$$2x + y + z = 3$$
$$\underline{x + y - z = 0}$$
$$3x + 2y \qquad = 3$$
Solve for y:
$$3(3) + 2y = 3$$
$$9 + 2y = 3$$
$$2y = -6$$
$$y = -3$$
Solve for z:
$$(3) + (-3) - z = 0$$
$$3 - 3 - z = 0$$
$$-z = 0$$
$$z = 0$$
The solution of the system is $(3, -3, 0)$.

9. Add the first equation to 2 times the second equation:
$$2x + 2y + z = 1$$
$$\underline{-2x + 2y + 4z = 6}$$
$$4y + 5z = 7 \quad (4)$$
Add the second equation to the third equation:
$$-x + y + 2z = 3$$
$$\underline{x + 2y + 4z = 0}$$
$$3y + 6z = 3 \quad (5)$$
Subtract 4 times the fifth equation from 3 times the fourth equation.
$$12y + 15z = 21$$
$$\underline{-12y - 24z = -12}$$
$$-9z = 9$$
$$z = -1$$
Solve for y:
$$3y + 6(-1) = 3$$
$$3y - 6 = 3$$
$$3y = 9$$
$$y = 3$$

Solve for x:
$$-x + (3) + 2(-1) = 3$$
$$-x + 3 - 2 = 3$$
$$-x + 1 = 3$$
$$x = -2$$
The solution of the system is $(-2, 3, -1)$.

11. Add the first and second equations:
$$2x - y + 2z = 6$$
$$\underline{3x + y - z = 5}$$
$$5x \quad\quad + z = 11 \quad (4)$$

Add 2 times the first equation to the third equation:
$$4x - 2y + 4z = 12$$
$$\underline{x + 2y + z = 3}$$
$$5x \quad\quad + 5z = 15 \quad (5)$$

Subtract the fourth equation from the fifth equation:
$$5x + 5z = 15$$
$$\underline{-5x \; - z = -11}$$
$$4z = 4$$
$$z = 1$$

Solve for x:
$$5x + (1) = 11$$
$$5x + 1 = 11$$
$$5x = 10$$
$$x = 2$$

Solve for y:
$$(2) + 2y + (1) = 3$$
$$2 + 2y + 1 = 3$$
$$2y + 3 = 3$$
$$2y = 0$$
$$y = 0$$
The solution of the system is $(2, 0, 1)$.

13. Add 2 times the first equation to 3 times the second equation:
$$2x \quad\quad - 6z = 12$$
$$\underline{3y + 6z = 6}$$
$$2x + 3y \quad\quad = 18 \quad (4)$$

Add 5 times the second equation to 2 times the third equation:
$$5y + 10z = 10$$
$$\underline{14x - 6y - 10z = 28}$$
$$14x - y \quad\quad = 38 \quad (5)$$

Add the fourth equation to 3 times the fifth equation:
$$2x + 3y = 18$$
$$\underline{42x - 3y = 114}$$
$$44x \quad\quad = 132$$
$$x \quad\quad = 3$$

Solve for y:
$$14(3) - y = 38$$
$$42 - y = 38$$
$$-y = -4$$
$$y = 4$$

Solve for z:
$$(3) - 3z = 6$$
$$-3z = 3$$
$$z = -1$$
The solution of the system is $(3, 4, -1)$.

15. Add the first equation to the second equation:
$$2x - y \quad\quad = -8$$
$$\underline{y + 3z = 22}$$
$$2x \quad\quad + 3z = 14 \quad (4)$$

Add the fourth equation to 3 times the third equation:
$$2x + 3z = 14$$
$$\underline{3x - 3z = -24}$$
$$5x \quad\quad = -10$$
$$x \quad\quad = -2$$

Solve for z:
$$(-2) - z = -8$$
$$-z = -6$$
$$z = 6$$

Solve for y:
$$2(-2) - y = -8$$
$$-4 - y = -8$$
$$-y = -4$$
$$y = 4$$
The solution of the system is $(-2, 4, 6)$.

17. Substitute the given points into
$$y = ax^2 + bx + c.$$

$$(1, 6): 6 = a(1)^2 + b(1) + c$$
$$(2, 11): 11 = a(2)^2 + b(2) + c$$
$$(3, 18): 18 = a(3)^2 + b(3) + c$$

Simplify these equations:
$$a + b + c = 6 \quad\quad (1)$$
$$4a + 2b + c = 11 \quad\quad (2)$$
$$9a + 3b + c = 18 \quad\quad (3)$$

Eliminate c by multiplying both sides of equation (1) by -1:
$$-a - b - c = -6 \quad\quad (4)$$

Adding the left sides and right sides of equations (2) and (4) gives:
$$3a + b = 5 \quad\quad (5)$$

Adding the left sides and right sides of equations (3) and (4) gives:

$8a + 2b = 12$ $\qquad$ (6)

Simplify:

$4a + b = 6$ $\qquad$ (7)

Eliminate b by multiplying equation (5) by -1 and add each side to the corresponding side of equation (7):

$a = 1$

Next, substitute 1 for a in equation (5):

$3(1) + b = 5$

$\qquad b = 2$

Then, substitute 1 for a and 2 for b in equation (1):

$a + b + c = 6$

$1 + 2 + c = 6$

$\qquad c = 3$

Therefore, $a = 1$, $b = 2$, and $c = 3$. So, the equation is $y = x^2 + 2x + 3$.

19. Substitute the given points into
$y = ax^2 + bx + c$.

$(1, 9): 9 = a(1)^2 + b(1) + c$

$(2, 7): 7 = a(2)^2 + b(2) + c$

$(4, -15): -15 = a(4)^2 + b(4) + c$

Simplify these equations:

$a + b + c = 9$ $\qquad$ (1)

$4a + 2b + c = 7$ $\qquad$ (2)

$16a + 4b + c = -15$ $\qquad$ (3)

Eliminate c by multiplying both sides of equation (1) by -1:

$-a - b - c = -9$ $\qquad$ (4)

Adding the left sides and right sides of equations (2) and (4) gives:

$3a + b = -2$ $\qquad$ (5)

Adding the left sides and right sides of equations (3) and (4) gives:

$15a + 3b = -24$ $\qquad$ (6)

Simplify:

$5a + b = -8$ $\qquad$ (7)

Eliminate b by multiplying equation (5) by -1 and add each side to the corresponding side of equation (7):

$2a = -6$

$\qquad a = -3$

Next, substitute -3 for a in equation (5):

$3(-3) + b = -2$

$\qquad b = 7$

Then, substitute -3 for a and 7 for b in equation (1):

$a + b + c = 9$

$-3 + 7 + c = 9$

$\qquad c = 5$

Therefore, $a = -3$, $b = 7$, and $c = 5$. So, the equation is $y = -3x^2 + 7x + 5$.

21. Substitute the given points into
$y = ax^2 + bx + c$.

$(2, 2): 2 = a(2)^2 + b(2) + c$

$(3, 11): 11 = a(3)^2 + b(3) + c$

$(4, 24): 24 = a(4)^2 + b(4) + c$

Simplify these equations:

$4a + 2b + c = 2$ $\qquad$ (1)

$9a + 3b + c = 11$ $\qquad$ (2)

$16a + 4b + c = 24$ $\qquad$ (3)

Eliminate c by multiplying both sides of equation (1) by -1:

$-4a - 2b - c = -2$ $\qquad$ (4)

Adding the left sides and right sides of equations (2) and (4) gives:

$5a + b = 9$ $\qquad$ (5)

Adding the left sides and right sides of equations (3) and (4) gives:

$12a + 2b = 22$ $\qquad$ (6)

Simplify:

$6a + b = 11$ $\qquad$ (7)

Eliminate b by multiplying equation (5) by -1 and add each side to the corresponding side of equation (7):

$a = 2$

Next, substitute 2 for a in equation (5):

$5(2) + b = 9$

$\qquad b = -1$

Then, substitute 2 for a and -1 for b in equation (1):

$4a + 2b + c = 2$

$4(2) + 2(-1) + c = 2$

$8 - 2 + c = 2$

$\qquad c = -4$

Therefore, $a = 2$, $b = -1$, and $c = -4$. So, the equation is $y = 2x^2 - x - 4$.

23. Substitute the given points into $y = ax^2 + bx + c$.

$(1, -3): -3 = a(1)^2 + b(1) + c$

$(3, 9): 9 = a(3)^2 + b(3) + c$

$(5, 29): 29 = a(5)^2 + b(5) + c$

Simplify these equations:

$a + b + c = -3 \qquad (1)$

$9a + 3b + c = 9 \qquad (2)$

$25a + 5b + c = 29 \qquad (3)$

Eliminate c by multiplying both sides of equation (1) by -1:

$-a - b - c = 3 \qquad (4)$

Adding the left sides and right sides of equations (2) and (4) gives:

$8a + 2b = 12 \qquad (5)$

Simplify:

$4a + b = 6 \qquad (6)$

Adding the left sides and right sides of equations (3) and (4) gives:

$24a + 4b = 32 \qquad (7)$

Simplify:

$6a + b = 8 \qquad (8)$

Eliminate b by multiplying equation (6) by -1 and add each side to the corresponding side of equation (8):

$2a = 2$

$a = 1$

Next, substitute 1 for a in equation (6):

$4(1) + b = 6$

$b = 2$

Then, substitute 1 for a and 2 for b in equation (1):

$a + b + c = -3$

$1 + 2 + c = -3$

$c = -6$

Therefore, $a = 1$, $b = 2$, and $c = -6$. So, the equation is $y = x^2 + 2x - 6$.

25. Substitute the given points into $y = ax^2 + bx + c$.

$(3, 7): 7 = a(3)^2 + b(3) + c$

$(4, 0): 0 = a(4)^2 + b(4) + c$

$(5, -11): -11 = a(5)^2 + b(5) + c$

Simplify these equations:

$9a + 3b + c = 7 \qquad (1)$

$16a + 4b + c = 0 \qquad (2)$

$25a + 5b + c = -11 \qquad (3)$

Eliminate c by multiplying both sides of equation (1) by -1:

$-9a - 3b - c = -7 \qquad (4)$

Adding the left sides and right sides of equations (2) and (4) gives:

$7a + b = -7 \qquad (5)$

Adding the left sides and right sides of equations (3) and (4) gives:

$16a + 2b = -18 \qquad (6)$

Simplify:

$8a + b = -9 \qquad (7)$

Eliminate b by multiplying equation (5) by -1 and add each side to the corresponding side of equation (7):

$a = -2$

Next, substitute -2 for a in equation (5):

$7(-2) + b = -7$

$b = 7$

Then, substitute -2 for a and 7 for b in equation (1):

$9a + 3b + c = 7$

$9(-2) + 3(7) + c = 7$

$-18 + 21 + c = 7$

$c = 4$

Therefore, $a = -2$, $b = 7$, and $c = 4$. So, the equation is $y = -2x^2 + 7x + 4$.

27. Substitute the given points into $y = ax^2 + bx + c$.

$(2, -5): -5 = a(2)^2 + b(2) + c$

$(4, 3): 3 = a(4)^2 + b(4) + c$

$(5, 13): 13 = a(5)^2 + b(5) + c$

Simplify these equations:

$4a + 2b + c = -5 \qquad (1)$

$16a + 4b + c = 3 \qquad (2)$

$25a + 5b + c = 13 \qquad (3)$

Eliminate c by multiplying both sides of equation (1) by -1:

$-4a - 2b - c = 5 \qquad (4)$

Adding the left sides and right sides of equations (2) and (4) gives:

$12a + 2b = 8 \qquad (5)$

Simplify:

$$6a + b = 4 \qquad (6)$$

Adding the left sides and right sides of equations (3) and (4) gives:

$$21a + 3b = 18 \qquad (7)$$

Simplify:

$$7a + b = 6 \qquad (8)$$

Eliminate b by multiplying equation (6) by -1 and add each side to the corresponding side of equation (8):

$$a = 2$$

Next, substitute 2 for a in equation (6):

$$6(2) + b = 4$$
$$b = -8$$

Then, substitute 2 for a and -8 for b in equation (1):

$$4a + 2b + c = -5$$
$$4(2) + 2(-8) + c = -5$$
$$8 - 16 + c = -5$$
$$c = 3$$

Therefore, $a = 2$, $b = -8$, and $c = 3$. So, the equation is $y = 2x^2 - 8x + 3$.

29. Substitute the given points into

$$y = ax^2 + bx + c.$$

$$(0,4): 4 = a(0)^2 + b(0) + c$$
$$(2,8): 8 = a(2)^2 + b(2) + c$$
$$(3,1): 1 = a(3)^2 + b(3) + c$$

Simplify these equations:

$$c = 4 \qquad (1)$$
$$4a + 2b + c = 8 \qquad (2)$$
$$9a + 3b + c = 1 \qquad (3)$$

Since $c = 4$, substitute 4 for c in equations (2) and (3):

$$4a + 2b + 4 = 8$$
$$9a + 3b + 4 = 1$$

Simplifying these equations gives:

$$4a + 2b = 4 \qquad (4)$$
$$9a + 3b = -3 \qquad (5)$$

To eliminate b, multiply both sides of equation (4) by -3 and both sides of equation (5) by 2:

$$-12a - 6b = -12 \qquad (6)$$
$$18a + 6b = -6 \qquad (7)$$

Adding the left and the right sides of equations (6) and (7) gives:

$$6a = -18$$
$$a = -3$$

Next, substitute -3 for a in equation (4):

$$4(-3) + 2b = 4$$
$$2b = 16$$
$$b = 8$$

Therefore, $a = -3$, $b = 8$, and $c = 4$. So, the equation is $y = -3x^2 + 8x + 4$.

31. Substitute the given points into

$$y = ax^2 + bx + c.$$

$$(0,-1): -1 = a(0)^2 + b(0) + c$$
$$(1,3): 3 = a(1)^2 + b(1) + c$$
$$(2,13): 13 = a(2)^2 + b(2) + c$$

Simplify these equations:

$$c = -1 \qquad (1)$$
$$a + b + c = 3 \qquad (2)$$
$$4a + 2b + c = 13 \qquad (3)$$

Since $c = -1$, substitute -1 for c in equations (2) and (3):

$$a + b + (-1) = 3$$
$$4a + 2b + (-1) = 13$$

Simplifying these equations gives:

$$a + b = 4 \qquad (4)$$
$$4a + 2b = 14 \qquad (5)$$

To eliminate b, multiply both sides of equation (4) by -2 and add each side to the corresponding side in equation (5):

$$2a = 6$$
$$a = 3$$

Next, substitute 3 for a in equation (4):

$$3 + b = 4$$
$$b = 1$$

Therefore, $a = 3$, $b = 1$, and $c = -1$. So, the equation is $y = 3x^2 + x - 1$.

33. Substitute the given points into

$$y = ax^2 + bx + c.$$

$$(1,1): 1 = a(1)^2 + b(1) + c$$
$$(2,4): 4 = a(2)^2 + b(2) + c$$
$$(3,9): 9 = a(3)^2 + b(3) + c$$

Simplify these equations:

$$a+b+c=1 \qquad (1)$$

$$4a+2b+c=4 \qquad (2)$$

$$9a+3b+c=9 \qquad (3)$$

Eliminate c by multiplying both sides of equation (1) by -1:

$$-a-b-c=-1 \qquad (4)$$

Adding the left sides and right sides of equations (2) and (4) gives:

$$3a+b=3 \qquad (5)$$

Adding the left sides and right sides of equations (3) and (4) gives:

$$8a+2b=8 \qquad (6)$$

Simplify:

$$4a+b=4 \qquad (7)$$

Eliminate b by multiplying equation (5) by -1 and add each side to the corresponding side of equation (7):

$$a=1$$

Next, substitute 1 for a in equation (5):

$$3(1)+b=3$$

$$b=0$$

Then, substitute 1 for a and 0 for b in equation (1):

$$a+b+c=1$$

$$1+0+c=1$$

$$c=0$$

Therefore, $a=1$, $b=0$, and $c=0$. So, the equation is $y=x^2$.

35. Substitute the given points into $y=ax^2+bx+c$.

$$(0,4): 4=a(0)^2+b(0)+c$$

$$(1,0): 0=a(1)^2+b(1)+c$$

$$(2,0): 0=a(2)^2+b(2)+c$$

Simplify these equations:

$$c=4 \qquad (1)$$

$$a+b+c=0 \qquad (2)$$

$$4a+2b+c=0 \qquad (3)$$

Since $c=4$, substitute 4 for c in equations (2) and (3):

$$a+b+4=0$$

$$4a+2b+4=0$$

Simplify these equations:

$$a+b=-4 \qquad (4)$$

$$4a+2b=-4 \qquad (5)$$

Eliminate b by multiplying equation (4) by -2 and add each side to the corresponding side of equation (5):

$$2a=4$$

$$a=2$$

Next, substitute 2 for a in equation (4):

$$2+b=-4$$

$$b=-6$$

Therefore, $a=2$, $b=-6$, and $c=4$. So, the equation is $y=2x^2-6x+4$.

37. Three possible points are $(2, 8)$, $(3, 4)$, and $(6, 4)$.

Substitute the given points into $y=ax^2+bx+c$.

$$(2,8): 8=a(2)^2+b(2)+c$$

$$(3,4): 4=a(3)^2+b(3)+c$$

$$(6,4): 4=a(6)^2+b(6)+c$$

Simplify these equations:

$$4a+2b+c=8 \qquad (1)$$

$$9a+3b+c=4 \qquad (2)$$

$$36a+6b+c=4 \qquad (3)$$

Eliminate c by multiplying both sides of equation (1) by -1:

$$-4a-2b-c=-8 \qquad (4)$$

Adding the left sides and right sides of equations (2) and (4) gives:

$$5a+b=-4 \qquad (5)$$

Adding the left sides and right sides of equations (3) and (4) gives:

$$32a+4b=-4 \qquad (6)$$

Simplify:

$$8a+b=-1 \qquad (7)$$

Eliminate b by multiplying equation (5) by -1 and add each side to the corresponding side of equation (7):

$$3a=3$$

$$a=1$$

Next, substitute 1 for a in equation (5):

$$5(1)+b=-4$$

$$b=-9$$

Then, substitute 1 for a and -9 for b in equation (1):

$$4a+2b+c=8$$

$$4(1)+2(-9)+c=8$$

$$4-18+c=8$$

$$c=22$$

Therefore, $a = 1$, $b = -9$, and $c = 22$. So, the equation is $y = x^2 - 9x + 22$.

39. Solve for a in $y = a(x-h)^2 + k$ by Substitute 5 for h and 8 for k since (5, 8) is the vertex, and substitute 4 for x and 6 for y since (4, 6) lies on the parabola:

$$y = a(x-h)^2 + k$$
$$6 = a(4-5)^2 + 8$$
$$6 = a + 8$$
$$a = -2$$

Since $a = -2$, $y = -2(x-5)^2 + 8$, or expanding the solution:

$$y = -2(x-5)(x-5) + 8$$
$$= -2(x^2 - 10x + 25) + 8$$
$$= -2x^2 + 20x - 50 + 8$$
$$= -2x^2 + 20x - 42$$

41. a.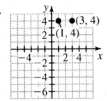

b. Answers may vary. Example:

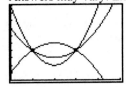

c. Answers may vary. Example:
$$y = (x-2)^2 + 3$$
$$y = -(x-2)^2 + 5$$
$$y = 2(x-2)^2 + 2$$

d. Answers may vary.

43. Answers may vary. Example:
$(0, 2)$, $(1, 1)$, and $(3, 5)$.
The equation of the parabola through these points is $y = x^2 - 2x + 2$.

45. Answers may vary. Example:
1. Select a pair of equations and eliminate a variable.
2. Select any other pair of equations and eliminate the same variable as in step 1.

3. The equations you found in steps 1 and 2 form a system of linear equations in two variables. Use elimination or substitution to solve this system.
4. Substitute the values of two of the variables you found in step 3 into one of the original equations that contains the third variable. Solve for the third variable.
5. Write your solution as an ordered triple.

47. Linear:
$$m = \frac{85-13}{7-3} = 18$$
$$y - 13 = 18(x-3)$$
$$y - 13 = 18x - 54$$
$$y = 18x - 41$$
$$L(x) = 18x - 41$$

Exponential:
$$13 = Ab^3$$
$$\frac{13}{b^3} = A$$
$$85 = Ab^7$$
$$\frac{85}{b^7} = A$$
$$\frac{13}{b^3} = \frac{85}{b^7}$$
$$13b^7 = 85b^3$$
$$b^4 \approx 6.54$$
$$b \approx 1.60$$
$$A = \frac{13}{(1.60)^3} = \frac{13}{4.096} \approx 3.17$$
$$E(x) = 3.17(1.60)^x$$

49. Linear:
$y = 2x + 2$ since the slope is 2 and the y-intercept is (0, 2)
Exponential:
$$y = 2(2)^x$$
Quadratic:
Answers may vary. Example: $y = 2x^2 + 2$.

51. Substitute the given points into
$$f(x) = ax^2 + bx + c.$$
$$(1,1): 1 = a(1)^2 + b(1) + c$$
$$(2,2): 2 = a(2)^2 + b(2) + c$$
$$(3,3): 3 = a(3)^2 + b(3) + c$$

Simplify these equations:

$a + b + c = 1$ (1)

$4a + 2b + c = 2$ (2)

$9a + 3b + c = 3$ (3)

Eliminate c by multiplying both sides of equation (1) by -1:

$-a - b - c = -1$ (4)

Adding the left sides and right sides of equations (2) and (4) gives:

$3a + b = 1$ (5)

Adding the left sides and right sides of equations (3) and (4) gives:

$8a + 2b = 2$ (6)

Simplify:

$4a + b = 1$ (7)

Eliminate b by multiplying equation (5) by -1 and add each side to the corresponding side of equation (7):

$a = 0$

Next, substitute 0 for a in equation (5):

$3(0) + b = 1$

$b = 1$

Then, substitute 0 for a and 1 for b in equation (1):

$a + b + c = 1$

$0 + 1 + c = 1$

$c = 0$

Therefore, $a = 0$, $b = 1$, and $c = 0$. So, the equation is $f(x) = x$, which is a linear function.

53. Since the parabola has a vertex of $(5, -7)$ it has the form: $f(x) = a(x - 5)^2 - 7$. Now substitute the point $(8, 11)$ and solve for a:

$11 = a((8) - 5)^2 - 7$

$11 = a(3)^2 - 7$

$11 = 9a - 7$

$18 = 9a$

$2 = a$

The equation of the parabola is:

$f(x) = 2(x - 5)^2 - 7$ or

$f(x) = 2x^2 - 20x + 43$.

55. $2x^2 - 10x + 7 = 0$

$x = \dfrac{-(-10) \pm \sqrt{(-10)^2 - 4(2)(7)}}{2(2)}$

$x = \dfrac{10 \pm \sqrt{44}}{4}$

$x = \dfrac{10 \pm 2\sqrt{11}}{4}$

$x = \dfrac{5 \pm \sqrt{11}}{2}$

This is a quadratic equation in one variable.

57. $f(2) = 2(2)^2 - 10(2) + 7$

$= 2(4) - 20 + 7$

$= 8 - 13$

$= -5$

This is a quadratic function.

59.

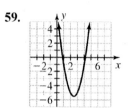

This is a quadratic function.

Homework 7.7

1. **a.** A quadratic function would be reasonable.

   b. A linear function would be reasonable.

   c. An exponential function would be reasonable.

   d. None of the mentioned types of functions would be reasonable for this scattergram.

3. By hand:
   Substitute the points $(14, 8.4)$, $(15, 13.6)$, and $(16, 20.1)$ into $f(t) = at^2 + bt + c$.

   $(14, 8.4): 8.4 = a(14)^2 + b(14) + c$

   $(15, 13.6): 13.6 = a(15)^2 + b(15) + c$

   $(16, 20.1): 20.1 = a(16)^2 + b(16) + c$

   Simplify these equations:

   $196a + 14b + c = 8.4$ (1)

   $225a + 15b + c = 13.6$ (2)

   $256a + 16b + c = 20.1$ (3)

   Eliminate c by multiplying both sides of equation (1) by -1:

   $-196a - 14b - c = -8.4$ (4)

Adding the left sides and right sides of equations (2) and (4) gives:

$29a + b = 5.2$ (5)

Adding the left sides and right sides of equations (3) and (4) gives:

$60a + 2b = 11.7$ (6)

Simplify:

$30a + b = 5.85$ (7)

Eliminate b by multiplying equation (5) by -1 and add each side to the corresponding side of equation (7):

$a = 0.65$

Next, substitute 0.65 for a in equation (5):

$29(0.65) + b = 5.2$

$b = -13.65$

Then, substitute 0.65 for a and -13.65 for b in equation (1):

$196(0.65) + 14(-13.65) + c = 8.4$

$127.4 - 191.1 + c = 8.4$

$c = 72.1$

Therefore, $a = 0.65$, $b = -13.65$, and $c = 72.10$. So, the equation of the parabola is

$f(t) = 0.65t^2 - 13.65t + 72.10$.

By regression:

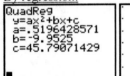 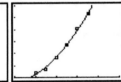

The equation of the parabola is

$f(t) = 0.52t^2 - 9.95t + 45.79$.

Answers may vary. Example:
The graphs of the two models are similar.

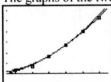

5. **a.** Substitute the points (2, 310), (7, 695), and (12, 1669) into $f(t) = at^2 + bt + c$.

$(2, 310): 310 = a(2)^2 + b(2) + c$

$(7, 695): 695 = a(7)^2 + b(7) + c$

$(12, 1669): 1669 = a(12)^2 + b(12) + c$

Simplify these equations:

$4a + 2b + c = 310$ (1)

$49a + 7b + c = 695$ (2)

$144a + 12b + c = 1669$ (3)

Eliminate c by multiplying both sides of equation (1) by -1:

$-4a - 2b - c = -310$ (4)

Adding the left sides and right sides of equations (2) and (4) gives:

$45a + 5b = 385$ (5)

Adding the left sides and right sides of equations (3) and (4) gives:

$140a + 10b = 1359$ (6)

Eliminate b by multiplying equation (5) by -2:

$-90a - 10b = -770$ (7)

Adding the left and right sides of equations (6) and (7) gives:

$50a = 589$

$a = 11.78$

Next, substitute 11.78 for a in equation (5):

$45(11.78) + 5b = 385$

$5b = 385 - 530.1$

$b = -29.02$

Then, substitute 11.78 for a and -29.02 for b in equation (1):

$4(11.78) + 2(-29.02) + c = 310$

$47.12 - 58.04 + c = 310$

$c = 320.92$

Therefore, $a = 11.78$, $b = -29.02$, and $c = 320.92$.
So, the equation of the parabola is

$f(t) = 11.78t^2 - 29.02t + 320.92$.

b.

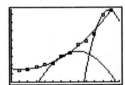

The equation

$f(t) = 11.78t^2 - 29.02t + 320.92$ for the years 2002, 2007, and 2012 better fits the data.

Answers may vary. Example:
This could have been guessed because the other equations would form parabolas that open down, and the data is better fit by a parabola that opens up.

7. a. By regression:

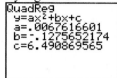

The equation is

$$f(t) = 0.0068t^2 - 0.13t + 6.49.$$

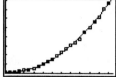

The model fits the data well.

b.   i.

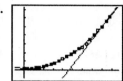

The quadratic function describes the population better for these years, as it represents all of the data well.

ii.

The linear function describes the population better for these years, as it is designed for these data.

iii. Answers may vary.

iv. It appears to be linear. Answers may vary.

9. a. By regression:

$$f(t) = 2.2t^2 - 31.65t + 115.51$$
$$g(t) = 6.11t - 42.75$$

b.

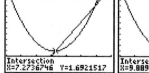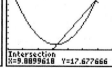

The points of intersection are (7.27, 1.69) and (9.89, 17.68). The point (7.27, 1.69) means that in 2007, the numbers of gigawatts of solar panels manufactured in China and installed in the world were both about 2 gigawatts. The point (9.89, 17.68) means that in 2010, the numbers of gigawatts of solar panels manufactured in China and installed in the world were both about 18 gigawatts.

11.

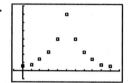

The data do not suggest a quadratic relationship based on the scattergram above. A quadratic function is not a reasonable function.

13. Answers may vary.

15. a. Use regression.

Linear:

$$f(t) = 0.70t + 8.86$$

Exponential:

$$f(t) = 12.09(1.029)^t$$

Quadratic:

$$f(t) = 0.017t^2 - 0.08t + 14.94$$

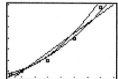

Both the exponential and quadratic models fit the data well. The linear model is not as good a fit.

b.

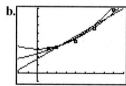

The exponential model provides the best estimates before 1976.

17. $\frac{1}{2}x - \frac{2}{3}y = 2$

$$3x - 4y = 12$$
$$3x = 12 + 4y$$
$$x = 4 + \frac{4}{3}y$$

$$\frac{4}{3}x + \frac{5}{2}y = 31$$
$$8x + 15y = 186$$
$$8\left(4 + \frac{4}{3}y\right) + 15y = 186$$
$$32 + \frac{32}{3}y + 15y = 186$$
$$\frac{77}{3}y = 154$$
$$y = 6$$

$$x = 4 + \frac{4}{3}(6) = 12$$
$(12, 6)$
This is a system of linear equations in two variables.

19. $\frac{1}{2}x - \frac{2}{3}y = 2$

$$-\frac{2}{3}y = 2 - \frac{1}{2}x$$
$$y = \frac{3}{4}x - 3$$

This is a linear equation in two variables.

21. $m = \frac{-7 + 2}{-2 + 5} = -\frac{5}{3}$

$$y = -\frac{5}{3}x + b$$
$$-2 = -\frac{5}{3}(-5) + b$$
$$-2 = \frac{25}{3} + b$$
$$b = -\frac{31}{3}$$

$$y = -\frac{5}{3}x - \frac{31}{3}$$
This is a linear equation in two variables.

Homework 7.8

1. a. The p-intercept is found when $t = 0$.

$$f(0) = 0.058(0)^2 - 0.98(0) + 15.43 = 15.43$$

The p-intercept is $(0, 15.43)$.

In 1990, about 15% percentage of households had outstanding student debt.

b. $f(19) = 0.058(19)^2 - 0.98(19) + 15.43$
$$= 17.748$$
About 18% of households had outstanding student debt in 2009. This involved interpolation because 2009 is between given ordered pairs.

c. $f(28) = 0.058(28)^2 - 0.98(28) + 15.43$
$$= 33.462$$
About 33% of households will have outstanding student debt in 2018. This involved extrapolation because 2018 is beyond 2010, the last year data were provided.

d. $36 = 0.058t^2 - 0.98t + 15.43$
$$0 = 0.058t^2 - 0.98t - 20.57$$
$$a = 0.058, \ b = -0.98, \ c = -20.57$$

$$t \approx \frac{-(-0.98) \pm \sqrt{(-0.98)^2 - 4(0.058)(-20.57)}}{2(0.058)}$$

$$t \approx \frac{0.98 \pm 2.39}{0.116}$$

$t \approx -12.16$ or $t \approx 29.05$
The model predicts that 36% of households will have outstanding student debt in $1990 + 29 = 2019$.

3. a. Use a calculator to find the quadratic regression equation for this set of data.

$$f(a) = -0.031a^2 + 4.03a - 47.05$$

```
QuadReg
y=ax²+bx+c
a=-.0313095238
b=4.031785714
c=-47.04770833
```

b. The a-intercepts occur when $f(a) = 0$. Use the quadratic formula to solve for a.

$$a = \frac{-4.03 \pm \sqrt{(4.03)^2 - 4(-0.031)(-47.05)}}{2(-0.031)}$$

$$a \approx \frac{-4.03 \pm \sqrt{10.41}}{-0.062}$$

$$a \approx \frac{-4.03 + 3.23}{-0.062} \quad \text{or} \quad a \approx \frac{-4.03 - 3.23}{-0.062}$$

$$a \approx 12.9 \qquad \text{or} \qquad a \approx 117.1$$

The a-intercepts are $(12.9, 0)$ and $(117.1, 0)$. Simplifying, $a \approx 13$ or $a \approx 117$. These values represent ages at which 0% own a home.

c. Model breakdown occurs for values less than 13 and greater than 117. These represent ages at which less than 0% own a home. The legal age for home ownership and human life expectancy also would factor into which values are reasonable.

d. The vertex represents the highest point of a downward-pointing parabola. The x-value of the vertex is given by

$$-\frac{b}{2a} = -\frac{4.03}{-0.062} = 65 \text{, or 65 years old.}$$

$$f(65) = -0.031(65)^2 + 4.03(65) - 47.05$$
$$= 83.93$$

So, approximately 84% of 65-year-olds own their own home.

e. Solve for a for $f(a) = 50$.

$$50 = -0.031(a)^2 + 4.03a - 47.05$$
$$0 = -0.031(a)^2 + 4.03a - 97.05$$

Use the quadratic formula to solve for a.

$$a = \frac{-4.03 \pm \sqrt{(4.03)^2 - 4(-0.031)(-97.05)}}{2(-0.031)}$$

$$a \approx \frac{-4.03 \pm \sqrt{4.21}}{-0.062}$$

$$a \approx \frac{-4.03 + 2.05}{-0.062} \quad \text{or} \quad a \approx \frac{-4.03 - 2.05}{-0.062}$$

$$a \approx 31.94 \qquad \text{or} \qquad a \approx 98.06$$

Simplifying, $a \approx 32$ or $a \approx 98$. These values represent ages in years at which 50% own a home.

5. a. $f(228) = 0.0068(228)^2 - 0.13(228) + 6.49$
$$\approx 330.34$$
In 2018, the U.S. population will be about 330.3 million people.

b. $0.0068t^2 - 0.13t + 6.49 = 335$
$$0.0068t^2 - 0.13t - 328.51 = 0$$
$$a = 0.0068, b = -0.13, c = -328.51$$

$$t = \frac{0.13 \pm \sqrt{(-0.13)^2 - 4(0.0068)(-328.51)}}{2(0.0068)}$$

$$t \approx \frac{0.13 \pm 2.992}{0.0136}$$

$$t \approx -210.44 \quad \text{or} \quad t \approx 229.56$$

The negative value for t indicates that the U.S. population was 325 million in $1790 - 210 = 1580$, so model breakdown has occurred. The U.S. population will be 325 million people in $1790 + 230 = 2020$.

c.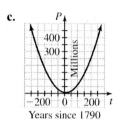
Years since 1790

d. The t-coordinate of the vertex for the model is given by

$$-\frac{b}{2a} = -\frac{-0.13}{2(0.0068)} = 9.56 \text{, which means}$$

that in $1790 + 9.56 \approx 1800$, the U.S. population will reach a minimum value. This minimum value is given by the model as:

$$f(9.56)$$
$$= 0.0068(9.56)^2 - 0.13(9.56) + 6.49$$
$$\approx 6.00$$

However, the census data for 1790 and 1800 are both lower than 6 million. Therefore, model breakdown occurs for years prior to 1800.

e. The graph should indicate that, for low values of t, the population should be small but slowly growing, while at large values of t, the population should level off at a maximum value (that is, there should not be an infinite-growth curve for a finite system). Between these two nearly horizontal curves, the growth should follow a quadratic or exponential function that closely models the data. Graphs may vary. Example:

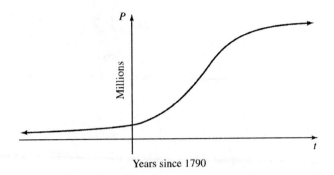

Years since 1790

7. a. By regression:

```
QuadReg
 y=ax²+bx+c
 a=.8807400802
 b=11.63440137
 c=⁻85.62591795
 ■
```

$f(t) = 0.88t^2 + 11.63t - 85.63$

b. $0.88t^2 + 11.63t - 85.63 = 755$

$0.88t^2 + 11.63t - 840.63 = 0$

$a = 0.88, b = 11.63, c = -840.63$

$$t = \frac{-11.63 \pm \sqrt{(11.63)^2 - 4(0.88)(-840.63)}}{2(0.88)}$$

$$t \approx \frac{-11.63 \pm 55.63}{1.76}$$

$t \approx 25$ or $t \approx -38.2$

We estimate that Barry Bonds reached 755 home runs in $1980 + 25 = 2005$.

c. $f(25) = 0.88(25)^2 + 11.63(25) - 85.63$
$= 755.12$

$f(26) = 0.88(26)^2 + 11.63(26) - 85.63$
$= 811.63$

$f(26) - f(25) = 811.63 - 755.12 = 56.51$

Yes, this result suggests that he hit enough home runs in 2006 to tie the record.

d. Answers may vary.

9. $f(t) = 2.2t^2 - 31.65t + 115.51$
$g(t) = 6.11t - 42.75$

$2.2t^2 - 31.65t + 115.51 = 6.11t - 42.75$

$2.2t^2 - 37.76t + 158.26 = 0$

$a = 2.2, b = -37.76, c = 158.26$

$$t = \frac{-(-37.76) \pm \sqrt{(-37.76)^2 - 4(2.2)(158.26)}}{2(2.2)}$$

$$t \approx \frac{37.76 \pm 5.76}{4.4}$$

$t \approx 7.27$ or $t \approx 9.89$

The number of gigawatts of solar panels manufactured in China was equal to the number of gigawatts of solar panels installed in the world in 2007 and 2010.

11. a. By regression:

```
QuadReg
 y=ax²+bx+c
 a=⁻.0805280201
 b=⁻2.060619878
 c=72.851705
 ■
```

$f(t) = -0.081t^2 - 2.06t + 72.85$

```
LinReg
 y=ax+b
 a=3.795588235
 b=10.11029412
 ■
```

$g(t) = 3.80t + 10.11$

b. $f(t) = -0.081t^2 - 2.06t + 72.85$
$g(t) = 3.80t + 10.11$

$-0.081t^2 - 2.06t + 72.85 = 3.80t + 10.11$

$-0.081t^2 - 5.86t + 62.74 = 0$

$a = -0.081, b = -5.86, c = 62.74$

$$t = \frac{5.86 \pm \sqrt{(-5.86)^2 - 4(-0.081)(62.74)}}{2(-0.081)}$$

$$t \approx \frac{5.86 \pm 7.39}{-0.162}$$

$t \approx -81.79$ or $t \approx 9.44$

As electronic voting machines did not exist in $1990 - 82 = 1908$, the positive root is the only meaningful result. In $1990 + 9 = 1999$, the percentage of voters using optical scan or other modern electrical system equaled the percentage of people using punch cards or lever machines. No, 1999 was not a major election year. 2000 is the closest election year.

c. $h(t) = 100 - (f + g)(t)$ represents the percentage of voters using the three other ways to vote.

d. $(f+g)(t) = \left(-0.081t^2 - 2.06t + 72.85\right)$
$$+ (3.80t + 10.11)$$
$$= -0.081t^2 + 1.74t + 82.96$$
$$h(t) = 100 - (f+g)(t)$$
$$= 100 - \left(-0.081t^2 + 1.74t + 82.96\right)$$
$$= 0.081t^2 - 1.74t + 17.04$$

e. $h(17) = 0.081(17)^2 - 1.74(17) + 17.04$
$$\approx 10.87$$
In 1990 + 17 = 2007, about 10.9% of registered voters used voting methods other than punch cards, lever machines, optical scan or other modern electronic systems.

13. $p = 250 - 5n$
$$R = pn = 250n - 5n^2$$
$$n = -\frac{b}{2a} = -\frac{250}{2(-5)} = 25$$
A group of 25 people would maximize the bus company's revenue.

15. $p = 28 - 0.2n$
$$R = pn = 28n - 0.2n^2$$
$$n = -\frac{b}{2a} = -\frac{28}{2(-0.2)} = 70$$
A party of 70 people would maximize the restaurant's revenue.

17. $160 = 2l + 2w$
$$80 = l + w$$
$$80 - w = l$$
$$A = lw = 80w - w^2$$
$$w = -\frac{b}{2a} = -\frac{80}{2(-1)} = 40$$
$$l = 80 - 40 = 40$$
The dimensions are 40 feet long by 40 feet wide. The area is 1600 square feet.

19. Answers may vary. Example:
The graph of f changes direction at the vertex, so that at $t = h$, a decreasing graph becomes increasing, and an increasing graph becomes decreasing. This is likely to reverse the trend represented by the model, and the model breaks down.
Graphs may vary.

21. Answers may vary.

23. a. Linear:
$$L(t) = 12.90t - 88.74$$
Exponential:
$$E(t) = 1.42(1.28)^t$$
Quadratic:
$$Q(t) = 1.23t^2 - 18.18t + 65.79$$

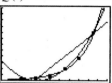

The linear model does not fit the data well; the quadratic model fits the data well; the exponential model fits the data the best.

b. The exponential model fits the data well and provides realistic results for years before 1994.

c. 7 billion = 7,000,000,000
1 MW = 1000 people
7,000,000 MW = 7,000,000,000 people
7000 thousand MW is the goal.
$$1.42(1.28)^t = 7000$$
$$1.28^t \approx 4929.58$$
$$\log_{1.28}\left(1.28^t\right) \approx \log_{1.28}(4929.58)$$
$$t \approx \frac{\log 4929.58}{\log 1.28} \approx 34.44$$
In 1990 + 34 = 2024, wind energy could meet the electricity needs of the entire world.

d. $1.23t^2 - 18.18t + 65.79 = 7000$
$$1.23t^2 - 18.18t - 6934.21 = 0$$
$$a = 1.23, b = -18.18, c = -6934.21$$
$$t = \frac{-(-18.18) \pm \sqrt{(-18.18)^2 - 4(1.23)(-6934.21)}}{2(1.23)}$$
$$t \approx \frac{18.18 \pm 185.60}{2.46}$$
$$t \approx -68.06 \text{ or } t \approx 82.84$$
Only the positive root is meaningful. In 1990 + 83 = 2073, wind energy could meet the electricity needs of the world.

e. Answers may vary. The exponential model predicts faster growth than the quadratic model.

25. Answers may vary.

27. Answers may vary.

29. Answers may vary.

31. Answers may vary.

Chapter 7 Review

1.

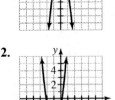

2.

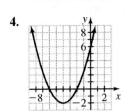

3.

4.

5. Since the parabola has a maximum point, $a < 0$. Since the vertex is in quadrant II, $h < 0$ and $k > 0$.

6. Vertex: $(2, -5)$

7. Vertex: $(2, 13)$

8. Vertex: $(1.43, -6.25)$

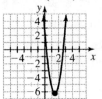

9. $\sqrt{72} = \sqrt{36 \cdot 2} = \sqrt{36}\sqrt{2} = 6\sqrt{2}$

10. $\sqrt{\dfrac{49}{100}} = \dfrac{\sqrt{49}}{\sqrt{100}} = \dfrac{7}{10}$

11. $3x^2 - 2x - 2 = 0$

$$x = \frac{2 \pm \sqrt{(-2)^2 - 4(3)(-2)}}{2(3)}$$

$$x = \frac{2 \pm \sqrt{28}}{6}$$

$$x = \frac{2 \pm 2\sqrt{7}}{6}$$

$$x = \frac{1 \pm \sqrt{7}}{3}$$

12. $5x^2 = 7$

$$x^2 = \frac{7}{5}$$

$$x = \pm\sqrt{\frac{7}{5}} = \pm\frac{\sqrt{7}}{\sqrt{5}} = \pm\frac{\sqrt{7}}{\sqrt{5}} \cdot \frac{\sqrt{5}}{\sqrt{5}} = \pm\frac{\sqrt{35}}{5}$$

13. $5(p - 3)^2 + 4 = 7$

$$5p^2 - 30p + 42 = 0$$

$$p = \frac{30 \pm \sqrt{(-30)^2 - 4(5)(42)}}{2(5)}$$

$$p = \frac{30 \pm \sqrt{60}}{10}$$

$$p = \frac{30 \pm 2\sqrt{15}}{10}$$

$$p = \frac{15 \pm \sqrt{15}}{5}$$

14. $(t+1)(t-7) = 4$

$t^2 - 6t - 11 = 0$

$x = \dfrac{6 \pm \sqrt{(-6)^2 - 4(1)(-11)}}{2(1)}$

$x = \dfrac{6 \pm \sqrt{80}}{2}$

$x = \dfrac{6 \pm 4\sqrt{5}}{2}$

$x = 3 \pm 2\sqrt{5}$

15. $\qquad 2x^2 = 4 - 5x$

$2x^2 + 5x - 4 = 0$

$x = \dfrac{-5 \pm \sqrt{5^2 - 4(2)(-4)}}{2(5)} = \dfrac{-5 \pm \sqrt{57}}{10}$

16. $\qquad 4x - x^2 = 1$

$x^2 - 4x + 1 = 0$

$x = \dfrac{4 \pm \sqrt{(-4)^2 - 4(1)(1)}}{2(1)}$

$x = \dfrac{4 \pm \sqrt{12}}{2}$

$x = \dfrac{4 \pm 2\sqrt{3}}{2}$

$x = 2 \pm \sqrt{3}$

17. $\qquad 5x^2 - 6x = 2$

$5x^2 - 6x - 2 = 0$

$x = \dfrac{6 \pm \sqrt{(-6)^2 - 4(5)(-2)}}{2(5)}$

$x = \dfrac{6 \pm \sqrt{76}}{10}$

$x = \dfrac{6 \pm 2\sqrt{19}}{10}$

$x = \dfrac{3 \pm \sqrt{19}}{5}$

18. $7x^2 - 20 = 0$

$7x^2 = 20$

$x^2 = \dfrac{20}{7}$

$x = \pm\sqrt{\dfrac{20}{7}}$

$x = \pm\dfrac{\sqrt{20}}{\sqrt{7}}$

$x = \pm\dfrac{\sqrt{20}}{\sqrt{7}} \cdot \dfrac{\sqrt{7}}{\sqrt{7}}$

$x = \pm\dfrac{\sqrt{140}}{7}$

$x = \pm\dfrac{2\sqrt{35}}{7}$

19. $(x+2)^2 + (x-3)^2 = 15$

$2x^2 - 2x - 2 = 0$

$x = \dfrac{2 \pm \sqrt{(-2)^2 - 4(2)(-2)}}{2(2)}$

$x = \dfrac{2 \pm \sqrt{20}}{4}$

$x = \dfrac{2 \pm 2\sqrt{5}}{4}$

$x = \dfrac{1 \pm \sqrt{5}}{2}$

20. $5(5x^2 - 8) = 9$

$25x^2 - 40 = 9$

$25x^2 = 49$

$x^2 = \dfrac{49}{25}$

$x = \pm\sqrt{\dfrac{49}{25}} = \pm\dfrac{7}{5}$

21. $\dfrac{3}{2}x^2 - \dfrac{3}{4}x = \dfrac{1}{2}$

$6x^2 - 3x - 2 = 0$

$x = \dfrac{3 \pm \sqrt{(-3)^2 - 4(6)(-2)}}{2(6)} = \dfrac{3 \pm \sqrt{57}}{12}$

22.
$$2.7x^2 - 5.1x = 9.8$$
$$2.7x^2 - 5.1x - 9.8 = 0$$
$$x = \frac{-(-5.1) \pm \sqrt{(-5.1)^2 - 4(2.7)(-9.8)}}{2(2.7)}$$
$$x \approx \frac{5.1 \pm 11.48}{5.4}$$
$$x \approx -1.18 \text{ or } x \approx 3.07$$

23.
$$1.7(x^2 - 2.3) = 3.4 - 2.8x$$
$$1.7x^2 - 3.91 = 3.4 - 2.8x$$
$$1.7x^2 + 2.8x - 7.31 = 0$$
$$x = \frac{-2.8 \pm \sqrt{2.8^2 - 4(1.7)(-7.31)}}{2(1.7)}$$
$$x \approx \frac{-2.8 \pm 7.59}{3.4}$$
$$x \approx -3.06 \text{ or } x \approx 1.41$$

24.
$$-2(x + 4)^2 = 9$$
$$-2x^2 - 16x - 41 = 0$$
$$x = \frac{16 \pm \sqrt{(-16)^2 - 4(-2)(-41)}}{2(-2)}$$
$$x = \frac{16 \pm \sqrt{-72}}{-4}$$
$$x = \frac{16 \pm 6i\sqrt{2}}{-4}$$
$$x = \frac{-8 \pm 3i\sqrt{2}}{2}$$

25.
$$2x^2 = 4x - 7$$
$$2x^2 - 4x + 7 = 0$$
$$x = \frac{4 \pm \sqrt{(-4)^2 - 4(2)(7)}}{2(2)}$$
$$x = \frac{4 \pm \sqrt{-40}}{4}$$
$$x = \frac{4 \pm 2i\sqrt{10}}{4}$$
$$x = \frac{2 \pm i\sqrt{10}}{2}$$

26.
$$x^2 + 6x - 4 = 0$$
$$x^2 + 6x = 4$$
$$x^2 + 6x + 9 = 4 + 9$$
$$(x + 3)^2 = 13$$
$$x + 3 = \pm\sqrt{13}$$
$$x = -3 \pm \sqrt{13}$$

27.
$$2t^2 = -3t + 6$$
$$2t^2 + 3t = 6$$
$$t^2 + \frac{3}{2}t = 3$$
$$t^2 + \frac{3}{2}t + \frac{9}{16} = 3 + \frac{9}{16}$$
$$\left(t + \frac{3}{4}\right)^2 = \frac{57}{16}$$
$$t + \frac{3}{4} = \pm\sqrt{\frac{57}{16}}$$
$$t + \frac{3}{4} = \pm\frac{\sqrt{57}}{4}$$
$$t = \frac{-3 \pm \sqrt{57}}{4}$$

28. Solve for x when $h(x) = 0$:
$$3x^2 + 2x - 2 = 0$$
$$x = \frac{-2 \pm \sqrt{2^2 - 4(3)(-2)}}{2(3)}$$
$$x = \frac{-2 \pm \sqrt{28}}{6}$$
$$x = \frac{-2 \pm 2\sqrt{7}}{6}$$
$$x = \frac{-1 \pm \sqrt{7}}{3}$$

The x-intercepts are $\left(\dfrac{-1 - \sqrt{7}}{3}, 0\right)$ and $\left(\dfrac{-1 + \sqrt{7}}{3}, 0\right)$.

29. Solve for x when $k(x) = 0$:
$$-5x^2 + 3x - 1 = 0$$
$$x = \frac{-3 \pm \sqrt{3^2 - 4(-5)(-1)}}{2(-5)} = \frac{-3 \pm \sqrt{-11}}{-10}$$

Since the square root of a negative number is not a real number, there are no real number solutions. Therefore, there are no x-intercepts.

30. Factoring:
$$x^2 - 2x - 8 = 0$$
$$(x - 4)(x + 2) = 0$$
$$x - 4 = 0 \quad \text{or} \quad x + 2 = 0$$
$$x = 4 \quad \text{or} \quad x = -2$$

Completing the square:

$$x^2 - 2x - 8 = 0$$

$$x^2 - 2x = 8$$

$$x^2 - 2x + 1 = 8 + 1$$

$$(x-1)^2 = 9$$

$$x - 1 = \pm\sqrt{9}$$

$$x - 1 = \pm 3$$

$$x - 1 = 3 \quad \text{or} \quad x - 1 = -3$$

$$x = 4 \quad \text{or} \quad x = -2$$

Using the quadratic formula:

$$x^2 - 2x - 8 = 0$$

$$x = \frac{-(-2) \pm \sqrt{(-2)^2 - 4(1)(-8)}}{2(1)}$$

$$x = \frac{2 \pm \sqrt{36}}{2}$$

$$x = \frac{2 \pm 6}{2}$$

$$x = \frac{2 + 6}{2} \quad \text{or} \quad x = \frac{2 - 6}{2}$$

$$x = \frac{8}{2} \quad \text{or} \quad x = \frac{-4}{2}$$

$$x = 4 \quad \text{or} \quad x = -2$$

31. $3x^2 - 5x + 4 = 0$

$$b^2 - 4ac = (-5)^2 - 4(3)(4) = -23 < 0$$

There are no real solutions. So, there are two imaginary solutions.

32. a. $3x^2 - 6x + 7 = 3$

$$3x^2 - 6x + 4 = 0$$

$$x = \frac{-(-6) \pm \sqrt{(-6)^2 - 4(3)(4)}}{2(3)} = \frac{6 \pm \sqrt{-12}}{6}$$

Since the square root of a negative is not a real number, there are no real number solutions. There is no such value for x.

b. $3x^2 - 6x + 7 = 4$

$$3x^2 - 6x + 3 = 0$$

$$x = \frac{-(-6) \pm \sqrt{(-6)^2 - 4(3)(3)}}{2(3)}$$

$$x = \frac{6 \pm \sqrt{0}}{6}$$

$$x = \frac{6}{6}$$

$$x = 1$$

There is one real solution.

c. $3x^2 - 6x + 7 = 5$

$$3x^2 - 6x + 2 = 0$$

$$x = \frac{-(-6) \pm \sqrt{(-6)^2 - 4(3)(2)}}{2(3)}$$

$$x = \frac{6 \pm \sqrt{12}}{6}$$

$$x = \frac{6 \pm 2\sqrt{3}}{6}$$

$$x = \frac{3 \pm \sqrt{3}}{3}$$

There are two real solutions.

d. Answers may vary.

33. $x = -0.6$ and $x = 1$

34. $x = -3$ or $x = 5$

35. $(-2.0, -1.3)$ and $(5.4, -3.8)$

36. Multiply the first equation by 2, and then subtract the second equation from the first equation.

$$2x + 4y - 6z = -8$$

$$\underline{2x - y + z = 3}$$

$$5y - 7z = -11 \quad (4)$$

Multiply the first equation by 3, and then subtract the third equation from the first equation:

$$3x + 6y - 9z = -12$$

$$\underline{3x + 2y + z = 10}$$

$$4y - 10z = -22 \quad (5)$$

Multiply (4) by 2 so that the right side equals the right side of (5) and then isolate:

$$10y - 14z = -22$$

$$4y - 10z = -22$$

$$4y - 10z = 10y - 14z$$

$$4z = 6y$$

$$y = \frac{2}{3}z$$

Solve for z:

$$5\left(\frac{2}{3}z\right) - 7z = -11$$

$$\frac{10}{3}z - 7z = -11$$

$$\frac{10z - 21z}{3} = -11$$

$$\frac{-11z}{3} = -11$$

$$z = 3$$

Solve for y:
$$y = \frac{2}{3}(3) = 2$$
Solve for x:
$$x + 2(2) - 3(3) = -4$$
$$x = 1$$
The solution of the system is $(1, 2, 3)$.

37. Isolate x in the first and second equations, and solve the third equation for x:
$$2x - 3z = -4$$
$$-3z = -4 - 2x$$
$$z = \frac{4 + 2x}{3}$$
$$3x + y = 0$$
$$y = -3x$$
$$x - 4(-3x) + 2\left(\frac{4 + 2x}{3}\right) = 17$$
$$x + 12x + \frac{8 + 4x}{3} = 17$$
$$3x + 36x + 8 + 4x = 51$$
$$43x = 43$$
$$x = 1$$
Solve for y:
$$y = -3(1) = -3$$
Solve for z:
$$z = \frac{4 + 2(1)}{3} = 2$$
The solution of the system is $(1, -3, 2)$.

38. Substitute the given points into
$$y = ax^2 + bx + c.$$
$$(2, 9): 9 = a(2)^2 + b(2) + c$$
$$(3, 18): 18 = a(3)^2 + b(3) + c$$
$$(5, 48): 48 = a(5)^2 + b(5) + c$$
Simplify these equations:
$$4a + 2b + c = 9 \qquad (1)$$
$$9a + 3b + c = 18 \qquad (2)$$
$$25a + 5b + c = 48 \qquad (3)$$
Eliminate c by multiplying both sides of equation (1) by -1:
$$-4a - 2b - c = -9 \qquad (4)$$
Adding the left sides and right sides of equations (2) and (4) gives:
$$5a + b = 9 \qquad (5)$$
Adding the left sides and right sides of equations (3) and (4) gives:
$$21a + 3b = 39 \qquad (6)$$

Simplify:
$$7a + b = 13 \qquad (7)$$
Eliminate b by multiplying equation (5) by -1 and add each side to the corresponding side of equation (7):
$$2a = 4$$
$$a = 2$$
Next, substitute 2 for a in equation (5):
$$5(2) + b = 9$$
$$b = -1$$
Then, substitute 2 for a and -1 for b in equation (1):
$$4a + 2b + c = 9$$
$$4(2) + 2(-1) + c = 9$$
$$8 - 2 + c = 9$$
$$c = 3$$
Therefore, $a = 2$, $b = -1$, and $c = 3$. So, the equation is $y = 2x^2 - x + 3$.

39. Substitute the given points into
$$y = ax^2 + bx + c.$$
$$(0, 5): 5 = a(0)^2 + b(0) + c$$
$$(2, 3): 3 = a(2)^2 + b(2) + c$$
$$(4, -15): -15 = a(4)^2 + b(4) + c$$
Simplify these equations:
$$c = 5 \qquad (1)$$
$$4a + 2b + c = 3 \qquad (2)$$
$$16a + 4b + c = -15 \qquad (3)$$
Since $c = 5$, substitute 5 for c in equations (2) and (3):
$$4a + 2b + (5) = 3$$
$$16a + 4b + (5) = -15$$
Simplifying these equations gives:
$$4a + 2b = -2 \qquad (4)$$
$$16a + 4b = -20 \qquad (5)$$
To eliminate b, multiply both sides of equation (4) by -2 and add each side to the corresponding side in equation (5):
$$8a = -16$$
$$a = -2$$
Next, substitute -2 for a in equation (4):
$$4a + 2b = -2$$
$$4(-2) + 2b = -2$$
$$2b = 6$$
$$b = 3$$
Therefore, $a = -2$, $b = 3$, and $c = 5$. So, the equation is $y = -2x^2 + 3x + 5$.

40. Linear:

$$\text{slope} = m = \frac{2-4}{1-0} = -2$$
$$y - \text{intercept} = (0, 4)$$
So, $y = -2x + 4$

Exponential:
As the value of x increases by 1, the value of y
is multiplied by ½, so the base $b = \dfrac{1}{2}$. The
y-intercept is $(0, 4)$. Therefore, the exponential
function is $y = 4\left(\dfrac{1}{2}\right)^x$

Quadratic:
Answers may vary. Example: $y = -2x^2 + 4$

41. Substitute the given points into
$y = ax^2 + bx + c$.

$$(0, 7): 7 = a(0)^2 + b(0) + c$$
$$(2, 1): 1 = a(2)^2 + b(2) + c$$
$$(5, 7): 7 = a(5)^2 + b(5) + c$$

Simplify these equations:
$$c = 7 \qquad\qquad (1)$$
$$4a + 2b + c = 1 \qquad (2)$$
$$25a + 5b + c = 7 \qquad (3)$$

Since $c = 7$, substitute 7 for c in equations (2)
and (3):
$$4a + 2b + (7) = 1$$
$$25a + 5b + (7) = 7$$
Simplifying these equations gives:
$$4a + 2b = -6 \qquad (4)$$
$$25a + 5b = 0 \qquad (5)$$
Simplify these equations:
$$2a + b = -3 \qquad (6)$$
$$5a + b = 0 \qquad (7)$$
To eliminate b, multiply both sides of
equation (6) by –1 and add each side to the
corresponding side in equation (7):
$$3a = 3$$
$$a = 1$$
Next, substitute 1 for a in equation (6):
$$2a + b = -3$$
$$2(1) + b = -3$$
$$b = -5$$
Therefore, $a = 1$, $b = -5$, and $c = 7$. So, the
equation is $y = x^2 - 5x + 7$.

42. a. Since the function is in the form
$h(t) = at^2 + bt + c$ and $a = -16 < 0$, the
vertex is the maximum point. Find the
$h(t)$-coordinate of the vertex.

$$h(0) = -16(0)^2 + 100(0) + 3 = 3$$

So, the h-intercept is $(0, 3)$. Next, find the
symmetric point by substituting 3 for $h(t)$
in the function and solve for t:

$$3 = -16t^2 + 100t + 3$$
$$0 = -16t^2 + 100t$$
$$0 = -4t(4t - 25)$$
$$-4t = 0 \quad \text{or} \quad 4t - 25 = 0$$
$$t = 0 \quad \text{or} \qquad 4t = 25$$
$$t = 0 \quad \text{or} \qquad t = \frac{25}{4} = 6.25$$

The symmetric points are $(0, 3)$ and
$(6.25, 3)$. Since the average of the
t-coordinates is $\dfrac{0 + 6.25}{2} = 3.125$, the
t-coordinate of the vertex is 3.125.
Substitute 3.125 for t in the function to find
the h-coordinate of the vertex:

$$h(3.125) = -16(3.125)^2 + 100(3.125) + 3$$
$$= 159.25$$

So the vertex is $(3.125, 159.25)$, which
means that the maximum height of the ball
is 159.25 feet and it is reached in
3.125 seconds.

b. Solve for t when $h(t) = 3$. From part (a), we
see that when $h(t) = 3$, t is either 0 or 6.25.
In this case, the fielder had 6.25 seconds to
get into position.

c.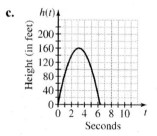

43. $$l + 2w = 180$$
$$l = 180 - 2w$$
$$A = lw = 180w - 2w^2$$
$$w = -\frac{180}{2(-2)} = 45$$
$$l = 180 - 2(45) = 90$$
$$A = (90)(45) = 4050$$

The rectangle should be 90 feet long and
45 feet wide. The area is 4050 square feet.

44. a. Draw a scatterplot.

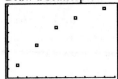

It looks like a parabola would fit the data well.
By regression:

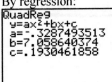

$$f(t) = -0.33t^2 + 7.06t + 0.19$$

b. The t-coordinate of the vertex for the model is given by

$$-\frac{b}{2a} = -\frac{7.06}{2(-0.33)} \approx 10.70$$

This maximum value is given by the model as:

$$f(10.7) = -0.33(10.7)^2 + 7.06(10.7) + 0.19$$
$$\approx 37.95$$

The largest percentage of military personnel who had done more than one tour of duty, about 38%, occurred in 2011.

c. Solve for t for $f(t) = 37$.

$$-0.33t^2 + 7.06t + 0.19 = 37$$
$$-0.33t^2 + 7.06t - 36.81 = 0$$
$$t = \frac{-7.06 \pm \sqrt{(7.06)^2 - 4(-0.33)(-36.81)}}{2(-0.33)}$$
$$t = \frac{-7.06 \pm 1.12}{-0.66}$$
$$t = 9 \text{ or } t \approx 12.39$$

In 2000 + 9 = 2009, and again in 2000 + 12 = 2013, 37% of military personnel served more than one tour of duty.

d. For $t = 6$,

$$f(6) = -0.33(6)^2 + 7.06(6) + 0.19 = 30.67$$

In 2006, the number of military personnel having served more than one tour of duty in Iraq and Afghanistan was about 31%.

45. a. Use the quadratic regression feature of a graphing calculator to find an equation of f.

QuadReg
y=ax²+bx+c
a=-.0270562771
b=-.5924242424
c=54.54069264

$$f(t) = -0.027t^2 - 0.59t + 54.54$$

Use the linear regression feature of a graphing calculator to find an equation of g.

LinReg
y=ax+b
a=2.346610762
b=-13.15583508

$$g(t) = 2.35t - 13.16$$

b. $-0.027t^2 - 0.59t + 54.54 = 2.35t - 13.16$
$$-0.027t^2 - 2.94t + 67.7 = 0$$
$$t = \frac{-(-2.94) \pm \sqrt{(-2.94)^2 - 4(-0.027)(67.7)}}{2(-0.027)}$$
$$t \approx \frac{2.94 \pm 3.994}{-0.054}$$
$$t \approx -128.41 \text{ or } t \approx 19.52$$

The Internet was not in existence in 1990 − 128 = 1862, so only the positive root is meaningful. In 1990 + 20 = 2010, the percentage of Americans who get their news every day from newspapers was equal to the percentage of Americans who get their news every day on the Internet.

$$g(19.52) = 2.35(19.52) - 13.16 \approx 33$$

That percentage is 33%.

Chapter 7 Test

1.

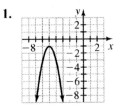

2. Since the vertex lies on the x-axis when $x > 0$, $h > 0$ and $k = 0$. Since the parabola is turned upward (has a minimum point), $a > 0$.

3. Answers may vary. Example:

$$y = -(x - 2)^2 - 7$$

4. Vertex: $(-1, 5)$

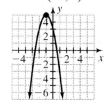

5. a. To find the x-intercepts, solve for x when $f(x) = 0$:

$$0 = x^2 - 2x - 8$$
$$0 = (x - 4)(x + 2)$$
$$x - 4 = 0 \quad \text{or} \quad x + 2 = 0$$
$$x = 4 \quad \text{or} \quad x = -2$$
$$(-2, 0), (4, 0)$$

b. Since the x-intercepts are symmetric points, the average of the x-coordinates for these points is the x-coordinate of the vertex, which is $\dfrac{4 + (-2)}{2} = 1$. Substitute 1 for x in the function to find the y-coordinate of the vertex:

$$y = (1)^2 - 2(1) - 8 = -9$$

So, the vertex is $(1, -9)$.

c.

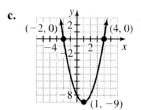

6. $\sqrt{32} = \sqrt{16 \cdot 2} = \sqrt{16}\sqrt{2} = 4\sqrt{2}$

7. $\sqrt{\dfrac{20}{75}} = \dfrac{\sqrt{20}}{\sqrt{75}} = \dfrac{2\sqrt{5}}{5\sqrt{3}} = \dfrac{2\sqrt{5}}{5\sqrt{3}} \cdot \dfrac{\sqrt{3}}{\sqrt{3}} = \dfrac{2\sqrt{15}}{15}$

8. $x^2 - 3x - 10 = 0$
$$(x - 5)(x + 2) = 0$$
$$x - 5 = 0 \quad \text{or} \quad x + 2 = 0$$
$$x = 5 \quad \text{or} \quad x = -2$$

9. $6x^2 = 100$
$$x^2 = \dfrac{100}{6}$$
$$x^2 = \dfrac{50}{3}$$
$$x = \pm\sqrt{\dfrac{50}{3}} = \pm\dfrac{\sqrt{50}}{\sqrt{3}} = \pm\dfrac{5\sqrt{2}}{\sqrt{3}} \cdot \dfrac{\sqrt{3}}{\sqrt{3}} = \pm\dfrac{5\sqrt{6}}{3}$$

10. $4(r - 3)^2 + 1 = 7$
$$4r^2 - 24r + 30 = 0$$
$$r = \dfrac{24 \pm \sqrt{(-24)^2 - 4(4)(30)}}{2(4)}$$
$$r = \dfrac{24 \pm \sqrt{96}}{8}$$
$$r = \dfrac{24 \pm 4\sqrt{6}}{8}$$
$$r = \dfrac{6 \pm \sqrt{6}}{2}$$

11. $\dfrac{5}{6}x^2 - \dfrac{1}{2}x = \dfrac{2}{3}$
$$5x^2 - 3x - 4 = 0$$
$$x = \dfrac{3 \pm \sqrt{(-3)^2 - 4(5)(-4)}}{2(5)} = \dfrac{3 \pm \sqrt{89}}{10}$$

12. $(x - 3)(x + 5) = 6$
$$x^2 + 5x - 3x - 15 = 6$$
$$x^2 + 2x - 21 = 0$$
$$x^2 + 2x = 21$$
$$x^2 + 2x + 1 = 21 + 1$$
$$(x + 1)^2 = 22$$
$$x + 1 = \pm\sqrt{22}$$
$$x = -1 \pm \sqrt{22}$$

13. $2x(x + 5) = 4x - 3$
$$2x^2 + 10x = 4x - 3$$
$$2x^2 + 6x + 3 = 0$$
$$x = \dfrac{-6 \pm \sqrt{6^2 - 4(2)(3)}}{2(2)}$$
$$x = \dfrac{-6 \pm \sqrt{12}}{4}$$
$$x = \dfrac{-6 \pm 2\sqrt{3}}{4}$$
$$x = \dfrac{-3 \pm \sqrt{3}}{2}$$

14.
$$3x^2 - 6x = 1$$
$$3x^2 - 6x - 1 = 0$$
$$x = \frac{6 \pm \sqrt{(-6)^2 - 4(3)(-1)}}{2(3)}$$
$$x = \frac{6 \pm \sqrt{48}}{6}$$
$$x = \frac{6 \pm 4\sqrt{3}}{6}$$
$$x = \frac{3 \pm 2\sqrt{3}}{3}$$

15.
$$3.7x^2 = 2.4 - 5.9x$$
$$3.7x^2 + 5.9x - 2.4 = 0$$
$$x = \frac{-5.9 \pm \sqrt{(5.9)^2 - 4(3.7)(-2.4)}}{2(3.7)}$$
$$x \approx \frac{-5.9 \pm 8.4}{7.4}$$
$$x \approx 0.34 \quad \text{or} \quad x \approx -1.93$$

16.
$$3x^2 - 6x = -5$$
$$3x^2 - 6x + 5 = 0$$
$$x = \frac{6 \pm \sqrt{(-6)^2 - 4(3)(5)}}{2(3)}$$
$$x = \frac{6 \pm \sqrt{-24}}{6}$$
$$x = \frac{6 \pm 2i\sqrt{6}}{6}$$
$$x = \frac{3 \pm i\sqrt{6}}{3}$$

17.
$$-2(p+4)^2 = 24$$
$$-2p^2 - 16p - 56 = 0$$
$$x = \frac{16 \pm \sqrt{(-16)^2 - 4(-2)(-56)}}{2(-2)}$$
$$x = \frac{16 \pm \sqrt{-192}}{-4}$$
$$x = \frac{16 \pm 8i\sqrt{3}}{-4}$$
$$x = -4 \pm 2i\sqrt{3}$$

18.
$$x^2 - 8x - 2 = 0$$
$$x^2 - 8x = 2$$
$$x^2 - 8x + 16 = 2 + 16$$
$$(x-4)^2 = 18$$
$$x - 4 = \pm\sqrt{18}$$
$$x - 4 = \pm 3\sqrt{2}$$
$$x = 4 \pm 3\sqrt{2}$$

19.
$$2(x^2 - 4) = -3x$$
$$x^2 - 4 = -\frac{3}{2}x$$
$$x^2 + \frac{3}{2}x = 4$$
$$x^2 + \frac{3}{2}x + \frac{9}{16} = 4 + \frac{9}{16}$$
$$\left(x + \frac{3}{4}\right)^2 = \frac{73}{16}$$
$$x + \frac{3}{4} = \pm\sqrt{\frac{73}{16}}$$
$$x + \frac{3}{4} = \pm\frac{\sqrt{73}}{4}$$
$$x = \frac{-3 \pm \sqrt{73}}{4}$$

20. Solve for x when $f(x) = 0$:
$$3x^2 - 8x + 1 = 0$$
$$x = \frac{-(-8) \pm \sqrt{(-8)^2 - 4(3)(1)}}{2(3)}$$
$$x = \frac{8 \pm \sqrt{52}}{6}$$
$$x = \frac{8 \pm 2\sqrt{13}}{6}$$
$$x = \frac{4 \pm \sqrt{13}}{3}$$

The x-intercepts are $\left(\dfrac{4 - \sqrt{13}}{3}, 0\right)$ and

$\left(\dfrac{4 + \sqrt{13}}{3}, 0\right)$.

21. $-2(x-3)^2 + 5 = 0$

$-2x^2 + 12x - 13 = 0$

$x = \dfrac{-12 \pm \sqrt{(12)^2 - 4(-2)(-13)}}{2(-2)}$

$x = \dfrac{-12 \pm \sqrt{40}}{-4}$

$x \approx \dfrac{-12 \pm 6.32}{-4}$

$x \approx 4.58$ or $x \approx 1.42$

The x-intercepts are $(1.42, 0)$ and $(4.58, 0)$.
The vertex is $(3, 5)$.

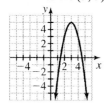

22. $b^2 - 4ac = 0$

$(-4)^2 - 4(a)(4a) = 0$

$16 - 16a^2 = 0$

$16a^2 = 16$

$a^2 = 1$

$a = \pm 1$

23. Substitute the given points into

$y = ax^2 + bx + c$.

$(1, 4): 4 = a(1)^2 + b(1) + c$

$(2, 9): 9 = a(2)^2 + b(2) + c$

$(3, 16): 16 = a(3)^2 + b(3) + c$

Simplify these equations:

$a + b + c = 4$ (1)

$4a + 2b + c = 9$ (2)

$9a + 3b + c = 16$ (3)

Eliminate c by multiplying both sides of
equation (1) by -1:

$-a - b - c = -4$ (4)

Adding the left sides and right sides of
equations (2) and (4) gives:

$3a + b = 5$ (5)

Adding the left sides and right sides of
equations (3) and (4) gives:

$8a + 2b = 12$ (6)

Simplify:

$4a + b = 6$ (7)

Eliminate b by multiplying equation (5) by -1
and add each side to the corresponding side of
equation (7):

$a = 1$

Next, substitute 1 for a in equation (5):

$3(1) + b = 5$

$b = 2$

Then, substitute 1 for a and 2 for b in
equation (1):

$a + b + c = 4$

$1 + 2 + c = 4$

$c = 1$

Therefore, $a = 1$, $b = 2$, and $c = 1$. So, the

equation is $y = x^2 + 2x + 1$ or $y = (x + 1)^2$.

24. Using the equation $y = a(x - h)^2 + k$,

substitute 5 for h and 3 for k since the vertex is
$(5, 3)$. Also, substitute 3 for x and 11 for y
since $(3, 11)$ lies on the parabola.

$y = a(x - h)^2 + k$

$11 = a(3 - 5)^2 + 3$

$11 = a(-2)^2 + 3$

$11 = 4a + 3$

$8 = 4a$

$a = 2$

So, the equation is $y = 2(x - 5)^2 + 3$ or

$y = 2x^2 - 20x + 53$.

25. a. $x^2 - 6x + 11 = 1$

$x^2 - 6x + 10 = 0$

$x = \dfrac{-(-6) \pm \sqrt{(-6)^2 - 4(1)(10)}}{2(1)} = \dfrac{6 \pm \sqrt{-4}}{2}$

Since the square root of a negative is not a
real number, there are no real number
solutions. There is no such value for x.

b. $x^2 - 6x + 11 = 2$

$x^2 - 6x + 9 = 0$

$x = \dfrac{-(-6) \pm \sqrt{(-6)^2 - 4(1)(9)}}{2(1)}$

$x = \dfrac{6 \pm \sqrt{0}}{2}$

$x = \dfrac{6}{2}$

$x = 3$

There is one real solution.

c. $x^2 - 6x + 11 = 3$

$x^2 - 6x + 8 = 0$

$x = \dfrac{-(-6) \pm \sqrt{(-6)^2 - 4(1)(8)}}{2(1)}$

$x = \dfrac{6 \pm \sqrt{4}}{2}$

$x = \dfrac{6 \pm 2}{2}$

$x = \dfrac{8}{2}$ or $x = \dfrac{4}{2}$

$x = 4$ or $x = 2$

There are two real solutions.

26. Multiply the first equation by –2, and then add the first and second equations:

$-2x - 8y - 6z = -4$

$\underline{2x + y + z = 10}$

$-7y - 5z = 6$

Add the first and third equations:

$x + 4y + 3z = 2$

$\underline{-x + y + 2z = 8}$

$5y + 5z = 10$

Isolate y:

$5y + 5z = 10$

$5y = 10 - 5z$

$y = 2 - z$

Solve for z:

$-7(2 - z) - 5z = 6$

$-14 + 7z - 5z = 6$

$2z = 20$

$z = 10$

Solve for y:

$y = 2 - 10 = -8$

Solve for x:

$x + 4(-8) + 3(10) = 2$

$x - 32 + 30 = 2$

$x = 4$

The solution of the system is $(4, -8, 10)$.

27. Add the first and second equations:

$2x - 3y = 4$

$\underline{ 3y + 2z = 2}$

$2x + 2z = 6$

Isolate z:

$2x + 2z = 6$

$2x = 6 - 2z$

$x = 3 - z$

Solve for z:

$x - z = -5$

$3 - z - z = -5$

$3 - 2z = -5$

$-2z = -8$

$z = 4$

Solve for y:

$3y + 2(4) = 2$

$3y + 8 = 2$

$3y = -6$

$y = -2$

Solve for x:

$x = 3 - 4 = -1$

The solution of the system is $(-1, -2, 4)$.

28. Using a graphing calculator, find that the maximum point of the parabola is approximately $(2.50, 103)$.

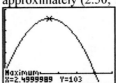

So, the maximum height reached by the ball is 103 feet at 2.5 seconds.

29. a. Draw a scatterplot for the data for f:

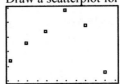

It looks like a parabola would fit the data.

By regression:

```
QuadReg
 y=ax²+bx+c
 a=-.0278708912
 b=2.535714114
 c=-15.92464858
```

$f(t) = -0.028t^2 + 2.54t - 15.92$

b. $f(31) = -0.028(31)^2 + 2.54(31) - 15.92$
   $= 35.912$
About 36% of 31-year-old Americans feel that they are taking a great risk by entering personal information into a pop-up ad.

c. $-0.028t^2 + 2.54t - 15.92 = 31$
   $-0.028t^2 + 2.54t - 46.92 = 0$
   $t = \dfrac{-2.54 \pm \sqrt{(2.54)^2 - 4(-0.028)(-46.92)}}{2(-0.028)}$
   $t \approx \dfrac{-2.54 \pm 1.094}{-0.056}$
   $t \approx 25.82$ or $t \approx 64.89$
About 31% of 26-year-old Americans and 31% of 65-year-old Americans feel that they are taking a great risk when entering personal information into a pop-up ad.

d. $-0.028t^2 + 2.54t - 15.92 = 0$
   $t = \dfrac{-2.54 \pm \sqrt{(2.54)^2 - 4(-0.028)(-15.92)}}{2(-0.028)}$
   $t \approx \dfrac{-2.54 \pm 2.16}{-0.056}$
   $t \approx 6.79$ or $t \approx 83.93$
The *t*-intercepts are therefore (6.79, 0) and (83.93, 0). These mean that there are no 7-year-old or 84-year-old Americans who feel that they are taking a great risk when entering personal information into a pop-up ad. Model breakdown has occurred.

e. $t = -\dfrac{b}{2a} = -\dfrac{2.54}{2(-0.028)} = 45.36$
   $f(45.36)$
   $= -0.028(45.36)^2 + 2.54(45.36) - 15.92$
   $= 41.68$
The vertex is (45.36, 41.68). This means that the age at which the maximum percentage, 42%, of Americans feel that they are taking a great risk is 45 years.

30. $p = 40 - 0.25x$
   $R = px = 40x - 0.25x^2$
   $x = -\dfrac{b}{2a} = -\dfrac{40}{2(-0.25)} = 80$
A group of 80 people would maximize the boat owner's revenue.

Cumulative Review of Chapters 1–7

1. $81x^2 - 49 = 0$
   $81x^2 = 49$
   $x^2 = \dfrac{49}{81}$
   $\sqrt{x^2} = \pm\sqrt{\dfrac{49}{81}}$
   $x = \pm\dfrac{7}{9}$

2. $\log_4(m + 3) = 2$
   $m + 3 = 4^2$
   $m + 3 = 16$
   $m = 13$

3. $5x^2 - 2x = 4$
   $5x^2 - 2x - 4 = 0$
   $x = \dfrac{2 \pm \sqrt{(-2)^2 - 4(5)(-4)}}{2(5)}$
   $x = \dfrac{2 \pm \sqrt{84}}{10}$
   $x = \dfrac{2 \pm 2\sqrt{21}}{10}$
   $x = \dfrac{1 \pm \sqrt{21}}{5}$

4. $2\left(3t^2 - 10\right) = -7t$
   $3t^2 - 10 = -\dfrac{7t}{2}$
   $3t^2 + \dfrac{7t}{2} - 10 = 0$
   $6t^2 + 7t - 20 = 0$
   $t = \dfrac{-7 \pm \sqrt{(7)^2 - 4(6)(-20)}}{2(6)}$
   $t = \dfrac{-7 \pm \sqrt{529}}{12}$
   $t = \dfrac{-7 \pm 23}{12}$
   $t = \dfrac{4}{3}$ or $t = -\dfrac{5}{2}$

5. $(2p-5)(3p+4) = 5p-2$

$6p^2 - 7p - 20 = 5p - 2$

$6p^2 - 12p - 18 = 0$

$6(p^2 - 2p - 3) = 0$

$6(p-3)(p+1) = 0$

$p = 3 \ \text{ or } \ p = -1$

6. $2(5x+2) - 1 = 9(x-3) - (3x-8)$

$10x + 4 - 1 = 9x - 27 - 3x + 8$

$10x + 3 = 6x - 19$

$4x = -22$

$x = -\dfrac{11}{2}$

7. $2x(3x-4) + 5 = 4 - x^2$

$6x^2 - 8x + 5 = 4 - x^2$

$7x^2 - 8x + 1 = 0$

$(7x-1)(x-1) = 0$

$x = \dfrac{1}{7} \ \text{ or } \ x = 1$

8. $3\ln(2w) + 2\ln(3w) = 8$

$\ln(2w)^3 + \ln(3w)^2 = 8$

$\ln(8w^3)(9w^2) = 8$

$\ln(72w^5) = 8$

$72w^5 = e^8$

$72w^5 = 2980.958$

$w^5 = 41.40$

$w = 2.1057$

9. $5b^6 + 4 = 82$

$5b^6 = 78$

$b^6 = \dfrac{78}{5}$

$b = \pm\left(\dfrac{78}{5}\right)^{\frac{1}{6}} \approx \pm 1.5807$

10. $3(2)^{4x-5} = 95$

$(2)^{4x-5} = \dfrac{95}{3}$

$\log(2)^{4x-5} = \log\left(\dfrac{95}{3}\right)$

$(4x-5)\log(2) = \log\left(\dfrac{95}{3}\right)$

$4x - 5 = \dfrac{\log\left(\dfrac{95}{3}\right)}{\log(2)}$

$4x = \dfrac{\log\left(\dfrac{95}{3}\right)}{\log(2)} + 5$

$x = \dfrac{\dfrac{\log\left(\dfrac{95}{3}\right)}{\log(2)} + 5}{4}$

$x \approx 2.4962$

11. $\log_b(65) = 4$

$b^{\log_b(65)} = b^4$

$65 = b^4$

$65^{1/4} = b$

$b \approx 2.8394$

12. $3e^x - 5 = 49$

$3e^x = 54$

$e^x = \dfrac{54}{3}$

$\ln(e^x) = \ln\left(\dfrac{54}{3}\right)$

$x\ln(e) = \ln\left(\dfrac{54}{3}\right)$

$x = \ln\left(\dfrac{54}{3}\right)$

$x \approx 2.8904$

13. $3\log_2\left(x^4\right) - 2\log_2\left(4x\right) = 5$

$$\log_2\left(x^4\right)^3 - \log_2\left(4x\right)^2 = 5$$

$$\log_2\left(x^{12}\right) - \log_2\left(16x^2\right) = 5$$

$$\log_2\left(\frac{x^{12}}{16x^2}\right) = 5$$

$$\frac{x^{12}}{16x^2} = 2^5$$

$$\frac{x^{10}}{16} = 2^5$$

$$\frac{x^{10}}{16} = 32$$

$$x^{10} = 512$$

$$x \approx 1.8661$$

14. $2x^2 - 6x = -5$

$$2x^2 - 6x + 5 = 0$$

$$x = \frac{6 \pm \sqrt{(-6)^2 - 4(2)(5)}}{2(2)}$$

$$x = \frac{6 \pm \sqrt{-4}}{4}$$

$$x = \frac{6 \pm 2i}{4}$$

$$x = \frac{3 \pm i}{2}$$

15. $2x^2 + 3x - 6 = 0$

$$2x^2 + 3x = 6$$

$$x^2 + \frac{3}{2}x = 3$$

$$x^2 + \frac{3}{2}x + \frac{9}{16} = 3 + \frac{9}{16}$$

$$\left(x + \frac{3}{4}\right)^2 = \frac{48}{16} + \frac{9}{16}$$

$$\left(x + \frac{3}{4}\right)^2 = \frac{57}{16}$$

$$x + \frac{3}{4} = \pm\sqrt{\frac{57}{16}}$$

$$x + \frac{3}{4} = \pm\frac{\sqrt{57}}{4}$$

$$x = -\frac{3}{4} \pm \frac{\sqrt{57}}{4}$$

$$x = \frac{-3 \pm \sqrt{57}}{4}$$

16. Substitute $y = 3x - 1$ in $2x - 3y = -11$.

$$2x - 3(3x - 1) = -11$$

$$2x - 9x + 3 = -11$$

$$-7x = -14$$

$$x = 2$$

Substitute 2 for x in one of the original equations to solve for y:

$$y = 3x - 1 = 3(2) - 1 = 5$$

The solution is $(2, 5)$.

17. In order to eliminate y, multiply $\frac{1}{2}x - y = \frac{5}{2}$

by $-\frac{3}{5}$. This gives $-\frac{3}{10}x + \frac{3}{5}y = -\frac{15}{10}$. Add the left sides and the right sides of the equations:

$$-\frac{3}{10}x + \frac{3}{5}y = -\frac{15}{10}$$

$$\frac{2}{5}x - \frac{3}{5}y = \frac{6}{5}$$

This yields:

$$-\frac{3}{10}x + \frac{2}{5}x = -\frac{15}{10} + \frac{6}{5}$$

$$-\frac{3}{10}x + \frac{4}{10}x = -\frac{15}{10} + \frac{12}{10}$$

$$\frac{1}{10}x = -\frac{3}{10}$$

$$\frac{10}{1} \cdot \frac{1}{10}x = -\frac{3}{10} \cdot \frac{10}{1}$$

$$x = -3$$

Substitute -3 for x in one of the original equations to solve for y:

$$\frac{1}{2}(-3) - y = \frac{5}{2}$$

$$-\frac{3}{2} - y = \frac{5}{2}$$

$$y = -\frac{3}{2} - \frac{5}{2} = -\frac{8}{2} = -4$$

So the solution is $(-3, -4)$.

18. Multiply the first equation by 2 and add the first and second equations:

$$4x - 2y + 6z = 2$$
$$\underline{3x + 2y - z = -6}$$
$$7x \qquad + 5z = -4$$

Multiply the first equation by -3 and add the first and third equations:

$$-6x + 3y - 9z = -3$$
$$\underline{4x - 3y + 2z = -7}$$
$$-2x \qquad - 7z = -10$$

Isolate z:
$$-2x - 7z = -10$$
$$-7z = 2x - 10$$
$$z = \frac{10 - 2x}{7}$$
Solve for x:
$$7x + 5\left(\frac{10 - 2x}{7}\right) = -4$$
$$7x + \frac{50 - 10x}{7} = -4$$
$$49x + 50 - 10x = -28$$
$$39x = -78$$
$$x = -2$$
Solve for z:
$$z = \frac{10 - 2(-2)}{7} = 2$$
Solve for y:
$$4(-2) - 2y + 6(2) = 2$$
$$-8 - 2y + 12 = 2$$
$$-2y = -2$$
$$y = 1$$
The solution of the system is $(-2, 1, 2)$.

19. $2(3x - 4) < 5 - 3(6x + 5)$
$$6x - 8 < 5 - 18x - 15$$
$$6x - 8 < -18x - 10$$
$$24x < -2$$
$$x < -\frac{1}{12}$$
$$\left(-\infty, -\frac{1}{12}\right)$$

20. $(2b^4 c^{-5})^3 (3b^{-1} c^{-2})^4$
$$= (2^3 b^{12} c^{-15})(3^4 b^{-4} c^{-8})$$
$$= (8 \cdot 81)(b^{12} b^{-4})(c^{-15} c^{-8})$$
$$= 648 b^8 c^{-23}$$
$$= \frac{648 b^8}{c^{23}}$$

21. $\left(\frac{6b^8 c^{-3}}{8b^{-4} c^{-1}}\right)^3 = \left(\frac{3b^{8-(-4)} c^{-3-(-1)}}{4}\right)^3$
$$= \left(\frac{3b^{12} c^{-2}}{4}\right)^3$$
$$= \left(\frac{3b^{12}}{4c^2}\right)^3$$
$$= \frac{27 b^{36}}{64 c^6}$$

22. $3 \log_b(4x) - 4 \log_b(x^3)$
$$= \log_b(4x)^3 - \log_b(x^3)^4$$
$$= \log_b(64x^3) - \log_b(x^{12})$$
$$= \log_b\left(\frac{64x^3}{x^{12}}\right)$$
$$= \log_b\left(\frac{64}{x^9}\right)$$

23. $3 \ln(x^7) + 2 \ln(x^5) = \ln(x^7)^3 + \ln(x^5)^2$
$$= \ln(x^{21}) + \ln(x^{10})$$
$$= \ln(x^{21})(x^{10})$$
$$= \ln(x^{31})$$

24. $(3x - 4y)^2 = 9x^2 - 24xy + 16y^2$

25. $(5p - 7q)(5p + 7q) = 25p^2 - 49q^2$

26. $-3x(x^2 - 5)(x^3 + 8)$
$$= -3x(x^5 + 8x^2 - 5x^3 - 40)$$
$$= -3x(x^5 - 5x^3 + 8x^2 - 40)$$
$$= -3x^6 + 15x^4 - 24x^3 + 120x$$

27. $(x^2 - 3x - 4)(x^2 + 4x - 5)$
$$= x^4 + 4x^3 - 5x^2 - 3x^3 - 12x^2 + 15x - 4x^2$$
$$- 16x + 20$$
$$= x^4 + x^3 - 21x^2 - x + 20$$

28. $f(x) = -2(x-5)^2 + 3$
$= -2(x-5)(x-5) + 3$
$= -2(x^2 - 5x - 5x + 25) + 3$
$= -2(x^2 - 10x + 25) + 3$
$= -2x^2 + 20x - 50 + 3$
$= -2x^2 + 20x - 47$

29. $(f+g)(x) = \left(2x^2 - 3x + 4\right) + \left(3x^2 - 5x - 2\right)$
$= 2x^2 + 3x^2 - 3x - 5x + 4 - 2$
$= 5x^2 - 8x + 2$
$(f+g)(2) = 5(2)^2 - 8(2) + 2 = 6$

30. $(f-g)(x) = \left(2x^2 - 3x + 4\right) - \left(3x^2 - 5x - 2\right)$
$= 2x^2 - 3x^2 - 3x + 5x + 4 + 2$
$= -x^2 + 2x + 6$
$(f-g)(-3) = -(-3)^2 + 2(-3) + 6 = -9$

31. $(g \cdot h)(x) = \left(3x^2 - 5x - 2\right)(4x - 2) \cdot$
$= 12x^3 - 6x^2 - 20x^2 + 10x - 8x + 4$
$= 12x^3 - 26x^2 + 2x + 4$
$(g \cdot h)(-1) = 12(-1)^3 - 26(-1)^2 + 2(-1) + 4$
$= -12 - 26 - 2 + 4$
$= -36$

32. $(g \circ k)(x) = g\big(k(x)\big)$
$= g(x-3)$
$= 3(x-3)^2 - 5(x-3) - 2$
$= 3\left(x^2 - 6x + 9\right) - 5(x-3) - 2$
$= 3x^2 - 18x + 27 - 5x + 15 - 2$
$= 3x^2 - 23x + 40$
$(g \circ k)(5) = 3(5)^2 - 23(5) + 40 = 0$

33.
$$\require{enclose}\begin{array}{r}3x^2 - x + 2 \\ 2x-3 \enclose{longdiv}{6x^3 - 11x^2 + 7x - 10} \\ \underline{-6x^3 + 9x^2 } \\ -2x^2 + 7x \\ \underline{2x^2 - 3x } \\ 4x - 10 \\ \underline{-4x + 6} \\ -4 \end{array}$$

$\dfrac{6x^3 - 11x^2 + 7x - 10}{2x - 3} = 3x^2 - x + 2 - \dfrac{4}{2x - 3}$

34.
$$\begin{array}{r|rrrr}\underline{-4} & 2 & 5 & -14 & -9 \\ & & -8 & 12 & 8 \\ \hline & 2 & -3 & -2 & -1 \end{array}$$

$\dfrac{2x^3 + 5x^2 - 14x - 9}{x + 4} = 2x^2 - 3x - 2 - \dfrac{1}{x + 4}$

35. $m^4 - 16n^4 = \left(m^2 - 4n^2\right)\left(m^2 + 4n^2\right)$
$= (m + 2n)(m - 2n)\left(m^2 + 4n^2\right)$

36. $x^3 - 13x^2 + 40x = x\left(x^2 - 13x + 40\right)$
$= x(x - 8)(x - 5)$

37. $8p^2 + 22pq - 21q^2 = (4p - 3q)(2p + 7q)$

38. $x^3 + 4x^2 - 9x - 36 = \left(x^2 - 9\right)(x + 4)$
$= (x - 3)(x + 3)(x + 4)$

39. Since the y-intercept is (0, 20) and as x increases by 1, $f(x)$ decreases by 3, so the slope is -3: $f(x) = -3x + 20$

40. As x increases by 1, $g(x)$ increases by a factor of 3, so the base $b = 3$. Substitute a point from the table into $g(x) = ab^x$ and solve for a.
$12 = a(3)^2$
$12 = 9a$
$a = \dfrac{4}{3}$
So, the equation is: $g(x) = \dfrac{4}{3}(3)^x$.

41. For the function k, as x increases by 1, $k(x)$ increases by 4, so the slope is 4.

42. $g(1) = 4$, so $x = 1$.

43. $k(3) = 4$, so $(f \circ k)(3) = f(k(3)) = f(4) = 8$.

44. $h(4) = 7$, so $h^{-1}(7) = 4$.

45.

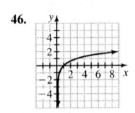

46.

47. First, change the form into $y = mx + b$

$$2x - 5y = 20$$
$$-5y = -2x + 20$$
$$y = \frac{2}{5}x - 4$$

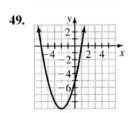

48.

49.

50.

51. The domain goes from −3 to 5, inclusive. The range goes from −2 to 4, inclusive. This is a function.

52. First, change the form of the equation given to $y = mx + b$.

$$3x - 4y = 5$$
$$-4y = -3x + 5$$
$$y = \frac{3}{4}x - \frac{5}{4}$$

The slope of the perpendicular line will be $-\frac{4}{3}$. So the equation of the perpendicular line will be $y = -\frac{4}{3}x + b$. Substitute the given point into this equation to solve for b.

$$6 = -\frac{4}{3}(-2) + b$$
$$6 = \frac{8}{3} + b$$
$$b = 6 - \frac{8}{3}$$
$$b = \frac{18}{3} - \frac{8}{3}$$
$$b = \frac{10}{3}$$

So, the equation is $y = -\frac{4}{3}x + \frac{10}{3}$ or $y = -1.33x + 3.33$.

53.
$$27 = ab^2$$
$$83 = ab^5$$
$$3.07 = b^3$$
$$b = 1.45$$
$$27 = a(1.45)^2$$
$$a = 12.84$$
$$y = 12.84(1.45)^x$$

54. Substitute the given points into $y = ax^2 + bx + c$.

$$(1, -1): -1 = a(1)^2 + b(1) + c$$
$$(2, 4): 4 = a(2)^2 + b(2) + c$$
$$(4, 20): 20 = a(4)^2 + b(4) + c$$

Simplify these equations:

$$a + b + c = -1 \qquad (1)$$
$$4a + 2b + c = 4 \qquad (2)$$
$$16a + 4b + c = 20 \qquad (3)$$

Eliminate c by multiplying both sides of equation (1) by -1:

$-a - b - c = 1 \qquad (4)$

Adding the left sides and right sides of equations (2) and (4) gives:

$3a + b = 5 \qquad (5)$

Adding the left sides and right sides of equations (3) and (4) gives:

$15a + 3b = 21 \qquad (6)$

Simplify:

$5a + b = 7 \qquad (7)$

Eliminate b by multiplying equation (5) by -1 and add each side to the corresponding side of equation (7):

$2a = 2$

$a = 1$

Next, substitute 1 for a in equation (5):

$3(1) + b = 5$

$\qquad b = 2$

Then, substitute 1 for a and 2 for b in equation (1):

$a + b + c = -1$

$1 + 2 + c = -1$

$\qquad c = -4$

Therefore, $a = 1$, $b = 2$, and $c = -4$. So, the equation is $y = x^2 + 2x - 4$.

55. **a.** Underline: Linear:

$\text{slope} = m = \dfrac{6 - 3}{1 - 0} = 3$

$y - \text{intercept} = (0, 3)$

So, $y = 3x + 3$

Exponential:

As the value of x increases by 1, the value of y is multiplied by 2, so the base $b = 2$. The y-intercept is (0, 3). Therefore, the exponential function is $y = 3(2)^x$.

Quadratic:

Answers may vary. Example: $y = 3x^2 + 3$.

b.

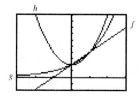

56. $-x^2 + 6x - 5 = 3$

$x^2 - 6x + 8 = 0$

$(x - 2)(x - 4) = 0$

$x - 2 = 0 \quad \text{or} \quad x - 4 = 0$

$\qquad x = 2 \quad \text{or} \qquad x = 4$

57. To find the x-intercepts, let $f(x) = 0$ and solve for x:

$-x^2 + 6x - 5 = 0$

$x^2 - 6x + 5 = 0$

$(x - 1)(x - 5) = 0$

$x - 1 = 0 \quad \text{or} \quad x - 5 = 0$

$\qquad x = 1 \quad \text{or} \qquad x = 5$

The x-intercepts are $(1, 0)$ and $(5, 0)$.

58.

59. $\log_2(16) = 4$ since $2^4 = 16$

60. $\log_5\left(\dfrac{1}{25}\right) = \log_5\left(\dfrac{1}{5^2}\right) = \log_5\left(5^{-2}\right) = -2$

61.

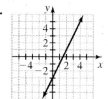

62. **a.** $f(x) = 3^x$

Replace $f(x)$ with y:

$y = 3^x$

Solve for x:

$\log_3 y = \log_3\left(3^x\right)$

$\log_3 y = x$

Replace x with $f^{-1}(y)$:

$f^{-1}(y) = \log_3 y$

Write in terms of x:

$f^{-1}(x) = \log_3 x$

b. $\left(f^{-1} \circ f\right)(x) = \log_3\left(3^x\right) = x$

c. $\left(f \circ f^{-1}\right)(x) = 3^{\log_3 x} = x$

63. a. $f(x) = \dfrac{2}{5}x + 1$

Replace $f(x)$ with y:

$y = \dfrac{2}{5}x + 1$

Solve for x:

$y - 1 = \dfrac{2}{5}x$

$\dfrac{5}{2}(y - 1) = x$

$x = \dfrac{5}{2}y - \dfrac{5}{2}$

Replace x with $f^{-1}(y)$:

$f^{-1}(y) = \dfrac{5}{2}y - \dfrac{5}{2}$

Write in terms of x:

$f^{-1}(x) = \dfrac{5}{2}x - \dfrac{5}{2}$

b. $\left(f^{-1} \circ f\right)(x) = \dfrac{5}{2}\left(f(x)\right) - \dfrac{5}{2}$

$= \dfrac{5}{2}\left(\dfrac{2}{5}x + 1\right) - \dfrac{5}{2}$

$= x + \dfrac{5}{2} - \dfrac{5}{2}$

$= x$

c. $\left(f \circ f^{-1}\right)(x) = \dfrac{2}{5}\left(f^{-1}(x)\right) + 1$

$= \dfrac{2}{5}\left(\dfrac{5}{2}x - \dfrac{5}{2}\right) + 1$

$= x - 1 + 1$

$= x$

64. Let x be the amount invested at 6%, so $12{,}000 - x$ is the amount invested at 11%.

$0.06x + (12000 - x)(0.11) = 845$

$0.06x + 1320 - 0.11x = 845$

$0.05x = 475$

$x = 9500$

$12000 - 9500 = 2500$

$9500 is invested at 6% and $2500 is invested at 11%.

65. $2010 - 1996 = 14$

$ab^0 = 480$

$a = 480$

$ab^{14} = 151$

$480b^{14} = 151$

$b^{14} = \dfrac{151}{480}$

$b = \left(\dfrac{151}{480}\right)^{1/14} \approx 0.92$

$480(0.92)^x \approx 80$

$(0.92)^x \approx \dfrac{1}{6}$

$x \approx \log_{0.92}\left(\dfrac{1}{6}\right)$

$x \approx \dfrac{\log\left(\dfrac{1}{6}\right)}{\log 0.92} \approx 21.49$

In $1996 + 21 = 2017$, there will be 80 cases of AIDS acquired at birth.

66. a. Linear:

$L(t) = 0.42t + 0.72$

Exponential:

$E(t) = 1.09(1.17)^t$

Quadratic:

$Q(t) = 0.03t^2 + 0.12t + 1.12$

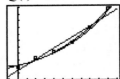

The linear model fits the data fairly well. The exponential and quadratic models both fit the data quite well.

b. $x = -\dfrac{b}{2a} = -\dfrac{0.12}{2(0.03)} = -2$

$Q(-2) = 0.03(-2)^2 + 0.12(-2) + 1.12 = 1$

The vertex is $(-2, 1)$, which means the admission rate was the lowest, 1 admission per 100,000 people, in 1998.

c. The exponential model works best for years before 1998. Answers may vary.

d.
$$f(t) = 1.09(1.17)^t$$
$$18 = 1.09(1.17)^t$$
$$16.51 \approx (1.17)^t$$
$$\log_{1.17}(16.51) \approx \log_{1.17}\left[(1.17)^t\right]$$
$$\log_{1.17}(16.51) \approx t$$
$$\frac{\log(16.51)}{\log(1.17)} \approx t$$
$$17.86 \approx t$$

The admission rate for opioid addiction will be 18 admissions per 100,000 people in 2018.

e. $18 = 0.03t^2 + 0.12t + 1.12$
$$0 = 0.03t^2 + 0.12t - 16.88$$
$$t = \frac{-0.12 \pm \sqrt{(0.12)^2 - 4(0.03)(-16.88)}}{2(0.03)}$$
$$t \approx \frac{-0.12 \pm 1.43}{0.06}$$
$$t \approx -25.83 \text{ or } t \approx 21.83$$

The admission rate for opioid addiction will be 18 admissions per 100,000 people in 2022.

f. Answers may vary. The exponential model predicts faster growth than the quadratic model.

67. a. A scattergram of the data shows that it is fairly linear.

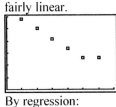

By regression:

```
LinReg
y=ax+b
a=-.6857142857
b=31.66666667
```

$$f(t) = -0.69t + 31.67$$

b. $f(37) = -0.69(37) + 31.67 = 6.14$

In 1980 + 37 = 2017, about 6% of union members will work in manufacturing.

c. $5 = -0.69t + 31.67$
$$-26.67 = -0.69t$$
$$38.65 \approx t$$
In 1980 + 39 = 2019, 5% of union members will work in manufacturing.

d.
$$f(t) = -0.69t + 31.67$$
$$y = -0.69x + 31.67$$
$$x = -0.69y + 31.67$$
$$0.69y = -x + 31.67$$
$$y = -1.45x + 45.90$$
$$f^{-1}(p) = -1.45p + 45.90$$

e. $f^{-1}(100) = -1.45(100) + 45.90 = -99.1$

In 1980 - 99 = 1881, 100% of union members worked in manufacturing. Model breakdown has likely occurred.

f. To find the t-intercept of the model, let $p = 0$ and solve for t.
$$-0.69t + 31.67 = 0$$
$$-0.69t = -31.67$$
$$t = 45.90$$
The t-intercept is (45.90, 0).
In 1980 + 46 = 2026, no union members will be working in manufacturing. Model breakdown has likely occurred.

g. For $t \le -99.1$ and $t > 45.90$, there is model breakdown.

68. a. $f(t) = 0.36t + 6.48$

$$g(t) = 0.18t + 9.16$$

b. $f(17) = 0.36(17) + 6.48 = 12.6$
$$g(17) = 0.18(17) + 9.16 = 12.22$$
In 2017, the average age of light trucks will be 12.6 years; in 2017, the average age of passenger cars will be 12.2 years.

c. For some value of t, $f(t) = g(t)$.
$$0.36t + 6.48 = 0.18t + 9.16$$
$$0.18t = 2.68$$
$$t \approx 14.89$$
$$f(14.89) = 0.36(14.89) + 6.48 = 11.84$$
$$g(14.89) = 0.18(14.89) + 9.16 = 11.84$$
In $2000 + 14.89 \approx 2015$, the average age of light trucks will be equal to the average age of passenger cars. That age will be about 11.8 years.

d. $0.36t + 6.48 > 0.18t + 9.16$

$\qquad 0.18t > 2.68$

$\qquad\quad t > 14.89$

The average age of light trucks will be greater than the average age of passenger cars after 2015.

e. $f(t) - g(t) = 0.36t + 6.48 - (0.18t + 9.16)$

$\qquad\qquad\quad = 0.18t - 2.68$

$(f - g)(18) = 0.18(18) - 2.68 = 0.56$

In 2018 the average age of trucks will be 0.56 year more than the average age of passenger cars.

f. $\dfrac{f(t) + g(t)}{2} = \dfrac{0.36t + 6.48 + 0.18t + 9.16}{2}$

$\qquad\qquad\quad = \dfrac{0.54t + 15.64}{2}$

$\qquad\qquad\quad = 0.27t + 7.82$

$\dfrac{f(18) + g(18)}{2} = 0.27(18) + 7.82 = 12.68$

Answers may vary.

Chapter 8
Rational Functions

1. $f(-1) = \dfrac{(-1)+1}{(-1)^2-9} = \dfrac{0}{1-9} = \dfrac{0}{-8} = 0$

$f(2) = \dfrac{(2)+1}{(2)^2-9} = \dfrac{3}{4-9} = \dfrac{3}{-5} = -\dfrac{3}{5}$

$f(3) = \dfrac{(3)+1}{(3)^2-9} = \dfrac{4}{9-9} = \dfrac{4}{0}$

$f(3)$ is undefined.

3. $f(-1) = \dfrac{(-1)^3-8}{2(-1)^2+3(-1)-1}$

$= \dfrac{-1-8}{2-3-1}$

$= \dfrac{-9}{-2}$

$= \dfrac{9}{2}$

$f(0) = \dfrac{(0)^3-8}{2(0)^2+3(0)-1} = \dfrac{0-8}{-1} = \dfrac{-8}{-1} = 8$

$f(3) = \dfrac{(3)^3-8}{2(3)^2+3(3)-1} = \dfrac{27-8}{2(9)+9-1} = \dfrac{19}{26}$

5. The only value that will make the function undefined (i.e., the denominator 0) is 0. Therefore, the domain is the set of all real numbers except 0.

7. Since there are no values that will make the function undefined, the domain is the set of all real numbers.

9. $x+3=0$

$x=-3$

Since -3 is the only value that will make the function undefined, the domain is the set of all real numbers except -3.

11. $2x+1=0$

$2x=-1$

$x=-\dfrac{1}{2}$

The domain is the set of all real numbers except $-\dfrac{1}{2}$.

13. $x^2-3x-10=0$

$(x-5)(x+2)=0$

$x-5=0 \quad \text{or} \quad x+2=0$

$x=5 \quad \text{or} \qquad x=-2$

The domain is the set of all real numbers except 5 and -2.

15. $\qquad 4x^2-25=0$

$(2x-5)(2x+5)=0$

$2x-5=0 \quad \text{or} \quad 2x+5=0$

$2x=5 \quad \text{or} \qquad 2x=-5$

$x=\dfrac{5}{2} \quad \text{or} \qquad x=-\dfrac{5}{2}$

The domain is the set of all real numbers except $\dfrac{5}{2}$ and $-\dfrac{5}{2}$.

17. $x^2+1=0$

$x^2=-1$

This equation has no real number solution. The domain is the set of all real numbers.

19. $2x^2-7x-15=0$

$(x-5)(2x+3)=0$

$x-5=0 \quad \text{or} \quad 2x+3=0$

$x=5 \quad \text{or} \qquad 2x=-3$

$x=5 \quad \text{or} \qquad x=-\dfrac{3}{2}$

The domain is the set of all real numbers except 5 and $-\dfrac{3}{2}$.

21. $x^2-3x+6=0$

$x=\dfrac{-(-3)\pm\sqrt{(-3)^2-4(1)(6)}}{6}$

$=\dfrac{3\pm\sqrt{9-24}}{6}$

$=\dfrac{3\pm\sqrt{-15}}{6}$

Since the solution is not real, the domain is the set of all real numbers.

23. $3x^2 - 2x - 7 = 0$

$$x = \frac{2 \pm \sqrt{4 - 4(3)(-7)}}{6}$$

$$= \frac{2 \pm \sqrt{4 + 84}}{6}$$

$$= \frac{2 \pm \sqrt{88}}{6}$$

$$= \frac{2 \pm \sqrt{4 \cdot 22}}{6}$$

$$= \frac{2 \pm 2\sqrt{22}}{6}$$

$$= \frac{1 \pm \sqrt{22}}{3}$$

The domain is the set of all real numbers

except $\dfrac{1 + \sqrt{22}}{3}$ and $\dfrac{1 - \sqrt{22}}{3}$.

25. $\qquad 4x^3 - 8x^2 - 9x + 18 = 0$

$$4x^2(x - 2) - 9(x - 2) = 0$$

$$(x - 2)(4x^2 - 9) = 0$$

$$(x - 2)(2x - 3)(2x + 3) = 0$$

$x - 2 = 0 \quad\text{or}\quad 2x - 3 = 0 \quad\text{or}\quad 2x + 3 = 0$

$\quad x = 2 \quad\text{or}\qquad 2x = 3 \quad\text{or}\qquad 2x = -3$

$\quad x = 2 \quad\text{or}\qquad x = \dfrac{3}{2} \quad\text{or}\qquad x = -\dfrac{3}{2}$

The domain is the set of all real numbers

except 2, $\dfrac{3}{2}$ and $-\dfrac{3}{2}$.

27. $f(x) = \dfrac{20x^7}{15x^4}$

$$= \frac{\cancel{5} \cdot 4 \cdot \cancel{x} \cdot \cancel{x} \cdot \cancel{x} \cdot \cancel{x} \cdot x \cdot x \cdot x}{\cancel{5} \cdot 3 \cdot \cancel{x} \cdot \cancel{x} \cdot \cancel{x} \cdot \cancel{x}}$$

$$= \frac{4x^3}{3}$$

29. $f(x) = \dfrac{4x - 28}{5x - 35} = \dfrac{4\cancel{(x - 7)}}{5\cancel{(x - 7)}} = \dfrac{4}{5}$

31. $f(x) = \dfrac{x^2 + 7x + 10}{x^2 - 7x - 18} = \dfrac{(x + 5)\cancel{(x + 2)}}{(x - 9)\cancel{(x + 2)}} = \dfrac{x + 5}{x - 9}$

33. $f(x) = \dfrac{x^2 - 49}{x^2 - 14x + 49} = \dfrac{\cancel{(x - 7)}(x + 7)}{\cancel{(x - 7)}(x - 7)} = \dfrac{x + 7}{x - 7}$

35. $f(x) = \dfrac{16x^2 - 25}{8x^2 - 22x + 15}$

$$= \frac{\cancel{(4x - 5)}(4x + 5)}{\cancel{(4x - 5)}(2x - 3)}$$

$$= \frac{4x + 5}{2x - 3}$$

37. $f(x) = \dfrac{x - 5}{5 - x} = \dfrac{\cancel{x - 5}}{-1\cancel{(x - 5)}} = \dfrac{1}{-1} = -1$

39. $f(x) = \dfrac{4x - 12}{18 - 6x} = \dfrac{4\cancel{(x - 3)}}{-6\cancel{(x - 3)}} = \dfrac{4}{-6} = -\dfrac{2}{3}$

41. $f(x) = \dfrac{6x - 18}{9 - x^2}$

$$= \frac{6(x - 3)}{(3 - x)(3 + x)}$$

$$= \frac{6\cancel{(x - 3)}}{-1\cancel{(x - 3)}(x + 3)}$$

$$= -\frac{6}{x + 3}$$

43. $f(x) = \dfrac{x^2 + 2x - 35}{-x^2 + 3x + 10}$

$$= \frac{x^2 + 2x - 35}{-1(x^2 - 3x - 10)}$$

$$= \frac{\cancel{(x - 5)}(x + 7)}{-1\cancel{(x - 5)}(x + 2)}$$

$$= -\frac{x + 7}{(x + 2)}$$

45. $f(x) = \dfrac{3x^3 + 21x^2 + 36x}{x^2 - 9}$

$$= \frac{3x(x^2 + 7x + 12)}{(x - 3)(x + 3)}$$

$$= \frac{3x(x + 4)\cancel{(x + 3)}}{(x - 3)\cancel{(x + 3)}}$$

$$= \frac{3x(x + 4)}{x - 3}$$

47. $f(x) = \dfrac{x^2 - 2x - 8}{4x^3 + 8x^2 - 9x - 18}$

$= \dfrac{(x-4)(x+2)}{4x^2(x+2) - 9(x+2)}$

$= \dfrac{(x-4)(x+2)}{(4x^2-9)(x+2)}$

$= \dfrac{(x-4)\cancel{(x+2)}}{(2x-3)(2x+3)\cancel{(x+2)}}$

$= \dfrac{x-4}{(2x-3)(2x+3)}$

49. $f(x) = \dfrac{x^3 + 8}{x^2 - 4}$

$= \dfrac{x^3 + 2^3}{(x-2)(x+2)}$

$= \dfrac{\cancel{(x+2)}(x^2 - 2x + 4)}{(x-2)\cancel{(x+2)}}$

$= \dfrac{x^2 - 2x + 4}{x-2}$

51. $f(x) = \dfrac{3x^2 + 7x - 6}{27x^3 - 8}$

$= \dfrac{(3x-2)(x+3)}{(3x)^3 - (2)^3}$

$= \dfrac{\cancel{(3x-2)}(x+3)}{\cancel{(3x-2)}(9x^2 + 6x + 4)}$

$= \dfrac{x+3}{9x^2 + 6x + 4}$

53. $\dfrac{x^2 - 6xy + 9y^2}{x^2 - 3xy} = \dfrac{(x-3y)\cancel{(x-3y)}}{x\cancel{(x-3y)}} = \dfrac{x-3y}{x}$

55. $\dfrac{6a^2 + ab - 2b^2}{3a^2 - 7ab - 6b^2} = \dfrac{\cancel{(3a+2b)}(2a-b)}{\cancel{(3a+2b)}(a-3b)} = \dfrac{2a-b}{a-3b}$

57. $\dfrac{p^3 - q^3}{p^2 - q^2} = \dfrac{\cancel{(p-q)}(p^2 + pq + q^2)}{\cancel{(p-q)}(p+q)}$

$= \dfrac{p^2 + pq + q^2}{p+q}$

59. $\left(\dfrac{f}{g}\right)(x) = \dfrac{x^2 + 2x - 8}{x^2 - 8x + 12}$

$= \dfrac{(x+4)\cancel{(x-2)}}{(x-6)\cancel{(x-2)}}$

$= \dfrac{x+4}{x-6}$

$\left(\dfrac{f}{g}\right)(3) = \dfrac{3+4}{3-6} = \dfrac{7}{-3} = -\dfrac{7}{3}$

61. $\left(\dfrac{h}{f}\right)(x) = \dfrac{3x^2 + 17x + 20}{x^2 + 2x - 8}$

$= \dfrac{(3x+5)\cancel{(x+4)}}{\cancel{(x+4)}(x-2)}$

$= \dfrac{3x+5}{x-2}$

$\left(\dfrac{h}{f}\right)(4) = \dfrac{3(4)+5}{4-2} = \dfrac{17}{2}$

63. $\left(\dfrac{f}{h}\right)(x) = \dfrac{3x^3 - x^2}{9x^2 - 1}$

$= \dfrac{x^2\cancel{(3x-1)}}{\cancel{(3x-1)}(3x+1)}$

$= \dfrac{x^2}{3x+1}$

$\left(\dfrac{f}{h}\right)(-2) = \dfrac{(-2)^2}{3(-2)+1} = \dfrac{4}{-6+1} = -\dfrac{4}{5}$

65. $\left(\dfrac{k}{g}\right)(x) = \dfrac{27x^3 + 1}{18x^3 + 12x^2 + 2x}$

$= \dfrac{\cancel{(3x+1)}(9x^2 - 3x + 1)}{\cancel{(3x+1)}(6x^2 + 2x)}$

$= \dfrac{9x^2 - 3x + 1}{2x(3x+1)}$

$\left(\dfrac{k}{g}\right)(-1) = \dfrac{9(-1)^2 - 3(-1) + 1}{6(-1)^2 + 2(-1)}$

$= \dfrac{9+3+1}{6-2}$

$= \dfrac{13}{4}$

67. a. Percentage of Americans who participated in SNAP in August of the year that is t years since 2000:

$$P(t) = \frac{F(t)}{U(t)} \cdot 100$$

$$= \frac{-1.03t^2 + 25.09t - 105.88}{0.0068t^2 + 2.71t + 277.9} \cdot \frac{100}{1}$$

$$= \frac{-103t^2 + 2509t - 10,588}{0.0068t^2 + 2.71t + 277.9}$$

The assumption that neither model will break down for a large domain of t has been assumed.

b. In the year 2012, $t = 12$:

$$P(12) \approx 0.151 \text{ or } 15.1\%$$

According to the model, about 15.1% of Americans participated in SNAP in August of 2012. The actual percent is:

$$\frac{47.0}{314.5} \approx 0.149 \text{ or } 14.9\%$$

The result of using the model is an overestimate.

c. $P(16) \approx 0.099$ or 9.9%

In August of 2016, 9.9% of Americans will participate in SNAP.

69. a. Using the quadratic regression feature of a graphing calculator, we get:

$$R(t) = 0.0048t^2 + 0.14t + 2.95$$

b. Using the quadratic regression feature of a graphing calculator, we get:

$$E(t) = 0.015t^2 + 0.055t + 6.05$$

c. $$P(t) = \frac{R(t)}{E(t)} \cdot 100$$

$$= \frac{0.0048t^2 + 0.14t + 2.95}{0.015t^2 + 0.055t + 6.05} \cdot \frac{100}{1}$$

$$= \frac{0.48t^2 + 14t + 295}{0.015t^2 + 0.055t + 6.05}$$

d. $2017 - 1980 = 37$

$$P(37) = \frac{0.48(37)^2 + 14(37) + 295}{0.015(37)^2 + 0.055(37) + 6.05}$$

$$\approx 51.37$$

In 2017, about 51.4% of voter-eligible Latinos will be registered to vote.

e.

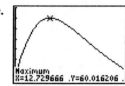

The minimum value for $P(t)$ between $t = 0$ and $t = 40$ is 60.02%, which occurs at $t = 12.73$. This means that the largest percentage of voter-eligible Latinos who were registered, about 60.0%, occurred in 1993.

f. The percentage of voter-registered Latinos who are registered to vote has decreased since 1992. The number of Latinos who are eligible to vote has been growing at a faster rate than the number of Latinos who are registered to vote.

71. Answers may vary. Examples:

$$f(x) = \frac{1}{(x-3)(x+3)}$$

$$g(x) = \frac{4x^2}{9 - x^2}$$

$$h(x) = \frac{3x^2 + 5}{\left(\dfrac{x}{3}\right)^2 - 1}$$

73. Domain: $-3, -2, -1, 0, 1, 2, 3$
Range: 50.4, 25.2, 16.8, 12.6, 10.08, 8.4, 7.2

75. The student is incorrect about 2 and 4, which are in the domain because they do not make the denominator 0. The only values that must be excluded are 5 and 1 because for these values the denominator equals 0.

77. Original expression:

$$\frac{2(3+4)+3}{(3+4)(3-1)} = \frac{2(7)+3}{7(2)} = \frac{14+3}{14} = \frac{17}{14}$$

Student's work:

$$\frac{5}{3-1} = \frac{5}{2}$$

The results are not the same, so the original expression is not equivalent to the student's expression.

79. a. $T(50) = \dfrac{420}{50} = 8.4$ hours

This represents the driving time, in hours.

b. $T(55) = \dfrac{420}{55} = 7.64$ hours

$T(60) = \dfrac{420}{60} = 7.00$ hours

$T(65) = \dfrac{420}{65} = 6.46$ hours

$T(70) = \dfrac{420}{70} = 6.00$ hours

c. T is decreasing for $s > 0$. This makes sense because the faster a person drives, the less time it will take to arrive at the destination, as shown in part (b).

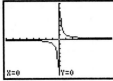

81. Answers may vary. Example:
In general, the denominator is set equal to zero. Any solution to this equation cannot be part of the domain of the function.

83. The domain is all real numbers.

85. The domain is all real numbers.

87. a. $\dfrac{x^2 + 5x + 6}{x + 3} = \dfrac{\cancel{(x+3)}(x+2)}{\cancel{x+3}} = x + 2$

b. $\begin{array}{r|rrr} -3 & 1 & 5 & 6 \\ & & -3 & -6 \\ \hline & 1 & 2 & 0 \end{array}$

$\dfrac{x^2 + 5x + 6}{x + 3} = x + 2$

c. Answers may vary.

89. $x^2 - 4x - 21 = 0$

$(x - 7)(x + 3) = 0$

$x - 7 = 0$ or $x + 3 = 0$

$x = 7$ or $x = -3$

The domain of f is all real numbers except 7 and -3.

The function $f(x)$ is a rational function in one variable.

91. $3x^3 + 5x^2 = 12x + 20$

$3x^3 + 5x^2 - 12x - 20 = 0$

$3x^3 - 12x + 5x^2 - 20 = 0$

$3x\left(x^2 - 4\right) + 5\left(x^2 - 4\right) = 0$

$\left(x^2 - 4\right)(3x + 5) = 0$

$(x - 2)(x + 2)(3x + 5) = 0$

$x - 2 = 0$ or $x + 2 = 0$ or $3x + 5 = 0$

$x = 2$ or $x = -2$ or $x = -\dfrac{5}{3}$

The solutions are 2, -2, and $-\dfrac{5}{3}$.

This is a cubic equation in one variable.

93. $8x^3 - 125$

$(2x)^3 - (5)^3$

$(2x - 5)\left(4x^2 + 10x + 25\right)$

This is a cubic expression in one variable.

Homework 8.2

1. $\dfrac{5}{x} \cdot \dfrac{2}{x} = \dfrac{10}{x^2}$

3. $\dfrac{7x^5}{2} \div \dfrac{5x^3}{6} = \dfrac{7x^5}{2} \cdot \dfrac{6}{5x^3}$

$= \dfrac{7 \cdot \cancel{x^3} \cdot x^2}{\cancel{2}} \cdot \dfrac{\cancel{2} \cdot 3}{5 \cdot \cancel{x^3}}$

$= \dfrac{21x^2}{5}$

5. $\dfrac{5p^3}{4p - 8} \cdot \dfrac{3p - 6}{10p^8} = \dfrac{5p^3}{4\cancel{(p-2)}} \cdot \dfrac{3\cancel{(p-2)}}{10p^8}$

$= \dfrac{15p^3}{40p^8}$

$= \dfrac{\cancel{5} \cdot 3 \cdot \cancel{p^3}}{\cancel{5} \cdot 8 \cdot \cancel{p^3} \cdot p^5} = \dfrac{3}{8p^5}$

7. $\dfrac{6x-18}{5x^5} \div \dfrac{5x-15}{x^3} = \dfrac{6x-18}{5x^5} \cdot \dfrac{x^3}{5x-15}$

$= \dfrac{6(x-3)}{5x^5} \cdot \dfrac{x^3}{5(x-3)}$

$= \dfrac{6x^3}{25x^5}$

$= \dfrac{6 \cdot x^3}{25 \cdot x^3 \cdot x^2}$

$= \dfrac{6}{25x^2}$

9. $\dfrac{4a^3}{9b^2} \cdot \dfrac{3b^5}{8a} = \dfrac{12a^3 b^5}{72ab^2}$

$= \dfrac{12 \cdot a \cdot a^2 \cdot b^3 \cdot b^2}{12 \cdot 6 \cdot a \cdot b^2}$

$= \dfrac{a^2 b^3}{6}$

11. $\dfrac{3y^5}{2x^4} \div \dfrac{15y^9}{16x^7} = \dfrac{3y^5}{2x^4} \cdot \dfrac{16x^7}{15y^9}$

$= \dfrac{48x^7 y^5}{30x^4 y^9}$

$= \dfrac{6 \cdot 8 \cdot x^4 \cdot x^3 \cdot y^5}{6 \cdot 5 \cdot x^4 \cdot y^5 \cdot y^4}$

$= \dfrac{8x^3}{5y^4}$

13. $\dfrac{r^2 +10r+21}{r-9} \cdot \dfrac{2r-18}{r^2-9}$

$= \dfrac{(r+7)(r+3)}{(r-9)} \cdot \dfrac{2(r-9)}{(r-3)(r+3)}$

$= \dfrac{2(r+7)}{r-3}$

15. $\dfrac{x^2 +3x+2}{3x-3} \div \dfrac{x^2 -x-6}{6x-6}$

$= \dfrac{x^2 +3x+2}{3x-3} \cdot \dfrac{6x-6}{x^2 -x-6}$

$= \dfrac{(x+2)(x+1)}{3(x-1)} \cdot \dfrac{6(x-1)}{(x+2)(x-3)}$

$= \dfrac{6(x+1)}{3(x-3)}$

$= \dfrac{2(x+1)}{x-3}$

17. $\dfrac{2x-12}{x+1} \cdot \dfrac{4x+4}{18-3x} = \dfrac{2(x-6)}{(x+1)} \cdot \dfrac{4(x+1)}{3(6-x)}$

$= \dfrac{8(x-6)}{3(6-x)}$

$= \dfrac{8(x-6)}{-3(x-6)}$

$= -\dfrac{8}{3}$

19. $\dfrac{2k^2 -32}{k^2 -2k-24} \div \dfrac{k+6}{k^2 -7k+6}$

$= \dfrac{2k^2 -32}{k^2 -2k-24} \cdot \dfrac{k^2 -7k+6}{k+6}$

$= \dfrac{2(k^2 -16)}{(k-6)(k+4)} \cdot \dfrac{(k-6)(k-1)}{(k+6)}$

$= \dfrac{2(k-4)(k+4)}{(k-6)(k+4)} \cdot \dfrac{(k-6)(k-1)}{(k+6)}$

$= \dfrac{2(k-4)(k-1)}{k+6}$

21. $\dfrac{2a^2 +3ab}{3a-6b} \cdot \dfrac{a^2 -4b^2}{2ab+3b^2}$

$= \dfrac{a(2a+3b)}{3(a-2b)} \cdot \dfrac{(a-2b)(a+2b)}{b(2a+3b)}$

$= \dfrac{a(a+2b)}{3b}$

23. $\dfrac{4-x}{x^2+10x+25} \div \dfrac{3x^2-9x-12}{25-x^2}$

$= \dfrac{4-x}{x^2+10x+25} \cdot \dfrac{25-x^2}{3x^2-9x-12}$

$= \dfrac{(4-x)}{(x+5)(x+5)} \cdot \dfrac{(5-x)(5+x)}{3(x^2-3x-4)}$

$= \dfrac{-1\cdot(x-4)}{(x+5)(x+5)} \cdot \dfrac{(5-x)(5+x)}{3(x-4)(x+1)}$

$= \dfrac{-(5-x)}{3(x+5)(x+1)}$

$= \dfrac{x-5}{3(x+5)(x+1)}$

25. $\dfrac{t^2-8t+16}{t^2-2t-3} \cdot \dfrac{3-t}{t^2-16}$

$= \dfrac{(t-4)(t-4)}{(t-3)(t+1)} \cdot \dfrac{-1\cdot(t-3)}{(t-4)(t+4)}$

$= \dfrac{-(t-4)}{(t+1)(t+4)}$

$= \dfrac{4-t}{(t+1)(t+4)}$

27. $\dfrac{-x^2+7x-10}{2x^2+5x-12} \div \dfrac{-x^2+4}{8x^2-18}$

$= \dfrac{-x^2+7x-10}{2x^2+5x-12} \cdot \dfrac{8x^2-18}{-x^2+4}$

$= \dfrac{-1(x^2-7x+10)}{(2x-3)(x+4)} \cdot \dfrac{2(4x^2-9)}{-1(x^2-4)}$

$= \dfrac{1(x-5)(x-2)}{(2x-3)(x+4)} \cdot \dfrac{2(2x-3)(2x+3)}{1(x-2)(x+2)}$

$= \dfrac{2(x-5)(2x+3)}{(x+4)(x+2)}$

29. $\dfrac{-4x-6}{36-x^2} \cdot \dfrac{4x+24}{6x^2+x-12}$

$= \dfrac{-2(2x+3)}{-1(x^2-36)} \cdot \dfrac{4(x+6)}{(3x-4)(2x+3)}$

$= \dfrac{-2(2x+3)}{-1(x+6)(x-6)} \cdot \dfrac{4(x+6)}{(3x-4)(2x+3)}$

$= \dfrac{8}{(x-6)(3x-4)}$

31. $\dfrac{9x^2-16}{x+2} \div \left(3x^2+5x-12\right)$

$= \dfrac{9x^2-16}{x+2} \cdot \dfrac{1}{3x^2+5x-12}$

$= \dfrac{(3x-4)(3x+4)}{(x+2)} \cdot \dfrac{1}{(3x-4)(x+3)}$

$= \dfrac{3x+4}{(x+2)(x+3)}$

33. $\dfrac{6m^2-17m-14}{m^2+6m+9} \cdot \dfrac{9-m^2}{4m^2-49}$

$= \dfrac{(2m-7)(3m+2)}{(m+3)(m+3)} \cdot \dfrac{(3-m)(3+m)}{(2m-7)(2m+7)}$

$= \dfrac{(3m+2)(3-m)}{(m+3)(2m+7)}$ or $-\dfrac{(3m+2)(m-3)}{(m+3)(2m+7)}$

35. $\dfrac{x^2-4x-32}{x^2+7x+12} \div \dfrac{x^2-2x-48}{x^2+3x-4}$

$= \dfrac{x^2-4x-32}{x^2+7x+12} \cdot \dfrac{x^2+3x-4}{x^2-2x-48}$

$= \dfrac{(x-8)(x+4)}{(x+3)(x+4)} \cdot \dfrac{(x+4)(x-1)}{(x-8)(x+6)}$

$= \dfrac{(x+4)(x-1)}{(x+3)(x+6)}$

37. $\dfrac{p^2+4pt-12t^2}{p^2+pt-12t^2} \cdot \dfrac{p^2+7pt+12t^2}{p^2-7pt+10t^2}$

$= \dfrac{(p+6t)(p-2t)}{(p+4t)(p-3t)} \cdot \dfrac{(p+4t)(p+3t)}{(p-2t)(p-5t)}$

$= \dfrac{(p+6t)(p+3t)}{(p-3t)(p-5t)}$

39. $\dfrac{2x^2-xy-3y^2}{3xy-5y^2} \div \dfrac{4x^2-9y^2}{3x^2-14xy+15y^2}$

$= \dfrac{2x^2-xy-3y^2}{3xy-5y^2} \cdot \dfrac{3x^2-14xy+15y^2}{4x^2-9y^2}$

$= \dfrac{(2x-3y)(x+y)}{y(3x-5y)} \cdot \dfrac{(3x-5y)(x-3y)}{(2x-3y)(2x+3y)}$

$= \dfrac{(x+y)(x-3y)}{y(2x+3y)}$

41. $\dfrac{3x^3 - 15x^2 + 18x}{x^2 + 16x + 64} \cdot \dfrac{x^2 - 64}{4x^4 - 28x^3 + 40x^2}$

$= \dfrac{3x\left(x^2 - 5x + 6\right)}{(x+8)(x+8)} \cdot \dfrac{(x-8)(x+8)}{4x^2\left(x^2 - 7x + 10\right)}$

$= \dfrac{3 \cdot \cancel{x}\,\cancel{(x-2)}(x-3)}{\cancel{(x+8)}(x+8)} \cdot \dfrac{(x-8)\,\cancel{(x+8)}}{4x \cdot \cancel{x}\,\cancel{(x-2)}(x-5)}$

$= \dfrac{3(x-3)(x-8)}{4x(x+8)(x-5)}$

43. $\dfrac{w^2 - 2w - 8}{12w^4 + 32w^3 - 12w^2} \div \dfrac{w^2 - 9w + 20}{12w^3 + 54w^2 + 54w}$

$= \dfrac{w^2 - 2w - 8}{12w^4 + 32w^3 - 12w^2} \cdot \dfrac{12w^3 + 54w^2 + 54w}{w^2 - 9w + 20}$

$= \dfrac{(w-4)(w+2)}{4w^2\left(3w^2 + 8w - 3\right)} \cdot \dfrac{6w\left(2w^2 + 9w + 9\right)}{(w-4)(w-5)}$

$= \dfrac{\cancel{(w-4)}(w+2)}{4w^2(3w-1)\,\cancel{(w+3)}} \cdot \dfrac{6w(2w+3)\,\cancel{(w+3)}}{\cancel{(w-4)}(w-5)}$

$= \dfrac{3 \cdot \cancel{2w}(w+2)(2w+3)}{2w \cdot \cancel{2w}(3w-1)(w-5)}$

$= \dfrac{3(w+2)(2w+3)}{2w(3w-1)(w-5)}$

45. $\dfrac{x^2 + 4x - 5}{x^3 + 6x^2 - 4x - 24} \cdot \dfrac{x^2 + 8x + 12}{x^2 + 10x + 25}$

$= \dfrac{(x+5)(x-1)}{x^2(x+6) - 4(x+6)} \cdot \dfrac{(x+2)(x+6)}{(x+5)(x+5)}$

$= \dfrac{(x+5)(x-1)}{(x+6)\left(x^2 - 4\right)} \cdot \dfrac{(x+2)(x+6)}{(x+5)(x+5)}$

$= \dfrac{\cancel{(x+5)}(x-1)}{\cancel{(x+6)}(x-2)\,\cancel{(x+2)}} \cdot \dfrac{\cancel{(x+2)}\,\cancel{(x+6)}}{\cancel{(x+5)}(x+5)}$

$= \dfrac{x-1}{(x-2)(x+5)}$

47. $\dfrac{18x^3 + 27x^2 - 8x - 12}{3x^2 - x - 2} \div \left(6x^2 + 5x - 6\right)$

$= \dfrac{18x^3 + 27x^2 - 8x - 12}{3x^2 - x - 2} \cdot \dfrac{1}{6x^2 + 5x - 6}$

$= \dfrac{9x^2(2x+3) - 4(2x+3)}{(3x+2)(x-1)} \cdot \dfrac{1}{(2x+3)(3x-2)}$

$= \dfrac{(2x+3)\left(9x^2 - 4\right)}{(3x+2)(x-1)} \cdot \dfrac{1}{(2x+3)(3x-2)}$

$= \dfrac{\cancel{(2x+3)}\,\cancel{(3x-2)}(3x+2)}{\cancel{(3x+2)}(x-1)} \cdot \dfrac{1}{\cancel{(2x+3)}\,\cancel{(3x-2)}}$

$= \dfrac{1}{x-1}$

49. $\dfrac{k^3 - 8}{k^3 + 27} \cdot \dfrac{k^2 - 9}{k^2 - 4}$

$= \dfrac{\cancel{(k-2)}\left(k^2 + 2k + 4\right)}{\cancel{(k+3)}\left(k^2 - 3k + 9\right)} \cdot \dfrac{(k-3)\,\cancel{(k+3)}}{\cancel{(k-2)}(k+2)}$

$= \dfrac{(k-3)\left(k^2 + 2k + 4\right)}{(k+2)\left(k^2 - 3k + 9\right)}$

51. $\dfrac{8x^3 - 27}{3x^2 - 6x + 12} \div \dfrac{8x^2 + 12x + 18}{6x^3 + 48}$

$= \dfrac{8x^3 - 27}{3x^2 - 6x + 12} \cdot \dfrac{6x^3 + 48}{8x^2 + 12x + 18}$

$= \dfrac{(2x-3)\left(4x^2 + 6x + 9\right)}{3\left(x^2 - 2x + 4\right)} \cdot \dfrac{6\left(x^3 + 8\right)}{2\left(4x^2 + 6x + 9\right)}$

$= \dfrac{(2x-3)\,\cancel{\left(4x^2 + 6x + 9\right)}}{3\,\cancel{\left(x^2 - 2x + 4\right)}} \cdot \dfrac{6(x+2)\,\cancel{\left(x^2 - 2x + 4\right)}}{2\,\cancel{\left(4x^2 + 6x + 9\right)}}$

$= \dfrac{\cancel{6}(2x-3)(x+2)}{\cancel{6}}$

$= (2x-3)(x+2)$

53. $\dfrac{a^2 + ab - 2b^2}{a^3 + b^3} \cdot \dfrac{a^2 + 2ab + b^2}{a^2 - b^2}$

$= \dfrac{(a+2b)\,\cancel{(a-b)}}{\cancel{(a+b)}\left(a^2 - ab + b^2\right)} \cdot \dfrac{\cancel{(a+b)}(a+b)}{\cancel{(a-b)}\,\cancel{(a+b)}}$

$= \dfrac{a + 2b}{a^2 - ab + b^2}$

55. $(f \cdot g)(x) = \dfrac{x^2 - 6x - 16}{x^2 + 3x - 40} \cdot \dfrac{x^2 - 64}{x^2 - 3x - 10}$

$= \dfrac{(x-8)(x+2)}{(x+8)(x-5)} \cdot \dfrac{(x-8)(x+8)}{(x-5)(x+2)}$

$= \dfrac{(x-8)(x-8)}{(x-5)(x-5)}$

$= \dfrac{(x-8)^2}{(x-5)^2}$

$= \left(\dfrac{x-8}{x-5}\right)^2$

$(f \cdot g)(6) = \left(\dfrac{6-8}{6-5}\right)^2 = \left(\dfrac{-2}{1}\right)^2 = 4$

57. $\left(\dfrac{g}{f}\right)(x) = \dfrac{x^2 - 64}{x^2 - 3x - 10} \div \dfrac{x^2 - 6x - 16}{x^2 + 3x - 40}$

$= \dfrac{x^2 - 64}{x^2 - 3x - 10} \cdot \dfrac{x^2 + 3x - 40}{x^2 - 6x - 16}$

$= \dfrac{(x-8)(x+8)}{(x-5)(x+2)} \cdot \dfrac{(x+8)(x-5)}{(x-8)(x+2)}$

$= \dfrac{(x+8)(x+8)}{(x+2)(x+2)}$

$= \left(\dfrac{x+8}{x+2}\right)^2$

$\left(\dfrac{g}{f}\right)(7) = \left(\dfrac{7+8}{7+2}\right)^2 = \left(\dfrac{15}{9}\right)^2 = \left(\dfrac{5}{3}\right)^2 = \dfrac{25}{9}$

59. $\left(\dfrac{f}{g}\right)(x) = \dfrac{1 - x^2}{x^2 - 3x - 28} \div \dfrac{x^2 - 8x + 7}{x^2 + 5x + 4}$

$= \dfrac{1 - x^2}{x^2 - 3x - 28} \cdot \dfrac{x^2 + 5x + 4}{x^2 - 8x + 7}$

$= \dfrac{-1 \cdot (x-1)(1+x)}{(x-7)(x+4)} \cdot \dfrac{(x+4)(x+1)}{(x-7)(x-1)}$

$= \dfrac{-(1+x)(x+1)}{(x-7)(x-7)}$

$= -\left(\dfrac{x+1}{x-7}\right)^2$

$\left(\dfrac{f}{g}\right)(4) = -\left(\dfrac{4+1}{4-7}\right)^2 = -\left(\dfrac{5}{-3}\right)^2 = -\dfrac{25}{9}$

61. $\left(\dfrac{20x^7}{x^2 - 9} \div \dfrac{x^2 - 14x + 24}{5x - 15}\right) \cdot \dfrac{x^2 + x - 6}{8x^{13}}$

$= \left(\dfrac{20x^7}{x^2 - 9} \cdot \dfrac{5x - 15}{x^2 - 14x + 24}\right) \cdot \dfrac{x^2 + x - 6}{8x^{13}}$

$= \dfrac{4 \cdot 5 x^7}{(x+3)(x-3)} \cdot \dfrac{5(x-3)}{(x-12)(x-2)} \cdot \dfrac{(x-2)(x+3)}{4 \cdot 2 \cdot x^7 \cdot x^6}$

$= \dfrac{25}{2x^6(x-12)}$

63. $\dfrac{12k^3}{k^2 - 4} \div \left(\dfrac{22k^6}{-6k + 12} \cdot \dfrac{k}{11k + 22}\right)$

$= \dfrac{12k^3}{k^2 - 4} \div \left(\dfrac{11 \cdot 2k^6}{-6(k-2)} \cdot \dfrac{k}{11(k+2)}\right)$

$= \dfrac{12k^3}{k^2 - 4} \div \dfrac{2k^7}{-6(k-2)(k+2)}$

$= \dfrac{12k^3}{k^2 - 4} \cdot \dfrac{-6(k-2)(k+2)}{2k^7}$

$= \dfrac{12k^3}{(k-2)(k+2)} \cdot \dfrac{-2 \cdot 3(k-2)(k+2)}{2 \cdot k^3 \cdot k^4}$

$= -\dfrac{36}{k^4}$

65. $\left(\left(\dfrac{x-4}{x+5}\right)^2 \cdot \left(\dfrac{x+5}{x-1}\right)^2\right) \div \left(\dfrac{x-4}{x-1}\right)^2$

$= \left(\left(\dfrac{x-4}{x+5}\right)^2 \cdot \left(\dfrac{x+5}{x-1}\right)^2\right) \cdot \left(\dfrac{x-1}{x-4}\right)^2$

$= \dfrac{(x-4)^2}{(x+5)^2} \cdot \dfrac{(x+5)^2}{(x-1)^2} \cdot \dfrac{(x-1)^2}{(x-4)^2}$

$= 1$

67. $\dfrac{63.8 \text{ inches}}{1} \cdot \dfrac{1 \text{ foot}}{12 \text{ inches}} \approx 5.32 \text{ feet}$

69. $\dfrac{15 \text{ gallons}}{1} \cdot \dfrac{4 \text{ quarts}}{1 \text{ gallon}} \cdot \dfrac{0.946 \text{ liter}}{1 \text{ quart}}$

$\approx 56.76 \text{ liters}$

71. $\dfrac{1.63 \text{ grams}}{1 \text{ pound}} \cdot \dfrac{1000 \text{ milligrams}}{1 \text{ gram}} \cdot \dfrac{1 \text{ pound}}{16 \text{ ounces}}$

$\approx \dfrac{101.88 \text{ milligrams}}{1 \text{ ounce}}$

73. $\dfrac{7533 \text{ liters}}{1 \text{ year}} \cdot \dfrac{1 \text{ year}}{365 \text{ days}} \cdot \dfrac{1 \text{ quart}}{0.946 \text{ liters}} \cdot \dfrac{1 \text{ gallon}}{4 \text{ quarts}}$

≈ 5.45 gallons per day

75. $\dfrac{67 \text{ miles}}{1 \text{ hour}} \cdot \dfrac{5280 \text{ feet}}{1 \text{ mile}} \cdot \dfrac{1 \text{ hour}}{3600 \text{ seconds}}$

≈ 98.27 feet per second

77. Substitute $x = 10$ into $\dfrac{x-2}{x+8} \div \dfrac{x+8}{x-5}$:

$\dfrac{10-2}{10+8} \div \dfrac{10+8}{10-5} = \dfrac{8}{18} \div \dfrac{18}{5} = \dfrac{8}{18} \cdot \dfrac{5}{18} = \dfrac{10}{81}$

Now, substitute $x = 10$ into $\dfrac{x-2}{x-5}$:

$\dfrac{10-2}{10-5} = \dfrac{8}{5}$

The two values are different, so the students' work is incorrect.

79. a. $\dfrac{1}{x} \div \dfrac{1}{x} = \dfrac{1}{x} \cdot \dfrac{x}{1} = \dfrac{x}{x} = 1$

b. $\dfrac{1}{x} \div \left(\dfrac{1}{x} \div \dfrac{1}{x} \right) = \dfrac{1}{x} \div \left(\dfrac{1}{x} \cdot \dfrac{x}{1} \right) = \dfrac{1}{x} \div 1 = \dfrac{1}{x}$

c. $\dfrac{1}{x} \div \left(\dfrac{1}{x} \div \left(\dfrac{1}{x} \div \dfrac{1}{x} \right) \right) = \dfrac{1}{x} \div \left(\dfrac{1}{x} \div 1 \right) = \dfrac{1}{x} \div \dfrac{1}{x} = 1$

d. $\dfrac{1}{x} \div \left(\dfrac{1}{x} \div \left(\dfrac{1}{x} \div \left(\dfrac{1}{x} \div \dfrac{1}{x} \right) \right) \right)$

$= \dfrac{1}{x} \div \left(\dfrac{1}{x} \div \left(\dfrac{1}{x} \div 1 \right) \right)$

$= \dfrac{1}{x} \div \left(\dfrac{1}{x} \div \dfrac{1}{x} \right)$

$= \dfrac{1}{x} \div 1$

$= \dfrac{1}{x}$

e. $\underbrace{\dfrac{1}{x} \div \left(\dfrac{1}{x} \div \left(\dfrac{1}{x} \div \cdots \div \left(\dfrac{1}{x} \div \left(\dfrac{1}{x} \div \dfrac{1}{x} \right) \right) \right) \right)}_{n \text{ division symbols}}$

The answer will be 1 if n is even, $\dfrac{1}{x}$, if n is odd.

81. Answers may vary.

83. $(f \cdot g)(x) = 8^x \cdot 2^x$

$= \left(2^3 \right)^x \cdot 2^x$

$= 2^{3x} \cdot 2^x$

$= 2^{3x+x}$

$= 2^{4x}$

$= \left(2^4 \right)^x$

$= 16^x$

$\left(\dfrac{f}{g} \right)(x) = \dfrac{8^x}{2^x}$

$= \dfrac{\left(2^3 \right)^x}{2^x}$

$= \dfrac{2^{3x}}{2^x}$

$= 2^{3x-x}$

$= 2^{2x}$

$= \left(2^2 \right)^x$

$= 4^x$

85. $(f \cdot g)(x) = 12(6)^x \cdot 3(2)^x$

$= 12 \cdot 3 (6)^x (2)^x$

$= 36 (6 \cdot 2)^x$

$= 36 (12)^x$

$\left(\dfrac{f}{g} \right)(x) = \dfrac{12(6)^x}{3(2)^x} = 4 \left(\dfrac{6}{2} \right)^x = 4(3)^x$

87. a. $\dfrac{2.5 \text{ yards}}{1} \cdot \dfrac{3 \text{ feet}}{1 \text{ yard}} \cdot \dfrac{12 \text{ inches}}{1 \text{ foot}} = 90$ inches

b. i. $f(w) = 12w$

$g(x) = 3x$

ii. $(f \circ g)(x) = f(g(x))$

$= f(3x)$

$= 12(3x)$

$= 36x$

iii. $(f \circ g)(2.5) = 36(2.5) = 90$

There are 90 inches in 2.5 yards. The two results are equal.

89. $4\log_b\left(2x^2\right) - 2\log_b\left(3x^3\right)$

$$= \log_b\left(2x^2\right)^4 - \log_b\left(3x^3\right)^2$$

$$= \log_b\left(16x^8\right) - \log_b\left(9x^6\right)$$

$$= \log_b\left(\frac{16x^8}{9x^6}\right)$$

$$= \log_b\left(\frac{16x^2 \cdot \cancel{x^6}}{9 \cdot \cancel{x^6}}\right)$$

$$= \log_b\left(\frac{16x^2}{9}\right)$$

This is a logarithmic expression in one variable.

91. $\log_2\left(x-3\right) + \log_2\left(x-2\right) = 3$

$$\log_2\left[\left(x-3\right)\left(x-2\right)\right] = 3$$

$$\left(x-3\right)\left(x-2\right) = 2^3$$

$$x^2 - 5x + 6 = 8$$

$$x^2 - 5x - 2 = 0$$

$$x = \frac{-(-5) \pm \sqrt{(-5)^2 - 4(1)(-2)}}{2(1)}$$

$$= \frac{5 \pm \sqrt{25+8}}{2}$$

$$x = \frac{5+\sqrt{33}}{2} \quad\text{or}\quad x = \cancel{\frac{5-\sqrt{33}}{2}}$$

The solution is $\dfrac{5+\sqrt{33}}{2} \approx 5.3723$.

This is a logarithmic equation in one variable.

93. $5(4)^x - 23 = 81$

$$5(4)^x = 104$$

$$(4)^x = \frac{104}{5}$$

$$x = \log_4\left(\frac{104}{5}\right)$$

$$x = \log_4\left(20.8\right)$$

$$x = \frac{\log_{10} 20.8}{\log_{10} 4} \approx 2.1893$$

This is an exponential equation in one variable.

Homework 8.3

1. $\dfrac{5}{x} + \dfrac{2}{x} = \dfrac{5+2}{x} = \dfrac{7}{x}$

3. $\dfrac{x}{x^2-9} + \dfrac{3}{x^2-9} = \dfrac{x+3}{x^2-9}$

$$= \frac{\cancel{(x+3)}}{\cancel{(x+3)}\left(x-3\right)}$$

$$= \frac{1}{x-3}$$

5. $\dfrac{6m^2}{m^2-4m+3} - \dfrac{4m^2+6m}{m^2-4m+3}$

$$= \frac{6m^2 - \left(4m^2+6m\right)}{m^2-4m+3}$$

$$= \frac{6m^2 - 4m^2 - 6m}{m^2-4m+3}$$

$$= \frac{2m^2 - 6m}{m^2-4m+3}$$

$$= \frac{2m\left(m-3\right)}{\left(m-1\right)\left(m-3\right)}$$

$$= \frac{2m}{m-1}$$

7. $\dfrac{3x^2+9x}{x^2+10x+21} - \dfrac{2x^2+x-15}{x^2+10x+21}$

$$= \frac{3x^2+9x - \left(2x^2+x-15\right)}{x^2+10x+21}$$

$$= \frac{3x^2+9x - 2x^2 - x + 15}{x^2+10x+21}$$

$$= \frac{x^2+8x+15}{x^2+10x+21}$$

$$= \frac{\cancel{(x+3)}\left(x+5\right)}{\cancel{(x+3)}\left(x+7\right)}$$

$$= \frac{x+5}{x+7}$$

9.
$$\frac{2}{x^6} - \frac{4}{x^2} = \frac{2}{x^6} - \frac{4}{x^2} \cdot \frac{x^4}{x^4}$$
$$= \frac{2}{x^6} - \frac{4x^4}{x^6}$$
$$= \frac{2 - 4x^4}{x^6}$$
$$= \frac{2\left(1 - 2x^4\right)}{x^6}$$

11.
$$\frac{3}{10x^6} + \frac{5}{12x^4} = \frac{3}{10x^6} \cdot \frac{6}{6} + \frac{5}{12x^4} \cdot \frac{5x^2}{5x^2}$$
$$= \frac{18}{60x^6} + \frac{25x^2}{60x^6}$$
$$= \frac{18 + 25x^2}{60x^6}$$

13.
$$\frac{7}{4a^2b} - \frac{5}{6ab^3} = \frac{7}{4a^2b} \cdot \frac{3b^2}{3b^2} - \frac{5}{6ab^3} \cdot \frac{2a}{2a}$$
$$= \frac{21b^3}{12a^2b^3} - \frac{10a}{12a^2b^3}$$
$$= \frac{21b^3 - 10a}{12a^2b^3}$$

15.
$$\frac{3}{x+1} + \frac{4}{x-2} = \frac{3}{x+1} \cdot \left(\frac{x-2}{x-2}\right) + \frac{4}{x-2} \cdot \left(\frac{x+1}{x+1}\right)$$
$$= \frac{3x-6}{(x+1)(x-2)} + \frac{4x+4}{(x+1)(x-2)}$$
$$= \frac{7x-2}{(x+1)(x-2)}$$

17.
$$\frac{6}{(x+4)(x-6)} - \frac{4}{(x-1)(x+4)}$$
$$= \frac{6}{(x+4)(x-6)} \cdot \left(\frac{x-1}{x-1}\right) - \frac{4}{(x-1)(x+4)} \cdot \left(\frac{x-6}{x-6}\right)$$
$$= \frac{6x-6}{(x+4)(x-6)(x-1)} - \frac{4x-24}{(x+4)(x-6)(x-1)}$$
$$= \frac{6x-6-4x+24}{(x+4)(x-6)(x-1)}$$
$$= \frac{2x+18}{(x+4)(x-6)(x-1)}$$
$$= \frac{2(x+9)}{(x+4)(x-6)(x-1)}$$

19.
$$\frac{5}{3t-6} - \frac{2}{5t+15}$$
$$= \frac{5}{3(t-2)} - \frac{2}{5(t+3)}$$
$$= \frac{5}{3(t-2)} \cdot \frac{5(t+3)}{5(t+3)} - \frac{2}{5(t+3)} \cdot \frac{3(t-2)}{3(t-2)}$$
$$= \frac{25(t+3)}{15(t-2)(t+3)} - \frac{6(t-2)}{15(t+3)(t-2)}$$
$$= \frac{25(t+3) - 6(t-2)}{15(t-2)(t+3)}$$
$$= \frac{25t + 75 - 6t + 12}{15(t-2)(t+3)}$$
$$= \frac{19t + 87}{15(t-2)(t+3)}$$

21.
$$\frac{3}{x^2 - 25} + \frac{5}{x^2 - 5x} = \frac{3}{(x+5)(x-5)} + \frac{5}{x(x-5)}$$
$$= \frac{3}{(x-5)(x+5)} \cdot \frac{x}{x} + \frac{5}{x(x-5)} \cdot \left(\frac{x+5}{x+5}\right)$$
$$= \frac{3x}{x(x-5)(x+5)} + \frac{5x+25}{x(x-5)(x+5)}$$
$$= \frac{8x+25}{x(x-5)(x+5)}$$

23.
$$\frac{2}{x^2 - 9} + \frac{3}{x^2 - 7x + 12}$$
$$= \frac{2}{(x+3)(x-3)} + \frac{3}{(x-4)(x-3)}$$
$$= \frac{2}{(x+3)(x-3)} \cdot \left(\frac{x-4}{x-4}\right) + \frac{3}{(x-4)(x-3)} \cdot \left(\frac{x+3}{x+3}\right)$$
$$= \frac{2x-8}{(x+3)(x-3)(x-4)} + \frac{3x+9}{(x+3)(x-3)(x-4)}$$
$$= \frac{5x+1}{(x+3)(x-3)(x-4)}$$

25.
$$2 + \frac{k-3}{k+1} = \frac{2}{1} \cdot \left(\frac{k+1}{k+1}\right) + \frac{k-3}{k+1}$$
$$= \frac{2k+2}{k+1} + \frac{k-3}{k+1}$$
$$= \frac{3k-1}{k+1}$$

27.
$$2 - \frac{2x+4}{x^2 + 3x + 2} = \frac{2}{1} - \frac{2\cancel{(x+2)}}{\cancel{(x+2)}(x+1)}$$
$$= \frac{2}{1} - \frac{2}{x+1} = \frac{2}{1} \cdot \left(\frac{x+1}{x+1}\right) - \frac{2}{x+1}$$
$$= \frac{2x+2}{x+1} - \frac{2}{x+1} = \frac{2x}{x+1}$$

29. $\dfrac{8}{x-6} - \dfrac{4}{6-x} = \dfrac{8}{x-6} + \dfrac{4}{x-6} = \dfrac{12}{x-6}$

31. $\dfrac{2x+1}{x^2-4x-21} + \dfrac{3}{14-2x}$

$= \dfrac{2x+1}{(x-7)(x+3)} + \dfrac{3}{-2(x-7)}$

$= \dfrac{2x+1}{(x-7)(x+3)} - \dfrac{3}{2(x-7)}$

$= \dfrac{2x+1}{(x-7)(x+3)} \cdot \dfrac{2}{2} - \dfrac{3}{2(x-7)} \cdot \left(\dfrac{x+3}{x+3} \right)$

$= \dfrac{4x+2}{2(x-7)(x+3)} - \dfrac{3x+9}{2(x-7)(x+3)}$

$= \dfrac{4x+2-3x-9}{2(x-7)(x+3)}$

$= \dfrac{x-7}{2(x-7)(x+3)}$

$= \dfrac{1}{2(x+3)}$

33. $\dfrac{-2c}{7-2c} - \dfrac{c+1}{4c^2-49}$

$= \dfrac{2c}{(2c-7)} - \dfrac{(c+1)}{(2c-7)(2c+7)}$

$= \dfrac{2c}{(2c-7)} \cdot \dfrac{(2c+7)}{(2c+7)} - \dfrac{(c+1)}{(2c-7)(2c+7)}$

$= \dfrac{2c(2c+7)-(c+1)}{(2c-7)(2c+7)}$

$= \dfrac{4c^2+14c-c-1}{(2c-7)(2c+7)}$

$= \dfrac{4c^2+13c-1}{(2c-7)(2c+7)}$

35. $\dfrac{2b}{a^2-b^2} + \dfrac{a}{ab-b^2}$

$= \dfrac{2b}{(a+b)(a-b)} + \dfrac{a}{b(a-b)}$

$= \dfrac{2b}{(a+b)(a-b)} \cdot \dfrac{b}{b} + \dfrac{a}{b(a-b)} \cdot \dfrac{(a+b)}{(a+b)}$

$= \dfrac{2b^2+a^2+ab}{b(a+b)(a-b)}$

$= \dfrac{a^2+ab+2b^2}{b(a+b)(a-b)}$

37. $\dfrac{x}{x^2+5x+6} - \dfrac{3}{x^2+7x+12}$

$= \dfrac{x}{(x+2)(x+3)} \cdot \dfrac{(x+4)}{(x+4)}$

$\quad - \dfrac{3}{(x+3)(x+4)} \cdot \dfrac{(x+2)}{(x+2)}$

$= \dfrac{x(x+4)-3(x+2)}{(x+2)(x+3)(x+4)}$

$= \dfrac{x^2+x-6}{(x+2)(x+3)(x+4)}$

$= \dfrac{(x-2)\cancel{(x+3)}}{(x+2)\cancel{(x+3)}(x+4)}$

$= \dfrac{x-2}{(x+2)(x+4)}$

39. $\dfrac{x-1}{x+2} + \dfrac{x+2}{x-1}$

$= \left(\dfrac{x-1}{x+2} \right)\left(\dfrac{x-1}{x-1} \right) + \left(\dfrac{x+2}{x-1} \right)\left(\dfrac{x+2}{x+2} \right)$

$= \dfrac{x^2-2x+1}{(x+2)(x-1)} + \dfrac{x^2+4x+4}{(x+2)(x-1)}$

$= \dfrac{2x^2+2x+5}{(x+2)(x-1)}$

41. $\dfrac{y-5}{y-3} - \dfrac{y+3}{y+5}$

$= \left(\dfrac{y-5}{y-3} \right)\left(\dfrac{y+5}{y+5} \right) - \left(\dfrac{y+3}{y+5} \right)\left(\dfrac{y-3}{y-3} \right)$

$= \dfrac{y^2-25}{(y-3)(y+5)} - \dfrac{y^2-9}{(y-3)(y+5)}$

$= \dfrac{y^2-25-y^2+9}{(y-3)(y+5)}$

$= \dfrac{-16}{(y-3)(y+5)}$

43. $\dfrac{x+4}{x^2-7x+10}-\dfrac{5}{x^2-25}$

$=\dfrac{x+4}{(x-2)(x-5)}-\dfrac{5}{(x-5)(x+5)}$

$=\dfrac{(x+4)}{(x-2)(x-5)}\cdot\dfrac{(x+5)}{(x+5)}$

$\quad-\dfrac{5}{(x-5)(x+5)}\cdot\dfrac{(x-2)}{(x-2)}$

$=\dfrac{(x+4)(x+5)-5(x-2)}{(x-2)(x-5)(x+5)}$

$=\dfrac{x^2+9x+20-5x+10}{(x-2)(x-5)(x+5)}$

$=\dfrac{x^2+4x+30}{(x-2)(x-5)(x+5)}$

45. $\dfrac{x+2}{(x-4)(x+3)^2}+\dfrac{x-1}{(x-4)(x+1)(x+3)}$

$=\left(\dfrac{x+2}{(x-4)(x+3)^2}\right)\left(\dfrac{x+1}{x+1}\right)$

$\quad+\left(\dfrac{x-1}{(x-4)(x+1)(x+3)}\right)\left(\dfrac{x+3}{x+3}\right)$

$=\dfrac{x^2+3x+2}{(x-4)(x+1)(x+3)^2}$

$\quad+\dfrac{x^2+2x-3}{(x-4)(x+1)(x+3)^2}$

$=\dfrac{2x^2+5x-1}{(x-4)(x+1)(x+3)^2}$

47. $\dfrac{c+2}{c^2-4}+\dfrac{3c}{c^2-2c}=\dfrac{c+2}{(c+2)(c-2)}+\dfrac{3c}{c(c-2)}$

$=\dfrac{1}{c-2}+\dfrac{3}{c-2}$

$=\dfrac{4}{c-2}$

49. $\dfrac{x-1}{4x^2+20x+25}-\dfrac{x+4}{6x^2+17x+5}$

$=\dfrac{x-1}{(2x+5)(2x+5)}-\dfrac{x+4}{(2x+5)(3x+1)}$

$=\left(\dfrac{x-1}{(2x+5)(2x+5)}\right)\left(\dfrac{3x+1}{3x+1}\right)$

$\quad-\left(\dfrac{x+4}{(2x+5)(3x+1)}\right)\left(\dfrac{2x+5}{2x+5}\right)$

$=\dfrac{3x^2-2x-1}{(3x+1)(2x+5)^2}-\dfrac{2x^2+13x+20}{(3x+1)(2x+5)^2}$

$=\dfrac{3x^2-2x-1-2x^2-13x-20}{(3x+1)(2x+5)^2}$

$=\dfrac{x^2-15x-21}{(3x+1)(2x+5)^2}$

51. $\dfrac{3x-1}{x^2+4x+4}+\dfrac{2x+1}{3x^2+5x-2}$

$=\dfrac{3x-1}{(x+2)(x+2)}-\dfrac{2x+1}{(3x-1)(x+2)}$

$=\left(\dfrac{3x-1}{(x+2)(x+2)}\right)\left(\dfrac{3x-1}{3x-1}\right)$

$\quad+\left(\dfrac{2x+1}{(3x-1)(x+2)}\right)\left(\dfrac{x+2}{x+2}\right)$

$=\dfrac{9x^2-6x+1}{(3x-1)(x+2)^2}+\dfrac{2x^2+5x+2}{(3x-1)(x+2)^2}$

$=\dfrac{9x^2-6x+1+2x^2+5x+2}{(3x-1)(x+2)^2}$

$=\dfrac{11x^2-x+3}{(3x-1)(x+2)^2}$

53. $\dfrac{3p}{p^2-2pq-24q^2}-\dfrac{2q}{p^2-3pq-18q^2}$

$=\dfrac{3p}{(p-6q)(p+4q)}-\dfrac{2q}{(p-6q)(p+3q)}$

$=\dfrac{3p}{(p-6q)(p+4q)}\cdot\dfrac{(p+3q)}{(p+3q)}$

$\quad-\dfrac{2q}{(p-6q)(p+3q)}\cdot\dfrac{(p+4q)}{(p+4q)}$

$=\dfrac{3p(p+3q)-2q(p+4q)}{(p-6q)(p+4q)(p+3q)}$

$=\dfrac{3p^2+9pq-2pq-8q^2}{(p-6q)(p+4q)(p+3q)}$

$=\dfrac{3p^2+7pq-8q^2}{(p-6q)(p+4q)(p+3q)}$

55. $\dfrac{x-1}{6x^2-24x}+\dfrac{5}{3x^3-6x^2-24x}$

$=\dfrac{x-1}{6x(x-4)}+\dfrac{5}{3x(x^2-2x-8)}$

$=\dfrac{x-1}{6x(x-4)}+\dfrac{5}{3x(x-4)(x+2)}$

$=\left(\dfrac{x-1}{6x(x-4)}\right)\left(\dfrac{x+2}{x+2}\right)$

$\quad+\left(\dfrac{5}{3x(x-4)(x+2)}\right)\left(\dfrac{2}{2}\right)$

$=\dfrac{x^2+x-2}{6x(x-4)(x+2)}+\dfrac{10}{6x(x-4)(x+2)}$

$=\dfrac{x^2+x+8}{6x(x-4)(x+2)}$

57. $\left(\dfrac{2}{x^2-4}+\dfrac{3}{x+2}\right)-\dfrac{1}{2x-4}$

$=\left(\dfrac{2}{(x+2)(x-2)}+\dfrac{3}{x+2}\right)-\dfrac{1}{2(x-2)}$

$=\left(\dfrac{2}{(x+2)(x-2)}+\left(\dfrac{3}{x+2}\right)\left(\dfrac{x-2}{x-2}\right)\right)$

$\quad-\dfrac{1}{2(x-2)}$

$=\left(\dfrac{2}{(x+2)(x-2)}+\dfrac{3x-6}{(x+2)(x-2)}\right)$

$\quad-\dfrac{1}{2(x-2)}$

$=\dfrac{3x-4}{(x+2)(x-2)}-\dfrac{1}{2(x-2)}$

$=\left(\dfrac{3x-4}{(x+2)(x-2)}\right)\left(\dfrac{2}{2}\right)-\left(\dfrac{1}{2(x-2)}\right)\left(\dfrac{x+2}{x+2}\right)$

$=\dfrac{6x-8}{2(x+2)(x-2)}-\dfrac{x+2}{2(x+2)(x-2)}$

$=\dfrac{5x-10}{2(x+2)(x-2)}$

$=\dfrac{5(x-2)}{2(x+2)(x-2)}$

$=\dfrac{5}{2(x+2)}$

59. $\dfrac{3}{t+1}-\left(\dfrac{2t-3}{t^2+6t+5}+\dfrac{2}{t+5}\right)$

$=\dfrac{3}{t+1}-\left(\dfrac{2t-3}{(t+5)(t+1)}+\dfrac{2}{t+5}\right)$

$=\dfrac{3}{t+1}-\left(\dfrac{2t-3}{(t+5)(t+1)}+\left(\dfrac{2}{t+5}\right)\left(\dfrac{t+1}{t+1}\right)\right)$

$=\dfrac{3}{t+1}-\left(\dfrac{2t-3}{(t+5)(t+1)}+\dfrac{2t+2}{(t+5)(t+1)}\right)$

$=\dfrac{3}{t+1}-\dfrac{4t-1}{(t+5)(t+1)}$

$=\left(\dfrac{3}{t+1}\right)\left(\dfrac{t+5}{t+5}\right)-\dfrac{4t-1}{(t+5)(t+1)}$

$=\dfrac{3t+15}{(t+1)(t+5)}-\dfrac{4t-1}{(t+1)(t+5)}$

$=\dfrac{3t+15-4t+1}{(t+1)(t+5)}$

$=\dfrac{-t+16}{(t+1)(t+5)}$

61. $f(x)+g(x)=\dfrac{x+3}{x-4}+\dfrac{x+4}{x-3}$

$=\left(\dfrac{x+3}{x-4}\right)\left(\dfrac{x-3}{x-3}\right)+\left(\dfrac{x+4}{x-3}\right)\left(\dfrac{x-4}{x-4}\right)$

$=\dfrac{x^2-9}{(x-4)(x+3)}+\dfrac{x^2-16}{(x-4)(x-3)}$

$=\dfrac{2x^2-25}{(x-4)(x-3)}$

63. $g(x)-f(x)$

$=\dfrac{x+4}{x-3}-\dfrac{x+3}{x-4}$

$=\left(\dfrac{x+4}{x-3}\right)\left(\dfrac{x-4}{x-4}\right)-\left(\dfrac{x+3}{x-4}\right)\left(\dfrac{x-3}{x-3}\right)$

$=\dfrac{x^2-16}{(x-3)(x-4)}-\dfrac{x^2-9}{(x-3)(x-4)}$

$=\dfrac{x^2-16-x^2+9}{(x-3)(x-4)}$

$=\dfrac{-7}{(x-3)(x-4)}$

65. $f(x) - g(x)$

$$= \frac{x-2}{x^2 - 2x - 8} - \frac{x+1}{3x+6}$$

$$= \frac{x-2}{(x-4)(x+2)} - \frac{x+1}{3(x+2)}$$

$$= \left(\frac{x-2}{(x-4)(x+2)}\right)\left(\frac{3}{3}\right) - \left(\frac{x+1}{3(x+2)}\right)\left(\frac{x-4}{x-4}\right)$$

$$= \frac{3x-6}{3(x-4)(x+2)} - \frac{x^2 - 3x - 4}{3(x-4)(x+2)}$$

$$= \frac{3x - 6 - x^2 + 3x + 4}{3(x-4)(x+2)}$$

$$= \frac{-x^2 + 6x - 2}{3(x-4)(x+2)}$$

67. The student did not multiply each expression by "1". The expression $\frac{2}{x+1}$ should have been multiplied by $\frac{x+2}{x+2}$ not $\frac{1}{x+2}$ and the expression $\frac{3}{x+2}$ should have been multiplied by $\frac{x+1}{x+1}$ not $\frac{1}{x+1}$. The correct addition is:

$$\frac{2}{x+1} + \frac{3}{x+2} = \frac{2}{x+1} \cdot \frac{x+2}{x+2} + \frac{3}{x+2} \cdot \frac{x+1}{x+1}$$

$$= \frac{2x+4}{(x+1)(x+2)} + \frac{3x+3}{(x+1)(x+2)}$$

$$= \frac{5x+7}{(x+1)(x+2)}$$

69. The student did not subtract the entire numerator in the second expression. The student only subtracted $5x$ and not the 1. The correct subtraction is:

$$\frac{9x}{x-3} - \frac{5x+1}{x-3} = \frac{9x - 5x - 1}{x-3} = \frac{4x-1}{x-3}$$

71. Answers may vary. Example:
The two rational expressions must have the same denominator in order to be added, so each expression must be multiplied by a fraction equal to "1," which introduces the necessary terms in each denominator that make them equal.

73. $(f+g)(x) = \left(6x^2 - 4x + 3\right) + \left(2x^2 - 7x - 5\right)$

$$= 6x^2 - 4x + 3 + 2x^2 - 7x - 5$$

$$= 8x^2 - 11x - 2$$

$(f-g)(x) = \left(6x^2 - 4x + 3\right) - \left(2x^2 - 7x - 5\right)$

$$= 6x^2 - 4x + 3 - 2x^2 + 7x + 5$$

$$= 4x^2 + 3x + 8$$

75. $(f+g)(x) = \left[2(5)^x\right] + \left[-3(5)^x\right]$

$$= 2(5)^x - 3(5)^x$$

$$= (2-3) \cdot (5)^x$$

$$= -(5)^x$$

$(f-g)(x) = \left[2(5)^x\right] - \left[-3(5)^x\right]$

$$= 2(5)^x + 3(5)^x$$

$$= (2+3) \cdot (5)^x$$

$$= 5 \cdot (5)^x$$

$$= 5^1 \cdot 5^x$$

$$= 5^{x+1}$$

77. $\dfrac{4x+5}{x+2} + \left(\dfrac{3x+15}{x^2 - 4} \cdot \dfrac{x^2 - 2x}{x^2 + 7x + 10}\right)$

$$= \frac{4x+5}{x+2} + \left(\frac{3(x+5)}{(x+2)(x-2)} \cdot \frac{x(x-2)}{(x+5)(x+2)}\right)$$

$$= \frac{4x+5}{x+2} + \frac{3x}{(x+2)^2}$$

$$= \left(\frac{4x+5}{x+2}\right)\left(\frac{x+2}{x+2}\right) + \frac{3x}{(x+2)^2}$$

$$= \frac{4x^2 + 13x + 10}{(x+2)^2} + \frac{3x}{(x+2)^2}$$

$$= \frac{4x^2 + 16x + 10}{(x+2)^2}$$

$$= \frac{2(2x^2 + 8x + 5)}{(x+2)^2}$$

79. $\dfrac{5x+5}{3x+6} \cdot \left(\dfrac{x^2+4x}{x^2+2x+1} + \dfrac{4}{x^2+2x+1} \right)$

$= \dfrac{5(x+1)}{3(x+2)} \cdot \left(\dfrac{x^2+4x+4}{x^2+2x+1} \right)$

$= \dfrac{5(x+1)}{3(x+2)} \cdot \dfrac{(x+2)(x+2)}{(x+1)(x+1)}$

$= \dfrac{5(x+2)}{3(x+1)}$

81.

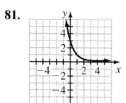

This is an exponential function.

83.

$5b^4 = 66$

$5b^4 - 66 = 0$

$5\left(b^4 - \dfrac{66}{5} \right) = 0$

$5\left(b^2 - \dfrac{\sqrt{330}}{5} \right)\left(b^2 + \dfrac{\sqrt{330}}{5} \right) = 0$

$b^2 - \dfrac{\sqrt{330}}{5} = 0$

$b^2 = \dfrac{\sqrt{330}}{5}$

$b = \pm\sqrt{\dfrac{\sqrt{330}}{5}}$

$b \approx \pm 1.9061$

This is a fourth-degree polynomial equation in one variable.

85. $y = ab^x$

$(3, 95): \ 95 = a \cdot b^3$

$a = \dfrac{95}{b^3}$

$(7, 2): \ 2 = a \cdot b^7$

$2 = \left(\dfrac{95}{b^3} \right) \cdot b^7$

$2 = 95b^4$

$\dfrac{2}{95} = b^4$

$b = \pm\sqrt[4]{\dfrac{2}{95}} \approx \pm 0.38$

$a \approx \dfrac{95}{\left(\sqrt[4]{\dfrac{2}{95}} \right)^3} \approx 1715.87$

$y \approx 1715.87(0.38)^x$

The equation is $y \approx 1715.98(0.38)^x$. It is an exponential equation in two variables.

Homework 8.4

1. $\dfrac{\dfrac{2}{x}}{\dfrac{3}{x}} = \dfrac{2}{x} \div \dfrac{3}{x} = \dfrac{2}{x} \cdot \dfrac{x}{3} = \dfrac{2}{3}$

3. $\dfrac{\dfrac{7}{x^2}}{\dfrac{21}{x^5}} = \dfrac{7}{x^2} \div \dfrac{21}{x^5} = \dfrac{7}{x^2} \cdot \dfrac{x^5}{21} = \dfrac{x^3}{3}$

5. $\dfrac{\dfrac{4a^2}{5b}}{\dfrac{6a}{15b^3}} = \dfrac{4a^2}{5b} \div \dfrac{6a}{15b^3}$

$= \dfrac{4a^2}{5b} \cdot \dfrac{15b^3}{6a}$

$= \dfrac{60a^2 b^3}{30ab}$

$= 2ab^2$

7. $\dfrac{\dfrac{3x-3}{2x+10}}{\dfrac{6x^2-6}{4x+20}} = \dfrac{3x-3}{2x+10} \div \dfrac{6x^2-6}{4x+20}$

$= \dfrac{3x-3}{2x+10} \cdot \dfrac{4x+20}{6x^2-6}$

$= \dfrac{3(x-1)}{2(x+5)} \cdot \dfrac{4(x+5)}{6(x^2-1)}$

$= \dfrac{12(x-1)(x+5)}{12(x+5)(x-1)(x+1)}$

$= \dfrac{1}{x+1}$

9. $\dfrac{\dfrac{x^2-49}{3x^2-9x}}{\dfrac{x^2-5x-14}{7x-21}} = \dfrac{x^2-49}{3x^2-9x} \cdot \dfrac{7x-21}{x^2-5x-14}$

$= \dfrac{(x+7)(x-7)}{3x(x-3)} \cdot \dfrac{7(x-3)}{(x-7)(x+2)}$

$= \dfrac{7(x+7)}{3x(x+2)}$

11. $\dfrac{\dfrac{25x^2-4}{9x^2-16}}{\dfrac{25x^2-20x+4}{9x^2-24x+16}}$

$= \dfrac{25x^2-4}{9x^2-16} \div \dfrac{25x^2-20x+4}{9x^2-24x+16}$

$= \dfrac{25x^2-4}{9x^2-16} \cdot \dfrac{9x^2-24x+16}{25x^2-20x+4}$

$= \dfrac{(5x-2)(5x+2)}{(3x-4)(3x+4)} \cdot \dfrac{(3x-4)(3x-4)}{(5x-2)(5x-2)}$

$= \dfrac{(5x+2)(3x-4)}{(3x+4)(5x-2)}$

13. $\dfrac{\dfrac{2}{x^3}-\dfrac{3}{x}}{\dfrac{5}{x^3}+\dfrac{4}{x^2}} = \dfrac{\left(\dfrac{2}{x^3}-\dfrac{3}{x}\right)}{\left(\dfrac{5}{x^3}+\dfrac{4}{x^2}\right)} \cdot \dfrac{x^3}{x^3} = \dfrac{2-3x^2}{5+4x}$

15. $\dfrac{4+\dfrac{3}{x}}{\dfrac{2}{x}-3} = \dfrac{\left(4+\dfrac{3}{x}\right)}{\left(\dfrac{2}{x}-3\right)} \cdot \dfrac{x}{x} = \dfrac{4x+3}{2-3x}$

17. $\dfrac{\dfrac{5}{2x^3}-4}{\dfrac{1}{6x^2}-3} = \dfrac{\left(\dfrac{5}{2x^3}-4\right)}{\left(\dfrac{1}{6x^2}-3\right)} \cdot \dfrac{6x^3}{6x^3}$

$= \dfrac{3(5-8x^3)}{1-18x^3}$ or $\dfrac{3(8x^3-5)}{18x^3-1}$

19. $\dfrac{\dfrac{a^2}{b}-b}{\dfrac{1}{b}-\dfrac{1}{a}} = \dfrac{\left(\dfrac{a^2}{b}-b\right)}{\left(\dfrac{1}{b}-\dfrac{1}{a}\right)} \cdot \dfrac{ab}{ab}$

$= \dfrac{a^3-ab^2}{a-b}$

$= \dfrac{a\left(a^2-b^2\right)}{a-b}$

$= \dfrac{a(a-b)(a+b)}{(a-b)}$

$= a(a+b)$

21. $\dfrac{\dfrac{1}{x}-\dfrac{8}{x^2}+\dfrac{15}{x^3}}{\dfrac{1}{x}-\dfrac{5}{x^2}} = \dfrac{\left(\dfrac{1}{x}-\dfrac{8}{x^2}+\dfrac{15}{x^3}\right)}{\left(\dfrac{1}{x}-\dfrac{5}{x^2}\right)} \cdot \dfrac{x^3}{x^3}$

$= \dfrac{x^2-8x+15}{x^2-5x}$

$= \dfrac{(x-5)(x-3)}{x(x-5)}$

$= \dfrac{x-3}{x}$

23. $\dfrac{\dfrac{x}{x-4}-\dfrac{2x}{x+1}}{\dfrac{x}{x+1}-\dfrac{2x}{x-4}}$

$= \dfrac{\left(\dfrac{x}{x-4}-\dfrac{2x}{x+1}\right)}{\left(\dfrac{x}{x+1}-\dfrac{2x}{x-4}\right)} \cdot \dfrac{(x-4)(x+1)}{(x-4)(x+1)}$

$= \dfrac{x(x+1)-2x(x-4)}{x(x-4)-2x(x+1)}$

$= \dfrac{x^2+x-2x^2+8x}{x^2-4x-2x^2-2x}$

$= \dfrac{-x^2+9x}{-x^2-6x}$

$= \dfrac{-x(x-9)}{-x(x+6)}$

$= \dfrac{x-9}{x+6}$

25.
$$\dfrac{p+\dfrac{2}{p-4}}{p-\dfrac{3}{p-4}} = \dfrac{\left(p+\dfrac{2}{p-4}\right)}{\left(p-\dfrac{3}{p-4}\right)} \cdot \dfrac{(p-4)}{(p-4)}$$

$$= \dfrac{p(p-4)+2}{p(p-4)-3}$$

$$= \dfrac{p^2-4p+2}{p^2-4p-3}$$

27.
$$\dfrac{\dfrac{1}{x+3}-\dfrac{1}{x}}{3} = \dfrac{\left(\dfrac{1}{x+3}-\dfrac{1}{x}\right)}{3} \cdot \left(\dfrac{x(x+3)}{x(x+3)}\right)$$

$$= \dfrac{x-(x+3)}{3x(x+3)}$$

$$= \dfrac{-3}{3x(x+3)}$$

$$= \dfrac{-1}{x(x+3)}$$

29.
$$\dfrac{\dfrac{3}{a+b}-\dfrac{3}{a-b}}{2ab}$$

$$= \dfrac{\left(\dfrac{3}{a+b}-\dfrac{3}{a-b}\right)}{2ab} \cdot \dfrac{(a+b)(a-b)}{(a+b)(a-b)}$$

$$= \dfrac{3(a-b)-3(a+b)}{2ab(a+b)(a-b)}$$

$$= \dfrac{3a-3b-3a-3b}{2ab(a+b)(a-b)}$$

$$= \dfrac{-6b}{2ab(a+b)(a-b)}$$

$$= \dfrac{-3}{a(a+b)(a-b)}$$

31.
$$\dfrac{\dfrac{1}{(x+2)^2}-\dfrac{1}{x^2}}{2} = \dfrac{\left(\dfrac{1}{(x+2)^2}-\dfrac{1}{x^2}\right)}{2} \cdot \dfrac{x^2(x+2)^2}{x^2(x+2)^2}$$

$$= \dfrac{x^2-(x+2)^2}{2x^2(x+2)^2}$$

$$= \dfrac{x^2-(x^2+4x+4)}{2x^2(x+2)^2}$$

$$= \dfrac{-4x-4}{2x^2(x+2)^2}$$

$$= \dfrac{-4(x+1)}{2x^2(x+2)^2}$$

$$= \dfrac{-2(x+1)}{x^2(x+2)^2}$$

33.
$$\dfrac{\dfrac{6}{2x-8}+\dfrac{10}{x^2-4x}}{\dfrac{1}{x^2-x-12}-\dfrac{2}{x^2-16}}$$

$$= \dfrac{\dfrac{6}{2(x-4)}+\dfrac{10}{x(x-4)}}{\dfrac{1}{(x-4)(x+3)}-\dfrac{2}{(x-4)(x+4)}}$$

$$= \dfrac{\dfrac{6}{2(x-4)}\cdot\dfrac{x}{x}+\dfrac{10}{x(x-4)}\cdot\dfrac{2}{2}}{\dfrac{1}{(x-4)(x+3)}\cdot\dfrac{(x+4)}{(x+4)}-\dfrac{2}{(x-4)(x+4)}\cdot\dfrac{(x+3)}{(x+3)}}$$

$$= \dfrac{\dfrac{6x+20}{2x(x-4)}}{\dfrac{x+4}{(x-4)(x+3)(x+4)}-\dfrac{2x+6}{(x-4)(x+3)(x+4)}}$$

$$= \dfrac{\dfrac{6x+20}{2x(x-4)}}{\dfrac{x+4-2x-6}{(x-4)(x+3)(x+4)}} \cdot \dfrac{2x(x-4)(x+3)(x+4)}{2x(x-4)(x+3)(x+4)}$$

$$= \dfrac{(6x+20)(x+3)(x+4)}{(-x-2)\cdot 2x}$$

$$= \dfrac{2(3x+10)(x+3)(x+4)}{-2x(x+2)}$$

$$= -\dfrac{(3x+10)(x+3)(x+4)}{x(x+2)}$$

35.
$$\dfrac{\dfrac{x+7}{x^2+7x+10}-\dfrac{6}{x^2+2x}}{\dfrac{x+1}{x^2+7x+10}+\dfrac{6}{x^2+5x}}$$

$$=\dfrac{\dfrac{x+7}{(x+5)(x+2)}-\dfrac{6}{x(x+2)}}{\dfrac{x+1}{(x+5)(x+2)}+\dfrac{6}{x(x+5)}}$$

$$=\dfrac{\dfrac{x+7}{(x+5)(x+2)}\cdot\dfrac{x}{x}-\dfrac{6}{x(x+2)}\cdot\dfrac{(x+5)}{(x+5)}}{\dfrac{x+1}{(x+5)(x+2)}\cdot\dfrac{x}{x}+\dfrac{6}{x(x+5)}\cdot\dfrac{(x+2)}{(x+2)}}$$

$$=\dfrac{\dfrac{x^2+7x}{x(x+5)(x+2)}-\dfrac{6x+30}{x(x+5)(x+2)}}{\dfrac{x^2+x}{x(x+5)(x+2)}+\dfrac{6x+12}{x(x+5)(x+2)}}$$

$$=\dfrac{\dfrac{x^2+7x-6x-30}{x(x+5)(x+2)}}{\dfrac{x^2+x+6x+12}{x(x+5)(x+2)}}$$

$$=\dfrac{\dfrac{x^2+x-30}{x(x+5)(x+2)}}{\dfrac{x^2+7x+12}{x(x+5)(x+2)}}$$

$$=\dfrac{x^2+x-30}{x(x+5)(x+2)}\cdot\dfrac{x(x+5)(x+2)}{x^2+7x+12}$$

$$=\dfrac{x^2+x-30}{x^2+7x+12}$$

$$=\dfrac{(x+6)(x-5)}{(x+3)(x+4)}$$

37. $\left(\dfrac{f}{g}\right)(x)=\dfrac{\dfrac{5x+10}{x^2-6x+9}}{\dfrac{4x+8}{x^2-4x+3}}$

$$=\dfrac{\dfrac{5(x+2)}{(x-3)(x-3)}}{\dfrac{4(x+2)}{(x-3)(x-1)}}$$

$$=\dfrac{5(x+2)}{(x-3)(x-3)}\cdot\dfrac{(x-3)(x-1)}{4(x+2)}$$

$$=\dfrac{5(x-1)}{4(x-3)}$$

39. $\left(\dfrac{f}{g}\right)(x)=\dfrac{\left(\dfrac{x}{2}-\dfrac{2}{x}\right)}{\left(\dfrac{3}{2}+\dfrac{3}{x}\right)}$

$$=\dfrac{\left(\dfrac{x}{2}-\dfrac{2}{x}\right)}{\left(\dfrac{3}{2}+\dfrac{3}{x}\right)}\cdot\dfrac{2x}{2x}$$

$$=\dfrac{x^2-4}{3x+6}$$

$$=\dfrac{(x-2)(x+2)}{3(x+2)}$$

$$=\dfrac{x-2}{3}$$

41. $\left(\dfrac{f}{g}\right)(x)$

$$=\dfrac{\dfrac{2x}{x^2-25}+\dfrac{x+5}{x-5}}{\dfrac{x-5}{x+5}+\dfrac{3x}{x^2-25}}$$

$$=\dfrac{\left(\dfrac{2x}{(x-5)(x+5)}+\dfrac{x+5}{x-5}\right)}{\left(\dfrac{x-5}{x+5}+\dfrac{3x}{(x-5)(x+5)}\right)}\cdot\dfrac{(x-5)(x+5)}{(x-5)(x+5)}$$

$$=\dfrac{2x+(x+5)(x+5)}{(x-5)(x-5)+3x}$$

$$=\dfrac{2x+x^2+10x+25}{x^2-10x+25+3x}$$

$$=\dfrac{x^2+12x+25}{x^2-7x+25}$$

43. The student must add the two expressions in the parentheses before taking the reciprocals. The correct simplification is:

$$\dfrac{x}{\dfrac{1}{x}+\dfrac{1}{2}}=x\div\left(\dfrac{1}{x}+\dfrac{1}{2}\right)$$

$$=x\div\left(\dfrac{1}{x}\cdot\dfrac{2}{2}+\dfrac{1}{2}\cdot\dfrac{x}{x}\right)$$

$$=x\div\left(\dfrac{2}{2x}+\dfrac{x}{2x}\right)$$

$$=x\div\dfrac{2+x}{2x}$$

$$=x\cdot\dfrac{2x}{2+x}$$

$$=\dfrac{2x^2}{x+2}$$

45. Method 1:

$$\frac{\dfrac{6}{x^2}-\dfrac{5}{x}}{\dfrac{2}{x^2}+\dfrac{3}{2x}} = \frac{\dfrac{6}{x^2}-\dfrac{5}{x}\cdot\dfrac{x}{x}}{\dfrac{2}{x^2}\cdot\dfrac{2}{2}+\dfrac{3}{2x}\cdot\dfrac{x}{x}}$$

$$=\frac{\dfrac{6}{x^2}-\dfrac{5x}{x^2}}{\dfrac{4}{2x^2}+\dfrac{3x}{2x^2}}$$

$$=\frac{\dfrac{6-5x}{x^2}}{\dfrac{4+3x}{2x^2}}$$

$$=\frac{6-5x}{x^2}\div\frac{4+3x}{2x^2}$$

$$=\frac{6-5x}{x^2}\cdot\frac{2x^2}{4+3x}$$

$$=\frac{2(6-5x)}{4+3x}\ \text{ or }\ -\frac{2(5x-6)}{3x+4}$$

Method 2:

$$\frac{\dfrac{6}{x^2}-\dfrac{5}{x}}{\dfrac{2}{x^2}+\dfrac{3}{2x}} = \frac{\left(\dfrac{6}{x^2}-\dfrac{5}{x}\right)}{\left(\dfrac{2}{x^2}+\dfrac{3}{2x}\right)}\cdot\frac{2x^2}{2x^2}$$

$$=\frac{12-10x}{4+3x}$$

$$=\frac{2(6-5x)}{4+3x}\ \text{ or }\ -\frac{2(5x-6)}{3x+4}$$

Answers may vary.

47. It implies $6\cdot\dfrac{1}{3}=2$, which is true.

49. A complex rational expression is a rational expression that has rational expressions in the numerator and in the denominator.
Answers may vary.

51. $\dfrac{8x^{-2}y^5}{6x^{-7}y^8}=\dfrac{4x^{-2-(-7)}}{3y^{8-5}}=\dfrac{4x^5}{3y^3}$

53. $\dfrac{x^{-1}+x^{-2}}{x^{-2}-x^{-1}}=\dfrac{\dfrac{1}{x}+\dfrac{1}{x^2}}{\dfrac{1}{x^2}-\dfrac{1}{x}}$

$$=\frac{\left(\dfrac{1}{x}+\dfrac{1}{x^2}\right)\cdot x^2}{\left(\dfrac{1}{x^2}-\dfrac{1}{x}\right)\cdot x^2}$$

$$=\frac{x+1}{1-x}\ \text{ or }\ -\frac{x+1}{x-1}$$

55. $\dfrac{2b^{-2}-4b}{8b^{-1}-6b}\cdot\dfrac{b^2}{b^2}=\dfrac{2-4b^3}{8b-6b^3}$

$$=\frac{2\left(1-2b^3\right)}{2b\left(4-3b^2\right)}$$

$$=\frac{1-2b^3}{b\left(4-3b^2\right)}\ \text{ or }\ \frac{2b^3-1}{b\left(3b^2-4\right)}$$

57. $(f\circ g)(x)=f\big(g(x)\big)$

$$=f\left(\frac{x}{x-1}\right)$$

$$=\frac{\left(\dfrac{x}{x-1}\right)+2}{3-\left(\dfrac{x}{x-1}\right)}$$

$$=\frac{\left(\dfrac{x}{x-1}\right)+2}{3-\left(\dfrac{x}{x-1}\right)}\cdot\frac{x-1}{x-1}$$

$$=\frac{x+2(x-1)}{3(x-1)-x}$$

$$=\frac{x+2x-2}{3x-3-x}$$

$$=\frac{3x-2}{2x-3}$$

59. **a.** $H_e=\dfrac{1}{\dfrac{1}{H_p}+\dfrac{1}{H_b}}=\dfrac{1}{\dfrac{H_b+H_p}{H_bH_p}}=\dfrac{H_bH_p}{H_b+H_p}$

b. $H_e=\dfrac{H_bH_p}{H_b+H_p}$

$$=\frac{(623)(87.4)}{623+87.4}$$

$$\approx 76.65\,\text{days}$$

61. $\dfrac{x-2}{x^2-2x-24} + \dfrac{x+4}{x^2-8x+12}$

$= \dfrac{x-2}{(x+4)(x-6)} + \dfrac{x+4}{(x-6)(x-2)}$

$= \dfrac{(x-2)}{(x+4)(x-6)} \cdot \dfrac{(x-2)}{(x-2)}$

$\quad + \dfrac{(x+4)}{(x-6)(x-2)} \cdot \dfrac{(x+4)}{(x+4)}$

$= \dfrac{\left(x^2-4x+4\right)+\left(x^2+8x+16\right)}{(x+4)(x-6)(x-2)}$

$= \dfrac{2x^2+4x+20}{(x+4)(x-6)(x-2)}$

This is a rational expression in one variable.

63. $\dfrac{x-2}{x^2-2x-24} \cdot \dfrac{x+4}{x^2-8x+12}$

$= \dfrac{(x-2)}{(x-6)(x+4)} \cdot \dfrac{(x+4)}{(x-6)(x-2)}$

$= \dfrac{1}{(x-6)^2}$

This is a rational expression in one variable.

65. $x^2-2x-24=0$

$(x-6)(x+4)=0$

$x-6=0 \quad \text{or} \quad x+4=0$

$\qquad x=6 \quad \text{or} \qquad x=-4$

The domain is all real numbers except 6 and −4.

The function $f(x)$ is a rational function.

Homework 8.5

1. $\dfrac{7}{x} = \dfrac{2}{x} + 1$

$x \cdot \dfrac{7}{x} = x\left(\dfrac{2}{x}+1\right)$

$7 = 2 + x$

$5 = x$

The solution is $x = 5$.

3. $\left(\dfrac{7}{4x} - \dfrac{5}{6}\right) = \dfrac{1}{12x}$

$12x \cdot \left(\dfrac{7}{4x} - \dfrac{5}{6}\right) = \dfrac{1}{12x} \cdot 12x$

$21 - 10x = 1$

$-10x = -20$

$x = 2$

The solution is $x = 2$.

5. $\dfrac{x-2}{x-7} = \dfrac{5}{x-7}$

$(x-7) \cdot \dfrac{x-2}{x-7} = \dfrac{5}{x-7} \cdot (x-7)$

$x-2 = 5$

$x = 7$

$x = 7$ makes each denominator equal 0. This equation has no solution.

7. $\dfrac{5}{4p-7} = \dfrac{2}{2p+3}$

$5(2p+3) = 2(4p-7)$

$10p+15 = 8p-14$

$2p = -29$

$p = -\dfrac{29}{2}$

The solution is $x = -\dfrac{29}{2}$.

9. $\dfrac{3}{x+1} + \dfrac{2}{5} = 1$

$5(x+1)\left(\dfrac{3}{x+1}+\dfrac{2}{5}\right) = 5(x+1)\cdot 1$

$15 + 2(x+1) = 5x+5$

$15 + 2x + 2 = 5x+5$

$2x+17 = 5x+5$

$12 = 3x$

$4 = x$

The solution is $x = 4$.

11.

$$\frac{1}{x-2}+\frac{1}{x+2}=\frac{4}{x^2-4}$$

$$\frac{1}{x-2}+\frac{1}{x+2}=\frac{4}{(x-2)(x+2)}$$

$$(x-2)(x+2)\left(\frac{1}{x-2}+\frac{1}{x+2}\right)=(x-2)(x+2)\frac{4}{(x-2)(x+2)}$$

$$x+2+x-2=4$$

$$2x=4$$

$$x=2$$

$x=2$ makes the denominators $x-2$ and x^2-2 equal 0. This equation has no solution.

13.

$$2+\frac{4}{k-2}=\frac{8}{k^2-2k}$$

$$2+\frac{4}{k-2}=\frac{8}{k(k-2)}$$

$$k(k-2)\left(2+\frac{4}{k-2}\right)=k(k-2)\frac{8}{k(k-2)}$$

$$2k(k-2)+4k=8$$

$$2k^2-4k+4k=8$$

$$2k^2=8$$

$$k^2=4$$

$$k=\pm2$$

$k=2$ makes the denominators $k-2$ and k^2-2k equal 0. The solution is $k=-2$.

15.

$$\frac{-48}{x^2-2x-15}-\frac{6}{x+3}=\frac{7}{x-5}$$

$$(x-5)(x+3)\cdot\left(\frac{-48}{(x-5)(x+3)}-\frac{6}{x+3}\right)=\frac{7}{x-5}\cdot(x-5)(x+3)$$

$$-48-6(x-5)=7(x+3)$$

$$-48-6x+30=7x+21$$

$$-6x-18=7x+21$$

$$-13x=39$$

$$x=-3$$

$x=-3$ makes the denominators $x^2-2x-15$ and $x-5$ equal 0. This equation has no solution.

17.

$$\frac{x^2-23}{2x^2-5x-3}+\frac{2}{x-3}=\frac{-1}{2x+1}$$

$$(2x+1)(x-3)\left(\frac{(x^2-23)}{(2x+1)(x-3)}+\frac{2}{x-3}\right)=\frac{-1(2x+1)(x-3)}{2x+1}$$

$$x^2-23+2(2x+1)=-(x-3)$$

$$x^2-23+4x+2=-x+3$$

$$x^2+5x-24=0$$

$$(x+8)(x-3)=0$$

$$x+8=0 \quad \text{or} \quad x-3=0$$

$$x=-8 \quad \text{or} \quad x=3$$

$x=3$ makes the denominators $2x^2-5x-3$ and $x-3$ equal 0. The solution is $x=-8$.

19.
$$\frac{w+7}{w^2-9} = \frac{-w+2}{w-3}$$
$$(w-3)(w+3)\frac{w+7}{(w-3)(w+3)} = \frac{(w-3)(w+3)(-w+2)}{w-3}$$
$$w+7 = (w+3)(-w+2)$$
$$w+7 = -w^2 - 3w + 2w + 6$$
$$w^2 + 2w + 1 = 0$$
$$(w+1)^2 = 0$$
$$w = -1$$

The solution is $w = -1$.

21.
$$x^2 \cdot \left(3 + \frac{2}{x}\right) = \frac{4}{x^2} \cdot x^2$$
$$3x^2 + 2x = 4$$
$$3x^2 + 2x - 4 = 0$$
$$x = \frac{-2 \pm \sqrt{2^2 - 4(3)(-4)}}{2(3)} = \frac{-2 \pm \sqrt{4 + 48}}{6} = \frac{-2 \pm \sqrt{52}}{6} = \frac{-2 \pm 2\sqrt{13}}{6} = \frac{-1 \pm \sqrt{13}}{3}$$

The solutions are $x = \frac{-1 \pm \sqrt{13}}{3}$.

23.
$$\frac{5}{r^2 - 3r + 2} - \frac{1}{r-2} = \frac{r+6}{3r-3}$$
$$\frac{5}{(r-2)(r-1)} - \frac{1}{r-2} = \frac{r+6}{3(r-1)}$$
$$3(r-2)(r-1)\left(\frac{5}{(r-2)(r-1)} - \frac{1}{r-2}\right) = 3(r-2)(r-1)\frac{r+6}{3(r-1)}$$
$$15 - 3(r-1) = (r-2)(r+6)$$
$$15 - 3r + 3 = r^2 + 6r - 2r - 12$$
$$18 - 3r = r^2 + 4r - 12$$
$$0 = r^2 + 7r - 30$$
$$0 = (r+10)(r-3)$$

The solutions are $r = 3$ and $r = -10$.

25.
$$\frac{2x}{x+1} - \frac{3}{2} = \frac{-2}{x+2}$$
$$2(x+1)(x+2)\left(\frac{2x}{x+1} - \frac{3}{2}\right) = \frac{2(x+1)(x+2)(-2)}{x+2}$$
$$4x(x+2) - 3(x+2)(x+1) = -4(x+1)$$
$$4x^2 + 8x - 3(x^2 + 3x + 2) = -4x - 4$$
$$4x^2 + 8x - 3x^2 - 9x - 6 = -4x - 4$$
$$x^2 + 3x - 2 = 0$$
$$x = \frac{-(3) \pm \sqrt{(3)^2 - 4(1)(-2)}}{2(1)} = \frac{-3 \pm \sqrt{17}}{2}$$

The solutions are $x = \frac{-3 \pm \sqrt{17}}{2}$.

27.
$$\frac{x-4}{x^2 - 7x + 12} - \frac{x+2}{x-3} = 0$$
$$\frac{x-4}{x^2 - 7x + 12} = \frac{x+2}{x-3}$$
$$\frac{x-4}{(x-4)(x-3)} = \frac{x+2}{x-3}$$
$$\frac{1}{x-3} = \frac{x+2}{x-3}$$
$$\left(\frac{x-3}{1}\right) \cdot \frac{1}{x-3} = \frac{x+2}{x-3} \cdot \left(\frac{x-3}{1}\right)$$
$$1 = x + 2$$
$$-1 = x$$

The solution is $x = -1$.

29.

$$\frac{t}{t-3} = 2 - \frac{5}{3-t}$$

$$(t-3)\cdot\frac{t}{t-3} = \left(2+\frac{5}{t-3}\right)\cdot(t-3)$$

$$t = 2(t-3)+5$$

$$t = 2t-6+5$$

$$-t = -1$$

$$t = 1$$

The solution is $t = 1$.

31.

$$\frac{12}{9-x^2}+\frac{3}{x+3} = \frac{-2}{x-3}$$

$$\frac{12}{(3-x)(3+x)}+\frac{3}{3+x} = \frac{2}{3-x}$$

$$(3-x)(3+x)\cdot\left(\frac{12}{(3-x)(3+x)}+\frac{3}{3+x}\right) = \frac{2}{3-x}\cdot(3-x)(3+x)$$

$$12+3(3-x) = 2(3+x)$$

$$12+9-3x = 6+2x$$

$$15 = 5x$$

$$3 = x$$

$x = 3$ makes the denominators $9-x^2$ and $x-3$ equal 0. This equation has no solution.

33.

$$\frac{x+2}{x-3}-\frac{x-3}{x+2} = \frac{5x}{x^2-x-6}$$

$$(x-3)(x+2)\left(\frac{x+2}{x-3}-\frac{x-3}{x+2}\right) = \frac{(x-3)(x+2)5x}{(x-3)(x+2)}$$

$$(x+2)^2-(x-3)^2 = 5x$$

$$x^2+4x+4-\left(x^2-6x+9\right) = 5x$$

$$10x-5 = 5x$$

$$5x = 5$$

$$x = 1$$

The solution is $x = 1$.

35.

$$\frac{2y}{y-2}-\frac{2y-5}{y^2-7y+10} = \frac{-4}{y-5}$$

$$(y-2)(y-5)\cdot\left(\frac{2y}{y-2}-\frac{2y-5}{(y-2)(y-5)}\right) = \frac{-4}{y-5}\cdot(y-2)(y-5)$$

$$2y(y-5)-(2y-5) = -4(y-2)$$

$$2y^2-10y-2y+5 = -4y+8$$

$$2y^2-8y-3 = 0$$

$$y = \frac{-(-8)\pm\sqrt{(-8)^2-4(2)(-3)}}{2(2)} = \frac{8\pm\sqrt{88}}{4} = \frac{8\pm2\sqrt{22}}{4} = \frac{4\pm\sqrt{22}}{2}$$

The solutions are $y = \frac{4\pm\sqrt{22}}{2}$.

37.

$$\frac{x-2}{(x-3)(x+1)} + \frac{x+5}{(x-1)(x+1)} = \frac{x+3}{(x-3)(x-1)}$$

$$\frac{x-2}{x^2-2x-3} + \frac{x+5}{x^2-1} = \frac{x+3}{x^2-4x+3}$$

$$(x-3)(x-1)(x+1) \cdot \left(\frac{x-2}{(x-3)(x+1)} + \frac{x+5}{(x-1)(x+1)} \right) = \frac{(x+3)(x-3)(x-1)(x+1)}{(x-3)(x-1)}$$

$$(x-2)(x-1) + (x+5)(x-3) = (x+3)(x+1)$$

$$x^2 - 3x + 2 + x^2 + 2x - 15 = x^2 + 4x + 3$$

$$x^2 - 5x - 16 = 0$$

$$x = \frac{-(-5) \pm \sqrt{(-5)^2 - 4(1)(-16)}}{2(1)} = \frac{5 \pm \sqrt{89}}{2}$$

The solutions are $x = \dfrac{5 \pm \sqrt{89}}{2}$.

39.

$$\frac{5}{x} - \frac{2}{x^2} = 4$$

$$x^2 \cdot \left(\frac{5}{x} - \frac{2}{x^2} \right) = 4 \cdot x^2$$

$$5x - 2 = 4x^2$$

$$4x^2 - 5x + 2 = 0$$

$$x = \frac{-(-5) \pm \sqrt{(-5)^2 - 4(4)(2)}}{2(4)} = \frac{5 \pm \sqrt{-7}}{8} = \frac{5 \pm i\sqrt{7}}{8}$$

The complex solutions are $x = \dfrac{5 \pm i\sqrt{7}}{8}$.

41.

$$\frac{2}{t-5} - \frac{3t}{t+5} = \frac{35}{t^2-25}$$

$$(t-5)(t+5) \cdot \left(\frac{2}{t-5} - \frac{3t}{t+5} \right) = \frac{35(t-5)(t+5)}{(t-5)(t+5)}$$

$$2(t+5) - 3t(t-5) = 35$$

$$2t + 10 - 3t^2 + 15t = 35$$

$$3t^2 - 17t + 25 = 0$$

$$t = \frac{-(-17) \pm \sqrt{(-17)^2 - 4(3)(25)}}{2(3)} = \frac{17 \pm \sqrt{-11}}{6} = \frac{17 \pm i\sqrt{11}}{6}$$

The complex solutions are $t = \dfrac{17 \pm i\sqrt{11}}{6}$.

43.

$$\frac{x-1}{3x-12} + \frac{-x+1}{x-5} = \frac{4x}{x^2-9x+20}$$

$$\frac{x-1}{3(x-4)} + \frac{-x+1}{x-5} = \frac{4x}{(x-4)(x-5)}$$

$$3(x-4)(x-5) \cdot \left(\frac{x-1}{3(x-4)} + \frac{-x+1}{x-5} \right) = \frac{3(x-4)(x-5)4x}{(x-4)(x-5)}$$

$$(x-5)(x-1) + 3(x-4)(-x+1) = 12x$$

$$x^2 - 6x + 5 + 3\left(-x^2 + 5x - 4\right) = 12x$$

$$x^2 - 6x + 5 - 3x^2 + 15x - 12 = 12x$$

$$-2x^2 - 3x - 7 = 0$$

$$2x^2 + 3x + 7 = 0$$

$$x = \frac{-3 \pm \sqrt{3^2 - 4(2)(7)}}{2(2)} = \frac{-3 \pm \sqrt{-47}}{4} = \frac{-3 \pm i\sqrt{47}}{4}$$

The complex solutions are $x = \dfrac{-3 \pm i\sqrt{47}}{4}$.

45.

$$4 = \frac{3}{x-5}$$

$$(x-5)4 = (x-5)\frac{3}{x-5}$$

$$4x - 20 = 3$$

$$4x = 23$$

$$x = \frac{23}{4}$$

The solution for $f(x) = 4$ is $x = \dfrac{23}{4}$.

47.

$$-1 = \frac{5}{x-1} + \frac{3}{x+1}$$

$$(x-1)(x+1) \cdot (-1) = (x-1)(x+1)\left(\frac{5}{x-1} + \frac{3}{x+1} \right)$$

$$-1(x^2 - 1) = 5(x+1) + 3(x-1)$$

$$-x^2 + 1 = 5x + 5 + 3x - 3$$

$$-x^2 + 1 = 8x + 2$$

$$0 = x^2 + 8x + 1$$

$$x = \frac{-8 \pm \sqrt{64 - 4(1)(1)}}{2}$$

$$= \frac{-8 \pm \sqrt{60}}{2}$$

$$= \frac{-8 \pm \sqrt{4 \cdot 15}}{2}$$

$$= \frac{-8 \pm 2\sqrt{15}}{2}$$

$$= -4 \pm \sqrt{15} \approx -0.13 \text{ or } -7.87$$

The solutions for $f(x) = -1$ are $x = -4 \pm \sqrt{15}$, or $x \approx -0.13$ and $x \approx -7.87$.

49.

$$0 = \frac{x-1}{x-5} - \frac{x+2}{x+3}$$

$$(x-5)(x+3) \cdot 0 = (x-5)(x+3)\left(\frac{x-1}{x-5} - \frac{x+2}{x+3} \right)$$

$$0 = (x+3)(x-1) - (x-5)(x+2)$$

$$0 = x^2 + 2x - 3 - (x^2 - 3x - 10)$$

$$0 = x^2 + 2x - 3 - x^2 + 3x + 10$$

$$0 = 5x + 7$$

$$-7 = 5x$$

$$-\frac{7}{5} = x$$

The x-intercept is $\left(-\dfrac{7}{5}, 0 \right)$.

51.

$$F = \frac{mv^2}{r}$$

$$r \cdot F = \frac{mv^2}{r} \cdot r$$

$$\frac{Fr}{F} = \frac{mv^2}{F}$$

$$r = \frac{mv^2}{F}$$

53.
$$F = \frac{-GMm}{r^2}$$
$$r^2 \cdot F = \frac{-GMm}{r^2} \cdot r^2$$
$$\frac{r^2 F}{-Gm} = \frac{-GMm}{-Gm}$$
$$M = -\frac{r^2 F}{Gm}$$

55.
$$P = \frac{A}{1+rt}$$
$$(1+rt) \cdot P = \left(\frac{A}{1+rt}\right) \cdot (1+rt)$$
$$P + Prt = A$$
$$\frac{Prt}{Pr} = \frac{A-P}{Pr}$$
$$t = \frac{A-P}{Pr}$$

57.
$$P(t) = \frac{0.48t^2 + 14t + 295}{0.015t^2 + 0.055t + 6.05}$$
$$50 = \frac{0.48t^2 + 14t + 295}{0.015t^2 + 0.055t + 6.05}$$
$$50\left(0.015t^2 + 0.055t + 6.05\right) = 0.48t^2 + 14t + 295$$
$$0.75t^2 + 2.75t + 302.5 = 0.48t^2 + 14t + 295$$
$$0.27t^2 - 11.25t + 7.5 = 0$$
$$t = \frac{-(-11.25) \pm \sqrt{(-11.25)^2 - 4(0.27)(7.5)}}{2(0.27)} \approx \frac{11.25 \pm 10.88}{0.54}$$

$t \approx 0.69$ or $t \approx 40.98$
Only the second solution is in the future, so discard the first solution.
Exactly half of voter-eligible Latinos will be registered to vote in $1980 + 41 = 2021$.

---

59. a. Using the linear regression feature of a graphing calculator, we get the equation:
$E(t) = 0.13t + 1.14$.

b. Using the linear regression feature of a graphing calculator, we get the equation:
$C(t) = 0.59t + 4.47$.

c. $P(t) = \dfrac{E(t)}{C(t)} \cdot 100$
$$= \frac{0.13t + 1.14}{0.59t + 4.47} \cdot \frac{100}{1}$$
$$= \frac{13t + 114}{0.59t + 4.47}$$

d. $P(17) = \dfrac{13(17) + 114}{0.59(17) + 4.47} \approx 23.10$

This means that in 2017, 23.1% of the total cost of health insurance will be paid by employees.

e. $\dfrac{13t + 114}{0.59t + 4.47} = 23$
$$13t + 114 = 23(0.59t + 4.47)$$
$$13t + 114 = 13.57t + 102.81$$
$$-0.57t = -11.19$$
$$t \approx 19.63$$
This means that employees will pay 23% of the total cost of health insurance in 2020.

f.
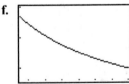

For the values of t between 7 and 20, P is decreasing.
This means that the percentage of the total cost of health insurance that employees pay is decreasing. This is possible because the rate of change of the amount of money employers pay for employees' health insurance is greater than the rate of change of the amount of money employees pay for health insurance.

61. The graphs of $y = \dfrac{5}{x-2}$ and $y = x^2 - 6x + 10$ intersect at approximately $(4.2,\ 2.3)$. The solution of the equation is $x \approx 4.2$.

63. The graphs of $y = \dfrac{5}{x-2}$ and $y = 4$ intersect at approximately $(3.3,\ 4)$. The solution of the equation is $x \approx 3.3$.

65. The graphs of $y = \dfrac{5}{x-2}$ and $y = -x^2 + x - 1$ intersect at approximately $(-0.6,\ -1.9)$. The solution of the system is approximately $(-0.6,\ -1.9)$.

67. $2 = \dfrac{0+a}{0+b}$ and $\dfrac{5}{2} = \dfrac{1+a}{1+b}$

$2 = \dfrac{a}{b}$ $\qquad 2(1+b)\dfrac{5}{2} = 2(1+b)\dfrac{1+a}{1+b}$

$2b = a$ $\qquad 5(1+b) = 2(1+a)$

By substituting $2b$ for a in the second equation we get:

$5(1+b) = 2(1+2b)$

$5 + 5b = 2 + 4b$

$b = -3$

By substituting -3 in the first equation for b we get:

$2 = \dfrac{a}{-3}$

$-6 = a$

The value for a is -6, and the value for b is -3, so $f(x) = \dfrac{x-6}{x-3}$.

69. A more efficient way would be to start by subtracting $\dfrac{4}{x+2}$ from both sides of the equation. Then, cross multiply:

$\dfrac{4}{x+2} - \dfrac{2}{x} = \dfrac{1}{x+2}$

$\dfrac{\cancel{4}}{\cancel{x+2}} - \dfrac{2}{x} - \dfrac{\cancel{4}}{\cancel{x+2}} = \dfrac{1}{x+2} - \dfrac{4}{x+2}$

$\dfrac{-2}{x} = \dfrac{-3}{x+2}$

$-2(x+2) = -3x$

$-2x - 4 = -3x$

$-4 = -x$

$x = 4$

71. Answers may vary. Example:
No, you cannot multiply a rational expression by the LCD. You must multiply it by the equivalent of 1, which could be the LCD divided by itself.
Yes, you can multiply both sides of a rational equation by the LCD, because you are multiplying both sides by the same amount, which does not change the equation.

73. I would tell the student that a solution to a rational equation should be a real number, not a rational expression.

75. $\dfrac{5}{x} + \dfrac{4}{x+1} - \dfrac{3}{x} = \dfrac{5}{x}\cdot\dfrac{(x+1)}{(x+1)} + \dfrac{4}{x+1}\cdot\dfrac{x}{x} - \dfrac{3}{x}\cdot\dfrac{(x+1)}{(x+1)}$

$= \dfrac{5(x+1) + 4x - 3(x+1)}{x(x+1)}$

$= \dfrac{5x + 5 + 4x - 3x - 3}{x(x+1)}$

$= \dfrac{6x + 2}{x(x+1)}$

$= \dfrac{2(3x+1)}{x(x+1)}$

77. $\dfrac{5}{x} + \dfrac{4}{x+1} = \dfrac{3}{x}$

$x(x+1) \cdot \left(\dfrac{5}{x} + \dfrac{4}{x+1} \right) = \dfrac{3}{x} \cdot x(x+1)$

$5(x+1) + 4x = 3(x+1)$

$5x + 5 + 4x = 3x + 3$

$6x = -2$

$x = -\dfrac{1}{3}$

The solution is $-\dfrac{1}{3}$.

79.

$$\frac{x+2}{x^2-5x+6} - \frac{x+1}{x^2-4} = \frac{4}{x^2-x-6}$$

$$\frac{x+2}{(x-2)(x-3)} - \frac{x+1}{(x-2)(x+2)} = \frac{4}{(x-3)(x+2)}$$

$$(x-2)(x+2)(x-3) \cdot \left(\frac{x+2}{(x-2)(x-3)} - \frac{x+1}{(x-2)(x+2)}\right) = \frac{4(x-2)(x+2)(x-3)}{(x-3)(x+2)}$$

$$(x+2)^2 - (x+1)(x-3) = 4(x-2)$$

$$x^2+4x+4 - \left(x^2-2x-3\right) = 4x-8$$

$$6x+7 = 4x-8$$

$$2x = -15$$

$$x = -\frac{15}{2}$$

The solution is $-\dfrac{15}{2}$.

81. $\dfrac{x+2}{x^2-5x+6} - \dfrac{x+1}{x^2-4} + \dfrac{4}{x^2-x-6}$

$$= \frac{x+2}{(x-2)(x-3)} - \frac{x+1}{(x-2)(x+2)} + \frac{4}{(x-3)(x+2)}$$

$$= \frac{x+2}{(x-2)(x-3)} \cdot \frac{(x+2)}{(x+2)} - \frac{x+1}{(x-2)(x+2)} \cdot \frac{(x-3)}{(x-3)} + \frac{4}{(x-3)(x+2)} \cdot \frac{(x-2)}{(x-2)}$$

$$= \frac{(x+2)^2 - (x+1)(x-3) + 4(x-2)}{(x-2)(x-3)(x+2)}$$

$$= \frac{x^2+4x+4 - \left(x^2-2x-3\right) + 4x-8}{(x-2)(x-3)(x+2)}$$

$$= \frac{10x-1}{(x-2)(x-3)(x+2)}$$

83.

$$2p^3 - p^2 = 8p - 4$$

$$2p^3 - p^2 - 8p + 4 = 0$$

$$2p^3 - 8p - p^2 + 4 = 0$$

$$2p\left(p^2-4\right) - 1 \cdot \left(p^2-4\right) = 0$$

$$\left(p^2-4\right)(2p-1) = 0$$

$$(p-2)(p+2)(2p-1) = 0$$

$$p = 2 \ \text{ or } \ p = -2 \ \text{ or } \ p = \frac{1}{2}$$

85. $2(4)^x + 3 = 106$

$$2(4)^x = 103$$

$$(4)^x = 51.5$$

$$x = \log_4 51.5 = \frac{\log_{10}(51.5)}{\log_{10}(4)} \approx 2.8433$$

87. $\log_3(5x-4) - \log_3(2x-3) = 2$

$$\log_3\left(\frac{5x-4}{2x-3}\right) = 2$$

$$3^2 = \frac{5x-4}{2x-3}$$

$$9 \cdot (2x-3) = \frac{5x-4}{2x-3} \cdot (2x-3)$$

$$18x - 27 = 5x - 4$$

$$13x = 23$$

$$x = \frac{23}{13}$$

89.

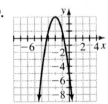

This is a quadratic function with vertex at $(-2, 3)$.

91.
$$-2(x+2)^2 + 3 = -15$$
$$-2(x+2)^2 = -18$$
$$(x+2)^2 = 9$$
$$\sqrt{(x+2)^2} = \sqrt{9}$$
$$|x+2| = 3$$
$$x+2 = 3 \quad \text{or} \quad x+2 = -3$$
$$x = 1 \quad \text{or} \qquad x = -5$$

This is a quadratic equation in one variable with solutions $x = 1$ and $x = -5$.

93.
$$-2(x+2)^2 + 3 = -2(x^2 + 4x + 4) + 3$$
$$= -2x^2 - 8x - 8 + 3$$
$$= -2x^2 - 8x - 5$$

This is a quadratic expression in one variable.

Homework 8.6

1. a. $C(n) = 1250 + 350n$

b. $M(n) = \dfrac{1250 + 350n}{n}$

c. $M(30) = \dfrac{1250 + 350(30)}{30} \approx \$391.67$

d.
$$400 = \dfrac{1250 + 350n}{n}$$
$$400n = 1250 + 350n$$
$$50n = 1250$$
$$n = 25$$

The minimum number of students needed to go on the trip is 25.

3. a. $T(n) = 500 + 50n$

b. $M(n) = \dfrac{500 + 50n}{n}$

c. $M(270) = \dfrac{500 + 50(270)}{270} \approx \$51.85$

If 270 people attend the reunion, the mean cost per person is $51.85.

d.
$$60 = \dfrac{500 + 50n}{n}$$
$$60n = 500 + 50n$$
$$10n = 500$$
$$n = 50$$

For the mean cost per person to be $60, 50 people would have to attend the reunion.

e.

f. As n gets very large, $M(n)$ decreases but never drops below 50. This makes sense in terms of the restaurant fees because each person pays $50 plus an equal share of $500 – the cost for the band. So the more people who attend (i.e., as n gets larger), the closer the mean cost per person gets to $50.

5. a. $C(n) = 90,000 + 7000n$

b. $B(n) = \dfrac{90,000 + 7000n}{n}$

c.
$$P(n) = \dfrac{90,000 + 7000n}{n} + 2000$$
$$= \dfrac{90,000 + 7000n}{n} + \dfrac{2000}{1} \cdot \dfrac{n}{n}$$
$$= \dfrac{90,000 + 7000n}{n} + \dfrac{2000n}{n}$$
$$= \dfrac{90,000 + 9000n}{n}$$

d. $P(40) = \dfrac{90,000 + 9000(40)}{40} = 11,250$

If the manufacturer produces and sells 40 cars, it would have to charge $11,250 for each car in order to make a profit of $2000 per car.

e. As the values of n get very large, $P(n)$ gets close to 9000. If a very large number of cars are produced and sold, the price per car can be just a little more than $9000 to ensure a profit of $2000 per car.

7. a. Using the linear regression feature of a graphing calculator, we get the equation:
$$I(t) = 441.53t + 3852.97$$

b. $M(t) = \dfrac{I(t)}{H(t)} = \dfrac{441.53t + 3852.97}{0.0013t + 0.093}$

c. $M(t)$ is measured in dollars per household. Since $I(t)$ is measured in billions of dollars and $H(t)$ is measured in billions of households, the units of the expressions on both sides of the equation are dollars per household.

d.
$$125,000 = \frac{441.53t + 3852.97}{0.0013t + 0.093}$$
$$125,000(0.0013t + 0.093) = 441.53t + 3852.97$$
$$162.5t + 11,625 = 441.53t + 3852.97$$
$$7772.03 = 279.03t$$
$$27.85 \approx t$$

The mean household income should reach $125,000 in the year $1990 + 28 = 2018$.

e.

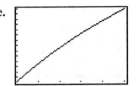

$M(t)$ is increasing over the interval $5 < t < 30$. This means that the mean annual income per household is steadily increasing between 1995 and 2020.

---

9. a. Using the linear regression feature of a graphing calculator, we get the equation $R(t) = 464.8t + 2795.2$.

b. $A(t) = \dfrac{R(t)}{E(t)} = \dfrac{464.8t + 2795.2}{0.026t^2 - 0.2t + 14.61}$

c. In the year 2018, $t = 28$, so:
$$A(28) = \frac{464.8(28) + 2795.2}{0.026(28)^2 - 0.2(28) + 14.61} \approx 537.85$$
The mean amount of money spent on books per student in the year 2018 will be about $537.85.

d.
$$525 = \frac{464.8t + 2795.2}{0.026t^2 - 0.2t + 14.61}$$
$$525\left(0.026t^2 - 0.2t + 14.61\right) = 464.8t + 2795.2$$
$$13.65t^2 - 105t + 7670.25 = 464.8t + 2795.2$$
$$13.65t^2 - 569.8t + 4875.05 = 0$$
$$t = \frac{-(-569.8) \pm \sqrt{(-569.8)^2 - 4(13.65)(4875.05)}}{2(13.65)} \approx \frac{569.8 \pm 241.9}{27.3}$$

$t \approx 12.01$ or $t \approx 29.7$
Discard the first solution, assuming model breakdown in the past.
The mean amount of money spent on books per student will reach $525 in the year $1990 + 30 = 2020$.

11. a. Using the linear regression feature of a graphing calculator, we get the equations:
$$M(t) = 16.98t + 395.66$$
$$E(t) = -29.87t + 1369.89$$

b. $(M + E)(t) = (16.98t + 395.66) + (-29.87t + 1369.89) = -12.89t + 1765.55$

The inputs of $M + E$ are the number of years since 1980, and the outputs are the total number of daily newspapers.

c. $P(t) = \left(\dfrac{M(t)}{M(t) + E(t)}\right) \cdot 100 = \dfrac{16.98t + 395.66}{-12.89t + 1765.55} \cdot \dfrac{100}{1} = \dfrac{1698t + 39,566}{-12.89t + 1765.55}$

d. $P(36) = \dfrac{1698(36) + 39{,}566}{-12.89(36) + 1765.55} \approx 77.37$

In the year 1980 + 36 = 2016, about 77.4% of dailies will be morning newspapers.

e. $\dfrac{1698t + 39{,}566}{-12.89t + 1765.55} = 80$

$1698t + 39{,}566 = 80(-12.89t + 1765.55)$

$1698t + 39{,}566 = -1031.2t + 141{,}244$

$2729.2t = 101{,}678$

$t \approx 37.26$

In the year 1980 + 37 = 2017, 80% of dailies will be morning newspapers.

13. a. Using the quadratic regression feature of a graphing calculator, we get the equations:

$W(t) = 0.31t^2 + 6.93t + 456.43$

$M(t) = 0.33t^2 - 2.96t + 482.26$

b. $(W + M)(t) = \left(0.31t^2 + 6.93t + 456.43\right) + \left(0.33t^2 - 2.96t + 482.26\right) = 0.64t^2 + 3.97t + 938.69$

The inputs of $W + M$ are the number of years since 1980, and the outputs are the total number (in thousands) of people who have earned a bachelor's degree.

c. $P(t) = \left(\dfrac{M(t)}{(W + M)(t)}\right) \cdot 100 = \dfrac{0.33t^2 - 2.96t + 482.26}{0.64t^2 + 3.97t + 938.69} \cdot \dfrac{100}{1} = \dfrac{33t^2 - 296t + 48{,}226}{0.64t^2 + 3.97t + 938.69}$

d.

$42.1 = \dfrac{33t^2 - 296t + 48{,}226}{0.64t^2 + 3.97t + 938.69}$

$42.1\left(0.64t^2 + 3.97t + 938.69\right) = 33t^2 - 296t + 48{,}226$

$26.944t^2 + 167.137t + 39{,}518.849 = 33t^2 - 296t + 48{,}226$

$0 = 6.06t^2 - 463.14t + 8707.15$

$t = \dfrac{-(-463.14) \pm \sqrt{(-463.14)^2 - 4(6.06)(8707.15)}}{2(6.06)}$

$t \approx \dfrac{463.14 \pm 58.63}{12.12}$

$t \approx 33.38 \text{ or } t \approx 43.05$

The possibility of model breakdown makes the larger solution unrealistic. Therefore, in the year 1980 + 33 = 2013, 42.1% of the people who earned bachelor's degrees were men.

e.

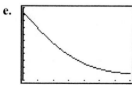

The graph of P is decreasing over the interval $0 < t < 35$. The number of bachelor's degrees earned by women has been increasing at a faster rate compared to men, which accounts for the decrease in the percentage of bachelor's degrees earned by men.

15.   Using $t = \dfrac{d}{s}$ gives the following:

$$t = \frac{85}{60} \approx 1.4 \text{ hours}$$

17. a. $T(a) = \dfrac{85}{a+70} + \dfrac{53}{a+65} = \dfrac{138a+9235}{(a+70)(a+65)}$

b. $T(5) = \dfrac{85}{5+70} + \dfrac{53}{5+65} = \dfrac{85}{75} + \dfrac{53}{70} \approx 1.89$

The driving time will be about 1.89 hours.

c.

$$1.8 = \frac{138a+9235}{(a+70)(a+65)}$$

$$1.8(a+70)(a+65) = 138a+9235$$

$$1.8\left(a^2+135a+4550\right) = 138a+9235$$

$$1.8a^2+243a+8190 = 138a+9235$$

$$1.8a^2+105a-1045 = 0$$

$$a = \frac{-105 \pm \sqrt{105^2 - 4(1.8)(-1045)}}{2(1.8)}$$

$$a = \frac{-105 \pm \sqrt{18{,}549}}{3.6}$$

$$a = \frac{-105 \pm \sqrt{18{,}549}}{3.6}$$

$$a = \frac{-105 - \sqrt{18{,}549}}{3.6} \quad \text{or} \quad a = \frac{-105 + \sqrt{18{,}549}}{3.6}$$

$$a \approx -67.00 \quad\quad \text{or} \quad a \approx 8.67$$

The value $a = -67.00$ represents driving under the speed limits by 67.00 mph—model breakdown has occurred. So, the student would have to drive about 8.67 mph over the speed limits for the driving time to be 1.8 hours.

19. a. $T(a) = \dfrac{83}{a+70} + \dfrac{37}{a+65}$

b. $T(0) = \dfrac{83}{0+70} + \dfrac{37}{0+65} = \dfrac{83}{70} + \dfrac{37}{65} \approx 1.19 + 0.57 \approx 1.76 \text{ hours}$

$T(10) = \dfrac{83}{10+70} + \dfrac{37}{10+65} = \dfrac{83}{80} + \dfrac{37}{75} \approx 1.04 + 0.49 \approx 1.53 \text{ hours}$

This means that if the student drives the speed limit, her driving time will be about 1.8 hours. If she drives 10 mph over the speed limit, her driving time will be about 1.5 hours.

c. $T(0) - T(10) = 0.23$

This shows the amount of time she will save if she drives 10 mph over the speed limit.

d.
$$1.6 = \frac{83}{a+70} + \frac{37}{a+65}$$

$$1.6(a+70)(a+65) = \left(\frac{83}{a+70} + \frac{37}{a+65}\right)(a+70)(a+65)$$

$$1.6a^2 + 216a + 7280 = 83a + 5395 + 37a + 2590$$

$$1.6a^2 + 96a - 705 = 0$$

$$a = \frac{-96 \pm \sqrt{(96)^2 - 4(1.6)(-705)}}{2(1.6)} = \frac{-96 \pm \sqrt{13728}}{3.2}$$

The only solution that makes sense in this context for exceeding the speed limit is the positive

value $\dfrac{-96 + \sqrt{13728}}{3.2} \approx 6.61$ miles per hour. This means that the student's driving time will be 1.6 hours if

she exceeds the speed limit by 6.61 miles per hour.

---

21. a. $T(a) = \dfrac{164}{a+70} + \dfrac{121}{a+65}$

$$= \left(\frac{164}{a+70}\right)\left(\frac{a+65}{a+65}\right) + \left(\frac{121}{a+65}\right)\left(\frac{a+70}{a+70}\right)$$

$$= \frac{164a+10,660}{(a+70)(a+65)} + \frac{121a+8470}{(a+70)(a+65)}$$

$$= \frac{285a+19,130}{(a+70)(a+65)}$$

b. $\dfrac{285(10)+19,130}{(10+70)(10+65)} = \dfrac{2850+19,130}{(80)(75)}$

$$= \frac{21,980}{6000}$$

$$\approx 3.66 \text{ hours}$$

23. a. Answers may vary. Example:
If all the salaries of a group of people are relatively close together, the mean salary will estimate the salary of the group well.

b. Answers may vary. Example:
If a few salaries of a group of people are very different from the rest of the group, the mean salary will not estimate the salary of the group well.

25. a. Linear:
$L(t) = 0.525t - 12.22$

Exponential:
$E(t) = 0.433(1.056)^t$

Quadratic:
$Q(t) = 0.01t^2 - 0.47t + 6.1$

The percentage of Americans who have shingles does not increase at a constant rate from year to year. So, the linear model would be the worst fit. The quadratic or exponential models provide a better fit of the data.

b. There is model breakdown for the linear model, because the model takes on negative values for ages less than 23 years. There is model breakdown also for the quadratic model, because it shows a decrease in the percent of Americans who have shingles, up to age 23. The exponential model is the best fit of the data.

c.
$$25 = 0.433(1.056)^t$$

$$\frac{25}{0.433} = (1.056)^t$$

$$\log\left(\frac{25}{0.433}\right) \approx t \log(1.056)$$

$$t \approx \frac{\log\left(\dfrac{25}{0.433}\right)}{\log(1.056)}$$

$$t \approx 74.44$$

So, 25% of people of age 74 have shingles.

d. The base of the exponential model is 1.056, which means that the number of Americans who have shingles increases by about 6% for each additional year that a person lives.

e. $f(t) = 0.433(1.056)^{85} = 44.46$

The model predicts that about 44.5% of Americans age 85 will have shingles. If the new vaccine reduces the number of shingles cases by 51%, then only about $44.46(1-0.51) = 21.79\%$ of people age 85 will have shingles.

27. $75x^3 - 50x^2 - 12x + 8$

$= 25x^2(3x-2) - 4(3x-2)$

$= (3x-2)(25x^2 - 4)$

$= (3x-2)(5x-2)(5x+2)$

This is a third-degree polynomial (cubic) expression in one variable.

29.
$$75x^3 - 50x^2 = 12x - 8$$
$$75x^3 - 50x^2 - 12x + 8 = 0$$
$$25x^2(3x-2) - 4(3x-2) = 0$$
$$(3x-2)(25x^2 - 4) = 0$$
$$(3x-2)(5x-2)(5x+2) = 0$$
$$x = \frac{2}{3} \text{ or } x = \frac{2}{5} \text{ or } x = -\frac{2}{5}$$

This is a third-degree polynomial (cubic) equation in one variable.

31. $(x^2 + 2x - 3)(3x^2 - x - 4)$

$= 3x^2(x^2 + 2x - 3) - x(x^2 + 2x - 3) -$

$\quad 4(x^2 + 2x - 3)$

$= 3x^4 + 6x^3 - 9x^2 - x^3 - 2x^2 + 3x -$

$\quad 4x^2 - 8x + 12$

$= 3x^4 + 5x^3 - 15x^2 - 5x + 12$

This is a fourth-degree polynomial expression in one variable.

Homework 8.7

1. $I = kt$

3. $w = \dfrac{k}{x+4}$

5. w varies inversely as r.

7. T varies directly as the square root of w.

9. $c = ku$

To find k, solve the following equation:

$12 = k(3)$

$4 = k$

$c = 4u$

11. $w = \dfrac{k}{\sqrt{t}}$

To find k, solve the following equation:

$3 = \dfrac{k}{\sqrt{16}}$

$3 = \dfrac{k}{4}$

$12 = k$

$w = \dfrac{12}{\sqrt{t}}$

13. $y = kx$

To find k, solve the following equation:

$12 = k(4)$

$3 = k$

$y = 3x$

Substitute 9 for x and solve for y to get the required value:

$y = 3(9) = 27$

15. $G = \dfrac{k}{r}$

To find k, solve the following equation:

$G = \dfrac{k}{r}$

$8 = \dfrac{k}{3}$

$24 = k$

$G = \dfrac{24}{r}$

Substitute 4 for r and solve for G to get the required value:

$G = \dfrac{24}{r} = \dfrac{24}{4} = 6$

17. $p = kx^2$

To find k, solve the following equation:

$6 = k(2)^2$

$6 = 4k$

$\dfrac{6}{4} = k$

$\dfrac{3}{2} = k$

$p = \dfrac{3}{2}x^2$

Substitute 24 for p and solve for x to get the required value:

$p = \dfrac{3}{2}x^2$

$24 = \dfrac{3}{2}x^2$

$48 = 3x^2$

$16 = x^2$

$x = \pm 4$

19. $I = \dfrac{k}{r+2}$

To find k, solve the following equation:

$I = \dfrac{k}{r+2}$

$9 = \dfrac{k}{3+2}$

$9 = \dfrac{k}{5}$

$45 = k$

$I = \dfrac{45}{r+2}$

Substitute 7 for I and solve for r to get the required value.

$I = \dfrac{45}{r+2}$

$7 = \dfrac{45}{r+2}$

$7(r+2) = 45$

$7r + 14 = 45$

$7r = 31$

$r = \dfrac{31}{7}$

21. As the value of w increases, the value of B will also increase for the given situation.

23. As the value of p increases, the value of w will decrease for the given situation.

25. If the GDP rises, the transaction demand will also rise.

27. A tall person will have less nerve conduction than a short person.

29. Bernice Fitz-Gibbon meant that as the number of cooks increase, the amount of creativity decreases.

31. Let c be the cost of tuition and h be the number of credit hours a student takes.

$c = kh$

Let $c = 1680$ and $h = 15$.

$1680 = k(15)$

$112 = k$

Therefore, $c = 112h$.

$c = 112(12) = \$1344$

33. Let t be the tension in the string and r be the radius of the circle.

$t = \dfrac{k}{r}$

Let $t = 80$ and $r = 60$.

$80 = \dfrac{k}{60}$

$4800 = k$

Therefore, $t = \dfrac{4800}{r}$.

$t = \dfrac{4800}{50} = 96$ newtons

35. Let d be the distance an object falls and t be the time in motion.

$d = kt^2$

Let $d = 144.9$ and $t = 3$.

$144.9 = k \cdot 3^2$

$144.9 = 9k$

$16.1 = k$

Therefore, $d = 16.1t^2$.

$d = 16.1(3.4)^2 = 186.116$ feet

37. Let i be the intensity of radiation and d be the distance from the machine.

$i = \dfrac{k}{d^2}$

Let $i = 90$ and $d = 2.5$.

$90 = \dfrac{k}{2.5^2}$

$90 = \dfrac{k}{6.25}$

$562.5 = k$

Therefore, $i = \dfrac{562.5}{d^2}$.

$$45 = \dfrac{562.5}{d^2}$$

$$45d^2 = 562.5$$

$$d^2 = 12.5$$

$$d \approx 3.54 \text{ meters}$$

39. a. $F = kw$

To find k solve the following equation:

$$50 = k(120)$$

$$\dfrac{5}{12} = k$$

The equation is $F = \dfrac{5}{12}w$.

b. $F = \dfrac{5}{12} \cdot 150 = 62.5$ pounds

c. Answers may vary. Example:
The carpet increases resistance to the applied force. So k would be larger for a carpeted floor than a wood floor.

41. a. $T = kd$

Let $T = 3$ and $d = 3313$.

$$3 = k(3313)$$

$$0.000906 \approx k$$

Therefore, $T \approx 0.000906d$.

b. $4 = 0.000906d$

$$d \approx 4415 \text{ feet}$$

c. For every additional foot away the lightning strikes, it takes another 0.000906 seconds to hear the thunder.

d. Answers may vary. Example:
The equation indicates that thunder travels 1 mile in about 4.78 seconds. A better rule of thumb is that, for every five seconds after you see the lightning, the thunder travels a mile further.

43. a. $w = f(d) = \dfrac{k}{d^2}$

$$200 = \dfrac{k}{4^2}$$

$$200 = \dfrac{k}{16}$$

$$3200 = k$$

Therefore, $w = f(d) = \dfrac{3200}{d^2}$.

b. If sea level is about 4 thousand miles from the center of the Earth, then 1 thousand miles above the surface would be a total of 5 thousand miles from the center of the Earth.

$$w = \dfrac{3200}{5^2} = 128 \text{ pounds}$$

c. $1 = \dfrac{3200}{d^2}$

$$d^2 = 3200$$

$$d = \sqrt{3200} \approx 56.569$$

An astronaut would weigh 1 pound at a distance of about 56.569 thousand miles, or 56,569 miles, from the center of Earth.

d. $f(239) = \dfrac{3200}{239^2} \approx 0.056$ pounds

Model breakdown as occurred. At the moon's surface, the gravitational effect of the moon is greater than that of Earth. This needs to be taken into account for the information to be accurate.

e. Answers may vary. Example:
The equation would have a slightly smaller value for k, because k is directly proportional to the astronaut's weight, and the distance is the same in both situations.

45. a.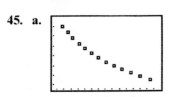

b. The model should be of the form

$$F(L) = \frac{k}{L}.$$

To find k take the average of all products LF from the table.
The sum of the products is 36,467.715.
Divide this sum by 13 (the total number of notes in the table) to get the average.

$$\frac{36,467.715}{13} \approx 2805.21$$

So $F(L) = \frac{2805.21}{L}$.

The graph below shows the equation graphed along with the scattergram of the data. By inspection, it appears the model fits the data extremely well.

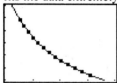

c. F varies inversely as L.

d. $F(7.58) = \dfrac{2805.21}{7.58} \approx 370.1$ hertz

e.
$$F = \frac{2805.21}{a} \qquad (1)$$

$$F = \frac{2805.21}{\frac{1}{2}a} \qquad (2)$$

$$F = 2805.21 \div \frac{1}{2}a$$

$$= 2805.21 \cdot \frac{2}{a}$$

$$= 2\left(\frac{2805.21}{a}\right)$$

Equation (2) is 2 times equation (1). So when we halved the effective length the frequency doubled.

---

47. a.

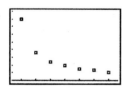

b. The model should be of the form $f(d) = \dfrac{k}{d}$. To find k take the average of all products dh from the table.

$$k = \frac{10 \cdot 16 + 20 \cdot 7.3 + 30 \cdot 4.8 + 40 \cdot 3.8 + 50 \cdot 3 + 60 \cdot 2.5 + 70 \cdot 2}{7} \approx 148.86$$

So $f(d) = \dfrac{148.86}{d}$

c. The apparent height varies inversely with the distance.

d. This makes sense in this case because the farther you are from the garage (i.e. the bigger d is) the smaller the apparent height $f(d)$ is.

e. $f(100) = \dfrac{148.86}{100} = 1.4886$ inches

f. $f(1) = \dfrac{148.86}{1} = 148.86$ inches

49. a. $T = k\sqrt{L}$

b. $\dfrac{T}{\sqrt{L}} = k$

c.

L	T	$\dfrac{T}{\sqrt{L}}$
5.0	0.50	0.22
10.0	0.63	0.20
15.0	0.88	0.23
20.0	1.00	0.22
25.0	1.13	0.23
32.5	1.25	0.22
45.0	1.50	0.22
60.0	1.75	0.23
85.0	2.00	0.22
110.0	2.25	0.21

A reasonable value for k is the average of all values in column 3, which is about 0.22.

d. $T = 0.22\sqrt{L}$

e.

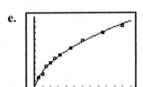

The model fits the data very well

f. $T = 0.22\sqrt{130} \approx 2.51$ seconds

51. $f(L) = 5L$
$k = 5$

53. $f(n) = \dfrac{2}{n}$
$k = 2$

55. $f(r) = 2\pi r$
$k = 2\pi$

57. a. f is an increasing function.
The number of albums sold increases as more money is spent on advertising.

b. The number of albums sold does not vary directly as the amount of money spent on advertising. "Varies directly" usually implies a linear relationship. A linear model would not fit this data well.

c. Answers may vary. Example:
The first statement indicates a direct, increasing linear relationship between the variables, while the second statement could refer to an increasing logarithmic, exponential, or quadratic function as well as an increasing linear function.

59. False. As a person gets older, the person doesn't necessarily keep getting taller. At some point everyone stops growing.

61. False. Coffee will get cooler as time passes, but only until it reaches room temperature.

63. $y = kx$
Solving this equation for x we get:
$\dfrac{y}{k} = x$
Yes, it follows that x varies directly as y. The variation constant is $\dfrac{1}{k}$.

65. The first student found an equation to convert kilometers (x) into miles (y). The second student found an equation to convert miles (x) into kilometers (y). One mile is approximately 1.61 kilometers and 1 kilometer is approximately 0.62 mile.

67. a. If y varies directly as x, x and y are linearly related. This relationship takes the form of $y = kx$, which is a linear relationship.

b. No. Just because w and t are linearly related doesn't mean they are directly related. The relationship may be either $w = kt$ or $w = \dfrac{k}{t}$.

69. $n = kt$
$310 = k(5)$
$62 = k$
$n = 62t$
The slope of the model is 62. This indicates that the typist can type 62 words per minute.

71. Answers may vary. Example:
$x + y = 2$
$x - y = 0$
$y = x$
$x + y = 2$
$x = 1$
$y = 1$

73. Answers may vary. Example:
$$\frac{x-2}{2x+2} - \frac{2x}{2x+2} = \frac{x-2-2x}{2x+2} = \frac{-2-x}{2x+2}$$

75. Answers may vary. Example:
$y = x^2$

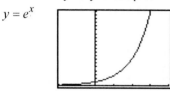

77. Answers may vary. Example:
$y = e^x$

Chapter 8 Review

1. $f(x) = \dfrac{5x-3}{2x^2 - 3x + 1}$

$f(0) = \dfrac{5(0)-3}{2(0)^2 - 3(0) + 1} = -3$

$f(2) = \dfrac{5(2)-3}{2(2)^2 - 3(2) + 1} = \dfrac{7}{3}$

2. The function will be defined for all x except where the denominator is 0.
$$4x^2 - 49 = 0$$
$$(2x - 7)(2x + 7) = 0$$
$$2x - 7 = 0 \quad \text{or} \quad 2x + 7 = 0$$
$$2x = 7 \quad \text{or} \quad 2x = -7$$
$$x = \frac{7}{2} \quad \text{or} \quad x = -\frac{7}{2}$$
The domain is the set of all real numbers except $\dfrac{7}{2}$ and $-\dfrac{7}{2}$.

3. The function will be defined for all x except where the denominator is 0.
$$12x^2 + 13x - 35 = 0$$
$$x = \frac{-13 \pm \sqrt{169 - 4(12)(-35)}}{24}$$
$$= \frac{-13 \pm \sqrt{169 + 1680}}{24}$$
$$= \frac{-13 \pm \sqrt{1849}}{24}$$
$$= \frac{-13 \pm 43}{24}$$

The solutions are $\dfrac{-13+43}{24} = \dfrac{5}{4}$ and $\dfrac{-13-43}{24} = -\dfrac{7}{3}$. The domain is the set of all real numbers except $\dfrac{5}{4}$ and $-\dfrac{7}{3}$.

4. The function will be defined for all x except where the denominator is 0.
$$9x^3 + 18x^2 - x - 2 = 0$$
$$9x^2(x + 2) - (x + 2) = 0$$
$$\left(9x^2 - 1\right)(x + 2) = 0$$
$$(3x + 1)(3x - 1)(x + 2) = 0$$
$$3x + 1 = 0 \quad \text{or} \quad 3x - 1 = 0 \quad \text{or} \quad x + 2 = 0$$
$$3x = -1 \quad \text{or} \quad 3x = 1 \quad \text{or} \quad x = -2$$
$$x = -\frac{1}{3} \quad \text{or} \quad x = \frac{1}{3} \quad \text{or} \quad x = -2$$
The domain is the set of all real numbers except $-\dfrac{1}{3}, \dfrac{1}{3},$ and -2.

5. $\dfrac{3x-12}{x^2 - 6x + 8} = \dfrac{3(x-4)}{(x-2)(x-4)} = \dfrac{3}{x-2}$

6. $\dfrac{16 - x^2}{2x^3 - 16x^2 + 32x} = \dfrac{(4+x)(4-x)}{2x(x^2 - 8x + 16)}$
$$= \dfrac{(4+x)(4-x)}{2x(x-4)(x-4)}$$
$$= \dfrac{-x-4}{2x(x-4)} \quad \text{or} \quad -\dfrac{x+4}{2x(x-4)}$$

7. $\dfrac{x+2}{x^3 + 8} = \dfrac{x+2}{(x+2)(x^2 - 2x + 4)} = \dfrac{1}{(x^2 - 2x + 4)}$

8. $\dfrac{6a^2 - 17ab + 5b^2}{3a^2 - 4ab + b^2} = \dfrac{(3a-b)(2a-5b)}{(3a-b)(a-b)} = \dfrac{2a-5b}{a-b}$

9. $\dfrac{f(x)}{g(x)} = \dfrac{x^2 + 3x - 28}{x^3 - x^2 - 12x}$
$$= \dfrac{(x+7)(x-4)}{x(x^2 - x - 12)}$$
$$= \dfrac{(x+7)(x-4)}{x(x-4)(x+3)}$$
$$= \dfrac{(x+7)}{x(x+3)}$$

$\dfrac{f(-2)}{g(-2)} = \dfrac{(-2+7)}{-2(-2+3)} = -\dfrac{5}{2}$

10. $\dfrac{3x+6}{2x-4} \cdot \dfrac{5x-10}{6x+12} = \dfrac{3(x+2)}{2(x-2)} \cdot \dfrac{5(x-2)}{6(x+2)} = \dfrac{15}{12} = \dfrac{5}{4}$

11. $\dfrac{x^2-49}{9-x^2} \cdot \dfrac{2x^3+8x^2-42x}{5x-35}$

$\qquad = \dfrac{(x-7)(x+7)}{-(x^2-9)} \cdot \dfrac{2x(x^2+4x-21)}{5(x-7)}$

$\qquad = \dfrac{(x-7)(x+7)}{-(x+3)(x-3)} \cdot \dfrac{2x(x+7)(x-3)}{5(x-7)}$

$\qquad = -\dfrac{2x(x+7)^2}{5(x+3)}$

12. $\dfrac{p^3-t^3}{p^2-t^2} \cdot \dfrac{p^2+6pt+5t^2}{p^2t+pt^2+t^3}$

$\qquad = \dfrac{(p-t)(p^2+pt+t^2)}{(p-t)(p+t)} \cdot \dfrac{(p+5t)(p+t)}{t(p^2+pt+t^2)}$

$\qquad = \dfrac{p+5t}{t}$

13. $\dfrac{x^2-4}{x^2+3x+2} \div \dfrac{4x^2-24x+32}{x^2-5x+4}$

$\qquad = \dfrac{x^2-4}{x^2+3x+2} \cdot \dfrac{x^2-5x+4}{4x^2-24x+32}$

$\qquad = \dfrac{(x-2)(x+2)}{(x+2)(x+1)} \cdot \dfrac{(x-4)(x-1)}{4(x^2-6x+8)}$

$\qquad = \dfrac{(x-2)(x+2)}{(x+2)(x+1)} \cdot \dfrac{(x-4)(x-1)}{4(x-4)(x-2)}$

$\qquad = \dfrac{(x-2)}{(x+1)} \cdot \dfrac{(x-1)}{4(x-2)}$

$\qquad = \dfrac{x-1}{4(x+1)}$

14. $\dfrac{4-x}{4x} \div \dfrac{16-x^2}{16x^2} = \dfrac{4-x}{4x} \cdot \dfrac{16x^2}{16-x^2}$

$\qquad = \dfrac{4-x}{4x} \cdot \dfrac{16x^2}{(4-x)(4+x)}$

$\qquad = \dfrac{4x}{4+x}$

15. $\dfrac{8x^3+4x^2-18x-9}{x^2-6x+9} \div \dfrac{4x^2+8x+3}{x^2-9}$

$\qquad = \dfrac{8x^3+4x^2-18x-9}{x^2-6x+9} \cdot \dfrac{x^2-9}{4x^2+8x+3}$

$\qquad = \dfrac{4x^2(2x+1)-9(2x+1)}{(x-3)(x-3)} \cdot \dfrac{(x-3)(x+3)}{(2x+1)(2x+3)}$

$\qquad = \dfrac{(2x+1)(4x^2-9)}{(x-3)(x-3)} \cdot \dfrac{(x-3)(x+3)}{(2x+1)(2x+3)}$

$\qquad = \dfrac{(2x+1)(2x-3)(2x+3)}{(x-3)(x-3)} \cdot \dfrac{(x-3)(x+3)}{(2x+1)(2x+3)}$

$\qquad = \dfrac{(2x-3)(x+3)}{x-3}$

16. $\dfrac{w}{w^2-5w+6} + \dfrac{3}{3-w}$

$\qquad = \dfrac{w}{(w-3)(w-2)} - \dfrac{3}{w-3}$

$\qquad = \dfrac{w}{(w-3)(w-2)} - \dfrac{3}{(w-3)} \cdot \dfrac{(w-2)}{(w-2)}$

$\qquad = \dfrac{w-3w+6}{(w-3)(w-2)}$

$\qquad = \dfrac{-2w+6}{(w-3)(w-2)}$

$\qquad = \dfrac{-2(w-3)}{(w-3)(w-2)}$

$\qquad = -\dfrac{2}{w-2}$

17.

$$\frac{x}{2x^3-3x^2-5x}+\frac{2}{x^3-x}=\frac{x}{x(2x^2-3x-5)}+\frac{2}{x(x^2-1)}$$

$$=\frac{x}{x(2x-5)(x+1)}+\frac{2}{x(x-1)(x+1)}$$

$$=\frac{x}{x(2x-5)(x+1)}\cdot\frac{(x-1)}{(x-1)}+\frac{2}{x(x-1)(x+1)}\cdot\frac{(2x-5)}{(2x-5)}$$

$$=\frac{x^2-x}{x(2x-5)(x+1)(x-1)}+\frac{4x-10}{x(2x-5)(x+1)(x-1)}$$

$$=\frac{x^2+3x-10}{x(2x-5)(x+1)(x-1)}$$

$$=\frac{(x-2)(x+5)}{x(2x-5)(x+1)(x-1)}$$

18.

$$\frac{x-1}{x^2-4}+\frac{x+3}{x^2-4x+4}=\frac{x-1}{(x-2)(x+2)}+\frac{x+3}{(x-2)(x-2)}$$

$$=\frac{(x-1)(x-2)}{(x-2)(x-2)(x+2)}+\frac{(x+2)(x+3)}{(x-2)(x-2)(x+2)}$$

$$=\frac{(x-1)(x-2)+(x+2)(x+3)}{(x-2)(x-2)(x+2)}$$

$$=\frac{x^2-3x+2+x^2+5x+6}{(x-2)^2(x+2)}$$

$$=\frac{2x^2+2x+8}{(x-2)^2(x+2)}$$

$$=\frac{2(x^2+x+4)}{(x-2)^2(x+2)}$$

19.

$$\frac{3}{4k-12}-\frac{k}{k^2-2k-3}=\frac{3}{4(k-3)}-\frac{k}{(k-3)(k+1)}$$

$$=\frac{3}{4(k-3)}\cdot\frac{(k+1)}{(k+1)}-\frac{k}{(k-3)(k+1)}\cdot\frac{(4)}{(4)}$$

$$=\frac{3k+3-4k}{4(k-3)(k+1)}$$

$$=\frac{-k+3}{4(k-3)(k+1)}$$

$$=\frac{-(k-3)}{4(k-3)(k+1)}$$

$$=\frac{-1}{4(k+1)}$$

20.
$$\frac{x+1}{25-x^2} - \frac{x-4}{2x^2-14x+20} = -\frac{x+1}{x^2-25} - \frac{x-4}{2x^2-14x+20}$$

$$= -\frac{x+1}{(x+5)(x-5)} - \frac{x-4}{2(x^2-7x+10)}$$

$$= -\frac{x+1}{(x+5)(x-5)} - \frac{x-4}{2(x-5)(x-2)}$$

$$= -\frac{(x+1)}{(x+5)(x-5)} \cdot \frac{2(x-2)}{2(x-2)} - \frac{(x-4)}{2(x-5)(x-2)} \cdot \frac{(x+5)}{(x+5)}$$

$$= -\frac{2(x^2-x-2)}{2(x+5)(x-5)(x-2)} - \frac{x^2+x-20}{2(x-5)(x-2)(x+5)}$$

$$= \frac{-2x^2+2x+4-x^2-x+20}{2(x+5)(x-5)(x-2)}$$

$$= \frac{-3x^2+x+24}{2(x+5)(x-5)(x-2)}$$

$$= -\frac{(x-3)(3x+8)}{2(x+5)(x-5)(x-2)}$$

21.
$$\frac{2m}{m^2-3mn-10n^2} - \frac{4n}{m^2+8mn+12n^2} = \frac{2m}{(m-5n)(m+2n)} - \frac{4n}{(m+6n)(m+2n)}$$

$$= \frac{2m(m+6n)}{(m-5n)(m+2n)(m+6n)} - \frac{4n(m-5n)}{(m+6n)(m+2n)(m-5n)}$$

$$= \frac{2m^2+12mn-4mn+20n^2}{(m+6n)(m+2n)(m-5n)}$$

$$= \frac{2m^2+8mn+20n^2}{(m+6n)(m+2n)(m-5n)}$$

$$= \frac{2\left(m^2+4mn+10n^2\right)}{(m+6n)(m+2n)(m-5n)}$$

22.
$$\frac{2}{x-5} - \left(\frac{x^2+5x+6}{3x^2-75} \div \frac{x^2+2x}{3x+15}\right) = \frac{2}{x-5} - \left(\frac{x^2+5x+6}{3x^2-75} \cdot \frac{3x+15}{x^2+2x}\right)$$

$$= \frac{2}{x-5} - \left(\frac{(x+3)(x+2)}{3(x^2-25)} \cdot \frac{3(x+5)}{x(x+2)}\right)$$

$$= \frac{2}{x-5} - \left(\frac{(x+3)(x+2)}{3(x+5)(x-5)} \cdot \frac{3(x+5)}{x(x+2)}\right)$$

$$= \frac{2}{x-5} - \frac{x+3}{x(x-5)}$$

$$= \frac{2}{x-5} \cdot \left(\frac{x}{x}\right) - \frac{x+3}{x(x-5)}$$

$$= \frac{2x-x-3}{x(x-5)}$$

$$= \frac{x-3}{x(x-5)}$$

23.
$$f(x)g(x) = \frac{x^2-x-2}{x^2+5x+6} \cdot \frac{x+3}{x+2} = \frac{(x-2)(x+1)}{(x+2)(x+3)} \cdot \frac{(x+3)}{(x+2)} = \frac{(x-2)(x+1)}{(x+2)^2}$$

24. $f(x) \div g(x) = \dfrac{x^2 - x - 2}{x^2 + 5x + 6} \div \dfrac{x+3}{x+2}$

$= \dfrac{x^2 - x - 2}{x^2 + 5x + 6} \cdot \dfrac{x+2}{x+3}$

$= \dfrac{(x-2)(x+1)}{(x+2)(x+3)} \cdot \dfrac{x+2}{x+3}$

$= \dfrac{(x-2)(x+1)}{(x+3)^2}$

25. $f(x) + g(x) = \dfrac{x^2 - x - 2}{x^2 + 5x + 6} + \dfrac{x+3}{x+2}$

$= \dfrac{(x-2)(x+1)}{(x+2)(x+3)} + \dfrac{x+3}{x+2}$

$= \dfrac{(x-2)(x+1)}{(x+2)(x+3)} + \dfrac{(x+3)}{(x+2)} \cdot \dfrac{(x+3)}{(x+3)}$

$= \dfrac{x^2 - x - 2}{(x+2)(x+3)} + \dfrac{x^2 + 6x + 9}{(x+2)(x+3)}$

$= \dfrac{2x^2 + 5x + 7}{(x+2)(x+3)}$

26. $f(x) - g(x)$

$= \dfrac{x^2 - x - 2}{x^2 + 5x + 6} - \dfrac{x+3}{x+2}$

$= \dfrac{x^2 - x - 2}{(x+2)(x+3)} - \dfrac{x+3}{x+2}$

$= \dfrac{x^2 - x - 2}{(x+2)(x+3)} - \dfrac{(x+3)}{(x+2)} \cdot \dfrac{(x+3)}{(x+3)}$

$= \dfrac{x^2 - x - 2}{(x+2)(x+3)} - \dfrac{x^2 + 6x + 9}{(x+2)(x+3)}$

$= \dfrac{x^2 - x - 2 - x^2 - 6x - 9}{(x+2)(x+3)}$

$= \dfrac{-7x - 11}{(x+2)(x+3)}$ or $-\dfrac{7x + 11}{(x+2)(x+3)}$

27. $\dfrac{100 \text{ meters}}{1} \cdot \dfrac{1 \text{ yard}}{0.914 \text{ meter}} \approx 109.41$ yards

The minimum length is 109.41 yards.

$\dfrac{110 \text{ meters}}{1} \cdot \dfrac{1 \text{ yard}}{0.914 \text{ meter}} \approx 120.35$ yards

The maximum length is 120.35 yards.

28. $\dfrac{8 \text{ gallons}}{1 \text{ hour}} \cdot \dfrac{4 \text{ quarts}}{1 \text{ gallon}} \cdot \dfrac{4 \text{ cups}}{1 \text{ quart}} \cdot \dfrac{1 \text{ hour}}{60 \text{ minutes}}$

≈ 2.13 cups per minute

29. $\dfrac{\dfrac{x-2}{x^2-9}}{\dfrac{x^2-4}{x+3}} = \dfrac{\dfrac{x-2}{(x-3)(x+3)}}{\dfrac{(x-2)(x+2)}{x+3}} = \dfrac{x-2}{(x-3)(x+3)} \div \dfrac{(x-2)(x+2)}{x+3} = \dfrac{\cancel{x-2}}{(x-3)\cancel{(x+3)}} \cdot \dfrac{\cancel{x+3}}{\cancel{(x-2)}(x+2)} = \dfrac{1}{(x-3)(x+2)}$

30. $\dfrac{\dfrac{4}{3x^4} - \dfrac{2}{6x^2}}{\dfrac{1}{2x} + \dfrac{1}{4}} = \left(\dfrac{\dfrac{4}{3x^4} - \dfrac{2}{6x^2}}{\dfrac{1}{2x} + \dfrac{1}{4}}\right) \cdot \dfrac{12x^4}{12x^4} = \dfrac{16 - 4x^2}{6x^3 + 3x^4} = \dfrac{-4(x^2-4)}{3x^3(2+x)} = \dfrac{-4(x+2)(x-2)}{3x^3(x+2)} = \dfrac{-4(x-2)}{3x^3}$

31.

$\dfrac{1}{x+5} - \dfrac{2}{x-2} = \dfrac{-14}{x^2 + 3x - 10}$

$\dfrac{1}{x+5} - \dfrac{2}{x-2} = \dfrac{-14}{(x+5)(x-2)}$

$(x+5)(x-2)\left(\dfrac{1}{x+5} - \dfrac{2}{x-2}\right) = (x+5)(x-2)\dfrac{-14}{(x+5)(x-2)}$

$x - 2 - 2(x+5) = -14$

$x - 2 - 2x - 10 = -14$

$-x - 12 = -14$

$-x = -2$

$x = 2$

$x = 2$ makes the denominators $x - 2$ and $x^2 + 3x - 10$ equal 0. This equation has no solution.

32.

$$\frac{x}{x+2} + \frac{3}{x+4} = \frac{14}{x^2+6x+8}$$

$$\frac{x}{x+2} + \frac{3}{x+4} = \frac{14}{(x+2)(x+4)}$$

$$(x+2)(x+4)\left(\frac{x}{x+2} + \frac{3}{x+4}\right) = (x+2)(x+4)\frac{14}{(x+2)(x+4)}$$

$$x(x+4) + 3(x+2) = 14$$

$$x^2 + 4x + 3x + 6 = 14$$

$$x^2 + 7x - 8 = 0$$

$$(x+8)(x-1) = 0$$

$$x+8 = 0 \quad \text{or} \quad x-1 = 0$$

$$x = -8 \quad \text{or} \quad x = 1$$

The solutions are $x = 1$ and $x = -8$.

33.

$$\frac{5}{x} + 3 = \frac{4}{x^2}$$

$$x^2\left(\frac{5}{x} + 3\right) = x^2\frac{4}{x^2}$$

$$5x + 3x^2 = 4$$

$$3x^2 + 5x - 4 = 0$$

$$x = \frac{-5 \pm \sqrt{5^2 - 4(3)(-4)}}{2(3)} = \frac{-5 \pm \sqrt{73}}{6}$$

The solutions are $x = \dfrac{-5 \pm \sqrt{73}}{6}$.

34.

$$\frac{x-3}{2x^2-7x-4} - \frac{5}{2x^2+3x+1} = \frac{x-1}{x^2-3x-4}$$

$$\frac{x-3}{(2x+1)(x-4)} - \frac{5}{(2x+1)(x+1)} = \frac{x-1}{(x-4)(x+1)}$$

$$(2x+1)(x-4)(x+1)\left(\frac{x-3}{(2x+1)(x-4)} - \frac{5}{(2x+1)(x+1)}\right) = (2x+1)(x-4)(x+1)\frac{x-1}{(x-4)(x+1)}$$

$$(x+1)(x-3) - 5(x-4) = (2x+1)(x-1)$$

$$x^2 - 2x - 3 - 5x + 20 = 2x^2 - x - 1$$

$$x^2 - 7x + 17 = 2x^2 - x - 1$$

$$-x^2 - 6x + 18 = 0$$

$$x^2 + 6x - 18 = 0$$

$$x = \frac{-6 \pm \sqrt{6^2 - 4(1)(-18)}}{2} = \frac{-6 \pm \sqrt{36+72}}{2} = \frac{-6 \pm \sqrt{108}}{2} = \frac{-6 \pm \sqrt{36 \cdot 3}}{2} = \frac{-6 \pm 6\sqrt{3}}{2} = -3 \pm 3\sqrt{3}$$

The solutions are $x = -3 \pm 3\sqrt{3}$.

35.

$$\frac{2x}{x+6} - \frac{4}{x-3} = \frac{-37}{x^2+3x-18}$$

$$\frac{2x}{x+6} - \frac{4}{x-3} = \frac{-37}{(x+6)(x-3)}$$

$$\frac{2x(x-3)}{(x+6)(x-3)} - \frac{4(x+6)}{(x+6)(x-3)} = \frac{-37}{(x+6)(x-3)}$$

$$\frac{2x(x-3) - 4(x+6)}{(x+6)(x-3)} = \frac{-37}{(x+6)(x-3)}$$

$$2x(x-3) - 4(x+6) = -37$$

$$2x^2 - 6x - 4x - 24 = -37$$

$$2x^2 - 10x = -13$$

$$2x^2 - 10x + 13 = 0$$

$$x = \frac{10 \pm \sqrt{(-10)^2 - 4(2)(13)}}{2(2)}$$

$$= \frac{10 \pm \sqrt{100 - 104}}{4}$$

$$= \frac{10 \pm \sqrt{-4}}{4}$$

$$= \frac{10 \pm 2i}{4}$$

$$= \frac{5 \pm i}{2}$$

The complex solutions are $x = \dfrac{5 \pm i}{2}$.

36.

$$0 = \frac{x-7}{x+1} - \frac{x+3}{x-4}$$

$$\frac{x-7}{x+1} = \frac{x+3}{x-4}$$

$$(x-4)(x-7) = (x+1)(x+3)$$

$$x^2 - 11x + 28 = x^2 + 4x + 3$$

$$25 = 15x$$

$$5 = 3x$$

$$x = \frac{5}{3}$$

The x – intercept is $\left(\dfrac{5}{3}, 0\right)$.

37.

$$S = \frac{a}{1-r}$$

$$S(1-r) = a$$

$$S - Sr = a$$

$$-Sr = a - S$$

$$Sr = S - a$$

$$r = \frac{S-a}{S}$$

38. H varies directly as the square of u

39. w varies inversely as $\log(t)$.

40. $y = k\sqrt{x}$

Solve the following equation to find k.

$$2 = k\sqrt{49}$$

$$2 = 7k$$

$$\frac{2}{7} = k$$

The equation is $y = \dfrac{2}{7}\sqrt{x}$.

41. $B = \dfrac{k}{r^3}$

Solve the following equation find k.

$$9 = \frac{k}{2^3}$$

$$9 = \frac{k}{8}$$

$$72 = k$$

The equation is $B = \dfrac{72}{r^3}$.

42. Let w be the number of inches of water and s be the number of inches of snow.

$$w = ks$$

Solve the following equation to find k.

$$2.24 = k(20)$$

$$0.112 = k$$

Using the equation $w = 0.112s$ we get:

$$w = 0.112(37) = 4.144$$

If 37 inches of snow melts, there will be 4.144 inches of water.

43. a. $m = kr^3$

b. $k = \dfrac{m}{r^3}$

c.

r	m	$\dfrac{m}{r^3}$
1.0	17.1	17.10
1.2	29.4	17.01
1.4	46.7	17.02
1.6	69.6	16.99
1.8	99.1	16.99
2.0	135.9	16.99

A reasonable value for k is the average of all values in the third column, which is approximately 17.02.

d. $m = 17.02r^3$

e. The model fits the data very well.

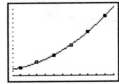

f. $m = 17.02(2.3)^3 \approx 207.08$ grams

44. a. $C(n) = 600 + 40n$

b. $M(n) = \dfrac{600 + 40n}{n}$

c. $M(270) = \dfrac{600 + 40(270)}{(270)} \approx 42.22$

The mean cost per person is $42.22 if 270 people use the room.

d. $50 = \dfrac{600 + 40n}{n}$

$50 \cdot n = \dfrac{600 + 40n}{n} \cdot n$

$50n = 600 + 40n$

$10n = 600$

$n = 60$

In order for the mean cost per person to be $50, 60 people must use the room.

45. a. $(C + R)(t)$

$= (1.13t + 39.29) + (-0.94t + 144.88)$

$= 0.19t + 184.17$

The inputs of $C + R$ are the number of years since 1970, and the outputs of $C + R$ are the total average annual per-person consumptions of chicken and red meat.

b. $P(t) = \left(\dfrac{C(t)}{(C + R)(t)} \right) \cdot 100$

$= \left(\dfrac{1.13t + 39.29}{0.19t + 184.17} \right) \cdot \dfrac{100}{1}$

$= \dfrac{113t + 3929}{0.19t + 184.17}$

c. For 2011, $t = 41$, so

$P(41) = \dfrac{113(41) + 3929}{0.19(41) + 184.17} \approx 44.6$

The model estimates that 44.6% of chicken and red meat consumed in 2011 was chicken.

Actual percentage:

$\dfrac{84.2}{84.2 + 104.3} \cdot 100 \approx 44.7$

The result from using the model is an underestimate.

d. $P(47) = \dfrac{113(47) + 3929}{0.19(47) + 184.17} \approx 47.85$

This means that in 2017, about 47.9% of the chicken and red meat consumed will be chicken.

e. $\dfrac{113t + 3929}{0.19t + 184.17} = 50$

$113t + 3929 = 50(0.19t + 184.17)$

$113t + 3929 = 9.5t + 9208.5$

$103.5t = 5279.5$

$t \approx 51.01$

In $1970 + 51 = 2021$, about half of the chicken and red meat consumed will be chicken.

46. a. $T(a) = \dfrac{75}{a+50} + \dfrac{40}{a+65}$

b. $T(5) = \dfrac{75}{(5)+50} + \dfrac{40}{(5)+65} \approx 1.94$

When the student drives 5 mph above the speed limit, the driving time is 1.94 hours.

c.
$$2 = \dfrac{75}{a+50} + \dfrac{40}{a+65}$$
$$(a+50)(a+65) \cdot 2 = (a+50)(a+65)\left(\dfrac{75}{a+50} + \dfrac{40}{a+65} \right)$$
$$2(a^2 + 115a + 3250) = 75(a+65) + 40(a+50)$$
$$2a^2 + 230a + 6500 = 75a + 4875 + 40a + 2000$$
$$2a^2 + 230a + 6500 = 115a + 6875$$
$$2a^2 + 115a - 375 = 0$$
$$a = \dfrac{-115 \pm \sqrt{115^2 - 4(2)(-375)}}{2(2)} = \dfrac{-115 \pm \sqrt{13225 + 3000}}{4} = \dfrac{-115 \pm \sqrt{16225}}{4}$$

The only answer that makes sense in this context is the positive value.

$\dfrac{-115 + \sqrt{16225}}{4} \approx 3.1$ mph above the speed limits.

The following graph shows the intersection of $y = 2$ and $y = \dfrac{75}{a+50} + \dfrac{40}{a+65}$.

X	Y1
0	2.1154
1	2.0766
2	2.0393
3	2.0033
4	1.9686
5	1.9351
6	1.9027

X=3

The table indicates that a value for a that is slightly in excess of 3 mph will make the driving time 2 hours.

Chapter 8 Test

1. $6x^2 + 11x - 10 = 0$

$x = \dfrac{-11 \pm \sqrt{11^2 - 4(6)(-10)}}{2(6)}$

$= \dfrac{-11 \pm \sqrt{121 + 240}}{12}$

$= \dfrac{-11 \pm \sqrt{361}}{12}$

$= \dfrac{-11 \pm 19}{12}$

$x = \dfrac{-11 + 19}{12} = \dfrac{2}{3}$ or $x = \dfrac{-11 - 19}{12} = -\dfrac{5}{2}$

The domain is the set of all real numbers except $\dfrac{2}{3}$ and $-\dfrac{5}{2}$.

2. $72 - 2x^2 = 0$

$36 - x^2 = 0$

$(6 - x)(6 + x) = 0$

$6 - x = 0$ or $6 + x = 0$

$6 = x$ or $x = -6$

The domain is the set of all real numbers except 6 and −6.

3. Since there is no value that will make the denominator 0, the domain is the set of all real numbers.

4. Answers may vary. Example:

$f(x) = \dfrac{4}{(x+3)(x-7)}$

5. $\dfrac{6-3x}{x^2 - 5x + 6} = \dfrac{-3(x-2)}{(x-2)(x-3)} = -\dfrac{3}{x-3}$

---

6. $\dfrac{9x^2-1}{18x^3-12x^2+2x}=\dfrac{(3x+1)(3x-1)}{2x(9x^2-6x+1)}$

$\qquad\qquad\qquad\ =\dfrac{(3x+1)(3x-1)}{2x(3x-1)(3x-1)}$

$\qquad\qquad\qquad\ =\dfrac{3x+1}{2x(3x-1)}$

7. $\dfrac{5x^4}{3x^2+6x+12}\cdot\dfrac{x^3-8}{15x^7}$

$\qquad =\dfrac{5x^4}{3(x^2+2x+4)}\cdot\dfrac{(x-2)(x^2+2x+4)}{3\cdot5\cdot x^4\cdot x^3}$

$\qquad =\dfrac{x-2}{9x^3}$

8. $\dfrac{p^2-4t^2}{p^2+6pt+9t^2}\div\dfrac{p^2-3pt+2t^2}{p^2+3pt}$

$\qquad =\dfrac{p^2-4t^2}{p^2+6pt+9t^2}\cdot\dfrac{p^2+3pt}{p^2-3pt+2t^2}$

$\qquad =\dfrac{(p-2t)(p+2t)}{(p+3t)(p+3t)}\cdot\dfrac{p(p+3t)}{(p-t)(p-2t)}$

$\qquad =\dfrac{(p-2t)(p+2t)p}{(p+3t)(p-t)(p-2t)}$

$\qquad =\dfrac{p(p+2t)}{(p+3t)(p-t)}$

9. $\dfrac{5x+12}{-2x^2-8x}-\dfrac{2x+1}{x^2+2x-8}$

$\qquad =\dfrac{5x+12}{-2x(x+4)}-\dfrac{2x+1}{(x-2)(x+4)}$

$\qquad =-\dfrac{(5x+12)}{2x(x+4)}\left(\dfrac{x-2}{x-2}\right)-\dfrac{(2x+1)}{(x-2)(x+4)}\left(\dfrac{2x}{2x}\right)$

$\qquad =\dfrac{-(5x+12)(x-2)}{2x(x+4)(x-2)}-\dfrac{2x(2x+1)}{2x(x+4)(x-2)}$

$\qquad =\dfrac{-(5x^2+2x-24)}{2x(x+4)(x-2)}-\dfrac{4x^2+2x}{2x(x+4)(x-2)}$

$\qquad =\dfrac{-5x^2-2x+24-4x^2-2x}{2x(x+4)(x-2)}$

$\qquad =\dfrac{-9x^2-4x+24}{2x(x+4)(x-2)}$

10. $\dfrac{x+2}{x^2-9}+\dfrac{3}{x^2+11x+24}$

$\qquad =\dfrac{x+2}{(x-3)(x+3)}+\dfrac{3}{(x+3)(x+8)}$

$\qquad =\dfrac{(x+2)(x+8)}{(x-3)(x+3)(x+8)}+\dfrac{3(x-3)}{(x-3)(x+3)(x+8)}$

$\qquad =\dfrac{x^2+10x+16+3x-9}{(x-3)(x+3)(x+8)}$

$\qquad =\dfrac{x^2+13x+7}{(x-3)(x+3)(x+8)}$

---

11. $\dfrac{3}{x^2-2x}\div\left(\dfrac{x}{5x-10}-\dfrac{x-1}{x^2-4}\right)=\dfrac{3}{x^2-2x}\div\left(\dfrac{x}{5(x-2)}-\dfrac{x-1}{(x+2)(x-2)}\right)$

$\qquad\qquad\qquad\qquad\qquad\qquad\quad =\dfrac{3}{x^2-2x}\div\left(\dfrac{x}{5(x-2)}\left(\dfrac{x+2}{x+2}\right)-\dfrac{(x-1)}{(x+2)(x-2)}\left(\dfrac{5}{5}\right)\right)$

$\qquad\qquad\qquad\qquad\qquad\qquad\quad =\dfrac{3}{x^2-2x}\div\left(\dfrac{x^2+2x}{5(x-2)(x+2)}-\dfrac{5x-5}{5(x-2)(x+2)}\right)$

$\qquad\qquad\qquad\qquad\qquad\qquad\quad =\dfrac{3}{x^2-2x}\div\left(\dfrac{x^2+2x-5x+5}{5(x-2)(x+2)}\right)$

$\qquad\qquad\qquad\qquad\qquad\qquad\quad =\dfrac{3}{x^2-2x}\div\dfrac{x^2-3x+5}{5(x-2)(x+2)}$

$\qquad\qquad\qquad\qquad\qquad\qquad\quad =\dfrac{3}{x(x-2)}\cdot\dfrac{5(x-2)(x+2)}{x^2-3x+5}$

$\qquad\qquad\qquad\qquad\qquad\qquad\quad =\dfrac{15(x+2)}{x(x^2-3x+5)}$

12. $f(x) - g(x)$

$$= \frac{x+1}{x-5} - \frac{x-2}{x+4}$$

$$= \left(\frac{x+1}{x-5}\right)\left(\frac{x+4}{x+4}\right) - \left(\frac{x-2}{x+4}\right)\left(\frac{x-5}{x-5}\right)$$

$$= \frac{x^2 + 5x + 4}{(x-5)(x+4)} - \frac{x^2 - 7x + 10}{(x-5)(x+4)}$$

$$= \frac{x^2 + 5x + 4 - x^2 + 7x - 10}{(x-5)(x+4)}$$

$$= \frac{12x - 6}{(x-5)(x+4)}$$

$$= \frac{6(2x-1)}{(x-5)(x+4)}$$

$$(f-g)(0) = \frac{6(2(0)-1)}{(0-5)(0+4)} = \frac{-6}{(-5)(4)} = \frac{6}{20} = \frac{3}{10}$$

13. $\dfrac{83.33 \text{ milligrams}}{1 \text{ ounce}} \cdot \dfrac{1 \text{ gram}}{1000 \text{ milligrams}} \cdot \dfrac{16 \text{ ounces}}{1 \text{ pound}}$

$= 1.33$ grams per pound

There are 1.33 grams of sodium in 1 pound of the chili.

14. $\dfrac{5 + \frac{2}{x}}{3 - \frac{4}{x-1}} = \left(\dfrac{5 + \frac{2}{x}}{3 - \frac{4}{x-1}}\right)\left(\dfrac{x(x-1)}{x(x-1)}\right)$

$$= \frac{5x(x-1) + 2(x-1)}{3x(x-1) - 4x}$$

$$= \frac{5x^2 - 5x + 2x - 2}{3x^2 - 3x - 4x}$$

$$= \frac{5x^2 - 3x - 2}{3x^2 - 7x}$$

$$= \frac{(5x+2)(x-1)}{x(3x-7)}$$

15. $\dfrac{2}{x-1} - \dfrac{5}{x+1} = \dfrac{4x}{x^2-1}$

$$\frac{2(x+1) - 5(x-1)}{(x+1)(x-1)} = \frac{4x}{(x+1)(x-1)}$$

$$2x + 2 - 5x + 5 = 4x$$

$$-3x + 7 = 4x$$

$$7 = 7x$$

$$1 = x$$

$x = 1$ makes the denominators $x - 1$ and $x^2 - 1$ equal 0. This equation has no solution.

16. $\dfrac{5}{w-3} = \dfrac{w}{w-2} + \dfrac{w}{w^2 - 5w + 6}$

$$\frac{5}{w-3} = \frac{w}{w-2} + \frac{w}{(w-3)(w-2)}$$

$$\frac{5(w-2)}{(w-3)(w-2)} = \frac{w(w-3)}{(w-3)(w-2)} + \frac{w}{(w-3)(w-2)}$$

$$\frac{5(w-2)}{(w-3)(w-2)} = \frac{w(w-3) + w}{(w-3)(w-2)}$$

$$5(w-2) = w(w-3) + w$$

$$5w - 10 = w^2 - 3w + w$$

$$5w - 10 = w^2 - 2w$$

$$w^2 - 7w + 10 = 0$$

$$(w-5)(w-2) = 0$$

$w = 5$ or $w = 2$

$w = 2$ makes the denominators $w - 2$ and $w^2 - 5w + 6$ equal 0. The solution is $w = 5$.

17. $5 = \dfrac{2}{x-4} + \dfrac{3}{x+1}$

$$(x-4)(x+1) \cdot 5 = (x-4)(x+1)\left(\frac{2}{x-4} + \frac{3}{x+1}\right)$$

$$5(x^2 - 3x - 4) = 2(x+1) + 3(x-4)$$

$$5x^2 - 15x - 20 = 2x + 2 + 3x - 12$$

$$5x^2 - 15x - 20 = 5x - 10$$

$$5x^2 - 20x - 10 = 0$$

$$x^2 - 4x - 2 = 0$$

$$x = \frac{4 \pm \sqrt{(-4)^2 - 4(1)(-2)}}{2(1)}$$

$$= \frac{4 \pm \sqrt{16 + 8}}{2}$$

$$= \frac{4 \pm \sqrt{24}}{2}$$

$$= \frac{4 \pm 2\sqrt{6}}{2}$$

$$= 2 \pm \sqrt{6}$$

The solutions are $x = 2 \pm \sqrt{6}$.

18. $f(x) = \dfrac{(x-5)(x+2)}{(x-1)(x+3)}$

$$f(-2) = \frac{((-2)-5)((-2)+2)}{((-2)-1)((-2)+3)} = \frac{(-5)(0)}{(-3)(1)} = 0$$

19. $f(x) = \dfrac{(x-5)(x+2)}{(x-1)(x+3)}$

$$f(1) = \frac{(1-5)(1+2)}{(1-1)(1+3)} = \frac{(-4)(3)}{(0)(4)} = \frac{-3}{0}$$

The solution is undefined.

20.
$$f(x) = \frac{(x-5)(x+2)}{(x-1)(x+3)}$$
$$0 = \frac{(x-5)(x+2)}{(x-1)(x+3)}$$
$$(x-1)(x+3) \cdot 0 = (x-1)(x+3)\frac{(x-5)(x+2)}{(x-1)(x+3)}$$
$$0 = (x-5)(x+2)$$
$$x-5 = 0 \quad \text{or} \quad x+2 = 0$$
$$x = 5 \quad \text{or} \quad x = -2$$
The solutions are $x = 5$ and $x = -2$.

21. $W = kt^2$

To find k solve the following equation:
$$3 = k(7)^2$$
$$3 = 49k$$
$$\frac{3}{49} = k$$

The equation is $W = \frac{3}{49}t^2$.

22. $y = \frac{k}{\sqrt{x}}$

To find k solve the following equation:
$$8 = \frac{k}{\sqrt{25}}$$
$$8 = \frac{k}{5}$$
$$40 = k$$

The equation is $y = \frac{40}{\sqrt{x}}$.

23. a. $C(n) = 200n + 10,000$

b. $B(n) = \frac{200n + 10,000}{n}$

c. $P(n) = \frac{200n + 10,000}{n} + 150$
$$= \frac{200n + 10,000}{n} + 150 \cdot \frac{n}{n}$$
$$= \frac{200n + 10,000}{n} + \frac{150n}{n}$$
$$= \frac{350n + 10,000}{n}$$

d. $P(100) = \frac{350(100) + 10,000}{100} = 450$

If the bike manufacturer makes and sells 100 bikes in a month, the price of each bike should be $450 to ensure that the manufacturer makes a profit of $150 per bike.

---

24. a. $T(a) = \frac{400}{a+70} + \frac{920}{a+75}$

b. $T(5) = \frac{400}{(5)+70} + \frac{920}{(5)+75} \approx 16.83$

When the student drives 5 mph above the speed limit, the trip takes about 16.8 hours.

c.
$$17 = \frac{400}{a+70} + \frac{920}{a+75}$$
$$(a+70)(a+75) \cdot 17 = (a+70)(a+75)\left(\frac{400}{a+70} + \frac{920}{a+75}\right)$$
$$17(a^2 + 145a + 5250) = 400(a+75) + 920(a+70)$$
$$17a^2 + 2465a + 89250 = 400a + 30000 + 920a + 64400$$
$$17a^2 + 2465a + 89250 = 1320a + 94400$$
$$17a^2 + 1145a - 5150 = 0$$
$$a = \frac{-1145 \pm \sqrt{1145^2 - 4(17)(-5150)}}{2(17)} = \frac{-1145 \pm \sqrt{1311025 + 350200}}{34} = \frac{-1145 \pm \sqrt{1661225}}{34}$$

The only solution that makes sense in this context is the positive solution: $\frac{-1145 + \sqrt{1661225}}{34} \approx 4.23$.

This means the student must drive 4.23 mph above the speed limits for the driving time to be 17 hours.

25. a. $F = \dfrac{k}{L^2}$

To find k solve the following equation:

$50 = \dfrac{k}{8^2}$

$50 = \dfrac{k}{64}$

$3200 = k$

The equation is $g(L) = F = \dfrac{3200}{L^2}$

b. $g(L) = F = \dfrac{3200}{6^2} \approx 88.9$ hertz

c. $200 = \dfrac{3200}{L^2}$

$200L^2 = 3200$

$L^2 = 16$

$L = 4$ cm

d. The graph is decreasing for $L > 0$.

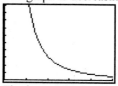

The longer the length of the prongs, the lower the frequency.

---

26. a. Using the quadratic regression feature of a graphing calculator, we get the equation:

$C(t) = -0.0027t^2 + 0.073t + 0.68$

b. By using the linear regression feature of a graphing calculator, we get the equation:

$P(t) = 1000t + 9000$

c. $M(t) = \dfrac{P(t)}{C(t)} = \dfrac{1000t + 9000}{-0.0027t^2 + 0.073t + 0.68}$

d. For 2017, $t = 17$.

$M(17) = \dfrac{1000(17) + 9000}{-0.0027(17)^2 + 0.073(17) + 0.68} \approx 22{,}793$

In 2017, there were about 22,793 patients per community center.

e. $21{,}000 = \dfrac{1000t + 9000}{-0.0027t^2 + 0.073t + 0.68}$

$21{,}000\left(-0.0027t^2 + 0.073t + 0.68\right) = 1000t + 9000$

$-56.7t^2 + 1533t + 14{,}280 = 1000t + 9000$

$-56.7t^2 + 533t + 5280 = 0$

$t = \dfrac{-533 \pm \sqrt{533^2 - 4(-56.7)(5280)}}{2(-56.7)} = \dfrac{-533 \pm 1217.2}{-113.4}$

$t \approx -6.03$ or $t \approx 15.43$

Considering the negative value to indicate model breakdown, in 2000 + 15 = 2015, the mean number of patients per community center will be 21,000 patients.

Chapter 9
Radical Functions

1. $x^{2/5} = \sqrt[5]{x^2}$

3. $\sqrt[4]{x^3} = x^{3/4}$

5. $\sqrt{w} = w^{1/2}$

7. $(2x+9)^{3/7} = \sqrt[7]{(2x+9)^3}$

9. $\sqrt[7]{(3k+2)^4} = (3k+2)^{4/7}$

11. $\sqrt{50} = \sqrt{25 \cdot 2} = \sqrt{25}\sqrt{2} = 5\sqrt{2}$

13. $\sqrt{x^8} = x^{8/2} = x^4$

15. $\sqrt{36x^6} = \sqrt{36} \cdot \sqrt{x^6} = 6x^3$

17. $\sqrt{5a^2b^{12}} = \sqrt{5} \cdot \sqrt{a^2b^{12}} = ab^6\sqrt{5}$

19. $\sqrt{x^9} = \sqrt{x^8 \cdot x} = \sqrt{x^8} \cdot \sqrt{x} = x^4\sqrt{x}$

21. $\sqrt{24x^5} = \sqrt{4 \cdot 6 \cdot x^4 \cdot x}$
$= \sqrt{4} \cdot \sqrt{6} \cdot \sqrt{x^4} \cdot \sqrt{x}$
$= 2x^2\sqrt{6x}$

23. $\sqrt{80x^3y^8} = \sqrt{16 \cdot 5 \cdot x^2 \cdot x \cdot y^8}$
$= \sqrt{16} \cdot \sqrt{5} \cdot \sqrt{x^2} \cdot \sqrt{x} \cdot \sqrt{y^8}$
$= 4xy^4\sqrt{5x}$

25. $\sqrt{200a^3b^5} = \sqrt{100 \cdot 2 \cdot a^2 \cdot a \cdot b^4 \cdot b}$
$= \sqrt{100} \cdot \sqrt{2} \cdot \sqrt{a^2} \cdot \sqrt{a} \cdot \sqrt{b^4} \cdot \sqrt{b}$
$= 10ab^2\sqrt{2ab}$

27. $\sqrt{(2x+5)^8} = (2x+5)^{8/2} = (2x+5)^4$

29. $\sqrt{(6t+3)^5} = \sqrt{(6t+3)^4(6t+3)}$
$= \sqrt{(6t+3)^4} \cdot \sqrt{6t+3}$
$= (6t+3)^2\sqrt{6t+3}$

31. $\sqrt[3]{27} = 3$

33. $\sqrt[6]{x^6} = x^{6/6} = x^1 = x$

35. $\sqrt[3]{8x^3} = \sqrt[3]{8} \cdot \sqrt[3]{x^3} = 2x$

37. $\sqrt[5]{-32x^{20}} = \sqrt[5]{-32} \cdot \sqrt[5]{x^{20}} = -2x^4$

39. $\sqrt[4]{81a^{12}b^{28}} = \sqrt[4]{81} \cdot \sqrt[4]{a^{12}} \cdot \sqrt[4]{b^{28}} = 3a^3b^7$

41. $\sqrt[6]{x^{17}} = \sqrt[6]{x^{12} \cdot x^5} = \sqrt[6]{x^{12}} \cdot \sqrt[6]{x^5} = x^2\sqrt[6]{x^5}$

43. $\sqrt[3]{-125a^{17}b^{12}} = \sqrt[3]{-125 \cdot a^{15} \cdot a^2 \cdot b^{12}}$
$= \sqrt[3]{-125} \cdot \sqrt[3]{a^{15}} \cdot \sqrt[3]{a^2} \cdot \sqrt[3]{b^{12}}$
$= -5a^5b^4\sqrt[3]{a^2}$

45. $\sqrt[5]{64x^{39}y^7}$
$= \sqrt[5]{32 \cdot 2 \cdot x^{35} \cdot x^4 \cdot y^5 \cdot y^2}$
$= \sqrt[5]{32} \cdot \sqrt[5]{2} \cdot \sqrt[5]{x^{35}} \cdot \sqrt[5]{x^4} \cdot \sqrt[5]{y^5} \cdot \sqrt[5]{y^2}$
$= 2x^7y\sqrt[5]{2x^4y^2}$

47. $\sqrt[5]{(6xy)^5} = (6xy)^{5/5} = (6xy)^1 = 6xy$

49. $\sqrt[4]{(3x+6)^4} = (3x+6)^{4/4} = (3x+6)^1 = 3x+6$

51. $\sqrt[5]{(4p+7)^{20}} = (4p+7)^{20/5} = (4p+7)^4$

53. $\sqrt[6]{(2x+9)^{31}} = \sqrt[6]{(2x+9)^{30}(2x+9)}$
$= \sqrt[6]{(2x+9)^{30}} \cdot \sqrt[6]{2x+9}$
$= (2x+9)^5\sqrt[6]{2x+9}$

55. $\sqrt[8]{x^6} = x^{6/8} = x^{3/4} = \sqrt[4]{x^3}$

57. $\sqrt[6]{x^4} = x^{4/6} = x^{2/3} = \sqrt[3]{x^2}$

59. $\sqrt[12]{(2m+7)^{10}} = (2m+7)^{10/12}$
$= (2m+7)^{5/6}$
$= \sqrt[6]{(2m+7)^5}$

61. $\sqrt[6]{x^{14}} = x^{14/6}$

$= x^{7/3}$

$= \sqrt[3]{x^7}$

$= \sqrt[3]{x^6 \cdot x}$

$= \sqrt[3]{x^6} \cdot \sqrt[3]{x}$

$= x^2 \sqrt[3]{x}$

63. $\sqrt[6]{27} = \sqrt[6]{3^3} = 3^{3/6} = 3^{1/2} = \sqrt{3}$

65. $\sqrt[4]{\sqrt[3]{p}} = \sqrt[4]{p^{1/3}} = p^{\frac{1}{3}\cdot\frac{1}{4}} = p^{1/12} = \sqrt[12]{p}$

67. $\sqrt[10]{16x^8} = \sqrt[10]{16} \cdot \sqrt[10]{x^8}$

$= \sqrt[10]{4^2} \cdot \sqrt[10]{x^8}$

$= 4^{2/10} \cdot x^{8/10}$

$= 4^{1/5} \cdot x^{4/5}$

$= \sqrt[5]{4} \cdot \sqrt[5]{x^4}$

$= \sqrt[5]{4x^4}$

69. $\sqrt[4]{\sqrt{ab}} = \sqrt[4]{ab^{1/2}} = ab^{\frac{1}{2}\cdot\frac{1}{4}} = ab^{1/8} = \sqrt[8]{ab}$

71. $f(-32) = \sqrt[5]{-32} = -2$

73. $g(2) = \sqrt[3]{3(2)+2} = \sqrt[3]{8} = 2$

75. $g(-7) = \sqrt[3]{3(-7)+2} = \sqrt[3]{-19} = -\sqrt[3]{19}$

77. $h(49) = 2\sqrt{49} - 5 = 2(7) - 5 = 14 - 5 = 9$

79.

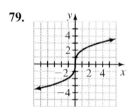

81. a.

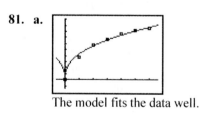

The model fits the data well.

b. $f(24) = 8.5\sqrt[5]{24^2} \approx 30.3$

The average temperature rise is about 30° Fahrenheit.

c. $f(45) = 8.5\sqrt[5]{45^2} \approx 38.97$

$90 + 39 = 129$

The temperature in the car is about 129° Fahrenheit.

d. $107 - 80 = 27$

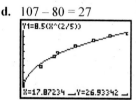

It would take about 18 minutes.

83. a. $f(3890) = \sqrt{9.8(3890)} \approx 195.25$

The speed of a tsunami in the Indian Ocean is about 195 meters per second.

b. $f(1000) = \sqrt{9.8(1000)} \approx 99$

$f(2000) = \sqrt{9.8(2000)} = 140$

$f(3000) = \sqrt{9.8(3000)} \approx 171$

f is an increasing function. This means that the speed of a tsunami increases as the depth of the ocean increases.

c. As a tsunami approaches the shore, the depth of the water decreases, which decreases the speed of the tsunami. At the same time, the height of the tsunami increases.

d. $195 \times 60 \times 60 = 702{,}000$ meters per hour

$702{,}000 \div 1609 \approx 436.3$ miles per hour

The speed of a tsunami in the Indian Ocean is about 436 miles per hour.

85. Answers may vary. Example:

$$\sqrt{x^{16}} = \left(x^{16}\right)^{1/2} = x^{16 \cdot \frac{1}{2}} = x^8$$

87. The exponent of x should be $\dfrac{3}{6}$, not $\dfrac{6}{3}$.

$$\sqrt[6]{x^3} = x^{3/6} = x^{1/2} = \sqrt{x}$$

89. $\sqrt[n]{\sqrt[n]{x}} = \sqrt[n]{x^{1/n}}$

$= \left(x^{1/n}\right)^{1/n}$

$= x^{(1/n)\cdot(1/n)}$

$= x^{1/n^2}$

$= \sqrt[n^2]{x}$

91. a. $\left(\sqrt{x}\right)^2 = \left(x^{1/2}\right)^2 = x^{2/2} = x^1 = x$

b. $\left(\sqrt[3]{x}\right)^3 = \left(x^{1/3}\right)^3 = x^{3/3} = x^1 = x$

c. $\left(\sqrt[n]{x}\right)^n = \left(x^{1/n}\right)^n = x^{n/n} = x^1 = x$

93. a. $\sqrt{16x^4y^6} = \sqrt{16} \cdot \sqrt{x^4} \cdot \sqrt{y^6} = 4x^2y^3$

b. $\sqrt{16x^4y^6} = \left(16x^4y^6\right)^{1/2}$

$= 16^{\frac{1}{2}} x^{4\cdot\frac{1}{2}} y^{6\cdot\frac{1}{2}}$

$= 4x^2y^3$

c. The answers are the same.

95.
$$\frac{2x}{x^2+x-6} - \frac{3x-1}{x^2+6x+9} = \frac{-3}{x+3}$$

$$\frac{2x}{(x+3)(x-2)} - \frac{3x-1}{(x+3)(x+3)} = \frac{-3}{x+3}$$

$$2x(x+3) - (3x-1)(x-2) = -3(x+3)(x-2)$$

$$2x^2 + 6x - 3x^2 + 7x - 2 = -3x^2 - 3x + 18$$

$$2x^2 + 6x - 3x^2 + 7x - 2 + 3x^2 + 3x - 18 = 0$$

$$2x^2 + 16x - 20 = 0$$

$$2(x^2 + 8x - 10) = 0$$

$$\frac{-b \pm \sqrt{b^2-4ac}}{2a} = \frac{-8 \pm \sqrt{8^2 - 4(1)(-10)}}{2(1)} = \frac{-8 \pm \sqrt{104}}{2} = -4 \pm \sqrt{26}$$

This is a rational equation in one variable.

---

97. $\dfrac{2x}{x^2+x-6} - \dfrac{3x-1}{x^2+6x+9}$

$= \dfrac{2x}{(x+3)(x-2)} - \dfrac{3x-1}{(x+3)(x+3)}$

$= \dfrac{2x(x+3) - (3x-1)(x-2)}{(x+3)(x+3)(x-2)}$

$= \dfrac{2x^2 + 6x - 3x^2 + 7x - 2}{(x+3)(x+3)(x-2)}$

$= \dfrac{-x^2 + 13x - 2}{(x+3)(x+3)(x-2)}$

$= -\dfrac{x^2 - 13x + 2}{(x+3)^2(x-2)}$

This is a rational expression in one variable.

99. $f(x) = \dfrac{2x}{x^2+x-6} = \dfrac{2x}{(x+3)(x-2)}$

The domain of $f(x)$ is all real numbers except $x = -3$ and $x = 2$.

This is a rational function.

Homework 9.2

1. $4\sqrt{x} + 5\sqrt{x} = (4+5)\sqrt{x} = 9\sqrt{x}$

3. $2\sqrt[3]{5x^2y} - 6\sqrt[3]{5x^2y} = (2-6)\sqrt[3]{5x^2y}$

$= -4\sqrt[3]{5x^2y}$

5. $3\sqrt{5a} + 2\sqrt{3b} - 6\sqrt{3b} + 7\sqrt{5a}$

   $= \left(3\sqrt{5a} + 7\sqrt{5a}\right) + \left(2\sqrt{3b} - 6\sqrt{3b}\right)$

   $= (3+7)\sqrt{5a} + (2-6)\sqrt{3b}$

   $= 10\sqrt{5a} - 4\sqrt{3b}$

7. $2\sqrt{x} + 5 - 7\sqrt[3]{x} - 9 + 5\sqrt[3]{x}$

   $= 2\sqrt{x} + (5-9) + (-7+5)\sqrt[3]{x}$

   $= 2\sqrt{x} - 4 - 2\sqrt[3]{x}$

9. $6\sqrt[3]{x-1} - 3\sqrt[3]{x-1} - 2\sqrt{x-1}$

   $= (6-3)\sqrt[3]{x-1} - 2\sqrt{x-1}$

   $= 3\sqrt[3]{x-1} - 2\sqrt{x-1}$

11. $3.7\sqrt[4]{x} - 1.1\sqrt[4]{x} - 4.2\sqrt[6]{x} + 4.2\sqrt[6]{x}$

    $= (3.7 - 1.1)\sqrt[4]{x} + (-4.2 + 4.2)\sqrt[6]{x}$

    $= 2.6\sqrt[4]{x}$

13. $3\left(7 - \sqrt{x} + 2\right) - \left(\sqrt{x} + 2\right)$

    $= 3 \cdot 7 - 3 \cdot \sqrt{x} + 3 \cdot 2 - \sqrt{x} - 2$

    $= 21 - 3\sqrt{x} + 6 - \sqrt{x} - 2$

    $= -3\sqrt{x} - \sqrt{x} + 21 + 6 - 2$

    $= (-3 - 1)\sqrt{x} + (21 + 6 - 2)$

    $= -4\sqrt{x} + 25$

15. $7\left(\sqrt[3]{x} + 1\right) - 7\left(\sqrt[3]{x} - 1\right)$

    $= 7 \cdot \sqrt[3]{x} + 7 \cdot 1 - 7 \cdot \sqrt[3]{x} - 7(-1)$

    $= 7\sqrt[3]{x} + 7 - 7\sqrt[3]{x} + 7$

    $= 7\sqrt[3]{x} - 7\sqrt[3]{x} + 7 + 7$

    $= 14$

17. $\sqrt{12b} + \sqrt{75b} = \sqrt{4 \cdot 3b} + \sqrt{25 \cdot 3b}$

    $= 2\sqrt{3b} + 5\sqrt{3b}$

    $= (2+5)\sqrt{3b}$

    $= 7\sqrt{3b}$

19. $\sqrt{18x^5} + 2x\sqrt{50x^3}$

    $= \sqrt{9 \cdot 2 \cdot x^4 \cdot x} + 2x\sqrt{25 \cdot 2 \cdot x^2 \cdot x}$

    $= \sqrt{9}\sqrt{x^4}\sqrt{2x} + 2x\sqrt{25}\sqrt{x^2}\sqrt{2x}$

    $= 3x^2\sqrt{2x} + 2x(5x)\sqrt{2x}$

    $= 3x^2\sqrt{2x} + 10x^2\sqrt{2x}$

    $= \left(3x^2 + 10x^2\right)\sqrt{2x}$

    $= 13x^2\sqrt{2x}$

21. $5\sqrt{4x^3} - x\sqrt{36x} = 5\sqrt{4 \cdot x^2 \cdot x} - x\sqrt{36 \cdot x}$

    $= 5\sqrt{4}\sqrt{x^2}\sqrt{x} - x\sqrt{36}\sqrt{x}$

    $= 5(2)x\sqrt{x} - 6x\sqrt{x}$

    $= 10x\sqrt{x} - 6x\sqrt{x}$

    $= (10 - 6)x\sqrt{x}$

    $= 4x\sqrt{x}$

23. $3\sqrt{81x^2} - 2\sqrt{100x^2} = 3 \cdot 9x - 2 \cdot 10x$

    $= 27x - 20x$

    $= 7x$

25. $a\sqrt{12b^3} + b\sqrt{75ba^2}$

    $= a\sqrt{4b^2 \cdot 3b} + b\sqrt{25a^2 \cdot 3b}$

    $= a\sqrt{4b^2}\sqrt{3b} + b\sqrt{25a^2}\sqrt{3b}$

    $= 2ab\sqrt{3b} + 5ab\sqrt{3b}$

    $= (2 + 5)ab\sqrt{3b}$

    $= 7ab\sqrt{3b}$

27. $\sqrt[3]{27x^5} - x\sqrt[3]{8x^2} = \sqrt[3]{27x^3 \cdot x^2} - x\sqrt[3]{8 \cdot x^2}$

    $= \sqrt[3]{27x^3}\sqrt[3]{x^2} - x\sqrt[3]{8}\sqrt[3]{x^2}$

    $= 3x\sqrt[3]{x^2} - 2x\sqrt[3]{x^2}$

    $= (3 - 2)x\sqrt[3]{x^2}$

    $= x\sqrt[3]{x^2}$

29. $y\sqrt[4]{16x^{11}y^4} - 3x\sqrt[4]{x^7y^8}$

    $= y\sqrt[4]{16 \cdot x^8 \cdot x^3 \cdot y^4} - 3x\sqrt[4]{x^4 \cdot x^3 \cdot y^8}$

    $= y\sqrt[4]{16x^8y^4}\sqrt[4]{x^3} - 3x\sqrt[4]{x^4y^8}\sqrt[4]{x^3}$

    $= y\left(2x^2y\right)\sqrt[4]{x^3} - 3x(xy^2)\sqrt[4]{x^3}$

    $= 2x^2y^2\sqrt[4]{x^3} - 3x^2y^2\sqrt[4]{x^3}$

    $= \left(2x^2y^2 - 3x^2y^2\right)\sqrt[4]{x^3}$

    $= -x^2y^2\sqrt[4]{x^3}$

31. $3\sqrt{x} \cdot 2\sqrt{x} = 3 \cdot 2 \cdot \sqrt{x} \cdot \sqrt{x}$

    $= 6\sqrt{x \cdot x}$

    $= 6\sqrt{x^2}$

    $= 6x$

33. $-2\sqrt{5x} \cdot 4\sqrt{3x} = -2 \cdot 4 \cdot \sqrt{5x} \cdot \sqrt{3x}$
$$= -8\sqrt{5x \cdot 3x}$$
$$= -8\sqrt{15x^2}$$
$$= -8\sqrt{15}\sqrt{x^2}$$
$$= -8x\sqrt{15}$$

35. $2\sqrt{7t}\left(\sqrt{7t} - \sqrt{2t}\right) = 2\sqrt{7t} \cdot \sqrt{7t} - 2\sqrt{7t} \cdot \sqrt{2t}$
$$= 2\sqrt{7t \cdot 7t} - 2\sqrt{7t \cdot 2t}$$
$$= 2\sqrt{49t^2} - 2\sqrt{14t^2}$$
$$= 2\sqrt{49}\sqrt{t^2} - 2\sqrt{14}\sqrt{t^2}$$
$$= 2 \cdot 7t - 2t\sqrt{14}$$
$$= 14t - 2t\sqrt{14}$$

37. $\left(2\sqrt{x} + 6\right)\left(5\sqrt{x} + 4\right)$
$$= 2\sqrt{x} \cdot 5\sqrt{x} + 6 \cdot 5\sqrt{x} + 2\sqrt{x} \cdot 4 + 6 \cdot 4$$
$$= 10\sqrt{x^2} + 30\sqrt{x} + 8\sqrt{x} + 24$$
$$= 10x + (30 + 8)\sqrt{x} + 24$$
$$= 10x + 38\sqrt{x} + 24$$

39. $\left(4\sqrt{x} + \sqrt{3}\right)\left(2\sqrt{x} - \sqrt{5}\right)$
$$= 4\sqrt{x} \cdot 2\sqrt{x} + \sqrt{3} \cdot 2\sqrt{x} - 4\sqrt{x} \cdot \sqrt{5} - \sqrt{3}\sqrt{5}$$
$$= 8\sqrt{x^2} + 2\sqrt{3x} - 4\sqrt{5x} - \sqrt{15}$$
$$= 8x + 2\sqrt{3x} - 4\sqrt{5x} - \sqrt{15}$$

41. $\left(5\sqrt{a} + \sqrt{b}\right)\left(\sqrt{a} - 2\sqrt{b}\right)$
$$= 5\sqrt{a} \cdot \sqrt{a} + \sqrt{b} \cdot \sqrt{a} - 5\sqrt{a} \cdot 2\sqrt{b} - \sqrt{b} \cdot 2\sqrt{b}$$
$$= 5\sqrt{a^2} + \sqrt{ab} - 10\sqrt{ab} - 2\sqrt{b^2}$$
$$= 5a + (1 - 10)\sqrt{ab} - 2b$$
$$= 5a - 2b - 9\sqrt{ab}$$

43. $\left(1 - \sqrt{w}\right)\left(1 + \sqrt{w}\right) = 1^2 - \left(\sqrt{w}\right)^2 = 1 - w$

45. $\left(7x + \sqrt{5}\right)\left(7x - \sqrt{5}\right) = \left(7x\right)^2 - \left(\sqrt{5}\right)^2$
$$= 49x^2 - 5$$

47. $\left(2\sqrt{a} - \sqrt{b}\right)\left(2\sqrt{a} + \sqrt{b}\right) = \left(2\sqrt{a}\right)^2 - \left(\sqrt{b}\right)^2$
$$= 2^2\left(\sqrt{a}\right)^2 - \left(\sqrt{b}\right)^2$$
$$= 4a - b$$

49. $\left(5 + 6\sqrt{x}\right)^2 = 5^2 + 2(5)\left(6\sqrt{x}\right) + \left(6\sqrt{x}\right)^2$
$$= 25 + 60\sqrt{x} + 6^2\left(\sqrt{x}\right)^2$$
$$= 25 + 60\sqrt{x} + 36x$$
$$= 36x + 60\sqrt{x} + 25$$

51. $\left(4\sqrt{x} - \sqrt{5}\right)^2$
$$= \left(4\sqrt{x}\right)^2 - 2\left(4\sqrt{x}\right)\left(\sqrt{5}\right) + \left(\sqrt{5}\right)^2$$
$$= 4^2\left(\sqrt{x}\right)^2 - 8\sqrt{5x} + 5$$
$$= 16x - 8\sqrt{5x} + 5$$

53. $\left(\sqrt{a} + 2\sqrt{b}\right)^2$
$$= \left(\sqrt{a}\right)^2 + 2\left(\sqrt{a}\right)\left(2\sqrt{b}\right) + \left(2\sqrt{b}\right)^2$$
$$= a + 4\sqrt{ab} + 2^2\left(\sqrt{b}\right)^2$$
$$= a + 4\sqrt{ab} + 4b$$

55. $\left(\sqrt{2x - 5} + 3\right)^2$
$$= \left(\sqrt{2x - 5}\right)^2 + 2\left(\sqrt{2x - 5}\right)(3) + (3)^2$$
$$= 2x - 5 + 6\sqrt{2x - 5} + 9$$
$$= 2x + 6\sqrt{2x - 5} + 4$$

57. $\sqrt{x}\sqrt[5]{x} = x^{\frac{1}{2}} \cdot x^{\frac{1}{5}}$
$$= x^{\frac{1}{2} + \frac{1}{5}}$$
$$= x^{\frac{5}{10} + \frac{2}{10}}$$
$$= x^{\frac{7}{10}}$$
$$= \sqrt[10]{x^7}$$

59. $\sqrt[5]{x^4}\sqrt[5]{x^3} = \sqrt[5]{x^4 \cdot x^3}$
$$= \sqrt[5]{x^7}$$
$$= \sqrt[5]{x^5 \cdot x^2}$$
$$= \sqrt[5]{x^5}\sqrt[5]{x^2}$$
$$= x\sqrt[5]{x^2}$$

61. $-5\sqrt{m}\left(\sqrt[4]{2m}-4\right)=-5\sqrt{m}\sqrt[4]{2m}-5\sqrt{m}\left(-4\right)$

$\qquad = -5\sqrt{m}\sqrt[4]{2}\sqrt[4]{m}+20\sqrt{m}$

$\qquad = -5\sqrt[4]{2}\;m^{\frac{1}{2}}\cdot m^{\frac{1}{4}}+20\sqrt{m}$

$\qquad = -5\sqrt[4]{2}\;m^{\frac{1}{2}+\frac{1}{4}}+20\sqrt{m}$

$\qquad = -5\sqrt[4]{2}m^{\frac{3}{4}}+20\sqrt{m}$

$\qquad = -5\sqrt[4]{2}\sqrt[4]{m^3}+20\sqrt{m}$

$\qquad = -5\sqrt[4]{2m^3}+20\sqrt{m}$

63. $\left(\sqrt[3]{x}+1\right)^2=\left(\sqrt[3]{x}\right)^2+2\left(\sqrt[3]{x}\right)(1)+(1)^2$

$\qquad = \sqrt[3]{x^2}+2\sqrt[3]{x}+1$

65. $\left(\sqrt[4]{k}-\sqrt[3]{k}\right)^2=\left(\sqrt[4]{k}\right)^2-2\left(\sqrt[4]{k}\right)\left(\sqrt[3]{k}\right)+\left(\sqrt[3]{k}\right)^2$

$\qquad = \sqrt[4]{k^2}-2k^{\frac{1}{4}}\cdot k^{\frac{1}{3}}+\sqrt[3]{k^2}$

$\qquad = k^{\frac{2}{4}}-2k^{\frac{1}{4}+\frac{1}{3}}+\sqrt[3]{k^2}$

$\qquad = k^{\frac{1}{2}}-2k^{\frac{7}{12}}+\sqrt[3]{k^2}$

$\qquad = \sqrt{k}-2\sqrt[12]{k^7}+\sqrt[3]{k^2}$

67. $\left(2\sqrt{x}-6\right)\left(3\sqrt[3]{x}+1\right)$

$\qquad = 2\sqrt{x}\cdot3\sqrt[3]{x}-6\cdot3\sqrt[3]{x}+2\sqrt{x}\cdot1-6\cdot1$

$\qquad = 6x^{\frac{1}{2}}\cdot x^{\frac{1}{3}}-18\sqrt[3]{x}+2\sqrt{x}-6$

$\qquad = 6x^{\frac{1}{2}+\frac{1}{3}}-18\sqrt[3]{x}+2\sqrt{x}-6$

$\qquad = 6x^{\frac{5}{6}}-18\sqrt[3]{x}+2\sqrt{x}-6$

$\qquad = 6\sqrt[6]{x^5}-18\sqrt[3]{x}+2\sqrt{x}-6$

69. $\left(3\sqrt[4]{x}+5\right)\left(3\sqrt[4]{x}-5\right)=\left(3\sqrt[4]{x}\right)^2-(5)^2$

$\qquad = 3^2\left(\sqrt[4]{x}\right)^2-25$

$\qquad = 9\sqrt{x}-25$

71. a. The flow rate, r, increases much more quickly as the value of d is increased. Answers may vary. Example: From the table, when the diameter doubles from 0.5 to 1.0, the flow rate quadruples.

In the equation, $r=30d^2\sqrt{P}$, the value of d is squared, so as d increases, the value of r increases by the squared value of d. If you consider the situation, the flow rate r

should increase at a greater rate as you increase the nozzle diameter d because more water is able to leave through the nozzle.

b. i. $r=30(0.5)^2\sqrt{100}=75$ gallons/minute

$\qquad r=30(1)^2\sqrt{100}=300$ gallons/minute

$\qquad r=30(1.5)^2\sqrt{100}=675$ gallons/minute

$\qquad r=30(2)^2\sqrt{100}=1200$ gallons/minute

$\qquad r=30(2.5)^2\sqrt{100}=1875$ gallons/minute

ii. $d=0.5$, error $=75-74=1$

$\qquad d=1$, error $=300-297=3$

$\qquad d=1.5$, error $=675-668=7$

$\qquad d=2$, error $=1200-1188=12$

$\qquad d=2.5$, error $=1875-1857=18$

The estimate for $d=2.5$ inches has the largest error at 18 gallons per minute.

iii. $d=0.5$, % error $=\dfrac{1}{74}\times100=1.35\%$

$\qquad d=1$, % error $=\dfrac{3}{297}\times100=1.01\%$

$\qquad d=1.5$, % error $=\dfrac{7}{668}\times100=1.05\%$

$\qquad d=2$, % error $=\dfrac{12}{1188}\times100=1.01\%$

$\qquad d=2.5$, % error $=\dfrac{18}{1857}\times100=0.97\%$

The estimate for $d=0.5$ inches has the largest percentage error at 1.35%.

c. $r=30(1.75)^2\sqrt{45}\approx616.3$

Yes, the minimum requirement of 500 gallons per minute is met and exceeded. The estimated flow rate is 616 gallons per minute.

73. The student squared each term individually instead of using FOIL.

$\left(x+\sqrt{7}\right)^2=\left(x+\sqrt{7}\right)\left(x+\sqrt{7}\right)$

$\qquad = (x)^2+2(x)\left(\sqrt{7}\right)+\left(\sqrt{7}\right)^2$

$\qquad = x^2+2x\sqrt{7}+7$

75. The student multiplied a number by a radical as if they were both radicals.

$7\left(2\sqrt{3}\right)=14\sqrt{3}$

77. $\dfrac{\sqrt{x}}{\sqrt[3]{x}} = \dfrac{x^{\frac{1}{2}}}{x^{\frac{1}{3}}} = x^{\frac{1}{2} - \frac{1}{3}} = x^{\frac{3}{6} - \frac{2}{6}} = x^{\frac{1}{6}} = \sqrt[6]{x}$

79. a. $\sqrt[4]{x}\sqrt[5]{x} = x^{\frac{1}{4}} \cdot x^{\frac{1}{5}}$

$= x^{\frac{1}{4} + \frac{1}{5}}$

$= x^{\frac{5}{20} + \frac{4}{20}}$

$= x^{\frac{9}{20}}$

$= \sqrt[20]{x^9}$

b. $\sqrt[k]{x}\sqrt[n]{x} = x^{\frac{1}{k}} \cdot x^{\frac{1}{n}}$

$= x^{\frac{1}{k} + \frac{1}{n}}$

$= x^{\frac{n}{n \cdot k} + \frac{k}{n \cdot k}}$

$= x^{\frac{n+k}{n \cdot k}}$

$= \sqrt[kn]{x^{k+n}}$

c. $\sqrt[4]{x}\sqrt[5]{x} \rightarrow \quad k = 4, n = 5$

$\sqrt[4]{x}\sqrt[5]{x} = \sqrt[4 \cdot 5]{x^{4+5}} = \sqrt[20]{x^9}$

The results are the same.

d. $\sqrt[3]{x}\sqrt[7]{x} = \sqrt[3 \cdot 7]{x^{3+7}} = \sqrt[21]{x^{10}}$

81. a.   **i.** true

    ii. false

    iii. true

    iv. false

    v. true

    vi. false

b. Answers may vary. Example:
The equations that are true involve multiplication and division. The equations that are false involve addition.

83. Answers may vary. Example:
If the indexes are the same, use the product property. If the indexes are different, change the radicals to exponents, then use the properties of exponents to solve. Finally, change the exponential expression back into a radical.

85. $3\sqrt{x} - 5\sqrt{x} = (3-5)\sqrt{x} = -2\sqrt{x}$

87. $\left(3\sqrt{x}\right)\left(-5\sqrt{x}\right) = -15\sqrt{x^2} = -15x$

89. $\log_b\left(x^2 + 3x - 40\right) - \log_b\left(x^2 - 64\right)$

$= \log_b\left[(x+8)(x-5)\right] - \log_b\left[(x+8)(x-8)\right]$

$= \log_b\left[\dfrac{(x+8)(x-5)}{(x+8)(x-8)}\right]$

$= \log_b\left(\dfrac{x-5}{x-8}\right)$

This is a logarithmic expression in one variable.

91. $\log_2(3x-4) - \log_2(2x-3) = 3$

$\log_2\left(\dfrac{3x-4}{2x-3}\right) = 3$

$2^3 = \dfrac{3x-4}{2x-3}$

$8 = \dfrac{3x-4}{2x-3}$

$8(2x-3) = 3x-4$

$16x - 24 = 3x - 4$

$13x = 20$

$x = \dfrac{20}{13}$

This is a logarithmic equation in one variable.

93. $2(3)^{5x-1} = 35$

$3^{5x-1} = 17.5$

$(5x-1)\ln(3) = \ln(17.5)$

$5x - 1 = \dfrac{\ln(17.5)}{(\ln 3)}$

$5x - 1 \approx 2.6053$

$5x \approx 3.6053$

$x \approx 0.7211$

This is an exponential equation in one variable.

Homework 9.3

1. $\dfrac{8}{\sqrt{x}} = \dfrac{8}{\sqrt{x}} \cdot \dfrac{\sqrt{x}}{\sqrt{x}} = \dfrac{8\sqrt{x}}{\left(\sqrt{x}\right)^2} = \dfrac{8\sqrt{x}}{x}$

3. $\dfrac{3}{\sqrt{5p}} = \dfrac{3}{\sqrt{5p}} \cdot \dfrac{\sqrt{5p}}{\sqrt{5p}} = \dfrac{3\sqrt{5p}}{\left(\sqrt{5p}\right)^2} = \dfrac{3\sqrt{5p}}{5p}$

5. $\dfrac{4}{3\sqrt{2x}} = \dfrac{4}{3\sqrt{2x}} \cdot \dfrac{\sqrt{2x}}{\sqrt{2x}} = \dfrac{4\sqrt{2x}}{3\left(\sqrt{2x}\right)^2} = \dfrac{4\sqrt{2x}}{3 \cdot 2x} = \dfrac{2\sqrt{2x}}{3x}$

7. $\dfrac{10}{\sqrt{8k}} = \dfrac{10}{\sqrt{8k}} \cdot \dfrac{\sqrt{8k}}{\sqrt{8k}} = \dfrac{10\sqrt{8k}}{\left(\sqrt{8k}\right)^2} = \dfrac{10\sqrt{4 \cdot 2k}}{8k} = \dfrac{10(2)\sqrt{2k}}{8k} = \dfrac{20\sqrt{2k}}{8k} = \dfrac{5\sqrt{2k}}{2k}$

9. $\sqrt{\dfrac{4}{x}} = \dfrac{\sqrt{4}}{\sqrt{x}} = \dfrac{2}{\sqrt{x}} \cdot \dfrac{\sqrt{x}}{\sqrt{x}} = \dfrac{2\sqrt{x}}{\left(\sqrt{x}\right)^2} = \dfrac{2\sqrt{x}}{x}$

11. $\sqrt{\dfrac{7}{2}} = \dfrac{\sqrt{7}}{\sqrt{2}} = \dfrac{\sqrt{7}}{\sqrt{2}} \cdot \dfrac{\sqrt{2}}{\sqrt{2}} = \dfrac{\sqrt{7 \cdot 2}}{\left(\sqrt{2}\right)^2} = \dfrac{\sqrt{14}}{2}$

13. $\sqrt{\dfrac{2y}{x}} = \dfrac{\sqrt{2y}}{\sqrt{x}} = \dfrac{\sqrt{2y}}{\sqrt{x}} \cdot \dfrac{\sqrt{x}}{\sqrt{x}} = \dfrac{\sqrt{2y \cdot x}}{\left(\sqrt{x}\right)^2} = \dfrac{\sqrt{2xy}}{x}$

15. $\sqrt{\dfrac{x}{12y}} = \dfrac{\sqrt{x}}{\sqrt{12y}} = \dfrac{\sqrt{x}}{\sqrt{12y}} \cdot \dfrac{\sqrt{12y}}{\sqrt{12y}} = \dfrac{\sqrt{x \cdot 12y}}{\left(\sqrt{12y}\right)^2} = \dfrac{\sqrt{x \cdot 4 \cdot 3y}}{12y} = \dfrac{2\sqrt{3xy}}{12y} = \dfrac{\sqrt{3xy}}{6y}$

17. $\dfrac{3}{\sqrt{x-4}} = \dfrac{3}{\sqrt{x-4}} \cdot \dfrac{\sqrt{x-4}}{\sqrt{x-4}} = \dfrac{3\sqrt{x-4}}{\left(\sqrt{x-4}\right)^2} = \dfrac{3\sqrt{x-4}}{x-4}$

19. $\dfrac{\sqrt{2a^3}}{\sqrt{3b}} = \dfrac{\sqrt{2a^3}}{\sqrt{3b}} \cdot \dfrac{\sqrt{3b}}{\sqrt{3b}} = \dfrac{\sqrt{6a^3b}}{\left(\sqrt{3b}\right)^2} = \dfrac{\sqrt{6 \cdot a^2 \cdot a \cdot b}}{3b} = \dfrac{a\sqrt{6ab}}{3b}$

21. $\dfrac{2}{\sqrt[3]{5}} = \dfrac{2}{\sqrt[3]{5}} \cdot \dfrac{\sqrt[3]{25}}{\sqrt[3]{25}} = \dfrac{2\sqrt[3]{25}}{\sqrt[3]{5 \cdot 25}} = \dfrac{2\sqrt[3]{25}}{\sqrt[3]{125}} = \dfrac{2\sqrt[3]{25}}{5}$

23. $\dfrac{5}{\sqrt[3]{4}} = \dfrac{5}{\sqrt[3]{4}} \cdot \dfrac{\sqrt[3]{2}}{\sqrt[3]{2}} = \dfrac{5\sqrt[3]{2}}{\sqrt[3]{4 \cdot 2}} = \dfrac{5\sqrt[3]{2}}{\sqrt[3]{8}} = \dfrac{5\sqrt[3]{2}}{2}$

25. $\dfrac{4}{5\sqrt[3]{x}} = \dfrac{4}{5\sqrt[3]{x}} \cdot \dfrac{\sqrt[3]{x^2}}{\sqrt[3]{x^2}} = \dfrac{4\sqrt[3]{x^2}}{5\sqrt[3]{x \cdot x^2}} = \dfrac{4\sqrt[3]{x^2}}{5\sqrt[3]{x^3}} = \dfrac{4\sqrt[3]{x^2}}{5x}$

27. $\dfrac{6}{\sqrt[3]{2x^2}} = \dfrac{6}{\sqrt[3]{2x^2}} \cdot \dfrac{\sqrt[3]{4x}}{\sqrt[3]{4x}} = \dfrac{6\sqrt[3]{4x}}{\sqrt[3]{2x^2 \cdot 4x}} = \dfrac{6\sqrt[3]{4x}}{\sqrt[3]{8x^3}} = \dfrac{6\sqrt[3]{4x}}{\sqrt[3]{8}\sqrt[3]{x^3}} = \dfrac{6\sqrt[3]{4x}}{2x} = \dfrac{3\sqrt[3]{4x}}{x}$

29. $\dfrac{7t}{\sqrt[4]{4t^3}} = \dfrac{7t}{\sqrt[4]{4t^3}} \cdot \dfrac{\sqrt[4]{4t}}{\sqrt[4]{4t}} = \dfrac{7t\sqrt[4]{4t}}{\sqrt[4]{4t^3 \cdot 4t}} = \dfrac{7t\sqrt[4]{4t}}{\sqrt[4]{16t^4}} = \dfrac{7t\sqrt[4]{4t}}{\sqrt[4]{16}\sqrt[4]{t^4}} = \dfrac{7t\sqrt[4]{4t}}{2t} = \dfrac{7\sqrt[4]{4t}}{2}$

31. $\dfrac{\sqrt[3]{x}}{\sqrt{x}} = \dfrac{\sqrt[3]{x}}{\sqrt{x}} \cdot \dfrac{\sqrt{x}}{\sqrt{x}} = \dfrac{x^{\frac{1}{3}} \cdot x^{\frac{1}{2}}}{\left(\sqrt{x}\right)^2} = \dfrac{x^{\frac{1}{3}+\frac{1}{2}}}{x} = \dfrac{x^{\frac{2}{6}+\frac{3}{6}}}{x} = \dfrac{x^{\frac{5}{6}}}{x} = \dfrac{\sqrt[6]{x^5}}{x}$

33. $\sqrt[5]{\dfrac{2}{x^3}} = \dfrac{\sqrt[5]{2}}{\sqrt[5]{x^3}} = \dfrac{\sqrt[5]{2}}{\sqrt[5]{x^3}} \cdot \dfrac{\sqrt[5]{x^2}}{\sqrt[5]{x^2}} = \dfrac{\sqrt[5]{2 \cdot x^2}}{\sqrt[5]{x^3 \cdot x^2}} = \dfrac{\sqrt[5]{2x^2}}{\sqrt[5]{x^5}} = \dfrac{\sqrt[5]{2x^2}}{x}$

35. $\sqrt[4]{\dfrac{4}{9x^2}} = \dfrac{\sqrt[4]{4}}{\sqrt[4]{9x^2}} = \dfrac{\sqrt[4]{4}}{\sqrt[4]{9x^2}} \cdot \dfrac{\sqrt[4]{9x^2}}{\sqrt[4]{9x^2}} = \dfrac{\sqrt[4]{4 \cdot 9x^2}}{\sqrt[4]{9x^2 \cdot 9x^2}} = \dfrac{\sqrt[4]{36x^2}}{\sqrt[4]{81x^4}} = \dfrac{\sqrt[4]{6^2 x^2}}{3x} = \dfrac{6^{2/4} x^{2/4}}{3x} = \dfrac{6^{1/2} x^{1/2}}{3x} = \dfrac{\sqrt{6x}}{3x}$

37. $\sqrt[5]{\dfrac{3w}{4x^4 y^2}} = \dfrac{\sqrt[5]{3w}}{\sqrt[5]{4x^4 y^2}} = \dfrac{\sqrt[5]{3w}}{\sqrt[5]{4x^4 y^2}} \cdot \dfrac{\sqrt[5]{8xy^3}}{\sqrt[5]{8xy^3}} = \dfrac{\sqrt[5]{3w}\sqrt[5]{8xy^3}}{\sqrt[5]{4x^4 y^2 \cdot 8xy^3}} = \dfrac{\sqrt[5]{3w}\sqrt[5]{8xy^3}}{\sqrt[5]{32x^5 y^5}} = \dfrac{\sqrt[5]{24wxy^3}}{2xy}$

39. $\dfrac{1}{5+\sqrt{3}} = \dfrac{1}{5+\sqrt{3}} \cdot \dfrac{5-\sqrt{3}}{5-\sqrt{3}} = \dfrac{1 \cdot 5 - 1 \cdot \sqrt{3}}{(5)^2 - \left(\sqrt{3}\right)^2} = \dfrac{5-\sqrt{3}}{25-3} = \dfrac{5-\sqrt{3}}{22}$

41. $\dfrac{2}{\sqrt{3}+\sqrt{7}} = \dfrac{2}{\sqrt{3}+\sqrt{7}} \cdot \dfrac{\sqrt{3}-\sqrt{7}}{\sqrt{3}-\sqrt{7}} = \dfrac{2 \cdot \sqrt{3} - 2 \cdot \sqrt{7}}{\left(\sqrt{3}\right)^2 - \left(\sqrt{7}\right)^2} = \dfrac{2\sqrt{3}-2\sqrt{7}}{3-7} = \dfrac{2\sqrt{3}-2\sqrt{7}}{-4} = \dfrac{\sqrt{7}-\sqrt{3}}{2}$

43. $\dfrac{1}{3\sqrt{r}-7} = \dfrac{1}{3\sqrt{r}-7} \cdot \dfrac{3\sqrt{r}+7}{3\sqrt{r}+7} = \dfrac{1 \cdot 3\sqrt{r} + 1 \cdot 7}{\left(3\sqrt{r}\right)^2 - (7)^2} = \dfrac{3\sqrt{r}+7}{9r-49}$

45. $\dfrac{\sqrt{x}}{\sqrt{x}-1} = \dfrac{\sqrt{x}}{\sqrt{x}-1} \cdot \dfrac{\sqrt{x}+1}{\sqrt{x}+1} = \dfrac{\sqrt{x} \cdot \sqrt{x} + \sqrt{x} \cdot 1}{\left(\sqrt{x}\right)^2 - (1)^2} = \dfrac{x+\sqrt{x}}{x-1}$

47. $\dfrac{3\sqrt{x}}{4\sqrt{x}-\sqrt{5}} = \dfrac{3\sqrt{x}}{4\sqrt{x}-\sqrt{5}} \cdot \dfrac{4\sqrt{x}+\sqrt{5}}{4\sqrt{x}+\sqrt{5}} = \dfrac{3\sqrt{x} \cdot 4\sqrt{x} + 3\sqrt{x} \cdot \sqrt{5}}{\left(4\sqrt{x}\right)^2 - \left(\sqrt{5}\right)^2} = \dfrac{12\left(\sqrt{x}\right)^2 + 3\sqrt{5x}}{4^2\left(\sqrt{x}\right)^2 - 5} = \dfrac{12x+3\sqrt{5x}}{16x-5}$

49. $\dfrac{\sqrt{x}}{\sqrt{x}-y} = \dfrac{\sqrt{x}}{\sqrt{x}-y} \cdot \dfrac{\sqrt{x}+y}{\sqrt{x}+y} = \dfrac{\sqrt{x} \cdot \sqrt{x} + \sqrt{x} \cdot y}{\left(\sqrt{x}\right)^2 - (y)^2} = \dfrac{x+y\sqrt{x}}{x-y^2}$

51. $\dfrac{\sqrt{x}-5}{\sqrt{x}+5} = \dfrac{\sqrt{x}-5}{\sqrt{x}+5} \cdot \dfrac{\sqrt{x}-5}{\sqrt{x}-5} = \dfrac{\left(\sqrt{x}\right)^2 - 2\left(\sqrt{x}\right)(5) + (5)^2}{\left(\sqrt{x}\right)^2 - (5)^2} = \dfrac{x-10\sqrt{x}+25}{x-25}$

53. $\dfrac{2\sqrt{x}+5}{3\sqrt{x}+1} = \dfrac{2\sqrt{x}+5}{3\sqrt{x}+1} \cdot \dfrac{3\sqrt{x}-1}{3\sqrt{x}-1} = \dfrac{2\sqrt{x}\cdot 3\sqrt{x}-2\sqrt{x}\cdot 1+5\cdot 3\sqrt{x}-5\cdot 1}{\left(3\sqrt{x}\right)^2-\left(1\right)^2} = \dfrac{6\left(\sqrt{x}\right)^2-2\sqrt{x}+15\sqrt{x}-5}{3^2\left(\sqrt{x}\right)^2-1} = \dfrac{6x+13\sqrt{x}-5}{9x-1}$

55. $\dfrac{6\sqrt{x}+\sqrt{5}}{3\sqrt{x}-\sqrt{7}} = \dfrac{6\sqrt{x}+\sqrt{5}}{3\sqrt{x}-\sqrt{7}} \cdot \dfrac{3\sqrt{x}+\sqrt{7}}{3\sqrt{x}+\sqrt{7}}$

$\qquad = \dfrac{6\sqrt{x}\cdot 3\sqrt{x}+6\sqrt{x}\cdot\sqrt{7}+\sqrt{5}\cdot 3\sqrt{x}+\sqrt{5}\cdot\sqrt{7}}{\left(3\sqrt{x}\right)^2-\left(\sqrt{7}\right)^2}$

$\qquad = \dfrac{18\left(\sqrt{x}\right)^2+6\sqrt{7}\cdot x+3\sqrt{5}\cdot x+\sqrt{5\cdot 7}}{3^2\left(\sqrt{x}\right)^2-7}$

$\qquad = \dfrac{18x+6\sqrt{7x}+3\sqrt{5x}+\sqrt{35}}{9x-7}$

57. $\dfrac{\sqrt{x}-\sqrt{y}}{\sqrt{x}+\sqrt{y}} = \dfrac{\sqrt{x}-\sqrt{y}}{\sqrt{x}+\sqrt{y}} \cdot \dfrac{\sqrt{x}-\sqrt{y}}{\sqrt{x}-\sqrt{y}} = \dfrac{\left(\sqrt{x}\right)^2-2\left(\sqrt{x}\right)\left(\sqrt{y}\right)+\left(\sqrt{y}\right)^2}{\left(\sqrt{x}\right)^2-\left(\sqrt{y}\right)^2} = \dfrac{x-2\sqrt{xy}+y}{x-y}$

59. $\dfrac{1}{\sqrt{x+1}-\sqrt{x}} = \dfrac{1}{\sqrt{x+1}-\sqrt{x}} \cdot \dfrac{\sqrt{x+1}+\sqrt{x}}{\sqrt{x+1}+\sqrt{x}} = \dfrac{\sqrt{x+1}+\sqrt{x}}{\left(\sqrt{x+1}\right)^2-\left(\sqrt{x}\right)^2} = \dfrac{\sqrt{x+1}+\sqrt{x}}{x+1-x} = \sqrt{x+1}+\sqrt{x}$

61. a. $\sqrt{\dfrac{3h}{2}} = \dfrac{\sqrt{3h}}{\sqrt{2}} = \dfrac{\sqrt{3h}}{\sqrt{2}} \cdot \dfrac{\sqrt{2}}{\sqrt{2}} = \dfrac{\sqrt{6h}}{\left(\sqrt{2}\right)^2} = \dfrac{\sqrt{6h}}{2}$

b. $\dfrac{\sqrt{6(1450)}}{2} = \dfrac{\sqrt{8700}}{2} \approx 46.64$

The horizon is about 47 miles from the top of the skyscraper.

c. $\dfrac{\sqrt{6(30,000)}}{2} = \dfrac{\sqrt{180,000}}{2} \approx 212.13$

The airplane is about 212 miles from the horizon.

63. a. Answers may vary. Example:

Cut the paper in half so that the dimensions are $\dfrac{x\sqrt{2}}{2}$ by x.

$\dfrac{x}{\dfrac{x\sqrt{2}}{2}} = \dfrac{x}{1} \cdot \dfrac{2}{x\sqrt{2}} = \dfrac{2x}{x\sqrt{2}} = \dfrac{2}{\sqrt{2}} = \dfrac{2}{\sqrt{2}} \cdot \dfrac{\sqrt{2}}{\sqrt{2}} = \dfrac{2\sqrt{2}}{\left(\sqrt{2}\right)^2} = \dfrac{2\sqrt{2}}{2} = \dfrac{\sqrt{2}}{1}$

b.
$$1 = x \cdot x\sqrt{2}$$
$$1 = x^2\sqrt{2}$$
$$x^2 = \frac{1}{\sqrt{2}}$$
$$x^2 = \frac{\sqrt{2}}{2}$$
$$x = \sqrt{\frac{\sqrt{2}}{2}} = \frac{\sqrt[4]{2}}{\sqrt{2}} \cdot \frac{\sqrt{2}}{\sqrt{2}} = \frac{2^{1/4} \cdot 2^{1/2}}{2} = \frac{\sqrt[4]{2^3}}{2} = \frac{\sqrt[4]{8}}{2} \approx 0.841$$

The page has a width of about 0.841 meters.

---

65. Student 1 did the work correctly.
Answers may vary. Example:
Student 2's error was to square the entire expression. This changes the value of the expression. To simplify the expression, you need to multiply by something equivalent to 1.

67. The student did not rationalize the denominator correctly.
Answers may vary. Example:
For the radicand in the denominator to be a perfect cube, $\sqrt[3]{x}$ needs to be multiplied by $\sqrt[3]{x^2}$ to yield $\sqrt[3]{x} \cdot \sqrt[3]{x^2} = \sqrt[3]{x^3}$.
$$\frac{5}{\sqrt[3]{x}} = \frac{5}{\sqrt[3]{x}} \cdot \frac{\sqrt[3]{x^2}}{\sqrt[3]{x^2}} = \frac{5\sqrt[3]{x^2}}{\sqrt[3]{x \cdot x^2}} = \frac{5\sqrt[3]{x^2}}{\sqrt[3]{x^3}} = \frac{5\sqrt[3]{x^2}}{x}$$

69. $\dfrac{\sqrt{x}}{3} = \dfrac{\sqrt{x}}{3} \cdot \dfrac{\sqrt{x}}{\sqrt{x}} = \dfrac{\sqrt{x^2}}{3\sqrt{x}} = \dfrac{x}{3\sqrt{x}}$

71.
$$\frac{\sqrt{x+2}-\sqrt{x}}{2} = \frac{\sqrt{x+2}-\sqrt{x}}{2} \cdot \frac{\sqrt{x+2}+\sqrt{x}}{\sqrt{x+2}+\sqrt{x}}$$
$$= \frac{\left(\sqrt{x+2}\right)^2 - \left(\sqrt{x}\right)^2}{2\left(\sqrt{x+2}+\sqrt{x}\right)}$$
$$= \frac{x+2-x}{2\left(\sqrt{x+2}+\sqrt{x}\right)}$$
$$= \frac{2}{2\left(\sqrt{x+2}+\sqrt{x}\right)}$$
$$= \frac{1}{\sqrt{x+2}+\sqrt{x}}$$

73.
$$\frac{\dfrac{1}{\sqrt{x}}-\dfrac{3}{x}}{\dfrac{2}{\sqrt{x}}+\dfrac{1}{x}} = \frac{\dfrac{1}{\sqrt{x}}\cdot\dfrac{\sqrt{x}}{\sqrt{x}}-\dfrac{3}{x}}{\dfrac{2}{\sqrt{x}}\cdot\dfrac{\sqrt{x}}{\sqrt{x}}+\dfrac{1}{x}}$$
$$= \frac{\dfrac{\sqrt{x}}{x}-\dfrac{3}{x}}{\dfrac{2\sqrt{x}}{x}+\dfrac{1}{x}}$$
$$= \frac{\dfrac{\sqrt{x}-3}{x}}{\dfrac{2\sqrt{x}+1}{x}}$$
$$= \frac{\sqrt{x}-3}{x} \div \frac{2\sqrt{x}+1}{x}$$
$$= \frac{\sqrt{x}-3}{x} \cdot \frac{x}{2\sqrt{x}+1}$$
$$= \frac{\sqrt{x}-3}{2\sqrt{x}+1}$$
$$= \frac{\sqrt{x}-3}{2\sqrt{x}+1} \cdot \frac{2\sqrt{x}-1}{2\sqrt{x}-1}$$
$$= \frac{\sqrt{x}\cdot 2\sqrt{x} - \sqrt{x}\cdot 1 - 3\cdot 2\sqrt{x} - 3(-1)}{\left(2\sqrt{x}\right)^2 - (1)^2}$$
$$= \frac{2x - \sqrt{x} - 6\sqrt{x} + 3}{4x - 1}$$
$$= \frac{2x - 7\sqrt{x} + 3}{4x - 1}$$

75.
$$x\sqrt{2} + 3\sqrt{5} = 9\sqrt{5}$$
$$x\sqrt{2} = 9\sqrt{5} - 3\sqrt{5}$$
$$x\sqrt{2} = 6\sqrt{5}$$
$$x = \frac{6\sqrt{5}}{\sqrt{2}} = \frac{6\sqrt{5}}{\sqrt{2}} \cdot \frac{\sqrt{2}}{\sqrt{2}} = \frac{6\sqrt{10}}{2} = 3\sqrt{10}$$

77. Answers may vary. Example:

(1) Determine the conjugate of the denominator.

(2) Multiply the original fraction by the fraction:
$$\frac{\text{conjugate}}{\text{conjugate}}$$

(3) Find the product of the fractions and simplify.

79. a. $A^3 + B^3 = (A+B)\left(A^2 - AB + B^2\right)$

b. $(A+B)\left(A^2 - AB + B^2\right)$
$$= A^3 - A^2B + AB^2 + A^2B - AB^2 + B^3$$
$$= A^3 + B^3$$
$(A+B)\left(A^2 - AB + B^2\right)$ is the factored form of $A^3 + B^3$.

c. $(x+2)\left(x^2 - 2x + 4\right)$
$$= x\left(x^2\right) + 2\left(x^2\right) - x(2x) - 2(2x)$$
$$\quad + x(4) + 2(4)$$
$$= x^3 + 2x^2 - 2x^2 - 4x + 4x + 8$$
$$= x^3 + 8$$
It follows the equation
$(A+B)\left(A^2 - AB + B^2\right) = A^3 + B^3$, where
$A = x$ and $B = 2$.

d. $\left(\sqrt[3]{x} + \sqrt[3]{2}\right)\left(\sqrt[3]{x^2} - \sqrt[3]{2x} + \sqrt[3]{4}\right)$
$$= \left(\sqrt[3]{x}\right)\left(\sqrt[3]{x^2}\right) + \left(\sqrt[3]{2}\right)\left(\sqrt[3]{x^2}\right)$$
$$\quad - \left(\sqrt[3]{x}\right)\left(\sqrt[3]{2x}\right) - \left(\sqrt[3]{2}\right)\left(\sqrt[3]{2x}\right)$$
$$\quad + \left(\sqrt[3]{x}\right)\left(\sqrt[3]{4}\right) + \left(\sqrt[3]{2}\right)\left(\sqrt[3]{4}\right)$$
$$= \sqrt[3]{x^3} + \sqrt[3]{2x^2} - \sqrt[3]{2x^2} - \sqrt[3]{4x}$$
$$\quad + \sqrt[3]{4x} + \sqrt[3]{8}$$
$$= x + 2$$
It follows the equation
$(A+B)\left(A^2 - AB + B^2\right) = A^3 + B^3$, where
$A = \sqrt[3]{x}$ and $B = \sqrt[3]{2}$.

e. $\dfrac{1}{\sqrt[3]{x} + \sqrt[3]{2}}$

$$= \frac{1}{\sqrt[3]{x} + \sqrt[3]{2}} \cdot \frac{\sqrt[3]{x^2} - \sqrt[3]{2x} + \sqrt[3]{4}}{\sqrt[3]{x^2} - \sqrt[3]{2x} + \sqrt[3]{4}}$$

$$= \frac{\sqrt[3]{x^2} - \sqrt[3]{2x} + \sqrt[3]{4}}{\sqrt[3]{x^3} + \sqrt[3]{2x^2} - \sqrt[3]{2x^2} - \sqrt[3]{4x} + \sqrt[3]{4x} + \sqrt[3]{8}}$$

$$= \frac{\sqrt[3]{x^2} - \sqrt[3]{2x} + \sqrt[3]{4}}{x + 2}$$

81. $(5x - 4)\left(3x^2 - 2x - 1\right)$
$$= 5x \cdot 3x^2 - 4 \cdot 3x^2 - 5x \cdot 2x + 4 \cdot 2x$$
$$\quad - 5x \cdot 1 + 4 \cdot 1$$
$$= 15x^3 - 12x^2 - 10x^2 + 8x - 5x + 4$$
$$= 15x^3 - 22x^2 + 3x + 4$$
This is a cubic (or third-degree) polynomial in one variable.

83. $24x^3 - 3000 = 24\left(x^3 - 125\right)$
$$= 24(x - 5)\left(x^2 + 5x + 25\right)$$
This is a cubic (or third-degree) polynomial in one variable.

85. $5x^2 - 3 = 4x - 1$
$$5x^2 - 3 - 4x + 1 = 0$$
$$5x^2 - 4x - 2 = 0$$
$$a = 5, b = -4, c = -2$$
$$x = \frac{4 \pm \sqrt{(-4)^2 - 4(5)(-2)}}{2(5)}$$
$$= \frac{4 \pm \sqrt{56}}{10}$$
$$= \frac{2 \pm \sqrt{14}}{5}$$
This is a quadratic equation in one variable.

Homework 9.4

1. $y = 2\sqrt{x}$

x	y
0	0
1	2
4	4
9	6
16	8

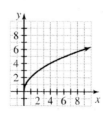

3. $y = -\sqrt{x}$

x	y
0	0
1	-1
4	-2
9	-3
16	-4

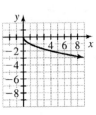

5. $y = \sqrt{x} + 3$

x	y
0	3
1	4
4	5
9	6
16	7

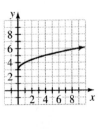

7. $y = 2\sqrt{x} - 5$

x	y
0	-5
1	-3
4	-1
9	1
16	3

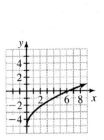

9. $y = -3\sqrt{x} + 4$

x	y
0	4
1	1
4	-2
9	-5
16	-8

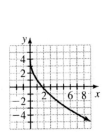

11. $y = \sqrt{x - 2}$

x	y
2	0
3	1
6	2
11	3
18	4

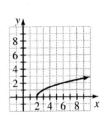

13. $y = -\sqrt{x + 2}$

x	y
-2	0
-1	-1
2	-2
7	-3
14	-4

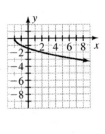

15. $y = \dfrac{1}{2}\sqrt{x - 4}$

x	y
4	0
5	$\dfrac{1}{2}$
8	1
13	$\dfrac{3}{2}$
20	2

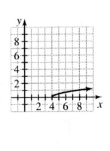

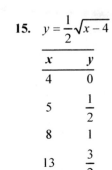

17. $y = \sqrt{x + 3} + 2$

x	y
-3	2
-2	3
1	4
6	5
13	6

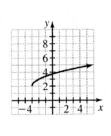

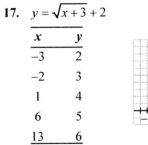

19. $y = -2\sqrt{x + 3} - 4$

x	y
-3	-4
-2	-6
1	-8
6	-10
13	-12

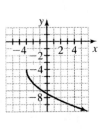

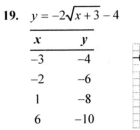

21. $y = 4\sqrt{x - 1} - 3$

x	y
1	-3
2	1
5	5
10	9
17	13

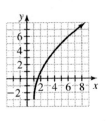

23. $\sqrt{x} + y = 4$

$\quad\quad y = -\sqrt{x} + 4$

x	y
0	4
1	3
4	2
9	1
16	0

25. $2y - 6\sqrt{x} = 8$

$\quad\quad 2y = 6\sqrt{x} + 8$

$\quad\quad\quad y = 3\sqrt{x} + 4$

x	y
0	4
1	7
4	10
9	13
16	16

27. $y = -2\sqrt{x}$

x	y
0	0
1	-2
4	-4
9	-6
16	-8

Domain: $x \geq 0$

Range: $y \leq 0$

29. $y = \sqrt{x + 2}$

x	y
-2	0
-1	1
2	2
7	3
14	4

Domain: $x \geq -2$

Range: $y \geq 0$

31. $y = \sqrt{x} + 2$

x	y
0	2
1	3
4	4
9	5
16	6

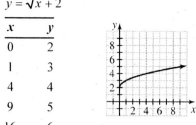

Domain: $x \geq 0$

Range: $y \geq 2$

33. $y = \sqrt{x - 5} - 3$

x	y
5	-3
6	-2
9	-1
14	0
21	1

Domain: $x \geq 5$

Range: $y \geq -3$

35. $y = 2\sqrt{x + 5} + 1$

x	y
-5	1
-4	3
-1	5
4	7
11	9

Domain: $x \geq -5$

Range: $y \geq 1$

37. $y = -\sqrt{x - 2} + 4$

x	y
2	4
3	3
6	2
11	1
18	0

Domain: $x \geq 2$

Range: $y \leq 4$

39. a. $f(x) = 2\sqrt{x-3}$

x	$f(x)$
3	0
4	2
7	4
12	6
19	8

b.

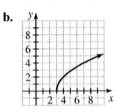

c. For each input-output pair, the output variable is equal to 2 times the square root of 3 less than the input variable.

41. $f(4) = 7\sqrt{4} - 3 = 7 \cdot 2 - 3 = 11$

43. $f(9c) = 7\sqrt{9c} - 3 = 7 \cdot 3\sqrt{c} - 3 = 21\sqrt{c} - 3$

45. $f + g = \left(5\sqrt{x} - 9\right) + \left(4\sqrt{x} + 1\right)$
$= 5\sqrt{x} - 9 + 4\sqrt{x} + 1$
$= 9\sqrt{x} - 8$

47. $f \cdot g = \left(5\sqrt{x} - 9\right)\left(4\sqrt{x} + 1\right)$
$= 5\sqrt{x} \cdot 4\sqrt{x} - 9 \cdot 4\sqrt{x} + 5\sqrt{x} \cdot 1 - 9 \cdot 1$
$= 20x - 36\sqrt{x} + 5\sqrt{x} - 9$
$= 20x - 31\sqrt{x} - 9$

49. $f - g = \left(2\sqrt{x} - 3\sqrt{5}\right) - \left(2\sqrt{x} + 3\sqrt{5}\right)$
$= 2\sqrt{x} - 3\sqrt{5} - 2\sqrt{x} - 3\sqrt{5}$
$= -6\sqrt{5}$

51. $\dfrac{f}{g} = \dfrac{2\sqrt{x} - 3\sqrt{5}}{2\sqrt{x} + 3\sqrt{5}}$

$= \dfrac{2\sqrt{x} - 3\sqrt{5}}{2\sqrt{x} + 3\sqrt{5}} \cdot \dfrac{2\sqrt{x} - 3\sqrt{5}}{2\sqrt{x} - 3\sqrt{5}}$

$= \dfrac{\left(2\sqrt{x}\right)^2 - 2\left(2\sqrt{x}\right)\left(3\sqrt{5}\right) + \left(3\sqrt{5}\right)^2}{\left(2\sqrt{x}\right)^2 - \left(3\sqrt{5}\right)^2}$

$= \dfrac{4x - 12\sqrt{5x} + 45}{4x - 45}$

53. $f + g = \left(\sqrt{x+1} - 2\right) + \left(\sqrt{x+1} + 2\right)$
$= \sqrt{x+1} - 2 + \sqrt{x+1} + 2$
$= 2\sqrt{x+1}$

55. $f \cdot g = \left(\sqrt{x+1} - 2\right)\left(\sqrt{x+1} + 2\right)$
$= \left(\sqrt{x+1}\right)^2 - (2)^2$
$= x + 1 - 4$
$= x - 3$

57. a.

The model fits the data well.

b. $t = 2005 - 1999 = 6$
$f(6) = 20.4\sqrt{6} + 21 \approx 71.0$
We estimate that in 2005, about 71% of e-mail was spam. This estimate involves interpolation, as it involves determining a value within existing data values.

c. $t = 2013 - 1999 = 14$
$f(14) = 20.4\sqrt{14} + 21 \approx 97.3$
We predict that in 2013, about 97% of e-mail will be spam. This estimate involves extrapolation, as it involves determining a value beyond existing values.

59. $f(-6) = 0$

61. $f(0) \approx 2.4$

63. $x = -6$

65. $x = 3$

67. a. $a < 0, h = 0,$ and $k > 0$

b. $a > 0, h < 0,$ and $k < 0$

c. $a > 0, h < 0,$ and $k > 0$

d. $a < 0, h > 0,$ and $k = 0$

69. Answers may vary. Example:

For the family of curves $y = a\sqrt{x-h} + k$,

$k = 2$, and $h = 0$.

Let $a = -4, -3, -2, -1, -\dfrac{1}{2}, \dfrac{1}{2}, 1, 2, 3$, and 4.

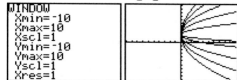

71. If $a < 0$, f has a maximum point at (h, k).

If $a > 0$, f has a minimum point at (h, k).

73. $f(x) = 2\sqrt{x+3} + 2$; $g(x) = -2\sqrt{x+3} + 2$

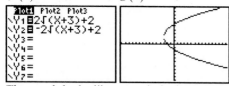

The graph looks like a parabola that opens to the right. The relation is not a function because it would fail the vertical line test.

75. Answers may vary. Example:
The y-coordinate of every point represented by $f(x)$ is the opposite of the y-coordinate of every point represented by $g(x)$ with the same x-coordinate. Therefore, for any x, the point $(x, g(x))$ is the same distance from the x-axis as the point $(x, f(x))$, but it is on the opposite side of the x-axis.

77. Answers may vary. Example:
The graph of f can be found by translating the graph of $f(x) = \sqrt{x}$ horizontally by $|h|$ units (left if $h < 0$ and right if $h > 0$), and vertically by $|k|$ units (up if $k > 0$ and down if $k < 0$).

Also, if $a > 0$, the graph is increasing and if $a < 0$, the graph is decreasing. The greater the absolute value of a, the steeper the slope of the graph.

79. $2x - 5y = 20$

$-5y = 20 - 2x$

$y = \dfrac{2}{5}x - 4$

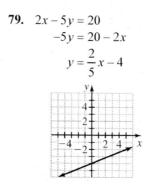

81. $y = 2\sqrt{x+3} - 4$

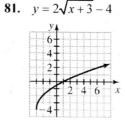

83. $y = 8\left(\dfrac{1}{2}\right)^x$

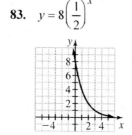

85. $f \circ g = f\big(g(x)\big)$

$= \sqrt{3(4x+5)+2} - 7$

$= \sqrt{12x+15+2} - 7$

$= \sqrt{12x+17} - 7$

87. $f \circ g = f\big(g(x)\big) = \sqrt{\left(x^2+2\right)+3} = \sqrt{x^2+5}$

89. $6x^2 - 5x - 6 = (3x+2)(2x-3)$

This is a quadratic polynomial in one variable.

91. $f(x) = 3x^2 - 2x + 4$

$6 = 3x^2 - 2x + 4$

$0 = 3x^2 - 2x - 2$

$\dfrac{-b \pm \sqrt{b^2 - 4ac}}{2a} = \dfrac{2 \pm \sqrt{(-2)^2 - 4(3)(-2)}}{2(3)}$

$= \dfrac{2 \pm \sqrt{28}}{6}$

$= \dfrac{1 \pm \sqrt{7}}{3}$

This is a quadratic function.

93. Use the points to write three equations.

$y = ax^2 + bx + c$

$4 = a(1)^2 + b(1) + c$

$4 = a + b + c$

$y = ax^2 + bx + c$

$14 = a(3)^2 + b(3) + c$

$14 = 9a + 3b + c$

$y = ax^2 + bx + c$

$25 = a(4)^2 + b(4) + c$

$25 = 16a + 4b + c$

Solve the system.

$14 = 9a + 3b + c$

$\underline{4 = a + b + c}$

$10 = 8a + 2b$

$5 = 4a + b$

$b = 5 - 4a$

$25 = 16a + 4(5 - 4a) + c$

$25 = 16a + 20 - 16a + c$

$c = 5$

$4 = a + 5 - 4a + 5$

$3a = 6$

$a = 2$

$b = 5 - 4(2) = -3$

$y = 2x^2 - 3x + 5$

This is a quadratic function.

Homework 9.5

1. $\sqrt{x} = 5$

$\left(\sqrt{x}\right)^2 = 5^2$

$x = 25$

Check $x = 25$:

$\sqrt{(25)} \overset{?}{=} 5$

$5 \overset{?}{=} 5$ true

The solution is $x = 25$.

3. $\sqrt{x} = -2$

$\left(\sqrt{x}\right)^2 = (-2)^2$

$x = 4$

Check $x = 4$:

$\sqrt{(4)} \overset{?}{=} -2$

$2 \overset{?}{=} -2$ false

There are no real solutions.

5. $\sqrt[3]{t} = -2$

$\left(\sqrt[3]{t}\right)^3 = (-2)^3$

$t = -8$

Check $t = -8$:

$\sqrt[3]{(-8)} \overset{?}{=} -2$

$-2 \overset{?}{=} -2$ true

The solution is $t = -8$.

7. $3\sqrt{x} - 1 = 5$

$3\sqrt{x} = 6$

$\sqrt{x} = 2$

$\left(\sqrt{x}\right)^2 = 2^2$

$x = 4$

Check $x = 4$:

$3\sqrt{(4)} - 1 \overset{?}{=} 5$

$5 \overset{?}{=} 5$ true

The solution is $x = 4$.

9. $\sqrt{x - 1} = 2$

$\left(\sqrt{x - 1}\right)^2 = 2^2$

$x - 1 = 4$

$x = 5$

Check $x = 5$:

$\sqrt{(5) - 1} \overset{?}{=} 2$

$2 \overset{?}{=} 2$ true

The solution is $x = 5$.

11. $\sqrt[4]{r + 2} = 2$

$\left(\sqrt[4]{r + 2}\right)^4 = (2)^4$

$r + 2 = 16$

$r = 14$

Check $r = 14$:

$\sqrt[4]{14 + 2} \overset{?}{=} 2$

$2 \overset{?}{=} 2$ true

The solution is $r = 14$.

13. $\sqrt{5x-7}+7=3$

$\sqrt{5x-7}=-4$

$\left(\sqrt{5x-7}\right)^2=(-4)^2$

$5x-7=16$

$5x=23$

$x=\dfrac{23}{5}$

Check $x=\dfrac{23}{5}$:

$\sqrt{5\left(\dfrac{23}{5}\right)-7}+7\overset{?}{=}3$

$11\overset{?}{=}3$ false

There are no real solutions.

15. $\sqrt[3]{2x-5}+3=7$

$\sqrt[3]{2x-5}=4$

$\left(\sqrt[3]{2x-5}\right)^3=(4)^3$

$2x-5=64$

$2x=69$

$x=\dfrac{69}{2}$

Check $x=\dfrac{69}{2}$:

$\sqrt[3]{2\left(\dfrac{69}{2}\right)-5}+3\overset{?}{=}7$

$7\overset{?}{=}7$ true

The solution is $x=\dfrac{69}{2}$.

17. $2-10\sqrt{6x+3}=-98$

$-10\sqrt{6x+3}=-100$

$\sqrt{6x+3}=10$

$\left(\sqrt{6x+3}\right)^2=(10)^2$

$6x+3=100$

$6x=97$

$x=\dfrac{97}{6}$

Check $x=\dfrac{97}{6}$:

$2-10\sqrt{6\left(\dfrac{97}{6}\right)+3}\overset{?}{=}-98$

$-98\overset{?}{=}-98$ true

The solution is $x=\dfrac{97}{6}$.

19. $\sqrt{3k+1}=\sqrt{2k+6}$

$\left(\sqrt{3k+1}\right)^2=\left(\sqrt{2k+6}\right)^2$

$3k+1=2k+6$

$k=5$

Check $k=5$:

$\sqrt{3(5)+1}\overset{?}{=}\sqrt{2(5+6)}$

$4\overset{?}{=}4$ true

The solution is $k=5$.

21. $\sqrt[4]{6x-3}=\sqrt[4]{2x+17}$

$\left(\sqrt[4]{6x-3}\right)^4=\left(\sqrt[4]{2x+17}\right)^4$

$6x-3=2x+17$

$4x=20$

$x=5$

Check $x=5$:

$\sqrt[4]{6(5)-3}\overset{?}{=}\sqrt[4]{2(5)+17}$

$\sqrt[4]{27}\overset{?}{=}\sqrt[4]{27}$ true

The solution is $x=5$.

23. $2\sqrt{1-x}-\sqrt{2x+5}=0$

$2\sqrt{1-x}=\sqrt{2x+5}$

$\left(2\sqrt{1-x}\right)^2=\left(\sqrt{2x+5}\right)^2$

$4(1-x)=2x+5$

$4-4x=2x+5$

$-6x=1$

$x=-\dfrac{1}{6}$

Check $x=-\dfrac{1}{6}$:

$2\sqrt{1-\left(-\dfrac{1}{6}\right)}-\sqrt{2\left(-\dfrac{1}{6}\right)+5}\overset{?}{=}0$

$2\sqrt{\dfrac{7}{6}}-\sqrt{\dfrac{28}{6}}\overset{?}{=}0$

$\sqrt{\dfrac{28}{6}}-\sqrt{\dfrac{28}{6}}\overset{?}{=}0$

$0\overset{?}{=}0$ true

The solution is $x=-\dfrac{1}{6}$.

25.
$$\sqrt{3w+3} = w-5$$
$$\left(\sqrt{3w+3}\right)^2 = (w-5)^2$$
$$3w+3 = w^2 - 10w + 25$$
$$w^2 - 13w + 22 = 0$$
$$(w-11)(w-2) = 0$$
$$w-11=0 \quad \text{or} \quad w-2=0$$
$$w=11 \quad \text{or} \qquad w=2$$

Check $w = 11$:
$$\sqrt{3(11)+3} \stackrel{?}{=} (11)-5$$
$$6 \stackrel{?}{=} 6 \quad \text{true}$$

Check $w = 2$:
$$\sqrt{3(2)+3} \stackrel{?}{=} (2)-5$$
$$3 \stackrel{?}{=} -3 \quad \text{false}$$
The solution is $w = 11$.

27.
$$\sqrt{12x+13} + 2 = 3x$$
$$\sqrt{12x+13} = 3x - 2$$
$$\left(\sqrt{12x+13}\right)^2 = (3x-2)^2$$
$$12x+13 = 9x^2 - 12x + 4$$
$$9x^2 - 24x - 9 = 0$$
$$3\left(3x^2 - 8x - 3\right) = 0$$
$$3(3x+1)(x-3) = 0$$
$$3x+1=0 \quad \text{or} \quad x-3=0$$
$$x = -\frac{1}{3} \quad \text{or} \qquad x = 3$$

Check $x = -\frac{1}{3}$:
$$\sqrt{12\left(-\frac{1}{3}\right)+13} + 2 \stackrel{?}{=} 3\left(-\frac{1}{3}\right)$$
$$5 \stackrel{?}{=} -1 \quad \text{false}$$

Check $x = 3$:
$$\sqrt{12(3)+13} + 2 \stackrel{?}{=} 3(3)$$
$$9 \stackrel{?}{=} 9 \quad \text{true}$$
The solution is $x = 3$.

29.
$$2 + \sqrt{10-x} = -x$$
$$\sqrt{10-x} = -x - 2$$
$$\left(\sqrt{10-x}\right)^2 = (-x-2)^2$$
$$10 - x = x^2 + 4x + 4$$
$$x^2 + 5x - 6 = 0$$
$$x = \frac{-5 \pm \sqrt{(5)^2 - 4(1)(-6)}}{2(1)}$$
$$= \frac{-5 \pm \sqrt{49}}{2}$$
$$= \frac{-5 \pm 7}{2}$$
$$x = \frac{-5-7}{2} \quad \text{or} \quad x = \frac{-5+7}{2}$$
$$x = \frac{-12}{2} \quad \text{or} \quad x = \frac{2}{2}$$
$$x = -6 \qquad \text{or} \quad x = 1$$

Check $x = -6$:
$$2 + \sqrt{10-(-6)} \stackrel{?}{=} -(-6)$$
$$2 + \sqrt{16} \stackrel{?}{=} 6$$
$$2 + 4 \stackrel{?}{=} 6$$
$$6 \stackrel{?}{=} 6 \quad \text{true}$$

Check $x = 1$:
$$2 + \sqrt{10-(1)} \stackrel{?}{=} -(1)$$
$$2 + \sqrt{9} \stackrel{?}{=} -1$$
$$2 + 3 \stackrel{?}{=} -1$$
$$5 \stackrel{?}{=} -1 \quad \text{false}$$
The solution is $x = -6$.

31.
$$\sqrt{r^2 - 5r + 1} = r - 3$$
$$\left(\sqrt{r^2 - 5r + 1}\right)^2 = (r-3)^2$$
$$r^2 - 5r + 1 = r^2 - 6r + 9$$
$$r = 8$$

Check $r = 8$:
$$\sqrt{(8)^2 - 5(8) + 1} \stackrel{?}{=} (8) - 3$$
$$5 \stackrel{?}{=} 5 \quad \text{true}$$
The solution is $r = 8$.

33.

$$\sqrt{x} - 1 = \sqrt{5 - x}$$
$$\left(\sqrt{x} - 1\right)^2 = \left(\sqrt{5 - x}\right)^2$$
$$x - 2\sqrt{x} + 1 = 5 - x$$
$$2x - 4 = 2\sqrt{x}$$
$$x - 2 = \sqrt{x}$$
$$(x - 2)^2 = \left(\sqrt{x}\right)^2$$
$$x^2 - 4x + 4 = x$$
$$x^2 - 5x + 4 = 0$$
$$(x - 4)(x - 1) = 0$$
$$x - 4 = 0 \quad \text{or} \quad x - 1 = 0$$
$$x = 4 \quad \text{or} \qquad x = 1$$

Check $x = 4$:

$$\sqrt{(4)} - 1 \stackrel{?}{=} \sqrt{5 - (4)}$$
$$1 \stackrel{?}{=} 1 \quad \text{true}$$

Check $x = 1$:

$$\sqrt{(1)} - 1 \stackrel{?}{=} \sqrt{5 - (1)}$$
$$0 \stackrel{?}{=} 2 \quad \text{false}$$

The solution is $x = 4$.

35.

$$\sqrt{x} - \sqrt{2x} = -1$$
$$\sqrt{2x} = \sqrt{x} + 1$$
$$\left(\sqrt{2x}\right)^2 = \left(\sqrt{x} + 1\right)^2$$
$$2x = x + 2\sqrt{x} + 1$$
$$x - 1 = 2\sqrt{x}$$
$$(x - 1)^2 = \left(2\sqrt{x}\right)^2$$
$$x^2 - 2x + 1 = 4x$$
$$x^2 - 6x + 1 = 0$$
$$x = \frac{-(-6) \pm \sqrt{(-6)^2 - 4(1)(1)}}{2(1)}$$
$$= \frac{6 \pm \sqrt{32}}{2}$$
$$= 3 \pm 2\sqrt{2}$$

Check $x = 3 + 2\sqrt{2} \quad (x \approx 5.83)$:

$$\sqrt{(5.83)} - \sqrt{2(5.83)} \stackrel{?}{=} -1$$
$$-1.00 \stackrel{?}{=} -1 \quad \text{true}$$

Check $3 - 2\sqrt{2} \quad (x \approx 0.172)$:

$$\sqrt{(0.172)} - \sqrt{2(0.172)} \stackrel{?}{=} -1$$
$$-0.172 \stackrel{?}{=} -1 \quad \text{false}$$

The solution is $x = 3 + 2\sqrt{2}$.

37.

$$\sqrt{x - 3} + \sqrt{x + 5} = 4$$
$$\sqrt{x - 3} = 4 - \sqrt{x + 5}$$
$$\left(\sqrt{x - 3}\right)^2 = \left(4 - \sqrt{x + 5}\right)^2$$
$$x - 3 = 16 - 8\sqrt{x + 5} + x + 5$$
$$8\sqrt{x + 5} = 24$$
$$\sqrt{x + 5} = 3$$
$$\left(\sqrt{x + 5}\right)^2 = 3^2$$
$$x + 5 = 9$$
$$x = 4$$

Check $x = 4$:

$$\sqrt{(4) - 3} + \sqrt{(4) + 5} \stackrel{?}{=} 4$$
$$4 \stackrel{?}{=} 4 \quad \text{true}$$

The solution is $x = 4$.

39.

$$\sqrt{2p - 1} + \sqrt{3p - 2} = 2$$
$$\sqrt{2p - 1} = 2 - \sqrt{3p - 2}$$
$$\left(\sqrt{2p - 1}\right)^2 = \left(2 - \sqrt{3p - 2}\right)^2$$
$$2p - 1 = 4 - 4\sqrt{3p - 2} + 3p - 2$$
$$4\sqrt{3p - 2} = p + 3$$
$$\left(4\sqrt{3p - 2}\right)^2 = (p + 3)^2$$
$$16(3p - 2) = p^2 + 6p + 9$$
$$48p - 32 = p^2 + 6p + 9$$
$$p^2 - 42p + 41 = 0$$
$$(p - 41)(p - 1) = 0$$
$$p - 41 = 0 \quad \text{or} \quad p - 1 = 0$$
$$p = 41 \quad \text{or} \qquad p = 1$$

Check $p = 41$:

$$\sqrt{2(41) - 1} + \sqrt{3(41) - 2} \stackrel{?}{=} 2$$
$$20 \stackrel{?}{=} 2 \quad \text{false}$$

Check $p = 1$:

$$\sqrt{2(1) - 1} + \sqrt{3(1) - 2} \stackrel{?}{=} 2$$
$$2 \stackrel{?}{=} 2 \quad \text{true}$$

The solution is $p = 1$.

41.
$$\sqrt{\sqrt{x}-2}=3$$
$$\left(\sqrt{\sqrt{x}-2}\right)^2=3^2$$
$$\sqrt{x}-2=9$$
$$\sqrt{x}=11$$
$$\left(\sqrt{x}\right)^2=11^2$$
$$x=121$$

Check $x=121$:

$$\sqrt{\sqrt{(121)}-2}\overset{?}{=}3$$
$$3\overset{?}{=}3 \quad\text{true}$$

The solution is $x=121$.

43.
$$\frac{1}{\sqrt{x+2}}=3-\sqrt{x+2}$$
$$\sqrt{x+2}\cdot\frac{1}{\sqrt{x+2}}=\sqrt{x+2}\left(3-\sqrt{x+2}\right)$$
$$1=3\sqrt{x+2}-x-2$$
$$3\sqrt{x+2}=x+3$$
$$\left(3\sqrt{x+2}\right)^2=(x+3)^2$$
$$9(x+2)=x^2+6x+9$$
$$9x+18=x^2+6x+9$$
$$x^2-3x-9=0$$
$$x=\frac{-(-3)\pm\sqrt{(-3)^2-4(1)(-9)}}{2(1)}=\frac{3\pm3\sqrt{5}}{2}$$

Check $\dfrac{3+3\sqrt{5}}{2}$ $(x\approx 4.8541)$:

$$\frac{1}{\sqrt{(4.8541)+2}}\overset{?}{=}3-\sqrt{(4.8541)+2}$$
$$0.382\overset{?}{=}0.382 \quad\text{true}$$

Check $\dfrac{3-3\sqrt{5}}{2}$ $(x\approx -1.8541)$:

$$\frac{1}{\sqrt{(-1.8541)+2}}\overset{?}{=}3-\sqrt{(-1.8541)+2}$$
$$2.62\overset{?}{=}2.62 \quad\text{true}$$

The solutions are $x=\dfrac{3+3\sqrt{5}}{2}$ and $x=\dfrac{3-3\sqrt{5}}{2}$.

45.
$$5.2\sqrt{x}-2.8=13.9$$
$$5.2\sqrt{x}=16.7$$
$$\sqrt{x}=\frac{16.7}{5.2}$$
$$\left(\sqrt{x}\right)^2=\left(\frac{16.7}{5.2}\right)^2$$
$$x\approx 10.31$$

47.
$$1.52-4.91\sqrt{3.18x-7.14}=-0.69$$
$$-4.91\sqrt{3.18x-7.14}=-2.21$$
$$\sqrt{3.18x-7.14}\approx 0.45$$
$$\left(\sqrt{3.18x-7.14}\right)^2\approx (0.45)^2$$
$$3.18x-7.14\approx 0.20$$
$$3.18x\approx 7.34$$
$$x\approx 2.31$$

49.
$$S=\sqrt{gd}$$
$$(S)^2=\left(\sqrt{gd}\right)^2$$
$$S^2=gd$$
$$\frac{S^2}{g}=d$$

51.
$$d=\sqrt{\frac{3h}{2}}$$
$$(d)^2=\left(\sqrt{\frac{3h}{2}}\right)^2$$
$$d^2=\frac{3h}{2}$$
$$2d^2=3h$$
$$\frac{2d^2}{3}=h$$

53.
$$v=\sqrt{\frac{2GM}{R}}$$
$$(v)^2=\left(\sqrt{\frac{2GM}{R}}\right)^2$$
$$v^2=\frac{2GM}{R}$$
$$Rv^2=2GM$$
$$R=\frac{2GM}{v^2}$$

55.

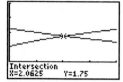

The solution is $x \approx 2.06$.

57.

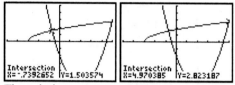

The solutions are $x \approx -0.74$ and $x \approx 4.97$.

59. The solutions are $x \approx -1.6$ and $x \approx 3.8$.

61. The solution is $x = -4$.

63. The solution is $(-3.2, 1.4)$.

65. $h(x) = 3\sqrt{-3x+4} - 15$

$$3\sqrt{-3x+4} - 15 = 0$$
$$3\sqrt{-3x+4} = 15$$
$$\sqrt{-3x+4} = 5$$
$$\left(\sqrt{-3x+4}\right)^2 = 5^2$$
$$-3x+4 = 25$$
$$-3x = 21$$
$$x = -7$$

Check $x = -7$:

$$3\sqrt{-3(-7)+4} - 15 \overset{?}{=} 0$$
$$0 \overset{?}{=} 0 \quad \text{true}$$

The x-intercept is $(-7, 0)$.

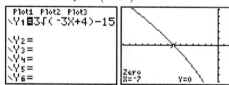

67. $f(x) = \sqrt{3x-2} - \sqrt{x+8}$

$$\sqrt{3x-2} - \sqrt{x+8} = 0$$
$$\sqrt{3x-2} = \sqrt{x+8}$$
$$\left(\sqrt{3x-2}\right)^2 = \left(\sqrt{x+8}\right)^2$$
$$3x-2 = x+8$$
$$2x = 10$$
$$x = 5$$

Check $x = 5$:

$$\sqrt{3(5)-2} - \sqrt{(5)+8} \overset{?}{=} 0$$
$$0 \overset{?}{=} 0 \quad \text{true}$$

The x-intercept is $(5, 0)$.

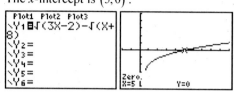

69. $h(x) = 2\sqrt{x+4} + 3\sqrt{x-5}$

$$2\sqrt{x+4} + 3\sqrt{x-5} = 0$$
$$2\sqrt{x+4} = -3\sqrt{x-5}$$
$$\left(2\sqrt{x+4}\right)^2 = \left(-3\sqrt{x-5}\right)^2$$
$$4(x+4) = 9(x-5)$$
$$4x+16 = 9x-45$$
$$-5x = -61$$
$$x = \frac{61}{5}$$

Check $x = \frac{61}{5}$:

$$2\sqrt{\left(\frac{61}{5}\right)+4} + 3\sqrt{\left(\frac{61}{5}\right)-5} \overset{?}{=} 0$$
$$2\sqrt{\left(\frac{81}{5}\right)} + 3\sqrt{\left(\frac{36}{5}\right)} \overset{?}{=} 0$$
$$\frac{36}{\sqrt{5}} \overset{?}{=} 0 \quad \text{false}$$

No real number solution. There are no x-intercepts.

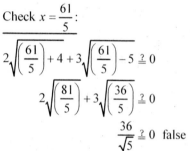

71. $f(x) = 3\sqrt{x} - 7$

$$3\sqrt{x} - 7 = -1$$
$$3\sqrt{x} = 6$$
$$\sqrt{x} = 2$$
$$\left(\sqrt{x}\right)^2 = 2^2$$
$$x = 4$$

Check $x = 4$:

$$3\sqrt{(4)} - 7 \overset{?}{=} -1$$
$$-1 \overset{?}{=} -1 \quad \text{true}$$

When $x = 4$, $f(x) = -1$.

73. $f(x) = -2\sqrt{x-4} + 5$

$$-2\sqrt{x-4} + 5 = -3$$
$$-2\sqrt{x-4} = -8$$
$$\sqrt{x-4} = 4$$
$$\left(\sqrt{x-4}\right)^2 = (4)^2$$
$$x - 4 = 16$$
$$x = 20$$

Check $x = 20$:

$$-2\sqrt{(20)-4} + 5 \stackrel{?}{=} -3$$
$$-3 \stackrel{?}{=} -3 \text{ true}$$

The solution is $x = 20$.

75. **a.** $f(t) = 20.4\sqrt{t} + 21$

$$f(4) = 20.4\sqrt{4} + 21 = 61.8$$

In $1999 + 4 = 2003$, about 62% of e-mails were spam.

b. $f(t) = 20.4\sqrt{t} + 21$

$$95 = 20.4\sqrt{t} + 21$$
$$74 = 20.4\sqrt{t}$$
$$\frac{74}{20.4} = \sqrt{t}$$
$$13.16 \approx t$$

In $1999 + 13 = 2012$, 95% of e-mails were spam.

c. $f(t) = 20.4\sqrt{t} + 21$

$$100 = 20.4\sqrt{t} + 21$$
$$79 = 20.4\sqrt{t}$$
$$\frac{79}{20.4} = \sqrt{t}$$
$$15.00 \approx t$$

In $1999 + 15 = 2014$, 100% of e-mails will be spam. Model breakdown has occurred.

d. $2011 - 1999 = 12$

$$f(t) = 20.4\sqrt{t} + 21$$
$$f(12) = 20.4\sqrt{12} + 21 \approx 91.67$$
$$0.9167 \cdot 72 \approx 66.0$$

We estimate that, on average, 66 spam e-mails were received daily by corporate employees in 2011.

e. $62 \dfrac{\text{e-mails}}{\text{day}} \cdot 5 \dfrac{\text{days}}{\text{week}} \cdot 50 \dfrac{\text{weeks}}{\text{year}}$

$$= 15,500 \dfrac{\text{e-mails}}{\text{year}}$$

$$\dfrac{\$1250}{15,500 \text{ e-mails}} \approx \$0.08 \text{ per e-mail}$$

We estimate that 8 cents in productivity is lost per e-mail.

77. **a.**

The model fits the data well.

b. $f(2) = 257\sqrt[4]{2+1} = 338.23$

The estimated charge is about \$338. It is an underestimate.

c. $385 = 257\sqrt[4]{n+1}$

$$\frac{385}{257} = \sqrt[4]{n+1}$$
$$5.03 \approx n+1$$
$$n \approx 4.03$$

Grade 4 students will pay this charge.

d. This charge is higher than what all other students pay, including those in higher grades. The table suggests that the function is increasing, and therefore the expected per-student charge would be between \$365 and \$410.

79. Answers may vary. Example:
In the third line, the student did not properly square $(x+3)$.

$$\sqrt{x^2 + 4x + 5} = x + 3$$
$$\left(\sqrt{x^2 + 4x + 5}\right)^2 = (x+3)^2$$
$$x^2 + 4x + 5 = x^2 + 6x + 9$$
$$-2x = 4$$
$$x = -2$$

81. $y = 3\sqrt{x} - 4$

$y = -2\sqrt{x} + 6$

Since the left hand sides are equal, set the right hand sides equal to each other and solve the resulting equation.

$3\sqrt{x} - 4 = -2\sqrt{x} + 6$

$5\sqrt{x} = 10$

$\sqrt{x} = 2$

$\left(\sqrt{x}\right)^2 = 2^2$

$x = 4$

Substitute this value into either original equation and solve for y.

$y = 3\sqrt{(4)} - 4 = 2$

The solution is $(4, 2)$.

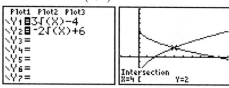

83.   Answers may vary. Example:

Let $a = 1$, $h = -1$, and $k = 3$.

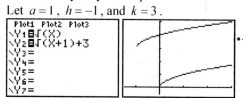

85.   Answers may vary. Example:

The left hand side was not squared properly in the second line. It should be:

$(2x)^2 - 2(2x)(x) + (-x)^2$

87. $3\sqrt{x} + 4 - 7\sqrt{x} + 1 = (3 - 7)\sqrt{x} + 5 = -4\sqrt{x} + 5$

89. $3\sqrt{x} + 4 - 7\sqrt{x} + 1 = -7$

$(3 - 7)\sqrt{x} + 5 = -7$

$-4\sqrt{x} + 5 = -7$

$-4\sqrt{x} = -12$

$\sqrt{x} = 3$

$\left(\sqrt{x}\right)^2 = (3)^2$

$x = 9$

91.
$$\left(\sqrt{p} + 3\right)\left(\sqrt{p} + 1\right) = 3$$
$$\sqrt{p} \cdot \sqrt{p} + 3 \cdot \sqrt{p} + \sqrt{p} \cdot 1 + 3 \cdot 1 = 3$$
$$p + 4\sqrt{p} + 3 = 3$$
$$p + 4\sqrt{p} = 0$$
$$p = 0$$

93.
$$\left(\sqrt{p} + 3\right)\left(\sqrt{p} + 1\right)$$
$$= \sqrt{p} \cdot \sqrt{p} + 3 \cdot \sqrt{p} + \sqrt{p} \cdot 1 + 3 \cdot 1$$
$$= p + 4\sqrt{p} + 3$$

95. $50 - 4(2)^x = -83$

$-4(2)^x = -133$

$2^x = 33.25$

$x \approx 5.0553$

97. $\sqrt{x + 3} - \sqrt{x - 2} = 1$

$\sqrt{x + 3} = 1 + \sqrt{x - 2}$

$\left(\sqrt{x + 3}\right)^2 = \left(1 + \sqrt{x - 2}\right)^2$

$x + 3 = 1 + 2\sqrt{x - 2} + x - 2$

$x + 3 = -1 + 2\sqrt{x - 2} + x$

$4 = 2\sqrt{x - 2}$

$2 = \sqrt{x - 2}$

$(2)^2 = \left(\sqrt{x - 2}\right)^2$

$4 = x - 2$

$6 = x$

99. $-3(2k - 5) + 1 = 2(4k + 3)$

$-6k + 15 + 1 = 8k + 6$

$-6k + 16 = 8k + 6$

$-14k = -10$

$k = \dfrac{5}{7}$

101. $\log_2 (5t - 1) = 5$

$5t - 1 = 2^5$

$5t - 1 = 32$

$5t = 33$

$t = \dfrac{33}{5}$

103. $\dfrac{3x^2 - x - 10}{x^3 - x^2 - x + 1} \div \dfrac{3x^2 - 12}{2x^2 + x - 3}$

$= \dfrac{(3x+5)(x-2)}{(x-1)(x-1)(x+1)} \div \dfrac{3(x-2)(x+2)}{(2x+3)(x-1)}$

$= \dfrac{(3x+5)(x-2)}{(x-1)(x-1)(x+1)} \cdot \dfrac{(2x+3)(x-1)}{3(x-2)(x+2)}$

$= \dfrac{(3x+5)(2x+3)}{3(x+1)(x-1)(x+2)}$

This is a rational expression in one variable.

105. $\dfrac{6}{b-2} + \dfrac{3b}{b^2 - 7b + 10}$

$= \dfrac{6(b-5)}{(b-2)(b-5)} + \dfrac{3b}{(b-2)(b-5)}$

$= \dfrac{3(3b-10)}{(b-2)(b-5)}$

This is a rational expression in one variable.

107. $\dfrac{6}{x-2} + \dfrac{3x}{x^2 - 7x + 10} = \dfrac{x}{x-5}$

$\dfrac{6}{x-2} \cdot \dfrac{x-5}{x-5} + \dfrac{3x}{(x-2)(x-5)} = \dfrac{x}{x-5} \cdot \dfrac{x-2}{x-2}$

$6x - 30 + 3x = x^2 - 2x$

$x^2 - 11x + 30 = 0$

$(x-5)(x-6) = 0$

$x = 5$ is not in the domain of the equation, so the only solution is $x = 6$.
This is a rational equation in one variable.

Homework 9.6

1. $(0,3)$ and $(4,5)$

Substitute the point $(0,3)$ into the equation $y = a\sqrt{x} + b$.

$3 = a\sqrt{0} + b$

$b = 3$

Substitute the point $(4,5)$ into the equation $y = a\sqrt{x} + 3$ and solve for a.

$5 = a\sqrt{4} + 3$

$2a = 2$

$a = 1$

The equation is $y = \sqrt{x} + 3$.

3. $(0,2)$ and $(9,6)$

Substitute the point $(0,2)$ into the equation $y = a\sqrt{x} + b$.

$2 = a\sqrt{0} + b$

$b = 2$

Substitute the point $(9,6)$ into the equation $y = a\sqrt{x} + 2$ and solve for a.

$6 = a\sqrt{9} + 2$

$3a = 4$

$a = \dfrac{4}{3} \approx 1.33$

The equation is $y = 1.33\sqrt{x} + 2$.

5. $(0, 4)$ and $(5, 7)$

Substitute the point $(0, 4)$ into the equation $y = a\sqrt{x} + b$.

$4 = a\sqrt{0} + b$

$b = 4$

Substitute the point $(5, 7)$ into the equation $y = a\sqrt{x} + 4$ and solve for a.

$7 = a\sqrt{5} + 4$

$3 = a\sqrt{5}$

$a \approx 1.34$

The equation is $y = 1.34\sqrt{x} + 4$.

7. $(0, 9)$ and $(3, 2)$

Substitute the point $(0, 9)$ into the equation $y = a\sqrt{x} + b$.

$9 = a\sqrt{0} + b$

$b = 9$

Substitute the point $(3, 2)$ into the equation $y = a\sqrt{x} + 9$ and solve for a.

$2 = a\sqrt{3} + 9$

$-7 = a\sqrt{3}$

$a \approx -4.04$

The equation is $y = -4.04\sqrt{x} + 9$.

9. $(1, 2)$ and $(4, 3)$

Substitute the points into the equation $y = a\sqrt{x} + b$.

$2 = a\sqrt{1} + b$

$3 = a\sqrt{4} + b$

Rewrite as:

$a + b = 2$

$2a + b = 3$

Substitute $b = 2 - a$ into the second equation.

$2a + (2 - a) = 3$

$2a + 2 - a = 3$

$a = 1$

Solve for b.

$b = 2 - 1 = 1$

The equation is $y = \sqrt{x} + 1$.

11. $(2, 4)$ and $(3, 5)$

Substitute the points into the equation

$y = a\sqrt{x} + b$.

$4 = a\sqrt{2} + b$

$5 = a\sqrt{3} + b$

Rewrite as:

$1.4142a + b = 4$

$1.7321a + b = 5$

Solve the resulting system. Multiply the first equation by -1 and add to the second equation.

$-1.4142a - b = -4$

$\underline{1.7321a + b = 5}$

$0.3179a = 1$

$a \approx 3.15$

Substitute the point $(2, 4)$ into the equation

$y = 3.15\sqrt{x} + b$ and solve for b.

$4 = 3.15\sqrt{2} + b$

$b \approx -0.45$

The equation is $y = 3.15\sqrt{x} - 0.45$.

13. $(2, 6)$ and $(5, 4)$

Substitute the points into the equation

$y = a\sqrt{x} + b$.

$6 = a\sqrt{2} + b$

$4 = a\sqrt{5} + b$

Rewrite as:

$1.4142a + b = 6$

$2.2361a + b = 4$

Solve the resulting system. Multiply the first equation by -1 and add to the second equation.

$-1.4142a - b = -6$

$\underline{2.2361a + b = 4}$

$0.8219a = -2$

$a \approx -2.43$

Substitute the point $(2, 6)$ into the equation

$y = -2.43\sqrt{x} + b$ and solve for b.

$6 = -2.43\sqrt{2} + b$

$b \approx 9.44$

The equation is $y = -2.43\sqrt{x} + 9.44$.

15. $(5, 7)$ and $(13, 21)$

Substitute the points into the equation

$y = a\sqrt{x} + b$.

$7 = a\sqrt{5} + b$

$21 = a\sqrt{13} + b$

Rewrite as:

$2.2361a + b = 7$

$3.6056a + b = 21$

Solve the resulting system. Multiply the first equation by -1 and add to the second equation.

$-2.2361a - b = -7$

$\underline{3.6056a + b = 21}$

$1.3695a = 14$

$a \approx 10.22$

Substitute the point $(5, 7)$ into the equation

$y = 10.22\sqrt{x} + b$ and solve for b.

$7 = 10.22\sqrt{5} + b$

$b \approx -15.85$

The equation is $y = 10.22\sqrt{x} - 15.85$.

17. $(7, 31)$ and $(10, 6)$

Substitute the points into the equation

$y = a\sqrt{x} + b$.

$31 = a\sqrt{7} + b$

$6 = a\sqrt{10} + b$

Rewrite as:

$2.6458a + b = 31$

$3.1623a + b = 6$

Solve the resulting system. Multiply the first equation by -1 and add to the second equation.

$-2.6458a - b = -31$

$\underline{3.1623a + b = 6}$

$0.5165a = 25$

$a \approx -48.40$

Substitute the point $(7, 31)$ into the equation

$y = -48.40\sqrt{x} + b$ and solve for b.

$31 = -48.40\sqrt{7} + b$

$b \approx 159.06$

The equation is $y = -48.40\sqrt{x} + 159.06$.

19. (15, 3) and (35, 18)
Substitute the points into the equation
$y = a\sqrt{x} + b$.

$3 = a\sqrt{15} + b$
$18 = a\sqrt{35} + b$

Rewrite as:
$3.873a + b = 3$
$5.9161a + b = 18$

Solve the resulting system. Multiply the first equation by -1 and add to the second equation.

$-3.873a - b = -3$
$\underline{5.9161a + b = 18}$
$2.0431a = 15$
$a \approx 7.34$

Substitute the point (15, 3) into the equation $y = 7.34\sqrt{x} + b$ and solve for b.

$3 = 7.34\sqrt{15} + b$
$b \approx -25.43$

The equation is $y = 7.34\sqrt{x} - 25.43$.

21. a. Create a scattergram of the data.

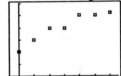

Answers may vary depending on the data points selected. Example:
Use the points (0, 10) and (6, 26).
Substitute the point (0, 10) into the equation $y = a\sqrt{x} + b$.

$10 = a\sqrt{0} + b$
$b = 10$

Substitute the point (6, 26) into the equation $y = a\sqrt{x} + 10$ and solve for a.

$26 = a\sqrt{6} + 10$
$16 = a\sqrt{6}$
$a = \dfrac{16}{\sqrt{6}} \approx 6.53$

$f(t) = 6.53\sqrt{t} + 10$

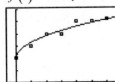

b. Answers may vary depending on the equation found in part (a). Example:
The n-intercept is (0, 10), which means

that there were 10 thousand American female troops in Iraq and Afghanistan in 2004.

c. Answers may vary depending on the equation found in part (a). Example:
$6.53\sqrt{t} + 10 = 28$
$6.53\sqrt{t} = 18$
$\sqrt{t} = \dfrac{18}{6.53}$
$t \approx 7.60$

In $2004 + 7.6 \approx 2012$, there were 28 thousand American female troops in Iraq and Afghanistan.

d. Answers may vary depending on the equation found in part (a). Example:
$t = 2012 - 2004 = 8$
$f(8) = 6.53\sqrt{8} + 10 \approx 28.5$
$28.5 - 15 = 13.5$

We estimate that there would have been 13.5 thousand American female troops in Iraq in 2012 if the withdrawal had not occurred.

23. a. Create a scattergram of the data.

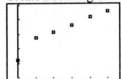

Answers may vary depending on the data points selected. Example:
Use the points (0, 56.0) and (5, 72.8).
Substitute the point (0, 56.0) into the equation $y = a\sqrt{x} + b$.

$56.0 = a\sqrt{0} + b$
$b = 56.0$

Substitute the point (5, 72.8) into the equation $y = a\sqrt{x} + 56.0$ and solve for a.

$a\sqrt{5} + 56.0 = 72.8$
$a\sqrt{5} = 16.8$
$a = \dfrac{16.8}{\sqrt{5}} \approx 7.51$

$f(t) = 7.51\sqrt{t} + 56.0$

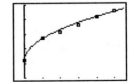

b. Answers may vary depending on the equation found in part (a). Example: The *M*-intercept is (0, 56.0), which means that the average monthly bill for pay-TV was $56 in 2006.

c. Answers may vary depending on the equation found in part (a). Example:

$$f(13) = 7.51\sqrt{13} + 56.0 \approx 83.08$$

This means that the average monthly bill for pay-TV will be about $83.08 in $2006 + 13 = 2019$.

d. Answers may vary depending on the equation found in part (a). Example:

$$7.51\sqrt{t} + 56.0 = 81$$
$$7.51\sqrt{t} = 25$$
$$\sqrt{t} = \frac{25}{7.51}$$
$$t = 11.08$$

This means that the average monthly bill for pay-TV will be $81 in $2006 + 11.08 \approx 2017$.

e. Answers may vary depending on the equation found in part (a). Example: Find the average monthly bill in 2017.
$$2017 - 2006 = 11$$

$$f(11) = 7.51\sqrt{11} + 56.0 \approx 80.91$$

$12 \cdot \$80.91 \cdot 100.9$ million ≈ 98.0 billion
We predict that the total annual revenue from all subscribers of pay-TV in 2017 will be $98 billion.

25. a. $f(18,000) = \sqrt{9.8(18,000)} = 420$

The average speed would be 420 meters per second.

b. No, this would suggest that the depth is much less because the speed is half of what it should be.

c. $203 = \sqrt{9.8t}$
$$t = 4205$$
$$210 = \sqrt{9.8t}$$
$$t = 4500$$
The average depth would be 4205 to 4500 meters.

d. Yes.

27. a. Create a scattergram of the data.

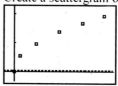

Answers may vary depending on the data points selected. Example:
Use the points $(0,0)$ and $(52.5, 1.94)$.

Substitute the point $(0,0)$ into the equation
$y = a\sqrt{x} + b$.
$$0 = a\sqrt{0} + b$$
$$b = 0$$

Substitute the point $(52.5, 1.94)$ into the equation $y = a\sqrt{x}$ and solve for *a*.
$$1.94 = a\sqrt{52.5}$$
$$a \approx 0.27$$
$$S(h) = 0.27\sqrt{h}$$

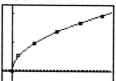

b.   i. Graph all three functions.

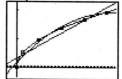

The model $S(h)$ appears to fit the best.

ii. $S(0) = 0; L(0) = 0.327; Q(0) = 0.165$

S models the situation best near 0 since it is the only model that passes through the origin.

iii. Zoom out.

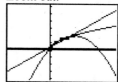

Q is not possible since it indicates that the falling time will reach 0 for larger drop heights.

iv. *S* models the situation the best and has no problems with larger *h*.

v. $T = \sqrt{\dfrac{2h}{32.2}} = \sqrt{\dfrac{2}{32.2}} \cdot \sqrt{h} \approx 0.249\sqrt{h}$

This is close to the model

$S(h) = 0.27\sqrt{h}$.

c. Answers may vary depending on the equation found in part (a). Example:

$0.27\sqrt{h} = 3$

$\sqrt{h} = \dfrac{3}{0.27}$

$\left(\sqrt{h}\right)^2 = \left(\dfrac{3}{0.27}\right)^2$

$h \approx 123.46$

According to the model, the height of the cliff is about 123.46 feet.

d. Answers may vary depending on the equation found in part (a). Example:

$S(1250) = 0.27\sqrt{1250} \approx 9.55$

It would take about 9.55 seconds for the baseball to reach the ground if it were dropped from the top of New York City's Empire State Building.

29. a. Create a scattergram of the data.

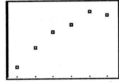

Answers may vary depending on the data points selected. Example:

Use the points $(2, 54.3)$ and $(4, 73)$.

Substitute the points into the equation

$y = a\sqrt{x} + b$.

$54.3 = a\sqrt{2} + b$

$73 = a\sqrt{4} + b$

Rewrite as:

$1.4142a + b = 54.3$

$2a + b = 73$

Solve the system of equations. Multiply the first equation by -1 and add to the second equation.

$-1.4142a - b = -54.3$

$\underline{2a + b = 73}$

$0.5858a = 18.7$

$a \approx 31.92$

Substitute the point $(2, 54.3)$ into the equation $y = 31.92\sqrt{x} + b$ and solve for b.

$54.3 = 31.92\sqrt{2} + b$

$b = 54.3 - 31.92\sqrt{2}$

$b \approx 9.16$

$f(n) = 31.92\sqrt{n} + 9.16$

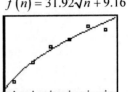

b. Answers may vary depending on the equation found in part (a). Example:

$f(7) = 31.92\sqrt{(7)} + 9.16 \approx 93.61$

About 93.6% of 7th births occurred despite the use of contraception.

c. Answers may vary depending on the equation found in part (a). Example:

$100 = 31.92\sqrt{n} + 9.16$

$31.92\sqrt{n} = 90.84$

$\sqrt{n} = \dfrac{90.84}{31.92}$

$\left(\sqrt{n}\right)^2 = \left(\dfrac{90.84}{31.92}\right)^2$

$n \approx 8.1$

All 8th births occurred despite the use of contraception. Model breakdown has likely occurred.

d. The higher the birth order, the higher the percent of births that happened despite the use of contraception.

Answers may vary. Example:

Perhaps couples without children are more careful in their use of contraception.

31. Sketches may vary. Example:

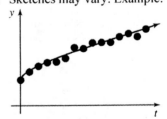

Increase the value of b to shift the graph up.

33. Answers may vary. Example:

Let $x = 0$ and solve for y.

$y = a\sqrt{x} + b = a\sqrt{0} + b = 0 + b = b$

When $x = 0$, $y = b$. The point $(0, b)$ is where the graph of the equation intercepts the y-axis, so it is the y-intercept.

35. a.

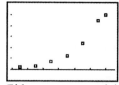

Either an exponential or quadratic function might fit the data well.

b. Use the regression feature.

Exponential: $f(t) = 1.09(1.37)^t$

Quadratic: $f(t) = 5.17t^2 - 105.43t + 556$

c.

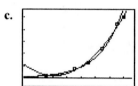

Both functions fit the data well.

d. The exponential function fits the data better before 1999.

e. Exponential model:

$$6000 = 1.09(1.37)^t$$

$$\frac{6000}{1.09} = 1.37^t$$

$$\ln\left(\frac{6000}{1.09}\right) = \ln\left(1.37^t\right)$$

$$\ln\left(\frac{6000}{1.09}\right) = t\ln(1.37)$$

$$t = \frac{\ln\left(\frac{6000}{1.09}\right)}{\ln(1.37)} \approx 27.36$$

The exponential model predicts that there will be 6000 U.S. communities with red-light cameras in $1990 + 27.36 \approx 2017$.

Quadratic model:

$$6000 = 5.17t^2 - 105.43t + 556$$

$$0 = 5.17t^2 - 105.43t - 5444$$

$$t = \frac{-(-105.43)\pm\sqrt{(-105.43)^2 - 4(5.17)(-5444)}}{2(5.17)}$$

$$= \frac{105.43\pm\sqrt{123{,}697.4049}}{10.34}$$

$t \approx -23.82$ or $t \approx 44.21$

The quadratic model predicts that there will be 6000 U.S. communities with red-light cameras in $1990 + 44.21 \approx 2034$. The exponential model increases at a much faster rate than the quadratic model.

f. **i.** $f(15) - f(14)$

$$= 1.09(1.37)^{15} - 1.09(1.37)^{14}$$

$$\approx 122.53 - 89.44$$

$$\approx 33.09$$

About 33 communities installed red-light cameras in 2005.

ii. $f(27) - f(26)$

$$= 1.09(1.37)^{27} - 1.09(1.37)^{26}$$

$$\approx 5356.51 - 3909.86$$

$$\approx 1446.65$$

About 1447 communities will install red-light cameras in 2017.

iii. $\dfrac{1446.65}{33.09} \cdot \$26\text{ million} \approx \$1137\text{ million}$

We predict that Redflex's revenue from sales of red-light cameras will be about $1137 million ($1.137 billion) in 2017.

We have assumed that the price of a red-light camera in 2017 will be equal to the price of one in 2005 and that the average number of red-light cameras sold per community in 2005 is equal to the average number of red-light cameras sold per community in 2017.

37. Answers may vary. Example:

$$\frac{4x-1}{2} = \frac{5x+3}{3}$$

$$12x - 3 = 10x + 6$$

$$2x = 9$$

$$x = \frac{9}{2}$$

39. Answers may vary. Example:

$$\frac{4x-1}{2} - \frac{5x+3}{3} = \frac{12x-3}{6} - \frac{10x+6}{6} = \frac{2x-9}{6}$$

41. Answers may vary. Example:

$$2\sqrt{x} - 5 = 4$$

$$2\sqrt{x} = 9$$

$$\sqrt{x} = \frac{9}{2}$$

$$x = \frac{81}{4}$$

43. Answers may vary. Example:

$$f(x) = 2(3)^x - 2.5$$

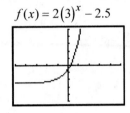

Chapter 9 Review

1. $x^{3/7} = \sqrt[7]{x^3}$

2. $\sqrt[5]{(3x+4)^7} = (3x+4)^{7/5}$

3. $\sqrt{8x^6} = \sqrt{4x^6 \cdot 2} = \sqrt{4x^6}\sqrt{2} = 2x^3\sqrt{2}$

4. $\sqrt{18x^7 y^{10}} = \sqrt{9x^6 y^{10} \cdot 2x}$
$= \sqrt{9x^6 y^{10}}\sqrt{2x}$
$= 3x^3 y^5 \sqrt{2x}$

5. $\sqrt[8]{x^6} = x^{\frac{6}{8}} = x^{\frac{3}{4}} = \sqrt[4]{x^3}$

6. $\sqrt[3]{24x^{10} y^{24}} = \sqrt[3]{8x^9 y^{24} \cdot 3x}$
$= \sqrt[3]{8x^9 y^{24}}\sqrt[3]{3x}$
$= 2x^3 y^8 \sqrt[3]{3x}$

7. $\sqrt[5]{(6x+11)^{27}} = \sqrt[5]{(6x+11)^{25} \cdot (6x+11)^2}$
$= \sqrt[5]{(6x+11)^{25}}\sqrt[5]{(6x+11)^2}$
$= (6x+11)^5 \sqrt[5]{(6x+11)^2}$

8. $5\sqrt{20x} - 2\sqrt{45x} + 7\sqrt{5x}$
$= 5\sqrt{4 \cdot 5x} - 2\sqrt{9 \cdot 5x} + 7\sqrt{5x}$
$= 5\sqrt{4}\sqrt{5x} - 2\sqrt{9}\sqrt{5x} + 7\sqrt{5x}$
$= 10\sqrt{5x} - 6\sqrt{5x} + 7\sqrt{5x}$
$= 11\sqrt{5x}$

9. $b\sqrt[3]{16a^5 b} + a\sqrt[3]{2a^2 b^4}$
$= b\sqrt[3]{8a^3 \cdot 2a^2 b} + a\sqrt[3]{b^3 \cdot 2a^2 b}$
$= b\sqrt[3]{8a^3}\sqrt[3]{2a^2 b} + a\sqrt[3]{b^3}\sqrt[3]{2a^2 b}$
$= 2ab\sqrt[3]{2a^2 b} + ab\sqrt[3]{2a^2 b}$
$= 3ab\sqrt[3]{2a^2 b}$

10. $5(4\sqrt{x} - \sqrt[3]{x}) - 2\sqrt[3]{x} + 8\sqrt{x}$
$= 20\sqrt{x} - 5\sqrt[3]{x} - 2\sqrt[3]{x} + 8\sqrt{x}$
$= (-5-2)\sqrt[3]{x} + (20+8)\sqrt{x}$
$= -7\sqrt[3]{x} + 28\sqrt{x}$

11. $3\sqrt{x}\left(\sqrt{x} - 7\right) = 3\sqrt{x} \cdot \sqrt{x} - 3\sqrt{x} \cdot 7$
$= 3\sqrt{x \cdot x} - 21\sqrt{x}$
$= 3\sqrt{x^2} - 21\sqrt{x}$
$= 3x - 21\sqrt{x}$

12. $\left(4\sqrt{x} - 3\right)\left(2\sqrt{x} + 1\right)$
$= 4\sqrt{x} \cdot 2\sqrt{x} - 3 \cdot 2\sqrt{x} + 4\sqrt{x} \cdot 1 - 3 \cdot 1$
$= 8\sqrt{x \cdot x} - 6\sqrt{x} + 4\sqrt{x} - 3$
$= 8\sqrt{x^2} - 2\sqrt{x} - 3$
$= 8x - 2\sqrt{x} - 3$

13. $\left(2\sqrt{a} - \sqrt{b}\right)\left(5\sqrt{a} + \sqrt{b}\right)$
$= 2\sqrt{a} \cdot 5\sqrt{a} - \sqrt{b} \cdot 5\sqrt{a} + 2\sqrt{a} \cdot \sqrt{b} - \sqrt{b} \cdot \sqrt{b}$
$= 10\sqrt{a \cdot a} - 5\sqrt{a \cdot b} + 2\sqrt{a \cdot b} - \sqrt{b \cdot b}$
$= 10\sqrt{a^2} - 3\sqrt{ab} - \sqrt{b^2}$
$= 10a - 3\sqrt{ab} - b$

14. $\left(5\sqrt{a} - 7\sqrt{b}\right)\left(5\sqrt{a} + 7\sqrt{b}\right)$
$= 5\sqrt{a} \cdot 5\sqrt{a} - 7\sqrt{b} \cdot 5\sqrt{a} + 5\sqrt{a} \cdot 7\sqrt{b} - 7\sqrt{b} \cdot 7\sqrt{b}$
$= 25\sqrt{a \cdot a} - 35\sqrt{a \cdot b} + 35\sqrt{a \cdot b} - 49\sqrt{b \cdot b}$
$= 25\sqrt{a^2} - 35\sqrt{ab} + 35\sqrt{ab} - 49\sqrt{b^2}$
$= 25a - 49b$

15. $\left(4\sqrt{x} + 3\right)^2 = \left(4\sqrt{x}\right)^2 + 2\left(4\sqrt{x}\right)(3) + (3)^2$
$= 16x + 24\sqrt{x} + 9$

16. $\left(2\sqrt[3]{x} - 5\right)^2 = \left(2\sqrt[3]{x}\right)^2 - 2\left(2\sqrt[3]{x}\right)(5) + (5)^2$
$= 4\sqrt[3]{x^2} - 20\sqrt[3]{x} + 25$

17. $\sqrt[4]{x}\sqrt[7]{x} = x^{\frac{1}{4}} \cdot x^{\frac{1}{7}}$
$= x^{\frac{1}{4} + \frac{1}{7}}$
$= x^{\frac{7}{28} + \frac{4}{28}}$
$= x^{\frac{11}{28}}$
$= \sqrt[28]{x^{11}}$

18. $\dfrac{\sqrt[4]{x}}{\sqrt[6]{x}} = \dfrac{x^{\frac{1}{4}}}{x^{\frac{1}{6}}} = x^{\frac{1}{4} - \frac{1}{6}} = x^{\frac{3}{12} - \frac{2}{12}} = x^{\frac{1}{12}} = \sqrt[12]{x}$

19. $\sqrt{\dfrac{3}{x}} = \dfrac{\sqrt{3}}{\sqrt{x}} = \dfrac{\sqrt{3}}{\sqrt{x}} \cdot \dfrac{\sqrt{x}}{\sqrt{x}} = \dfrac{\sqrt{3 \cdot x}}{\sqrt{x \cdot x}} = \dfrac{\sqrt{3x}}{x}$

20. $\dfrac{5t}{\sqrt[3]{t}} = \dfrac{5t}{\sqrt[3]{t}} \cdot \dfrac{\sqrt[3]{t^2}}{\sqrt[3]{t^2}}$

$= \dfrac{5t\sqrt[3]{t^2}}{\sqrt[3]{t \cdot t^2}}$

$= \dfrac{5t\sqrt[3]{t^2}}{\sqrt[3]{t^3}}$

$= \dfrac{5t\sqrt[3]{t^2}}{t}$

$= 5\sqrt[3]{t^2}$

21. $\sqrt[5]{\dfrac{7y}{27x^2}} = \dfrac{\sqrt[5]{7y}}{\sqrt[5]{27x^2}}$

$= \dfrac{\sqrt[5]{7y}}{\sqrt[5]{27x^2}} \cdot \dfrac{\sqrt[5]{9x^3}}{\sqrt[5]{9x^3}}$

$= \dfrac{\sqrt[5]{7y \cdot 9x^3}}{\sqrt[5]{27x^2 \cdot 9x^3}}$

$= \dfrac{\sqrt[5]{63x^3 y}}{\sqrt[5]{243x^5}}$

$= \dfrac{\sqrt[5]{63x^3 y}}{3x}$

22. $\dfrac{\sqrt{a}}{\sqrt{a} - 2\sqrt{b}} = \dfrac{\sqrt{a}}{\sqrt{a} - 2\sqrt{b}} \cdot \dfrac{\sqrt{a} + 2\sqrt{b}}{\sqrt{a} + 2\sqrt{b}}$

$= \dfrac{\sqrt{a}\left(\sqrt{a} + 2\sqrt{b}\right)}{\left(\sqrt{a}\right)^2 - \left(2\sqrt{b}\right)^2}$

$= \dfrac{a + 2\sqrt{ab}}{a - 4b}$

23. $\dfrac{5\sqrt{x} - 4}{2\sqrt{x} + 3}$

$= \dfrac{5\sqrt{x} - 4}{2\sqrt{x} + 3} \cdot \dfrac{2\sqrt{x} - 3}{2\sqrt{x} - 3}$

$= \dfrac{5\sqrt{x} \cdot 2\sqrt{x} - 4 \cdot 2\sqrt{x} - 5\sqrt{x} \cdot 3 + 4 \cdot 3}{\left(2\sqrt{x}\right)^2 - (3)^2}$

$= \dfrac{10\sqrt{x \cdot x} - 8\sqrt{x} - 15\sqrt{x} + 12}{4x - 9}$

$= \dfrac{10x - 23\sqrt{x} + 12}{4x - 9}$

24. $y = -\sqrt{x - 5} + 3$

x	y
5	3
6	2
9	1
14	0
21	−1

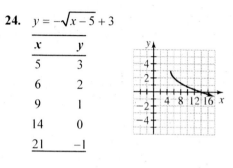

25. $y = 2\sqrt{x + 4} - 1$

x	y
−4	−1
−3	1
0	3
5	5
12	7

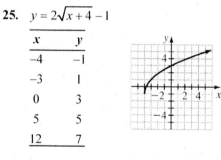

26. $(f + g)(x) = \left(3\sqrt{x} + 5\right) + \left(2 - 4\sqrt{x}\right)$

$= 3\sqrt{x} + 5 + 2 - 4\sqrt{x}$

$= -\sqrt{x} + 7$

27. $(f - g)(x) = \left(3\sqrt{x} + 5\right) - \left(2 - 4\sqrt{x}\right)$

$= 3\sqrt{x} + 5 - 2 + 4\sqrt{x}$

$= 7\sqrt{x} + 3$

28. $(f \cdot g)(x) = \left(3\sqrt{x} + 5\right)\left(2 - 4\sqrt{x}\right)$

$= 3\sqrt{x} \cdot 2 + 5 \cdot 2 - 3\sqrt{x} \cdot 4\sqrt{x} - 5 \cdot 4\sqrt{x}$

$= 6\sqrt{x} + 10 - 12x - 20\sqrt{x}$

$= -12x - 14\sqrt{x} + 10$

29. $\left(\dfrac{f}{g}\right)(x) = \dfrac{3\sqrt{x}+5}{2-4\sqrt{x}}$

$$= \dfrac{3\sqrt{x}+5}{2-4\sqrt{x}} \cdot \dfrac{2+4\sqrt{x}}{2+4\sqrt{x}}$$

$$= \dfrac{3\sqrt{x}\cdot 2 + 5\cdot 2 + 3\sqrt{x}\cdot 4\sqrt{x} + 5\cdot 4\sqrt{x}}{(2)^2 - \left(4\sqrt{x}\right)^2}$$

$$= \dfrac{6\sqrt{x}+10+12x+20\sqrt{x}}{4-16x}$$

$$= \dfrac{12x+26\sqrt{x}+10}{4-16x}$$

$$= \dfrac{2\left(6x+13\sqrt{x}+5\right)}{2\left(2-8x\right)}$$

$$= \dfrac{6x+13\sqrt{x}+5}{2-8x}$$

30. $\sqrt{2x+1}+4 = 7$

$\sqrt{2x+1} = 3$

$\left(\sqrt{2x+1}\right)^2 = 3^2$

$2x+1 = 9$

$2x = 8$

$x = 4$

Check $x = 4$:

$\sqrt{2(4)+1}+4 \overset{?}{=} 7$

$7 \overset{?}{=} 7$ true

The solution is $x = 4$.

31. $\sqrt{2p-4}-p = -2$

$\sqrt{2p-4} = p-2$

$\left(\sqrt{2p-4}\right)^2 = (p-2)^2$

$2p-4 = p^2-4p+4$

$p^2-6p+8 = 0$

$(p-4)(p-2) = 0$

$p-4 = 0$ or $p-2 = 0$

$p = 4$ or $p = 2$

Check $p = 4$:

$\sqrt{2(4)-4}-4 \overset{?}{=} -2$

$-2 \overset{?}{=} -2$ true

Check $p = 2$:

$\sqrt{2(2)-4}-2 \overset{?}{=} -2$

$-2 \overset{?}{=} -2$ true

The solutions are $p = 4$ and $p = 2$.

32. $\sqrt{x}+6 = x$

$\sqrt{x} = x-6$

$\left(\sqrt{x}\right)^2 = (x-6)^2$

$x = x^2-12x+36$

$x^2-13x+36 = 0$

$(x-9)(x-4) = 0$

$x-9 = 0$ or $x-4 = 0$

$x = 9$ or $x = 4$

Check $x = 9$:

$\sqrt{9}+6 \overset{?}{=} 9$

$9 \overset{?}{=} 9$ true

Check $x = 4$:

$\sqrt{4}+6 \overset{?}{=} 4$

$8 \overset{?}{=} 4$ false

The solution is $x = 9$.

33. $\sqrt{13x+4} = \sqrt{5x+20}$

$\left(\sqrt{13x+4}\right)^2 = \left(\sqrt{5x+20}\right)^2$

$13x+4 = 5x+20$

$8x = 16$

$x = 2$

Check $x = 2$:

$\sqrt{13(2)+4} \overset{?}{=} \sqrt{5(2)+20}$

$\sqrt{30} \overset{?}{=} \sqrt{30}$ true

The solution is $x = 2$.

34. $\sqrt{w+2}+\sqrt{w+9} = 7$

$\sqrt{w+2} = 7-\sqrt{w+9}$

$\left(\sqrt{w+2}\right)^2 = \left(7-\sqrt{w+9}\right)^2$

$w+2 = 49-14\sqrt{w+9}+w+9$

$14\sqrt{w+9} = 56$

$\sqrt{w+9} = 4$

$\left(\sqrt{w+9}\right)^2 = 4^2$

$w+9 = 16$

$w = 7$

Check $w = 7$:

$\sqrt{(7)+2}+\sqrt{(7)+9} \overset{?}{=} 7$

$7 \overset{?}{=} 7$ true

The solution is $w = 7$.

35.
$$3.57 + 2.99\sqrt{8.06x - 6.83} = 14.55$$
$$2.99\sqrt{8.06x - 6.83} = 10.98$$
$$\sqrt{8.06x - 6.83} \approx 3.67$$
$$8.06x - 6.83 \approx 13.47$$
$$8.06x \approx 20.3$$
$$x \approx 2.52$$

36.

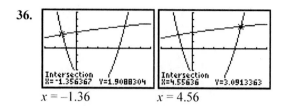

$$x = -1.36 \qquad\qquad x = 4.56$$

37. $f(x) = \sqrt{4x - 7} - \sqrt{2x + 1}$
$$\sqrt{4x - 7} - \sqrt{2x + 1} = 0$$
$$\sqrt{4x - 7} = \sqrt{2x + 1}$$
$$\left(\sqrt{4x - 7}\right)^2 = \left(\sqrt{2x + 1}\right)^2$$
$$4x - 7 = 2x + 1$$
$$2x = 8$$
$$x = 4$$

Check $x = 4$:
$$\sqrt{4(4) - 7} - \sqrt{2(4) + 1} \overset{?}{=} 0$$
$$0 \overset{?}{=} 0 \quad \text{true}$$

The x-intercept is $(4, 0)$.

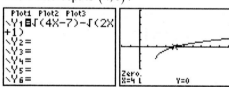

38. Sketches may vary. Example:

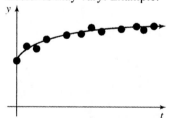

Increase b to raise the y-intercept and decrease a to lower the rate of increase.

39. The equation is of the form $y = a\sqrt{x} + b$.

The y-intercept is $(0, 3)$ so $b = 3$.

Substitute the point $(4, 8)$ into the equation $y = a\sqrt{x} + 3$ and solve for a.
$$8 = a\sqrt{4} + 3$$
$$2a = 5$$
$$a = \frac{5}{2} = 2.5$$

The equation is $y = 2.5\sqrt{x} + 3$.

40. $(3, 7)$ and $(5, 4)$

Substitute the points into the equation $y = a\sqrt{x} + b$.
$$7 = a\sqrt{3} + b$$
$$4 = a\sqrt{5} + b$$
Rewrite as:
$$1.7321a + b = 7$$
$$2.2361a + b = 4$$
Solve the system of equations. Multiply the first equation by -1 and add to the second equation.
$$-1.7321a - b = -7$$
$$\underline{2.2361a + b = 4}$$
$$0.504a = -3$$
$$a \approx -5.95$$
Substitute the point $(3, 7)$ into the equation $y = -5.95\sqrt{x} + b$ and solve for b.
$$7 = -5.95\sqrt{3} + b$$
$$b = 7 + 5.95\sqrt{3}$$
$$b \approx 17.31$$
The equation is $y = -5.95\sqrt{x} + 17.31$.

41. a. Create a scattergram of the data.

Answers may vary depending on the data points selected. Example:

Use the points $(0, 3.8)$ and $(12, 9.6)$.

The y-intercept is $(0, 3.8)$ so $b = 3.8$.

Substitute the point $(12, 9.6)$ into the equation $y = a\sqrt{x} + 3.8$ and solve for a.

$$9.6 = a\sqrt{12} + 3.8$$
$$5.8 = a\sqrt{12}$$
$$a = \frac{5.8}{\sqrt{12}} \approx 1.67$$
$$f(t) = 1.67\sqrt{t} + 3.8$$

b. The D-intercept is $(0, 3.8)$. This means that in 1992, the average credit card debt per household was $3.8 thousand.

c. $f(26) = 1.67\sqrt{26} + 3.8 \approx 12.32$

This means that in $1992 + 26 = 2018$, the average credit card debt per household will be about $12.3 thousand.

d. $13 = 1.67\sqrt{t} + 3.8$
$$9.2 = 1.67\sqrt{t}$$
$$\sqrt{t} = \frac{9.2}{1.67}$$
$$t \approx 30.35$$

This means that the average credit card debt per household will be $13 thousand in $1992 + 30.35 \approx 2022$.

Chapter 9 Test

1. $\sqrt{32x^9 y^{12}} = \sqrt{16x^8 y^{12} \cdot 2x}$
$$= \sqrt{16x^8 y^{12}} \sqrt{2x}$$
$$= 4x^4 y^6 \sqrt{2x}$$

2. $\sqrt[3]{64x^{22} y^{14}} = \sqrt[3]{64x^{21} y^{12} \cdot xy^2}$
$$= \sqrt[3]{64x^{21} y^{12}} \sqrt[3]{xy^2}$$
$$= 4x^7 y^4 \sqrt[3]{xy^2}$$

3. $\sqrt[4]{(2x+8)^{27}} = \sqrt[4]{(2x+8)^{24} \cdot (2x+8)^3}$
$$= \sqrt[4]{(2x+8)^{24}} \sqrt[4]{(2x+8)^3}$$
$$= (2x+8)^6 \sqrt[4]{(2x+8)^3}$$

4. $\dfrac{4\sqrt[3]{x}}{6\sqrt[5]{x}} = \dfrac{2x^{\frac{1}{3}}}{3x^{\frac{1}{5}}}$
$$= \frac{2}{3} x^{\frac{1}{3} - \frac{1}{5}}$$
$$= \frac{2}{3} x^{\frac{5}{15} - \frac{3}{15}}$$
$$= \frac{2}{3} x^{\frac{2}{15}}$$
$$= \frac{2\sqrt[15]{x^2}}{3}$$

5. $\dfrac{\sqrt{x}+1}{2\sqrt{x}-3} = \dfrac{\sqrt{x}+1}{2\sqrt{x}-3} \cdot \dfrac{2\sqrt{x}+3}{2\sqrt{x}+3}$
$$= \frac{\sqrt{x} \cdot 2\sqrt{x} + 1 \cdot 2\sqrt{x} + \sqrt{x} \cdot 3 + 1 \cdot 3}{\left(2\sqrt{x}\right)^2 - (3)^2}$$
$$= \frac{2x + 2\sqrt{x} + 3\sqrt{x} + 3}{4x - 9}$$
$$= \frac{2x + 5\sqrt{x} + 3}{4x - 9}$$

6. $4\sqrt{12x^3} - 2x\sqrt{75x} + \sqrt{3x^3}$
$$= 4\sqrt{4x^2 \cdot 3x} - 2x\sqrt{25 \cdot 3x} + \sqrt{x^2 \cdot 3x}$$
$$= 4 \cdot 2x\sqrt{3x} - 2x \cdot 5\sqrt{3x} + x\sqrt{3x}$$
$$= 8x\sqrt{3x} - 10x\sqrt{3x} + x\sqrt{3x}$$
$$= -x\sqrt{3x}$$

7. $3\sqrt{x}\left(6\sqrt{x} - 5\right) = 3\sqrt{x} \cdot 6\sqrt{x} - 3\sqrt{x} \cdot 5$
$$= 18\sqrt{x^2} - 15\sqrt{x}$$
$$= 18x - 15\sqrt{x}$$

8. $\left(2 + 4\sqrt{x}\right)\left(3 - 5\sqrt{x}\right)$
$$= 2 \cdot 3 + 4\sqrt{x} \cdot 3 - 2 \cdot 5\sqrt{x} - 4\sqrt{x} \cdot 5\sqrt{x}$$
$$= 6 + 12\sqrt{x} - 10\sqrt{x} - 20\sqrt{x^2}$$
$$= -20x + 2\sqrt{x} + 6$$

9. $\left(3\sqrt{a} - 5\sqrt{b}\right)\left(3\sqrt{a} + 5\sqrt{b}\right) = \left(3\sqrt{a}\right)^2 - \left(5\sqrt{b}\right)^2$
$$= 9a - 25b$$

10. $\left(4\sqrt[5]{x} - 3\right)^2 = \left(4\sqrt[5]{x}\right)^2 - 2\left(4\sqrt[5]{x}\right)(3) + (3)^2$
$$= 16\sqrt[5]{x^2} - 24\sqrt[5]{x} + 9$$

11.
$$\frac{\sqrt[n]{x}}{\sqrt[k]{x}} = \frac{x^{\frac{1}{n}}}{x^{\frac{1}{k}}}$$

$$= x^{\frac{1}{n} - \frac{1}{k}}$$

$$= x^{\frac{k}{kn} - \frac{n}{kn}}$$

$$= x^{\frac{k-n}{kn}}$$

$$= \sqrt[kn]{x^{k-n}}$$

12. $y = -2\sqrt{x+3} + 1$

x	y
-3	1
-2	-1
1	-3
6	-5
13	-7

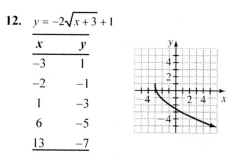

13. **a.** We need $a < 0$ and $k \geq 0$, or we need $a > 0$ and $k \leq 0$. In either case, h can be any real number.

b. $f(x) = a\sqrt{x-h} + k$

$$a\sqrt{x-h} + k = 0$$

$$a\sqrt{x-h} = -k$$

$$\sqrt{x-h} = -\frac{k}{a}$$

$$\left(\sqrt{x-h}\right)^2 = \left(-\frac{k}{a}\right)^2$$

$$x - h = \frac{k^2}{a^2}$$

$$x = h + \frac{k^2}{a^2} = \frac{a^2 h + k^2}{a^2}$$

The x-intercept is $\left(\dfrac{a^2 h + k^2}{a^2}, 0\right)$.

14. $(f+g)(x) = \left(7 - 3\sqrt{x}\right) + \left(4 + 5\sqrt{x}\right)$

$$= 7 - 3\sqrt{x} + 4 + 5\sqrt{x}$$

$$= 2\sqrt{x} + 11$$

15. $(f-g)(x) = \left(7 - 3\sqrt{x}\right) - \left(4 + 5\sqrt{x}\right)$

$$= 7 - 3\sqrt{x} - 4 - 5\sqrt{x}$$

$$= 3 - 8\sqrt{x}$$

16. $(f \cdot g)(x) = \left(7 - 3\sqrt{x}\right)\left(4 + 5\sqrt{x}\right)$

$$= 7 \cdot 4 - 3\sqrt{x} \cdot 4 + 7 \cdot 5\sqrt{x} - 3\sqrt{x} \cdot 5\sqrt{x}$$

$$= 28 - 12\sqrt{x} + 35\sqrt{x} - 15x$$

$$= -15x + 23\sqrt{x} + 28$$

17. $\left(\dfrac{f}{g}\right)(x) = \dfrac{7 - 3\sqrt{x}}{4 + 5\sqrt{x}}$

$$= \frac{7 - 3\sqrt{x}}{4 + 5\sqrt{x}} \cdot \frac{4 - 5\sqrt{x}}{4 - 5\sqrt{x}}$$

$$= \frac{7 \cdot 4 - 3\sqrt{x} \cdot 4 - 7 \cdot 5\sqrt{x} + 3\sqrt{x} \cdot 5\sqrt{x}}{(4)^2 - \left(5\sqrt{x}\right)^2}$$

$$= \frac{28 - 12\sqrt{x} - 35\sqrt{x} + 15x}{16 - 25x}$$

$$= \frac{15x - 47\sqrt{x} + 28}{16 - 25x}$$

18. $2\sqrt{x} + 3 = 13$

$$2\sqrt{x} = 10$$

$$\sqrt{x} = 5$$

$$\left(\sqrt{x}\right)^2 = 5^2$$

$$x = 25$$

Check $x = 25$:

$$2\sqrt{(25)} + 3 \stackrel{?}{=} 13$$

$$13 \stackrel{?}{=} 13 \text{ true}$$

The solution is $x = 25$.

19. $3\sqrt{5x - 4} = 27$

$$\sqrt{5x - 4} = 9$$

$$\left(\sqrt{5x - 4}\right)^2 = 9^2$$

$$5x - 4 = 81$$

$$5x = 85$$

$$x = 17$$

Check $x = 17$:

$$3\sqrt{5(17) - 4} \stackrel{?}{=} 27$$

$$27 \stackrel{?}{=} 27 \text{ true}$$

The solution is $x = 17$.

20.
$$3 - 2\sqrt{x} + \sqrt{9-x} = 0$$
$$\sqrt{9-x} = 2\sqrt{x} - 3$$
$$\left(\sqrt{9-x}\right)^2 = \left(2\sqrt{x}-3\right)^2$$
$$9 - x = 4x - 12\sqrt{x} + 9$$
$$12\sqrt{x} = 5x$$
$$\left(12\sqrt{x}\right)^2 = \left(5x\right)^2$$
$$144x = 25x^2$$
$$25x^2 - 144x = 0$$
$$x\left(25x - 144\right) = 0$$
$$x = 0 \quad \text{or} \quad 25x - 144 = 0$$
$$x = 0 \quad \text{or} \quad 25x = 144$$
$$x = 0 \quad \text{or} \quad x = \frac{144}{25}$$

Check $x = 0$:
$$3 - 2\sqrt{(0)} + \sqrt{9 - (0)} \stackrel{?}{=} 0$$
$$6 \stackrel{?}{=} 0 \quad \text{false}$$

Check $x = \dfrac{144}{25}$:
$$3 - 2\sqrt{\left(\frac{144}{25}\right)} + \sqrt{9 - \left(\frac{144}{25}\right)} \stackrel{?}{=} 0$$
$$0 \stackrel{?}{=} 0 \quad \text{true}$$

The solution is $x = \dfrac{144}{25}$.

21.
$$f(8) = 6 - 4\sqrt{(8) + 1}$$
$$= 6 - 4\sqrt{9}$$
$$= 6 - 4(3)$$
$$= 6 - 12$$
$$= -6$$

22.
$$-2 = 6 - 4\sqrt{x+1}$$
$$-8 = -4\sqrt{x+1}$$
$$2 = \sqrt{x+1}$$
$$4 = x + 1$$
$$3 = x$$

23.
$$f(x) = 3\sqrt{2x-4} - 2\sqrt{2x+1}$$
$$3\sqrt{2x-4} - 2\sqrt{2x+1} = 0$$
$$3\sqrt{2x-4} = 2\sqrt{2x+1}$$
$$\left(3\sqrt{2x-4}\right)^2 = \left(2\sqrt{2x+1}\right)^2$$
$$9(2x-4) = 4(2x+1)$$
$$18x - 36 = 8x + 4$$
$$10x = 40$$
$$x = 4$$

Check $x = 4$:
$$3\sqrt{2(4)-4} - 2\sqrt{2(4)+1} \stackrel{?}{=} 0$$
$$0 \stackrel{?}{=} 0 \quad \text{true}$$

The x-intercept is $(4, 0)$.

24.
$$\sqrt{x+4} = x^2 - 4x + 5$$
$$x \approx 0.9 \text{ and } x \approx 3.3$$

25.
$$\sqrt{x+4} = 1$$
$$x = -3$$

26. Sketches may vary. Example:

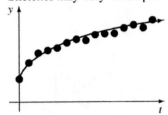

Decrease b to lower the y-intercept and increase a to increase the rate of increase.

27. Substitute the points $(2, 4)$ and $(5, 6)$ into the equation $y = a\sqrt{x} + b$.
$$4 = a\sqrt{2} + b$$
$$6 = a\sqrt{5} + b$$
Rewrite as:
$$1.4142a + b = 4$$
$$2.2361a + b = 6$$
Solve the system of equations. Multiply the first equation by -1 and add to the second equation.
$$-1.4142a - b = -4$$
$$\underline{2.2361a + b = 6}$$
$$0.8219a = 2$$
$$a \approx 2.43$$

Substitute the point $(2, 4)$ into the equation $y = 2.43\sqrt{x} + b$ and solve for b.
$$4 = 2.43\sqrt{2} + b$$
$$b = 4 - 2.43\sqrt{2} \approx 0.56$$

The equation is $y = 2.43\sqrt{x} + 0.56$.

28. a. Create a scattergram of the data.

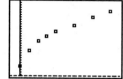

Answers may vary depending on the data points selected. Example:

Use the points $(0, 20.5)$ and $(60, 43.4)$.

The y-intercept is $(0, 20.5)$ so $b = 20.5$.

Substitute the point $(60, 43.4)$ into the

equation $y = a\sqrt{x} + 20.5$ and solve for a.

$$43.4 = a\sqrt{60} + 20.5$$
$$a\sqrt{60} = 22.9$$
$$a \approx 2.96$$
$$f(t) = 2.96\sqrt{t} + 20.5$$

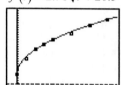

b. Answers may vary depending on the equation found in part (a). Example:

$t = 6$ years $= 6 \times 12$ months $= 72$ months

$$f(72) = 2.96\sqrt{72} + 20.5 \approx 45.6$$

According to the model, the median height of 6-year-old boys is about 45.6 inches.

c. Answers may vary depending on the equation found in part (a). Example:

$f(t) = 3$ feet $= 3 \times 12$ inches $= 36$ inches

$$36 = 2.96\sqrt{t} + 20.5$$
$$2.96\sqrt{t} = 15.5$$
$$\sqrt{t} = \frac{15.5}{2.96}$$
$$\left(\sqrt{t}\right)^2 = \left(\frac{15.5}{2.96}\right)^2$$
$$t \approx 27.42$$

The median height of 27-month-old boys is 3 feet.

d. Answers may vary depending on the equation found in part (a). Example:

The h-intercept is $(0, 20.5)$. The median height of boys at birth is 20.5 inches.

Chapter 10
Sequences and Series

1. $11 - 3 = 8, 19 - 11 = 8, 27 - 19 = 8, 35 - 27 = 8$

   The sequence has a common difference of 8. It is arithmetic.

3. $5 - 1 = 4, 7 - 5 = 2$

   The sequence does not have a common difference. It is not arithmetic.

5. $-13 - (-20) = 7, -6 - (-13) = 7, 1 - (-6) = 7,$
   $8 - 1 = 7$

   The sequence has a common difference of 7. It is arithmetic.

7. $44 - 4 = 40, 444 - 44 = 400$

   The sequence does not have a common difference. It is not arithmetic.

9. The sequence has a common difference of 6.
   $a_n = 5 + (n-1) \cdot 6 = 5 + 6n - 6 = 6n - 1$

11. The sequence has a common difference of -11.
    $$a_n = -4 + (n-1) \cdot (-11)$$
    $$= -4 + (-11n) + 11$$
    $$= -11n + 7$$

13. The sequence has a common difference of -6.
    $$a_n = 100 + (n-1) \cdot (-6)$$
    $$= 100 - 6n + 6$$
    $$= -6n + 106$$

15. The sequence has a common difference of 2.
    $a_n = 1 + (n-1) \cdot 2 = 1 + 2n - 2 = 2n - 1$

17. The sequence has a common difference of 3.
    $a_{37} = 5 + (37-1) \cdot 3 = 5 + 36 \cdot 3 = 5 + 108 = 113$

19. The sequence has a common difference of -9.
    $$a_{45} = 200 + (45-1) \cdot (-9)$$
    $$= 200 + 44 \cdot (-9)$$
    $$= 200 + (-396)$$
    $$= -196$$

21. The sequence has a common difference of 1.6.
    $$a_{96} = 4.1 + (96-1) \cdot 1.6$$
    $$= 4.1 + 95 \cdot 1.6$$
    $$= 4.1 + 152$$
    $$= 156.1$$

23. The sequence has a common difference of 1.
    $a_{400} = 1 + (400-1) \cdot 1 = 400$

25. The sequence has a common difference of 5.
    $$533 = 3 + (n-1) \cdot 5$$
    $$533 = 3 + 5n - 5$$
    $$533 = 5n - 2$$
    $$535 = 5n$$
    $$107 = n$$
    533 is the 107th term.

27. The sequence has a common difference of 8.
    $$695 = 7 + (n-1) \cdot 8$$
    $$695 = 7 + 8n - 8$$
    $$695 = 8n - 1$$
    $$696 = 8n$$
    $$87 = n$$
    695 is the 87th term.

29. The sequence has a common difference of 8.
    $$2469 = -27 + (n-1) \cdot 8$$
    $$2469 = -27 + 8n - 8$$
    $$2469 = 8n - 35$$
    $$2504 = 8n$$
    $$313 = n$$
    2469 is the 313th term.

31. The sequence has a common difference of -4.
    $$-14{,}251 = 29 + (n-1) \cdot (-4)$$
    $$-14{,}251 = 29 - 4n + 4$$
    $$-14{,}251 = 33 - 4n$$
    $$-14{,}284 = -4n$$
    $$3571 = n$$
    $-14{,}251$ is the 3571st term.

33. The sequence has a common difference of -5.
    $$-493 = -8 + (n-1) \cdot (-5)$$
    $$-493 = -8 - 5n + 5$$
    $$-493 = -3 - 5n$$
    $$-490 = -5n$$
    $$98 = n$$
    -493 is the 98th term.

35. $15 - 8 = 7, 22 - 15 = 7, 29 - 22 = 7,$
$36 - 29 = 7$
The sequence has a common difference of 7. It is arithmetic.
$2537 = 8 + (n-1) \cdot 7$
$2537 = 8 + 7n - 7$
$2537 = 7n + 1$
$2536 = 7n$
$\quad n \approx 362.3$
Since n is not a counting number, 2537 is not a term in the sequence.

37. Since $d = 9$ and $a_1 = 8,$
$\quad f(n) = 8 + (n-1) \cdot 9 = 8 + 9n - 9 = 9n - 1$

39. a. $a_n = 27,500 + (n-1) \cdot 800$
$\quad\quad = 27,500 + 800n - 800$
$\quad\quad = 800n + 26,700$

   b. $a_{22} = 26,700 + 800 \cdot 22 = \$44,300$
The salary for the 22nd year will be $44,300.

   c. $50,000 = 26,700 + 800n$
$23,300 = 800n$
$29.125 = n$
The salary will first be above $50,000 in the 30th year.

41. a. $a_n = 35 + \dfrac{1}{6}n$

   b. $a_1 = 35 + \dfrac{1}{6} \cdot 1 \approx 35.17$

$a_2 = 35 + \dfrac{1}{6} \cdot 2 \approx 35.33$

$a_3 = 35 + \dfrac{1}{6} \cdot 3 = 35.5$

$a_4 = 35 + \dfrac{1}{6} \cdot 4 \approx 35.67$

These values represent the number of hours the instructor would work if she had 1, 2, 3, or 4 students, respectively.

   c. $a_{130} = 35 + \dfrac{1}{6} \cdot 130 \approx 56.67$
She works 56.7 hours per week.

   d. $60 = 35 + \dfrac{1}{6}n$

$25 = \dfrac{1}{6}n$

$150 = n$
The greatest number is 150 students.

43. a. The band collects $0.3(6) = \$1.80$ per cover charge, so the common difference is 1.8.
$a_n = -50 + 0.3(6)n$
$\quad = -50 + 1.8n$

   b. $256 = -50 + 1.8n$
$306 = 1.8n$
$170 = n$
170 people paid the cover charge.

   c. Total number of people who pay is
$200 - 18 - 11 - 6 = 165.$
$a_{165} = -50 + 1.8(165) = 247$
Their maximum profit is $247.00.

   d. $0 = -50 + 0.3(6)n$
$0 = -50 + 1.8n$
$50 = 1.8n$
$\quad n \approx 27.8$
Little Muddy will lose money for values of n less than or equal to 27.

45. a. Create a scattergram of the data.

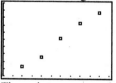

The points appear to be linear.
Use linear regression to find an equation.

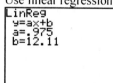

$f(t) = 0.98t + 12.11$

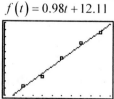

The equation fits the points well.

b.

X	Y1	
13	24.85	
14	25.83	
15	26.81	
16	27.79	
17	28.77	
18	29.75	
19	30.73	

X=19

$$f(13) = 24.85$$
$$f(14) = 25.83$$
$$f(15) = 26.81$$
$$f(16) = 27.79$$
$$f(17) = 28.77$$

From 2013 through 2017, the pharmaceutical industry spent (in millions of dollars) about 24.9, 25.8, 26.8, 27.8, and 28.8, respectively, on government and politics.

c. $$30 = 0.98t + 12.11$$
$$17.89 = 0.98t$$
$$t \approx 18.3$$

We predict that the pharmaceutical industry's spending on government and politics will be \$30 million in $2000 + 18.3 \approx 2018$.

47. a. $$a_n = 0.9 + (n-1) \cdot 0.2$$
$$= 0.9 + 0.2n - 0.2$$
$$= 0.2n + 0.7$$

b. $$a_{13} = 0.2(13) + 0.7 = 3.3$$
The postage is \$3.30.

c. $$a_{16} = 0.2(16) + 0.7 = 3.9$$
No, \$3.90 is a better deal.

d.   5 pounds is equivalent to 80 ounces.
$$a_{80} = 0.2(80) + 0.7 = 16.7$$
Using the formula, the postage is \$16.70.

49. Equation 1: $$500 = a_1 + (41-1)d$$
$$500 = a_1 + 40d$$
$$500 - 40d = a_1$$

Equation 2: $$500 = a_1 + (81-1)d$$
$$500 = a_1 + 80d$$
$$500 - 80d = a_1$$

Solve the system of linear equations by setting the left side of each equation equal to each other.
$$500 - 40d = 500 - 80d$$
$$40d = 0$$
$$d = 0$$

Use equation 1 and $d = 0$ to find a_1.
$$500 - 40 \cdot 0 = a_1$$
$$500 = a_1$$
$$a_{990} = 500 + (990 - 1) \cdot 0 = 500$$

51. $$f(1) = 4(1) - 2 = 2$$
$$f(2) = 4(2) - 2 = 6$$
$$f(3) = 4(3) - 2 = 10$$
Yes, this sequence is arithmetic.
Answers may vary. Example:
The sequence has a common difference equal to the slope, 4.

53. $$f(1) = 1^2 = 1$$
$$f(2) = 2^2 = 4$$
$$f(3) = 3^2 = 9$$
No, this sequence is not arithmetic.
Answers may vary. Example:
There is no common difference.

55.   The student's work is not correct.
Answers may vary. Example:
The sequence is not arithmetic because there is no common difference.

57. a. $7 - 5 = 2, 9 - 7 = 2, 11 - 9 = 2, 13 - 11 = 2$
The arithmetic sequence has a common difference of 2.

b. $$m = \frac{7-5}{2-1} = \frac{2}{1} = 2$$
$$m = \frac{9-7}{3-2} = \frac{2}{1} = 2$$
$$m = \frac{11-9}{4-3} = \frac{2}{1} = 2$$
$$m = \frac{13-11}{5-4} = \frac{2}{1} = 2$$
The slope of the line containing the points given is $m = 2$.

c.   The common difference of the arithmetic sequence is equal to the slope of the function.
Answers may vary. Example:
The form of the equation of an arithmetic sequence is the same as the form of a line in slope–intercept form, where m takes the place of d.

59.
$$2\sqrt{x+3}-1=5$$
$$2\sqrt{x+3}=6$$
$$\sqrt{x+3}=3$$
$$x+3=9$$
$$x=6$$
Check:
$$2\sqrt{(6)+3}-1\overset{?}{=}5$$
$$2\sqrt{9}-1\overset{?}{=}5$$
$$2(3)-1\overset{?}{=}5$$
$$6-1\overset{?}{=}5$$
$$5\overset{?}{=}5 \quad \text{true}$$
This is a radical equation in one variable.

61. $f(x)=2\sqrt{x+3}-1$

This is a radical function.

63. $\left(4\sqrt{x}-5\right)\left(3\sqrt{x}-2\right)$
$$=12\cdot x-8\sqrt{x}-15\sqrt{x}+10$$
$$=12x-23\sqrt{x}+10$$
This is a radical expression in one variable.

Homework 10.2

1. $\dfrac{28}{4}=7, \dfrac{196}{28}=7, \dfrac{1372}{196}=7, \dfrac{9604}{1372}=7$
The sequence has a common ratio of 7. It is geometric.

3. $6-13=-7, -1-6=-7, -8-(-1)=-7,$
$-15-(-8)=-7$
The sequence has a common difference of -7. It is arithmetic.

5. $\dfrac{4}{3}\approx1.33, \dfrac{6}{4}=1.5$
The sequence has no common ratio.
$4-3=1, 6-4=2$
The sequence has no common difference. The sequence is neither arithmetic nor geometric.

7. $\dfrac{40}{200}=\dfrac{1}{5}, \dfrac{8}{40}=\dfrac{1}{5}, \dfrac{\frac{8}{5}}{8}=\dfrac{1}{5}, \dfrac{\frac{8}{25}}{\frac{8}{5}}=\dfrac{1}{5}$

The sequence has a common ratio of $\dfrac{1}{5}$. It is geometric.

9. The sequence has a common ratio of 2.
$$a_n=3(2)^{n-1}=3(2)^n(2)^{-1}=\dfrac{3(2)^n}{2}=\dfrac{3}{2}(2)^n$$
$$a_n=3(2)^{n-1} \text{ or } a_n=\dfrac{3}{2}(2)^n$$

11. The sequence has a common ratio of $\dfrac{1}{4}$.
$$a_n=800\left(\dfrac{1}{4}\right)^{n-1}$$
$$=800\left(\dfrac{1}{4}\right)^n\left(\dfrac{1}{4}\right)^{-1}$$
$$=800\left(\dfrac{1}{4}\right)^n(4)$$
$$=3200\left(\dfrac{1}{4}\right)^n$$
$$a_n=800\left(\dfrac{1}{4}\right)^{n-1} \text{ or } a_n=3200\left(\dfrac{1}{4}\right)^n$$

13. The sequence has a common ratio of $\dfrac{1}{2}$.
$$a_n=100\left(\dfrac{1}{2}\right)^{n-1}$$
$$=100\left(\dfrac{1}{2}\right)^n\left(\dfrac{1}{2}\right)^{-1}$$
$$=100\left(\dfrac{1}{2}\right)^n(2)$$
$$=200\left(\dfrac{1}{2}\right)^n$$
$$a_n=100\left(\dfrac{1}{2}\right)^{n-1} \text{ or } a_n=200\left(\dfrac{1}{2}\right)^n$$

15. The sequence has a common ratio of 4.
$$a_n=1(4)^{n-1}=4^n(4)^{-1}=4^n\left(\dfrac{1}{4}\right)=\dfrac{1}{4}(4)^n$$
$$a_n=1(4)^{n-1}=4^{n-1} \text{ or } a_n=\dfrac{1}{4}(4)^n$$

17. The sequence has a common ratio of 5.
$$a_{34} = 4(5)^{34-1} = 4(5)^{33} \approx 4.6566 \times 10^{23}$$

19. The sequence has a common ratio of $\frac{1}{2}$.

$$a_{27} = 80\left(\frac{1}{2}\right)^{27-1} = 80\left(\frac{1}{2}\right)^{26} \approx 1.1921 \times 10^{-6}$$

21. The sequence has a common ratio of 2.
$$a_{23} = 8(2)^{23-1}$$
$$= 8(2)^{22}$$
$$= 33,554,432 \approx 3.3554 \times 10^{7}$$

23. The sequence has a common ratio of $\frac{1}{2}$.

$$0.46875 = 240\left(\frac{1}{2}\right)^{n-1}$$

$$0.001953125 = \left(\frac{1}{2}\right)^{n-1}$$

$$\log(0.001953125) = \log\left(\frac{1}{2}\right)^{n-1}$$

$$\log(0.001953125) = (n-1)\log\left(\frac{1}{2}\right)$$

$$\frac{\log(0.001953125)}{\log\left(\frac{1}{2}\right)} = n-1$$

$$\frac{\log(0.001953125)}{\log\left(\frac{1}{2}\right)} + 1 = n$$

$$10 = n$$

25. Use the sequence 0.00224, 0.0112, 0.056, 0.28, 1.4,…,109,375. This sequence has a common ratio of 5.
$$109,375 = 0.00224(5)^{n-1}$$
$$48,828,125 = 5^{n-1}$$
$$\log(48,828,125) = \log(5^{n-1})$$
$$\log(48,828,125) = (n-1)\log(5)$$
$$\frac{\log(48,828,125)}{\log(5)} = n-1$$
$$\frac{\log(48,828,125)}{\log(5)} + 1 = n$$
$$12 = n$$

27. The sequence has a common ratio of 2.
$$3,407,872 = 13(2^{n-1})$$
$$262,144 = 2^{n-1}$$
$$\log(262,144) = \log(2^{n-1})$$
$$\log(262,144) = (n-1)\log(2)$$
$$\frac{\log(262,144)}{\log(2)} = n-1$$
$$\frac{\log(262,144)}{\log(2)} + 1 = n$$
$$19 = n$$

29. The sequence has a common ratio of 3.
$$28,697,814 = 2(3)^{n-1}$$
$$14,348,907 = 3^{n-1}$$
$$\log(14,348,907) = \log(3^{n-1})$$
$$\log(14,348,907) = (n-1)\log(3)$$
$$\frac{\log(14,348,907)}{\log(3)} = n-1$$
$$\frac{\log(14,348,907)}{\log(3)} + 1 = n$$
$$16 = n$$

31. The sequence has a common ratio of 3.
$$f(n) = 8(3)^{n-1}$$

33. No. Answers may vary. Example: The sequence is geometric with a common ratio of 2 and $a_1 = 13$. Therefore, all terms must be divisible by 13. 9,238,946 is not divisible by 13, so it is not a term in the geometric sequence.

35. **a.** $a_n = 27,000(1.04)^{n-1}$

    b. $a_{10} = 27,000(1.04)^{10-1}$
    $$= 27,000(1.04)^{9} \approx \$38,429.42$$
    The person's salary will be \$38,429.42 for the 10th year.

c.
$$50,000 = 27,000(1.04)^{n-1}$$

$$\frac{50}{27} = 1.04^{n-1}$$

$$\log\left(\frac{50}{27}\right) = \log\left(1.04^{n-1}\right)$$

$$\log\left(\frac{50}{27}\right) = (n-1)\log(1.04)$$

$$\frac{\log\left(\dfrac{50}{27}\right)}{\log(1.04)} = n-1$$

$$\frac{\log\left(\dfrac{50}{27}\right)}{\log(1.04)} + 1 = n$$

$$16.7 \approx n$$

The salary will first exceed \$50,000 in the 17th year.

37. a. $2, 4, 8, 16, 32$

b. The sequence has a common ratio of 2.
$$a_n = 2(2)^{n-1} = 2^1 \cdot 2^{n-1} = 2^{1+n-1} = 2^n$$

c. $a_8 = 2^8 = 256$ ancestors

d. $a_{35} = 2^{35} \approx 3.436 \times 10^{10}$
$$\approx 34.36 \text{ billion ancestors}$$
Model breakdown has occurred.
Answers may vary. Example:
The number of ancestors is much higher than the world's current population. One assumption is that no ancestor is related to any other ancestor, and this assumption is likely false.

39. a. Create a scattergram of the data.

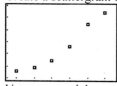

Use exponential regression to find an equation.

$$f(t) = 26.81(1.144)^t$$

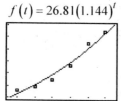

The equation fits the points well.

b.

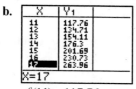

$f(11) \approx 117.76$
$f(12) \approx 134.71$
$f(13) \approx 154.11$
$f(14) \approx 176.30$
$f(15) \approx 201.69$

From 2011 through 2015, the federal student loan origination volumes (in billions of dollars) will be about 117.8, 134.7, 154.1, 176.3, and 201.7, respectively.

c.
$$250 = 26.81(1.144)^t$$

$$\frac{250}{26.81} = (1.144)^t$$

$$\log\left(\frac{250}{26.81}\right) = \log(1.144)^t$$

$$\log\left(\frac{250}{26.81}\right) = t\log(1.144)$$

$$t = \frac{\log\left(\dfrac{250}{26.81}\right)}{\log(1.144)} \approx 16.60$$

We predict that the origination volume of federal student loans will be \$250 billion in $2000 + 16.60 \approx 2017$.

41. a. $a_n = 5(3)^{n-1}$

b. $a_5 = 5(3)^{5-1} = 5(3)^4 = 405$
a_5 is equal to 405 students.

c. $a_{11} = 5(3)^{11-1} = 5(3)^{10} = 295,245$
Yes, model breakdown has occurred.
Answers may vary. Example:
No campus has 295,245 students.

d. Answers may vary. Example:
One of the assumptions was that a student would tell the rumor to 3 other students who have not heard the rumor yet. This assumption is reasonable for the first several days. However, as the number of students who heard the rumor grows larger, it is unlikely that those students would each know 3 other students who had not heard the rumor yet.

43. The sequence is geometric. The common ratio is 5.

45. The sequence is arithmetic. The common difference is the slope which is 7.

47. The student is using the formula for a geometric sequence but the sequence is arithmetic.
The arithmetic sequence has a common difference of 4.
$a_{17} = 2 + (17 - 1) \cdot 4 = 66$

49. a. $\dfrac{14}{7} = 2, \dfrac{28}{14} = 2, \dfrac{56}{28} = 2, \dfrac{112}{56} = 2$
The series is geometric with a common ratio of 2.

b. $y = ab^x$
$7 = a \cdot b^1$
$7 = ab$
$a = \dfrac{7}{b}$

Substitute $a = \dfrac{7}{b}$ into $14 = ab^2$.

$14 = \left(\dfrac{7}{b}\right)b^2$
$14 = 7b$
$2 = b$
So the base is 2.

c. The common ratio for the sequence and the base of the function are the same.
Answers may vary. Example:
The form of the equation of a geometric sequence is the same as the form of an exponential function, where the common ratio of the sequence is the base of the function.

51. The sequence has a common difference of 5.
$a_n = 14 + (n - 1) \cdot 5 = 14 + 5n - 5 = 5n + 9$

53. The sequence has a common ratio of $\dfrac{1}{2}$.

$a_n = 448\left(\dfrac{1}{2}\right)^{n-1}$

$= 448\left(\dfrac{1}{2}\right)^{n}\left(\dfrac{1}{2}\right)^{-1}$

$= 448\left(\dfrac{1}{2}\right)^{n}(2)$

$a_n = 896\left(\dfrac{1}{2}\right)^{n}$

$a_n = 448\left(\dfrac{1}{2}\right)^{n-1}$ or $a_n = 896\left(\dfrac{1}{2}\right)^{n}$

55. The sequence has a common ratio of 5.
$a_9 = 2(5)^{9-1} = 2(5)^8 = 781,250$

57. The sequence has a common difference of -5.
$a_{99} = 17 + (99 - 1) \cdot (-5) = -473$

59. The sequence has a common difference of 3.
$367 = 4 + (n - 1) \cdot 3$
$367 = 4 + 3n - 3$
$367 = 1 + 3n$
$366 = 3n$
$122 = n$

61. The sequence is geometric with a common ratio of $\dfrac{1}{4}$ and $a_1 = 8192$.

$$0.0078125 = 8192\left(\dfrac{1}{4}\right)^{n-1}$$

$$\dfrac{0.0078125}{8192} = \left(\dfrac{1}{4}\right)^{n-1}$$

$$\log\left(\dfrac{0.0078125}{8192}\right) = \log\left(\dfrac{1}{4}\right)^{n-1}$$

$$\log\left(\dfrac{0.0078125}{8192}\right) = (n-1)\log\left(\dfrac{1}{4}\right)$$

$$\dfrac{\log\left(\dfrac{0.0078125}{8192}\right)}{\log\left(\dfrac{1}{4}\right)} = n - 1$$

$$\dfrac{\log\left(\dfrac{0.0078125}{8192}\right)}{\log\left(\dfrac{1}{4}\right)} + 1 = n$$

$$11 = n$$

63. $-3(4)^x = -44$

$$4^x = \frac{44}{3}$$

$$\log\left(4^x\right) = \log\left(\frac{44}{3}\right)$$

$$x \log\left(4\right) = \log\left(\frac{44}{3}\right)$$

$$x = \frac{\log\left(\frac{44}{3}\right)}{\log\left(4\right)} \approx 1.9372$$

Check:

$$-3(4)^{1.937234559} \overset{?}{=} -44$$

$$-3\left(\frac{44}{3}\right) \overset{?}{=} -44$$

$$-44 \overset{?}{=} -44 \quad \text{true}$$

This is an exponential equation in one variable.

65.

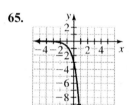

This is an exponential function.

67. $2\log_b\left(5x^3\right) - 3\log_b\left(2x^7\right)$

$$= \log_b\left(5x^3\right)^2 - \log_b\left(2x^7\right)^3$$

$$= \log_b\left(25x^6\right) - \log_b\left(8x^{21}\right)$$

$$= \log_b\left(\frac{25x^6}{8x^{21}}\right)$$

$$= \log_b\left(\frac{25}{8x^{15}}\right)$$

This is a logarithmic expression in one variable.

Homework 10.3

1. $S_{90} = \dfrac{90\left(2 + 447\right)}{2} = 20{,}205$

3. $S_{108} = \dfrac{108(13 + 548)}{2} = 30{,}294$

5. $S_{72} = \dfrac{72\left(37 + \left(-1099\right)\right)}{2} = -38{,}232$

7.   The series is arithmetic with a common difference of 8.

$$S_{74} = \frac{74(5 + 589)}{2} = 21{,}978$$

9.   The series is arithmetic with a common difference of –4.

$$S_{101} = \frac{101\left(93 + \left(-307\right)\right)}{2} = -10{,}807$$

11.   The series is arithmetic with a common difference of 0.

$$S_{117} = \frac{117(4 + 4)}{2} = 468$$

13.   The series is arithmetic with a common difference of 10.

$$a_{125} = 3 + (125 - 1) \cdot 10 = 3 + 124 \cdot 10 = 1243$$

$$S_{125} = \frac{125(3 + 1243)}{2} = 77{,}875$$

15.   The series is arithmetic with a common difference of 11.

$$a_{81} = 8 + (81 - 1) \cdot 11 = 8 + 80 \cdot 11 = 888$$

$$S_{81} = \frac{81(8 + 888)}{2} = 36{,}288$$

17.   The series is arithmetic with a common difference of –13.

$$a_{152} = -15 + (152 - 1) \cdot \left(-13\right) = -1978$$

$$S_{152} = \frac{152\left(-15 + \left(-1978\right)\right)}{2} = -151{,}468$$

19.   The series is arithmetic with a common difference of 3.

$$a_{137} = -40 + (137 - 1) \cdot 3 = 368$$

$$S_{137} = \frac{137(-40 + 368)}{2} = 22{,}468$$

21.   The series is arithmetic with a common difference of 6.

$$247 = 19 + (n - 1) \cdot 6$$

$$247 = 19 + 6n - 6$$

$$247 = 13 + 6n$$

$$234 = 6n$$

$$39 = n$$

$$S_{39} = \frac{39(19 + 247)}{2} = 5187$$

23. The series is arithmetic with a common difference of -8.
$$-900 = 900 + (n-1) \cdot (-8)$$
$$-900 = 900 - 8n + 8$$
$$-900 = 908 - 8n$$
$$-1808 = -8n$$
$$226 = n$$
$$S_{226} = \frac{226(900 + (-900))}{2} = 0$$

25. The series is arithmetic with a common difference of 3.
$$340 = 4 + (n-1) \cdot 3$$
$$340 = 4 + 3n - 3$$
$$340 = 3n + 1$$
$$339 = 3n$$
$$113 = n$$
$$S_{113} = \frac{113(4 + 340)}{2} = 19,436$$

27. The series is arithmetic with a common difference of 1.
$$10,000 = 1 + (n-1) \cdot 1$$
$$10,000 = 1 + n - 1$$
$$10,000 = n$$
$$a_{10,000} = 10,000$$
$$S_{10,000} = \frac{10,000(1 + 10,000)}{2} = 50,005,000$$

29. a. $a_{28} = 28,500 + (28-1) \cdot 1100$
$$= 28,500 + 27 \cdot 1100$$
$$= \$58,200$$

b. $S_{28} = \dfrac{28(28,500 + 58,200)}{2} = \$1,213,800$

31. Company A:
$$a_{20} = 35,000 + (20-1) \cdot 700 = \$48,300$$
$$S_{20} = \frac{20(35,000 + 48,300)}{2} = \$833,000$$
Company B:
$$a_{20} = 27,000 + (20-1) \cdot 1500 = \$55,500$$
$$S_{20} = \frac{20(27,000 + 55,500)}{2} = \$825,000$$
Your total earnings for 20 years would be greater at Company A by \$8000.

33. a. $a_{30} = 20 + (30-1) \cdot 4 = 136$
There are 136 seats in the 30th row, the back row.

b. $S_{30} = \dfrac{30(20 + 136)}{2} = 2340$
There are 2340 seats in the auditorium.

35. a. Since $t = 0$ corresponds to the year 2000, according to the model:
$$f(0) = 0.98(0) + 12.11 = 12.11$$
The model estimates that the pharmaceutical industry spent approximately \$12.1 million on government and politics in 2000.

b. Since $t = 17$ corresponds to the year 2017, according to the model:
$$f(17) = 0.98(17) + 12.11 = 28.77$$
The model predicts that in 2017, the pharmaceutical industry will spend approximately \$28.8 million on government and politics.

c. $S_{18} = \dfrac{18(12.1 + 28.8)}{2} = 368.1$
From 2000 through 2017 (including both 2000 and 2017), the pharmaceutical industry will spend a total of approximately \$368.1 million, or \$3.681 billion, on government and politics.

37. a. $a_{26} = 24,800 + (26-1) \cdot 1200 = \$54,800$
$$S_{26} = \frac{26(24,800 + 54,800)}{2} = \$1,034,800$$

b. $a_1 = 0$
$$a_{26} = 54,800 - 24,800 = 30,000$$
$$n = 26$$
$$S_{26} = \frac{26(0 + 30,000)}{2} = 390,000$$
Alternatively, subtract the total with no raises, or \$24,800(26) = \$644,800, from the total with raises.
\$1,034,800 - \$644,800 = \$390,000
The total amount of money earned from raises in 26 years is \$390,000.

c. mean $= \dfrac{1,034,800}{26} = \$39,800$
The mean salary over the 26 years is \$39,800. For the first 13 years this mean will be greater than the yearly salary. For the last 13 years, the mean will be less than the yearly salary.

d. The taxable income for the first 5 years is $20,550, $21,750, $22,950, $24,150, and $25,350. The taxable income does not exceed $25,000 until the 5th year. Therefore, the total income taxed at the lower rate will be:

$20,550 + 21,750 + 22,950 + 24,150$
$+ 22(25,000) = \$639,400$

The taxable income at the higher rate is an arithmetic sequence whose first term is $350 \ (25,350 - 25,000)$ and whose common difference is $1200. The number of terms in this sequence is 22 because taxable income did not exceed $25,000 until the 5th year.

$a_{22} = 350 + (22 - 1) \cdot 1200 = \$25,550$

$S_{22} = \dfrac{22(350 + 25,550)}{2} = \$284,900$

The estimated income tax will be
$\$639,400(0.15016) + \$284,900(0.1704)$
$\approx \$144,559.26$

39. S_n is positive.
Answers may vary. Example:
It will be the sum of n positive numbers.

41. S_n is positive.
Answers may vary. Example:
It will be the sum of three relatively small negative numbers and many larger positive numbers.

43. Yes, the series is arithmetic.
Answers may vary. Example:
The common difference is the slope, 7.

45. The arithmetic sequence has a common difference of 16.
$a_{15} = 8 + (15 - 1) \cdot 16 = 232$

47. The arithmetic series has a common difference of 16.
$a_{15} = 8 + (15 - 1) \cdot 16 = 232$
$S_{15} = \dfrac{15(8 + 232)}{2} = 1800$

49. $\dfrac{x-5}{x^2-9} + \dfrac{x+3}{x^2-8x+15}$

$= \dfrac{x-5}{(x-3)(x+3)} + \dfrac{x+3}{(x-3)(x-5)}$

$= \dfrac{(x-5)}{(x-3)(x+3)} \cdot \dfrac{(x-5)}{(x-5)} + \dfrac{(x+3)}{(x-3)(x-5)} \cdot \dfrac{(x+3)}{(x+3)}$

$= \dfrac{(x-5)(x-5)}{(x-3)(x+3)(x-5)} + \dfrac{(x+3)(x+3)}{(x-3)(x-5)(x+3)}$

$= \dfrac{(x-5)(x-5) + (x+3)(x+3)}{(x-3)(x+3)(x-5)}$

$= \dfrac{x^2 - 5x - 5x + 25 + x^2 + 3x + 3x + 9}{(x-3)(x+3)(x-5)}$

$= \dfrac{2x^2 - 4x + 34}{(x-3)(x+3)(x-5)}$

$= \dfrac{2(x^2 - 2x + 17)}{(x-3)(x+3)(x-5)}$

This is a rational expression in one variable.

51. $\dfrac{x-5}{x^2-9} \cdot \dfrac{x+3}{x^2-8x+15}$

$= \dfrac{x-5}{(x-3)(x+3)} \cdot \dfrac{x+3}{(x-3)(x-5)}$

$= \dfrac{(x-5)}{(x-3)(x+3)} \cdot \dfrac{(x+3)}{(x-3)(x-5)}$

$= \dfrac{(x-5)}{(x-3)} \cdot \dfrac{1}{(x-3)(x-5)}$

$= \dfrac{1}{(x-3)^2}$

This is a rational expression in one variable.

53. $\dfrac{x-5}{x^2-9} + \dfrac{x+3}{x^2-8x+15} = \dfrac{2}{x-5}$

$\dfrac{x-5}{(x-3)(x+3)} + \dfrac{x+3}{(x-3)(x-5)} = \dfrac{2}{x-5}$

$(x-5)(x-5) + (x+3)(x+3) = 2(x-3)(x+3)$

$x^2 - 10x + 25 + x^2 + 6x + 9 = 2x^2 - 18$

$2x^2 - 4x + 34 = 2x^2 - 18$

$-4x + 34 = -18$

$-4x = -52$

$x = 13$

Check:

$$\frac{(13)-5}{(13)^2-9}+\frac{(13)+3}{(13)^2-8(13)+15}\overset{?}{=}\frac{2}{(13)-5}$$

$$0.25\overset{?}{=}0.25 \text{ true}$$

This is a rational equation in one variable.

Homework 10.4

1. $S_{13}=\dfrac{5(1-2^{13})}{1-2}=40{,}955$

3. $S_{12}=\dfrac{6(1-1.3^{12})}{1-1.3}\approx 445.9617$

5. $S_{13}=\dfrac{13(1-0.8^{13})}{1-0.8}\approx 61.4266$

7. $S_{10}=\dfrac{2.3(1-0.9^{10})}{1-0.9}\approx 14.9804$

9. The series is geometric with a common ratio of 5.

$$S_{13}=\frac{2(1-5^{13})}{1-5}=610{,}351{,}562$$

11. The series is geometric with a common ratio of 0.3.

$$S_{11}=\frac{600(1-0.3^{11})}{1-0.3}\approx 857.1413$$

13. The series is geometric with a common ratio of $\dfrac{2}{3}$.

$$S_{10}=\frac{3\left(1-\left(\frac{2}{3}\right)^{10}\right)}{1-\frac{2}{3}}\approx 8.8439$$

15. The series is geometric with a common ratio of 4.

$$67{,}108{,}864=1(4)^{n-1}$$
$$\log(67{,}108{,}864)=\log\left(4^{n-1}\right)$$
$$\log(7{,}108{,}864)=(n-1)\log(4)$$
$$\frac{\log(67{,}108{,}864)}{\log(4)}=n-1$$
$$\frac{\log(67{,}108{,}864)}{\log(4)}+1=n$$
$$14=n$$
$$S_{14}=\frac{1(1-4^{14})}{1-4}=89{,}478{,}485$$

17. The series is geometric with a common ratio of 1.2.

$$21.4990848=5(1.2)^{n-1}$$
$$4.29981696=1.2^{n-1}$$
$$\log(4.29981696)=\log\left(1.2^{n-1}\right)$$
$$\log(4.29981696)=(n-1)\log(1.2)$$
$$\frac{\log(4.29981696)}{\log(1.2)}=n-1$$
$$\frac{\log(4.29981696)}{\log(1.2)}+1=n$$
$$9=n$$
$$S_9=\frac{5(1-1.2^9)}{1-1.2}\approx 103.9945$$

19. The series is geometric with a common ratio of $\dfrac{1}{2}$.

$$4.8828125=10{,}000\left(\frac{1}{2}\right)^{n-1}$$
$$0.00048828125=\left(\frac{1}{2}\right)^{n-1}$$
$$\log\left(0.00048828125\right)=\log\left[\left(\frac{1}{2}\right)^{n-1}\right]$$
$$\log\left(0.00048828125\right)=(n-1)\log\left(\frac{1}{2}\right)$$
$$\frac{\log\left(0.00048828125\right)}{\log\left(\frac{1}{2}\right)}=n-1$$
$$\frac{\log\left(0.00048828125\right)}{\log\left(\frac{1}{2}\right)}+1=n$$
$$12=n$$
$$S_{12}=\frac{10{,}000\left(1-\left(\frac{1}{2}\right)^{12}\right)}{1-\frac{1}{2}}\approx 19{,}995.1172$$

21. The series is arithmetic with a common difference of 0.

$$S_{100}=\frac{100(1+1)}{2}=100$$

23. The series is geometric with a common ratio of $\frac{1}{3}$.

$$\frac{4}{729} = 324\left(\frac{1}{3}\right)^{n-1}$$

$$\frac{4}{729 \cdot 324} = \left(\frac{1}{3}\right)^{n-1}$$

$$\frac{1}{59,049} = \left(\frac{1}{3}\right)^{n-1}$$

$$\log\left(\frac{1}{59,049}\right) = \log\left[\left(\frac{1}{3}\right)^{n-1}\right]$$

$$\log\left(\frac{1}{59,049}\right) = (n-1)\log\left(\frac{1}{3}\right)$$

$$\frac{\log\left(\frac{1}{59,049}\right)}{\log\left(\frac{1}{3}\right)} = n-1$$

$$\frac{\log\left(\frac{1}{59,049}\right)}{\log\left(\frac{1}{3}\right)} + 1 = n$$

$$11 = n$$

$$S_{11} = \frac{324\left[1-\left(\frac{1}{3}\right)^{11}\right]}{1-\left(\frac{1}{3}\right)} \approx 485.9973$$

25. $S_{20} = \frac{\$23,500(1-1.04^{20})}{1-1.04} \approx \$699,784.85$

The person's total earnings after 20 years of work will be $699,784.85.

27. Company A:

$S_{30} = \frac{\$26,000(1-1.05^{30})}{1-1.05} \approx \$1,727,410.04$

Company B:

$S_{30} = \frac{\$31,000(1-1.03^{30})}{1-1.03} \approx \$1,474,837.89$

The earnings at Company A after 30 years will be $252,572.15 more than Company B.

29. Recall, the number of ancestors n generations back is a geometric series with a common ratio of 2.

$$S_n = \frac{2\left(1-2^{10}\right)}{1-2} = 2046 \text{ ancestors}$$

31. a. The entrepreneur's name would be taken off the list in the 11th round. The amount of money sent to the entrepreneur each round is a geometric series with a common ratio of 8.

$$S_{10} = \frac{40(1-8^{10})}{1-8} = 6,135,667,560$$

The entrepreneur could receive as much as approximately $6.14 billion.

b.

$$7,000,000,000 = \frac{8(1-8^n)}{1-8}$$

$$7,000,000,000 = -\frac{8}{7}(1-8^n)$$

$$6,125,000,000 = 8^n - 1$$

$$6,125,000,001 = 8^n$$

$$\log(6,125,000,001) = \log\left(8^n\right)$$

$$\log(6,125,000,001) = n\log(8)$$

$$\frac{\log(6,125,000,001)}{\log(8)} = n$$

$$10.84 \approx n$$

There will be ten full rounds and part of an 11th round.

c. The money from the first 10 rounds will go to the entrepreneur. The chain letter runs out of people (and money!) to complete the 11th round. All the money from the 11th round would go to the first eight people besides the entrepreneur. This amount (in billions) is:

$$\frac{35-6.136}{8} \approx 3.61$$

So, nine people will receive money from the chain letters. The entrepreneur will receive approximately $6.14 billion. The other eight people will receive an average of $3.61 billion each.

33. a. Create a scattergram of the data.

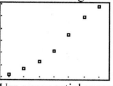

Use exponential regression to find an equation.

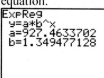

$$f(t) = 927.46(1.35)^t$$

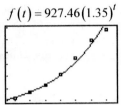

The equation fits the points well.

b. $f(1) = 927.46(1.35)^1 \approx 1252.07$

This means that 1252 Nevaehs were born in 2001.

c. $f(17) = 927.46(1.35)^{17} \approx 152,394.49$

This means that 152,394 Nevaehs will be born in 2017.

d. Find the sum of

$f(1), f(2), \cdots, f(13), f(17)$.

$$S_{17} = \frac{1252.07(1 - 1.35^{17})}{1 - 1.35} \approx 584,229.49$$

This means that 584,229 Nevaehs will be born from 2001 to 2017, inclusive.

35. S_n is positive.

Answers may vary. Example:
It is the sum of all positive values.

37. The series is arithmetic.

Answers may vary. Example:
$f(x)$ is linear. The common difference of the series is the slope of $f(x)$ which is -1.

39. a. The series is geometric with a common ratio of 2.

$$2560 = 5(2)^{n-1}$$
$$512 = 2^{n-1}$$
$$\log(512) = \log\left(2^{n-1}\right)$$
$$\log(512) = (n-1)\log(2)$$
$$\frac{\log(512)}{\log(2)} = n - 1$$
$$\frac{\log(512)}{\log(2)} + 1 = n$$
$$10 = n$$
$$S_{10} = \frac{5(1 - 2^{10})}{1 - 2} = 5115$$

b.
$$a_n = a_1 r^{n-1}$$
$$\frac{a_n}{a_1} = r^{n-1}$$
$$\log\left(\frac{a_n}{a_1}\right) = \log\left(r^{n-1}\right)$$
$$\log\left(\frac{a_n}{a_1}\right) = (n-1)\log(r)$$
$$\frac{\log\left(\frac{a_n}{a_1}\right)}{\log(r)} = (n-1)$$
$$\frac{\log\left(\frac{a_n}{a_1}\right)}{\log(r)} + 1 = n$$
$$\frac{\log\left(\frac{a_n}{a_1}\right)}{\log(r)} + \frac{\log(r)}{\log(r)} = n$$
$$\frac{\log\left(\frac{a_n}{a_1}\right) + \log(r)}{\log(r)} = n$$
$$\frac{\log\left(\frac{a_n r}{a_1}\right)}{\log(r)} = n$$

c. $$S_n = \frac{a_1\left(1 - r^n\right)}{1 - r}$$
$$S_n = \frac{a_1\left(1 - r^{\log(a_n r/a_1)/\log(r)}\right)}{1 - r}$$

d. $$S_{10} = \frac{5\left(1 - 2^{\log(2560 \cdot 2/5)/\log(2)}\right)}{1 - 2}$$
$$= \frac{5\left(1 - 2^{10}\right)}{1 - 2}$$
$$= 5115$$

e. Answers may vary.

41. The series is arithmetic with a common difference of 6.

$$351 = 3 + (n-1)\cdot 6$$
$$351 = 3 + 6n - 6$$
$$351 = 6n - 3$$
$$354 = 6n$$
$$59 = n$$
$$S_{59} = \frac{59(3 + 351)}{2} = 10,443$$

43. The series is geometric with a common ratio of 0.9.

$$3.486784401 = 10(0.9)^{n-1}$$
$$0.3486784401 = 0.9^{n-1}$$
$$\log(0.3486784401) = \log\left(0.9^{n-1}\right)$$
$$\log(0.3486784401) = (n-1)\log(0.9)$$
$$\frac{\log(0.3486784401)}{\log(0.9)} = n-1$$
$$\frac{\log(0.3486784401)}{\log(0.9)} + 1 = n$$
$$11 = n$$
$$S_{11} = \frac{10\left(1-0.9^{11}\right)}{1-0.9} \approx 68.6189$$

45. Answers may vary. Example:

$$y = 4x^2 - 8x + 6$$

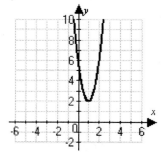

47. Answers may vary. Example:

$$\frac{x^2 - 6x + 9}{x^2 + 7x + 10} \div \frac{x-3}{x+2} = \frac{(x-3)(x-3)}{(x+5)(x+2)} \cdot \frac{x+2}{x-3}$$
$$= \frac{x-3}{x+5}$$

49. Answers may vary. Example:

$$x^2 - 9x + 14 = 0$$
$$(x-2)(x-7) = 0$$
$$x-2 = 0 \quad \text{or} \quad x-7 = 0$$
$$x = 2 \quad \text{or} \quad x = 7$$

51. Answers may vary. Example:

$$y = 3(2)^x$$

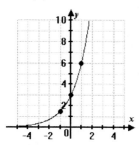

53. Answers may vary. Example:

$$y = x + 3$$
$$4x + 2y = 15$$

Solve by substitution:

$$4x + 2(x+3) = 15$$
$$4x + 2x + 6 = 15$$
$$6x = 9$$
$$x = \frac{3}{2}$$

Substitute $x = \frac{3}{2}$ into $y = x + 3$:

$$y = x + 3 = \frac{3}{2} + 3 = \frac{9}{2}$$

The solution is $\left(\frac{3}{2}, \frac{9}{2}\right)$.

Chapter 10 Review

1. $\dfrac{40}{160} = \dfrac{1}{4}, \dfrac{10}{40} = \dfrac{1}{4}, \dfrac{2.5}{10} = \dfrac{1}{4}, \dfrac{0.625}{2.5} = \dfrac{1}{4}$

The sequence is geometric with a common ratio of $\dfrac{1}{4}$.

2. $24 - 13 = 11, 35 - 24 = 11, 46 - 35 = 11,$
$57 - 46 = 11$
The series is arithmetic with a common difference of 11.

3. $95 - 101 = -6, 89 - 95 = -6,$
$83 - 89 = -6, 77 - 83 = -6$
The sequence is arithmetic with a common difference of -6.

4. $\dfrac{\frac{7}{5}}{7} = \dfrac{1}{5}, \dfrac{\frac{7}{25}}{\frac{7}{5}} = \dfrac{1}{5}, \dfrac{\frac{7}{125}}{\frac{7}{25}} = \dfrac{1}{5}, \dfrac{\frac{7}{625}}{\frac{7}{125}} = \dfrac{1}{5}$

The series is geometric with a common ratio of $\dfrac{1}{5}$.

5. The sequence is geometric with a common ratio of 3.

$$a_n = 2(3)^{n-1} = 2(3)^n (3)^{-1} = 2(3)^n \left(\frac{1}{3}\right) = \frac{2}{3}(3)^n$$

$$a_n = 2(3)^{n-1} \text{ or } a_n = \frac{2}{3}(3)^n$$

6. The sequence is arithmetic with a common difference of -5.

$$a_n = 9 + (n-1) \cdot (-5) = 9 - 5n + 5 = -5n + 14$$

7. The sequence is geometric with a common ratio of $\frac{1}{2}$.

$$a_n = 200\left(\frac{1}{2}\right)^{n-1}$$
$$= 200\left(\frac{1}{2}\right)^n \left(\frac{1}{2}\right)^{-1}$$
$$= 200\left(\frac{1}{2}\right)^n (2)$$
$$= 400\left(\frac{1}{2}\right)^n$$

$$a_n = 200\left(\frac{1}{2}\right)^{n-1} \text{ or } a_n = 400\left(\frac{1}{2}\right)^n$$

8. The sequence is arithmetic with a common difference of 2.7.
$$a_n = 3.2 + (n-1)\cdot 2.7$$
$$= 3.2 + 2.7n - 2.7$$
$$= 2.7n + 0.5$$

9. The sequence is geometric with a common ratio of 2.
$$a_{47} = 6(2)^{47-1} = 6(2)^{46} \approx 4.2221 \times 10^{14}$$

10. The sequence is geometric with a common ratio of $\frac{1}{4}$.
$$a_9 = 768\left(\frac{1}{4}\right)^{9-1} = 768\left(\frac{1}{4}\right)^8 \approx 1.1719 \times 10^{-2}$$

11. The sequence is arithmetic with a common difference of -3.
$$a_{98} = 87 + (98-1)\cdot(-3) = -204$$

12. The sequence is arithmetic with a common difference of 2.6.
$$a_{87} = 2.3 + (87-1)\cdot 2.6 = 225.9$$

13. The sequence is arithmetic with a common difference of 4.
$$2023 = 7 + (n-1)\cdot 4$$
$$2023 = 7 + 4n - 4$$
$$2023 = 3 + 4n$$
$$2020 = 4n$$
$$505 = n$$
2023 is the 505th term.

14. The sequence is arithmetic with a common difference of -8.
$$-107 = 501 + (n-1)\cdot(-8)$$
$$-107 = 501 - 8n + 8$$
$$-107 = 509 - 8n$$
$$-616 = -8n$$
$$77 = n$$
-107 is the 77th term.

15. The sequence is geometric with a common ratio of 3.
$$470{,}715{,}894{,}135 = 5(3)^{n-1}$$
$$94{,}143{,}178{,}827 = 3^{n-1}$$
$$\log(94{,}143{,}178{,}827) = \log\left(3^{n-1}\right)$$
$$\log(94{,}143{,}178{,}827) = (n-1)\log(3)$$
$$\frac{\log(94{,}143{,}178{,}827)}{\log(3)} = n-1$$
$$\frac{\log(94{,}143{,}178{,}827)}{\log(3)} + 1 = n$$
$$24 = n$$
$470{,}715{,}894{,}135$ is the 24th term.

16. <u>Equation 1:</u>
$$52 = a_1 + (5-1)d$$
$$52 = a_1 + 4d$$
$$52 - 4d = a_1$$

<u>Equation 2:</u>
$$36 = a_1 + (9-1)d$$
$$36 = a_1 + 8d$$
$$36 - 8d = a_1$$

Solve the system of linear equations by setting the left side of each equation equal to each other.
$$52 - 4d = 36 - 8d$$
$$4d = -16$$
$$d = -4$$
Use equation 1 and $d = -4$ to find a_1.
$$52 - 4\cdot(-4) = a_1$$
$$68 = a_1$$
$$a_{69} = 68 + (69-1)\cdot(-4) = -204$$

17. $S_{43} = \dfrac{43(52 + -200)}{2} = -3182$

18. $S_{22} = \dfrac{4(1 - 1.7^{22})}{1 - 1.7} \approx 671{,}173.0723$

19. The series is geometric with a common ratio of 2.

$$1{,}610{,}612{,}736 = 3(2)^{n-1}$$
$$536{,}870{,}912 = 2^{n-1}$$
$$\log(536{,}870{,}912) = \log\left(2^{n-1}\right)$$
$$\log(536{,}870{,}912) = (n-1)\log(2)$$
$$\frac{\log(536{,}870{,}912)}{\log(2)} = n-1$$
$$\frac{\log(536{,}870{,}912)}{\log(2)} + 1 = n$$
$$30 = n$$
$$S_{30} = \frac{3(1-2^{30})}{1-2} = 3{,}221{,}225{,}469$$

20. The series is arithmetic with a common difference of 6.
$$1200 = 30 + (n-1)\cdot 6$$
$$1200 = 30 + 6n - 6$$
$$1200 = 24 + 6n$$
$$1176 = 6n$$
$$196 = n$$
$$S_{196} = \frac{196(30+1200)}{2} = 120{,}540$$

21. The series is arithmetic with a common difference of –4.
$$a_{33} = 11 + (33-1)\cdot(-4) = -117$$
$$S_{33} = \frac{33\left(11+(-117)\right)}{2} = -1749$$

22. The series is geometric with a common ratio of $\frac{1}{3}$.

$$a_n = a_1 r^{n-1}$$
$$a_{13} = 531{,}441\left(\frac{1}{3}\right)^{13-1} = 1$$

$$S_{13} = \frac{531{,}441\left(1-\left(\frac{1}{3}\right)^{13}\right)}{1-\frac{1}{3}} = 797{,}161$$

23. A geometric series with a common ratio of 5, since the ratio of any two consecutive terms is 5.
$$\frac{4(5)^x}{4(5)^{x-1}} = \frac{4(5)^x}{4(5)^x(5)^{-1}} = \frac{1}{(5)^{-1}} = 5$$

24. An arithmetic sequence with a common difference of –9, since the difference between two consecutive terms is –9.
$$\left(-9(x)+40\right)-\left(-9(x-1)+40\right)$$
$$=\left(-9x+40\right)-\left(-9x+9+40\right)$$
$$=\left(-9x+40\right)-\left(-9x+49\right)$$
$$=-9x+40+9x-49$$
$$=-9$$

25. a. <u>Company A:</u>
$$a_{25} = 28{,}000(1.04)^{25-1} \approx \$71{,}772.52$$
<u>Company B:</u>
$$a_{25} = 34{,}000 + (24-1)\cdot 1500 = \$70{,}000$$

b. <u>Company A:</u>
$$S_{25} = \frac{28{,}000(1-1.04^{25})}{1-1.04} \approx \$1{,}166{,}085.43$$
<u>Company B:</u>
$$S_{25} = \frac{25(34{,}000+70{,}000)}{2} = \$1{,}300{,}000$$

c. Answers may vary. Example:
You could earn more money in the early years at Company A, but receive smaller raises.

26. a. Create a scattergram of the data.

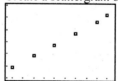

Use linear regression to find an equation.

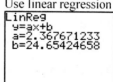

$$f(t) = 2.37t + 24.65$$

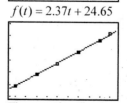

The equation fits the points well.

b. The slope of the graph of
$f(t) = 2.37t + 24.65$ is 2.37.
The slope indicates that the spending on pets in the United States increased by \$2.37 billion per year.

c. $f(18) = 2.37(18) + 24.65 = 67.31$

The spending on pets in 2018 is predicted to be $67.31 billion.

d. Since f is linear we can treat the sum of all the terms as an arithmetic series with a common difference of 2.37. Since $f(0) = 2.37(0) + 24.65 = 24.65$, $a_0 = 24.65$. There are 19 terms, so we find S_{19}.

$$S_{19} = \frac{19(24.65 + 67.31)}{2} = 873.62$$

The total spending on pets from 2000 through 2018 is estimated to be $873.6 billion.

Chapter 10 Test

1. $\frac{6}{3} = 2, \frac{12}{6} = 2, \frac{24}{12} = 2, \frac{48}{24} = 2$

It is a geometric sequence with a common ratio of 2.

2. $19 - 20 = -1, 17 - 19 = -2$

$\frac{19}{20} = 0.95, \frac{17}{19} \approx 0.89$

The sequence has neither a common difference nor a common ratio, so it is none of these.

3. $\frac{35}{7} = 5, \frac{175}{35} = 5, \frac{875}{175} = 5, \frac{4375}{875} = 5$

It is a geometric series with a common ratio of 5.

4. It is an arithmetic series with a common difference of -8.

5. The sequence is arithmetic with a common difference of -6.

$a_n = 31 + (n-1) \cdot (-6) = 31 - 6n + 6 = 37 - 6n$

6. The sequence is geometric with a common ratio of 6.

$a_n = 6(4)^{n-1} = 6(4)^n (4)^{-1} = 6(4)^n \left(\frac{1}{4}\right) = \frac{3}{2}(4)^n$

$a_n = 6(4)^{n-1}$ or $a_n = \frac{3}{2}(4)^n$

7. The sequence is arithmetic with a common difference of 3.

$a_{87} = 4 + (87 - 1) \cdot 3 = 262$

8. The sequence is geometric with a common ratio of $\frac{1}{2}$.

$a_{16} = 6144 \left(\frac{1}{2}\right)^{16-1} = \frac{3}{16}$ or 0.1875

9. The sequence is arithmetic with a common difference of 4.

$1789 = -27 + (n-1) \cdot 4$

$1789 = -27 + 4n - 4$

$1789 = -31 + 4n$

$1820 = 4n$

$455 = n$

1789 is the 455th term in the sequence.

10. The sequence is geometric with a common ratio of 1.1.

$$428.717762 = 200(1.1)^{n-1}$$

$$2.14358881 = 1.1^{n-1}$$

$$\log(2.14358881) = \log\left(1.1^{n-1}\right)$$

$$\log(2.14358881) = (n-1)\log(1.1)$$

$$\frac{\log(2.14358881)}{\log(1.1)} = n - 1$$

$$\frac{\log(2.14358881)}{\log(1.1)} + 1 = n$$

$$9 = n$$

428.717762 is the 9th term in the sequence.

11. The series is geometric with a common ratio of $\frac{1}{3}$.

$$S_{20} = \frac{27\left(1 - \left(\frac{1}{3}\right)^{20}\right)}{1 - \frac{1}{3}} \approx 40.5000$$

12. The series is geometric with a common ratio of 2.

$$2{,}147{,}483{,}648 = 4(2)^{n-1}$$

$$536{,}870{,}912 = 2^{n-1}$$

$$\log(536{,}870{,}912) = \log\left(2^{n-1}\right)$$

$$\log(536{,}870{,}912) = (n-1)\log(2)$$

$$\frac{\log(536{,}870{,}912)}{\log(2)} = n - 1$$

$$\frac{\log(536{,}870{,}912)}{\log(2)} + 1 = n$$

$$30 = n$$

$$S_{30} = \frac{4(1 - 2^{30})}{1 - 2} = 4{,}294{,}967{,}292 \approx 4.2950 \times 10^9$$

13. The series is arithmetic with a common difference of -4.

$$-78 = 50 + (n-1) \cdot (-4)$$
$$-78 = 50 - 4n + 4$$
$$-78 = 54 - 4n$$
$$-132 = -4n$$
$$33 = n$$

$$S_{33} = \frac{33(50 + (-78))}{2} = -462$$

14. The series is arithmetic with a common difference of 14.

$$a_{400} = 19 + (400-1) \cdot 14 = 5605$$

$$S_{400} = \frac{400(19 + 5605)}{2}$$
$$= 1,124,800$$
$$= 1.1248 \times 10^6$$

15. $(7+2) + (7 \cdot 2 + 2^2) + (7 \cdot 3 + 2^3) + (7 \cdot 4 + 2^4)$

$\quad + (7 \cdot 5 + 2^5) + \ldots + (7 \cdot 20 + 2^{20})$

$= (7 + 7 \cdot 2 + 7 \cdot 3 + 7 \cdot 4 + 7 \cdot 5 + \ldots + 7 \cdot 20)$

$\quad + (2 + 2^2 + 2^3 + 2^4 + 2^5 + \ldots + 2^{20})$

$= \dfrac{20(7 + 140)}{2} + \dfrac{2(1 - 2^{20})}{1 - 2}$

$= 1470 + 2,097,150$

$= 2,098,620$

16. $f(1) = 3(1)^2 + 1 = 4$

$f(2) = 3(2)^2 + 1 = 13$

$f(3) = 3(3)^2 + 1 = 28$

The series is neither geometric nor arithmetic.
Answers may vary. Example:
The series has neither a common difference nor a common ratio,

17. S_n is negative.

Answers may vary. Example:
Most of the terms of the series will be negative. The sum of negative numbers is negative.

18. a. Create a scattergram of the data.

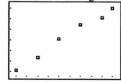

Use linear regression to find an equation.

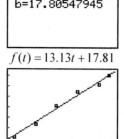

$f(t) = 13.13t + 17.81$

The equation fits the points well.

b. $f(1) = 13.13(1) + 17.81 = 30.94$

The model estimates that in 2001, online retail sales were about $31 billion.

c. $f(18) = 13.13(18) + 17.81 = 254.15$

The model predicts that in 2018, online retail sales will be about $254 billion.

d. Since f is linear we can treat the sum of all the terms as an arithmetic series with a common difference of 13.13 and $a_1 = 30.94$

$$S_{18} = \frac{18(30.94 + 254.15)}{2} = 2565.81$$

Total online retail sales from 2001 through 2018 will be about $2566 billion, or about $2.6 trillion.

19. a. $a_n = 32(1.03)^{n-1}$

b.
$$40 = 32(1.03)^{n-1}$$
$$1.25 = 1.03^{n-1}$$
$$\log(1.25) = \log\left(1.03^{n-1}\right)$$
$$\log(1.25) = (n-1)\log(1.03)$$
$$\frac{\log(1.25)}{\log(1.03)} = n-1$$
$$\frac{\log(1.25)}{\log(1.03)} + 1 = n$$
$$8.5 \approx n$$

The salary will first be above $40,000 in the 9th year.

c. $a_{25} = 32(1.03)^{24} \approx 65.04941$ (in thousands)

The salary in the 25th year will be $65,049.41.

d. $S_{25} = \dfrac{32(1 - 1.03^{25})}{1 - 1.03} \approx 1166.69646$

(in thousands) The salary for the sum of the 1st through 25th years will be $1,166,696.46.

Cumulative Review of Chapters 1–10

1. $\quad 6x^2 + 13x = 5$

$6x^2 + 13x - 5 = 0$

$(3x - 1)(2x + 5) = 0$

$3x - 1 = 0 \quad \text{or} \quad 2x + 5 = 0$

$3x = 1 \quad \text{or} \qquad 2x = -5$

$x = \dfrac{1}{3} \quad \text{or} \qquad x = -\dfrac{5}{2}$

2. $\log_3(4x - 7) = 4$

$4x - 7 = 3^4$

$4x - 7 = 81$

$4x = 88$

$x = 22$

3. $\quad (t + 3)(t - 4) = 5$

$t^2 + 3t - 4t - 12 = 5$

$t^2 - t - 17 = 0$

$t = \dfrac{-(-1) \pm \sqrt{(-1)^2 - 4(1)(-17)}}{2(1)} = \dfrac{1 \pm \sqrt{69}}{2}$

4.

$\dfrac{1}{w^2 - w - 6} - \dfrac{w}{w + 2} = \dfrac{w - 2}{w - 3}$

$\dfrac{1}{(w - 3)(w + 2)} - \dfrac{w}{w + 2} = \dfrac{w - 2}{w - 3}$

$\dfrac{1}{(w - 3)(w + 2)} - \dfrac{w(w - 3)}{(w - 3)(w + 2)} = \dfrac{(w - 2)(w + 2)}{(w - 3)(w + 2)}$

$\dfrac{1 - w^2 + 3w}{(w - 3)(w + 2)} = \dfrac{w^2 - 4}{(w - 3)(w + 2)}$

$\dfrac{w^2 - 4 - 1 + w^2 - 3w}{(w - 3)(w + 2)} = 0$

$\dfrac{2w^2 - 3w - 5}{(w - 3)(w + 2)} = 0$

$2w^2 - 3w - 5 = 0$

$(2w - 5)(w + 1) = 0$

$2w - 5 = 0 \quad \text{or} \quad w + 1 = 0$

$2w = 5 \quad \text{or} \qquad w = -1$

$w = \dfrac{5}{2} \quad \text{or} \qquad w = -1$

5. $5(3x - 2)^2 + 7 = 17$

$5(3x - 2)^2 = 10$

$(3x - 2)^2 = 2$

$3x - 2 = \pm\sqrt{2}$

$3x = 2 \pm \sqrt{2}$

$x = \dfrac{2 \pm \sqrt{2}}{3}$

6. $\log_6(3x) + \log_6(x - 1) = 1$

$\log_6[3x(x - 1)] = 1$

$\log_6(3x^2 - 3x) = 1$

$3x^2 - 3x = 6^1$

$3x^2 - 3x = 6$

$3x^2 - 3x - 6 = 0$

$3(x^2 - x - 2) = 0$

$3(x - 2)(x + 1) = 0$

$(x - 2)(x + 1) = 0$

$x - 2 = 0 \quad \text{or} \quad x + 1 = 0$

$x = 2 \quad \text{or} \qquad x = -1$

Check $x = 2$:

$\log_6[3(2)] + \log_6[(2) - 1] \overset{?}{=} 1$

$\log_6 6 + \log_6 1 \overset{?}{=} 1$

$1 + 0 \overset{?}{=} 1$

$1 \overset{?}{=} 1 \quad \text{true}$

Check $x = -1$:

$\log_6[3(-1)] + \log_6[(-1) - 1] \overset{?}{=} 1$

$\log_6(-3) + \log_6(-2) \overset{?}{=} 1 \quad \text{false}$

We cannot take the logarithm of a negative number. The solution is $x = 2$.

7. $20 - 4x = 7(2x + 9)$

$20 - 4x = 14x + 63$

$-4x = 14x + 43$

$-18x = 43$

$x = -\dfrac{43}{18} = -2\dfrac{7}{18}$

8. $\sqrt{x+1} - \sqrt{2x-5} = 1$

$\qquad \sqrt{x+1} = 1 + \sqrt{2x-5}$

Square both sides.

$\left(\sqrt{x+1}\right)^2 = \left(1 + \sqrt{2x-5}\right)^2$

$\quad x+1 = 1 + \sqrt{2x-5} + \sqrt{2x-5} + (2x-5)$

$\quad x+1 = 1 + 2\sqrt{2x-5} + 2x - 5$

$\quad x+1 = 2\sqrt{2x-5} + 2x - 4$

$\quad -x+5 = 2\sqrt{2x-5}$

$\quad \dfrac{-x+5}{2} = \sqrt{2x-5}$

Square both sides again.

$\left(\dfrac{-x+5}{2}\right)^2 = \left(\sqrt{2x-5}\right)^2$

$\dfrac{x^2 - 5x - 5x + 25}{4} = 2x - 5$

$\dfrac{x^2 - 10x + 25}{4} = 2x - 5$

$x^2 - 10x + 25 = 8x - 20$

$x^2 - 18x + 45 = 0$

$(x-3)(x-15) = 0$

$x - 3 = 0 \quad \text{or} \quad x - 15 = 0$

$x = 3 \quad \text{or} \qquad x = 15$

Check $x = 3$:

$\sqrt{4} - \sqrt{1} \overset{?}{=} 1$

$\quad 2 - 1 \overset{?}{=} 1$

$\qquad 1 \overset{?}{=} 1 \ \text{true}$

Check $x = 15$:

$\sqrt{(15)+1} - \sqrt{2(15)-5} \overset{?}{=} 1$

$\qquad \sqrt{16} - \sqrt{25} \overset{?}{=} 1$

$\qquad\qquad 4 - 5 \overset{?}{=} 1$

$\qquad\qquad\qquad -1 \overset{?}{=} 1 \ \text{false}$

The solution is $x = 3$.

9. $2b^7 - 3 = 51$

$\quad 2b^7 = 54$

$\qquad b^7 = 27$

$\qquad b = 27^{1/7}$

$\qquad b = \sqrt[7]{27} \approx 1.6013$

10. $6(3)^x - 5 = 52$

$\quad 6(3)^x = 57$

$\qquad 3^x = \dfrac{19}{2}$

$x = \log_3\left(\dfrac{19}{2}\right) = \dfrac{\log\left(\dfrac{19}{2}\right)}{\log(3)} \approx 2.0492$

11. $\qquad 5e^x = 98$

$\qquad e^x = \dfrac{98}{5}$

$\ln\left(e^x\right) = \ln\left(\dfrac{98}{5}\right)$

$\qquad x = \ln\left(\dfrac{98}{5}\right) \approx 2.9755$

12. $\qquad 3x^2 - 5x + 1 = 0$

$\qquad 3x^2 - 5x = -1$

$\quad 3\left(x^2 - \dfrac{5}{3}x\right) = -1$

$3\left(x^2 - \dfrac{5}{3}x + \dfrac{25}{36}\right) = -1 + 3 \cdot \dfrac{25}{36}$

$\quad 3\left(x - \dfrac{5}{6}\right)^2 = -1 + \dfrac{75}{36}$

$\quad 3\left(x - \dfrac{5}{6}\right)^2 = \dfrac{39}{36}$

$\qquad \left(x - \dfrac{5}{6}\right)^2 = \dfrac{13}{36}$

$\qquad x - \dfrac{5}{6} = \pm\sqrt{\dfrac{13}{36}}$

$\qquad x = \dfrac{5}{6} \pm \dfrac{\sqrt{13}}{6} = \dfrac{5 \pm \sqrt{13}}{6}$

13. $2x^2 = 4x - 3$

$2x^2 - 4x + 3 = 0$

$x = \dfrac{-(-4) \pm \sqrt{(-4)^2 - 4(2)(3)}}{2(2)}$

$\quad = \dfrac{4 \pm \sqrt{-8}}{4}$

$\quad = \dfrac{4 \pm i\sqrt{8}}{4}$

$\quad = \dfrac{4 \pm 2i\sqrt{2}}{4}$

$\quad = \dfrac{2 \pm i\sqrt{2}}{2}$

14. $2x + 4y = 0$

$5x + 3y = 7$

Multiply the first equation by -3 and the second equation by 4.

$-6x - 12y = 0$

$20x + 12y = 28$

Add the two equations and solve the result for x.

$14x = 28$

$x = 2$

Substitute this result for x in the first equation.

$2(2) + 4y = 0$

$4 + 4y = 0$

$4y = -4$

$y = -1$

The solution of the system is $(2, -1)$.

15. $y = 3x + 9$

$4x + 2y = -2$

Substitute the first equation for y in the second equation.

$4x + 2(3x + 9) = -2$

$4x + 6x + 18 = -2$

$10x = -20$

$x = -2$

Substitute this value for x into the first equation.

$y = 3x + 9 = 3(-2) + 9 = -6 + 9 = 3$

The solution of the system is $(-2, 3)$.

16. $2x - 3y + 4z = 19$

$5x + y - 5z = -6$

$3x - y + 2z = 13$

Multiply the second equation by 3 and add to first equation.

$2x - 3y + 4z = 19$

$15x + 3y - 15z = -18$

$17x - 11z = 1$

Add the second and third equations.

$5x + y - 5z = -6$

$3x - y + 2z = 13$

$\overline{8x - 3z = 7}$

$8x - 3z = 7 \quad \rightarrow \quad -88x + 33z = -77$

$17x - 11z = 1 \quad \rightarrow \quad 51x - 33z = 3$

$\overline{-37x = -74}$

$x = 2$

Substitute $x = 2$ into $8x - 3z = 7$ and solve for z.

$8(2) - 3z = 7$

$16 - 3z = 7$

$-3z = -9$

$z = 3$

Substitute $x = 2$ and $z = 3$ into $2x - 3y + 4z = 19$ and solve for y.

$2(2) - 3y + 4(3) = 19$

$4 - 3y + 12 = 19$

$-3y + 16 = 19$

$-3y = 3$

$y = -1$

The solution of the system is $(2, -1, 3)$.

17. $5 - 2(3x - 5) + 1 \geq 2 - 4x$

$5 - 6x + 10 + 1 \geq 2 - 4x$

$16 - 6x \geq 2 - 4x$

$-2x \geq -14$

$x \leq 7$

Interval: $(-\infty, 7]$

18. $\left(3b^{-2}c^{-3}\right)^4 \left(6b^{-5}c^2\right)^2 = 81b^{-8}c^{-12} \cdot 36b^{-10}c^4$

$= 2916b^{-8+(-10)}c^{-12+4}$

$= 2916b^{-18}c^{-8}$

$= \dfrac{2916}{b^{18}c^8}$

19. $\dfrac{8b^{1/2}c^{-4/3}}{10b^{3/4}c^{-7/3}} = \dfrac{4}{5}b^{1/2-3/4}c^{-4/3-(-7/3)}$

$= \dfrac{4}{5}b^{2/4-3/4}c^{-4/3+7/3}$

$= \dfrac{4}{5}b^{-1/4}c^{3/3}$

$= \dfrac{4c}{5b^{1/4}}$

20. $3y\sqrt{8x^3} - 2x\sqrt{18xy^2}$

$= 3y\sqrt{4x^2 \cdot 2x} - 2x\sqrt{9 \cdot 2x \cdot y^2}$

$= 3y\sqrt{4x^2}\sqrt{2x} - 2x\sqrt{9}\sqrt{2x}\sqrt{y^2}$

$= 3y \cdot 2x\sqrt{2x} - 2x \cdot 3y\sqrt{2x}$

$= 6xy\sqrt{2x} - 6xy\sqrt{2x}$

$= 0$

21. $\sqrt{12x^7y^{14}} = \sqrt{4x^6 \cdot 3x \cdot y^{14}}$
$$= \sqrt{4x^6}\sqrt{3x}\sqrt{y^{14}}$$
$$= 2x^3 y^7 \sqrt{3x}$$

22. $\sqrt[3]{\dfrac{4}{x}} = \dfrac{\sqrt[3]{4}}{\sqrt[3]{x}} = \dfrac{\sqrt[3]{4}}{\sqrt[3]{x}} \cdot \dfrac{\sqrt[3]{x^2}}{\sqrt[3]{x^2}} = \dfrac{\sqrt[3]{4 \cdot x^2}}{\sqrt[3]{x \cdot x^2}} = \dfrac{\sqrt[3]{4x^2}}{x}$

23. $\dfrac{3\sqrt{x} - \sqrt{y}}{2\sqrt{x} + \sqrt{y}}$

$= \dfrac{3\sqrt{x} - \sqrt{y}}{2\sqrt{x} + \sqrt{y}} \cdot \dfrac{2\sqrt{x} - \sqrt{y}}{2\sqrt{x} - \sqrt{y}}$

$= \dfrac{3\sqrt{x} \cdot 2\sqrt{x} - 3\sqrt{x} \cdot \sqrt{y} - 2\sqrt{x} \cdot \sqrt{y} + \sqrt{y} \cdot \sqrt{y}}{\left(2\sqrt{x}\right)^2 - \left(\sqrt{y}\right)^2}$

$= \dfrac{6x - 3\sqrt{xy} - 2\sqrt{xy} + y}{4x - y}$

$= \dfrac{6x - 5\sqrt{xy} + y}{4x - y}$

24. $2\ln\left(x^4\right) + 3\ln\left(x^9\right) = \ln\left[\left(x^4\right)^2\right] + \ln\left[\left(x^9\right)^3\right]$
$$= \ln\left(x^8\right) + \ln\left(x^{27}\right)$$
$$= \ln\left(x^8 x^{27}\right)$$
$$= \ln\left(x^{8+27}\right)$$
$$= \ln\left(x^{35}\right)$$

25. $4\log_b\left(x^5\right) - 5\log_b\left(2x\right)$

$= \log_b\left(x^5\right)^4 - \log_b\left(2x\right)^5$

$= \log_b\left(x^{20}\right) - \log_b\left(32x^5\right)$

$= \log_b\left(\dfrac{x^{20}}{32x^5}\right)$

$= \log_b\left(\dfrac{x^{20-5}}{32}\right)$

$= \log_b\left(\dfrac{x^{15}}{32}\right)$

26. $(3a - 5b)^2 = (3a)^2 - 2(3a)(5b) + (5b)^2$
$$= 9a^2 - 30ab + 25b^2$$

27. $\left(3\sqrt{k} - 4\right)\left(2\sqrt{k} + 7\right)$
$$= 3\sqrt{k} \cdot 2\sqrt{k} - 4 \cdot 2\sqrt{k} + 3\sqrt{k} \cdot 7 - 4 \cdot 7$$
$$= 6k - 8\sqrt{k} + 21\sqrt{k} - 28$$
$$= 6k + 13\sqrt{k} - 28$$

28. $\left(2x^2 - x + 3\right)\left(x^2 + 2x - 1\right)$
$$= 2x^2\left(x^2 + 2x - 1\right) - x\left(x^2 + 2x - 1\right)$$
$$\quad + 3\left(x^2 + 2x - 1\right)$$
$$= 2x^4 + 4x^3 - 2x^2 - x^3 - 2x^2 + x + 3x^2$$
$$\quad + 6x - 3$$
$$= 2x^4 + 3x^3 - x^2 + 7x - 3$$

29.

$$
\begin{array}{r}
3x^2 - x + 4 \\
3x-2\overline{)9x^3 - 9x^2 + 14x - 13} \\
\underline{-9x^3 + 6x^2} \\
-3x^2 + 14x \\
\underline{3x^2 - 2x} \\
12x - 13 \\
\underline{-12x + 8} \\
-5
\end{array}
$$

$\dfrac{9x^3 - 9x^2 + 14x - 13}{3x - 2} = 3x^2 - x + 4 - \dfrac{5}{3x - 2}$

30. $\dfrac{x^3 - 27}{2x^2 - 3x + 1} \div \dfrac{2x^3 + 6x^2 + 18x}{4x^2 - 1}$

$= \dfrac{x^3 - 27}{2x^2 - 3x + 1} \cdot \dfrac{4x^2 - 1}{2x^3 + 6x^2 + 18x}$

$= \dfrac{(x-3)\left(x^2 + 3x + 9\right)}{(2x-1)(x-1)} \cdot \dfrac{(2x-1)(2x+1)}{2x\left(x^2 + 3x + 9\right)}$

$= \dfrac{(x-3)\left(x^2 + 3x + 9\right)}{(2x-1)(x-1)} \cdot \dfrac{(2x-1)(2x+1)}{2x\left(x^2 + 3x + 9\right)}$

$= \dfrac{(x-3)(2x+1)}{2x(x-1)}$

31. $\dfrac{3x}{x^2-10x+25} - \dfrac{x+2}{x^2-7x+10}$

$= \dfrac{3x}{(x-5)(x-5)} - \dfrac{x+2}{(x-5)(x-2)}$

$= \dfrac{3x(x-2)}{(x-5)(x-5)(x-2)} - \dfrac{(x+2)(x-5)}{(x-5)(x-5)(x-2)}$

$= \dfrac{3x^2-6x}{(x-5)^2(x-2)} - \dfrac{x^2-3x-10}{(x-5)^2(x-2)}$

$= \dfrac{3x^2-6x-x^2+3x+10}{(x-5)^2(x-2)}$

$= \dfrac{2x^2-3x+10}{(x-5)^2(x-2)}$

32. $\dfrac{4x-x^2}{6x^2+10x-4} \cdot \dfrac{7-21x}{x^2-8x+16}$

$= \dfrac{-x(x-4)}{2(3x-1)(x+2)} \cdot \dfrac{-7(3x-1)}{(x-4)(x-4)}$

$= \dfrac{-x(x-4)}{2(3x-1)(x+2)} \cdot \dfrac{-7(3x-1)}{(x-4)(x-4)}$

$= \dfrac{7x}{2(x+2)(x-4)}$

33. $\dfrac{1}{x^2+12x+27} + \dfrac{x+2}{x^3+x^2-9x-9}$

$= \dfrac{1}{(x+9)(x+3)} + \dfrac{x+2}{x^2(x+1)-9(x+1)}$

$= \dfrac{1}{(x+9)(x+3)} + \dfrac{x+2}{(x+1)(x^2-9)}$

$= \dfrac{1}{(x+9)(x+3)} + \dfrac{x+2}{(x+1)(x+3)(x-3)}$

$= \dfrac{(x-3)(x+1)}{(x+9)(x+3)(x-3)(x+1)}$

$\quad + \dfrac{(x+2)(x+9)}{(x+1)(x+9)(x-3)(x+9)}$

$= \dfrac{x^2-3x+x-3+x^2+2x+9x+18}{(x+9)(x+3)(x-3)(x+1)}$

$= \dfrac{2x^2+9x+15}{(x+9)(x+3)(x-3)(x+1)}$

34. $\dfrac{\dfrac{x+2}{x^2-64}}{\dfrac{x^2+4x+4}{3x+24}} = \dfrac{x+2}{x^2-64} \cdot \dfrac{3x+24}{x^2+4x+4}$

$= \dfrac{(x+2)}{(x-8)(x+8)} \cdot \dfrac{3(x+8)}{(x+2)(x+2)}$

$= \dfrac{(x+2)}{(x-8)(x+8)} \cdot \dfrac{3(x+8)}{(x+2)(x+2)}$

$= \dfrac{1}{(x-8)(x+8)} \cdot \dfrac{3(x+8)}{(x+2)}$

$= \dfrac{3}{(x-8)(x+2)}$

35. $f(x) = -3(x+3)^2 - 7$

$= -3\left(x^2+6x+9\right) - 7$

$= -3x^2-18x-27-7$

$= -3x^2-18x-34$

$f(x) = -3x^2-18x-34$

36. $(f \circ g)(x) = f(g(x))$

$= 2(x-2)^2 - 4(x-2) + 3$

$= 2\left(x^2-4x+4\right) - 4x+8+3$

$= 2x^2-8x+8-4x+11$

$= 2x^2-12x+19$

37. $4x^3 - 8x^2 - 25x + 50$

$= 4x^2(x-2) - 25(x-2)$

$= (x-2)\left(4x^2-25\right)$

$= (x-2)(2x-5)(2x+5)$

38. $2x^3 - 4x^2 - 30x = 2x\left(x^2-2x-15\right)$

$= 2x(x-5)(x+3)$

39. $6w^2 + 2wy - 20y^2 = 2\left(3w^2+wy-10y^2\right)$

$= 2(3w-5y)(w+2y)$

40. $100p^2 - 1 = (10p-1)(10p+1)$

41. $f(2) = 3$

42. When $f(x) = 3$, $x = 0$ or $x = 2$.

43. The graph is quadratic so the function is of the form $f(x) = a(x-h)^2 + k$. The vertex is $(h,k) = (1,4)$ so we have

$f(x) = a(x-1)^2 + 4$.

Choosing another point on the graph, $(0,3)$, we can find the value of a.

$3 = a(0-1)^2 + 4$

$3 = a + 4$

$-1 = a$

Thus, the function is

$f(x) = -(x-1)^2 + 4 = -x^2 + 2x + 3$

44. Domain of f:
the set of all real numbers or $(-\infty, \infty)$

45. Range of f: $\{y \mid y \le 4\}$ or $(-\infty, 4]$

46. $y = -3(x-4)^2 + 3$

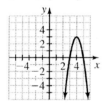

47. $y = 2\sqrt{x+5} - 4$

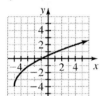

48. $y = 15\left(\dfrac{1}{3}\right)^x$

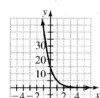

49. $y = 2x^2 + 5x - 1$

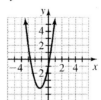

50. $2x(x-3) + y = 5(x+1)$

$2x^2 - 6x + y = 5x + 5$

$\qquad y = -2x^2 + 11x + 5$

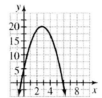

51. $(-3,2)$ and $(2,-5)$

$m = \dfrac{-5-2}{2-(-3)} = -\dfrac{7}{5}$

Using the slope m and the point $(-3,2)$, we get:

$y = mx + b$

$2 = -\dfrac{7}{5}(-3) + b$

$b = -\dfrac{11}{5}$

The equation of the line is $y = -\dfrac{7}{5}x - \dfrac{11}{5}$ or

$7x + 5y = -11$.

52. $(3,95)$ and $(6,12)$

We want to fit the model $y = a(b)^x$.

Substitute both points into the equation.

$95 = a(b)^3$

$12 = a(b)^6$

Divide the second equation by the first equation.

$\dfrac{12}{95} = \dfrac{ab^6}{ab^3}$

$\dfrac{12}{95} = b^3$

$b = \sqrt[3]{\dfrac{12}{95}} \approx 0.50$

Substitute the point $(3,95)$ into the equation

$$y = a\left(\sqrt[3]{\frac{12}{95}}\right)^x.$$

$$95 = a\left(\sqrt[3]{\frac{12}{95}}\right)^3$$

$$a = \frac{95}{\left(\sqrt[3]{\frac{12}{95}}\right)^3} \approx 752.08$$

The equation is $y = 752.08(0.50)^x$.

53. $(2,1), (3,6),$ and $(4,15)$

Substitute the points into the equation

$y = ax^2 + bx + c$.

$$1 = a(2)^2 + b(2) + c$$

$$6 = a(3)^2 + b(3) + c$$

$$15 = a(4)^2 + b(4) + c$$

Rewrite as:

$$4a + 2b + c = 1$$

$$9a + 3b + c = 6$$

$$16a + 4b + c = 15$$

Multiply the first equation by -1 and add to both the second and third equations.

$$4a + 2b + c = 1$$

$$5a + b = 5$$

$$12a + 2b = 14$$

Multiply the second equation by -2 and add to the third equation.

$$4a + 2b + c = 1$$

$$5a + b = 5$$

$$2a = 4$$

Solve the third equation for a.

$$2a = 4$$

$$a = 2$$

Substitute this value into the second equation and solve for b.

$$5(2) + b = 5$$

$$10 + b = 5$$

$$b = -5$$

Substitute the values for a and b into the first equation and solve for c.

$$4(2) + 2(-5) + c = 1$$

$$8 - 10 + c = 1$$

$$c = 3$$

The equation is $y = 2x^2 - 5x + 3$.

54. $(2,5)$ and $(6,17)$

Substitute the points into the equation

$y = a\sqrt{x} + b$.

$$5 = a\sqrt{2} + b$$

$$17 = a\sqrt{6} + b$$

Rewrite as:

$$1.4142a + b = 5$$

$$2.4495a + b = 17$$

Multiply the first equation by -1 and add to the second equation.

$$1.4142a + b = 5$$

$$1.0353a = 12$$

Solve the second equation for a.

$$1.0353a = 12$$

$$a \approx 11.59$$

Substitute this value into the first equation and solve for b.

$$1.4142(11.59) + b = 5$$

$$b \approx -11.39$$

The equation is approximately

$y = 11.59\sqrt{x} - 11.39$.

55. a. <u>Linear:</u> $f(x) = mx + b$

The y-intercept is $(0,2)$ so $b = 2$.

$$m = \frac{4-2}{1-0} = 2$$

$$f(x) = 2x + 2$$

<u>Exponential:</u> $g(x) = a \cdot b^x$

The y-intercept is $(0,2)$ so $a = 2$.

Now substitute the point $(1,4)$.

$$4 = 2(b)^1$$

$$4 = 2b$$

$$2 = b$$

$$g(x) = 2(2)^x$$

<u>Quadratic:</u> $h(x) = ax^2 + bx + c$

Answers may vary. Example:

Let $a = 2$ so we have

$h(x) = 2x^2 + bx + c$.

Substitute the point $(0,2)$.

$$2 = 2(0)^2 + b(0) + c$$

$$2 = c$$

$$h(x) = 2x^2 + bx + 2$$

Substitute the point $(1,4)$.

$$4 = 2(1)^2 + b(1) + 2$$
$$4 = 2 + b + 2$$
$$4 = b + 4$$
$$0 = b$$
$$h(x) = 2x^2 + 2$$

b.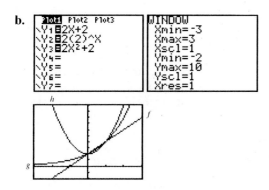

56. $\log_3(81) = \log_3(3^4) = 4\log_3(3) = 4 \cdot 1 = 4$

57. $\log_b(\sqrt{b}) = \log_b(b^{1/2}) = \frac{1}{2}\log_b(b) = \frac{1}{2} \cdot 1 = \frac{1}{2}$

58. $y = g(x) = \log_2(x)$
Switch x and y, and solve for y.
$$x = \log_2(y)$$
$$y = 2^x$$
$$g^{-1}(x) = 2^x$$

59. $y = f(x) = -4x - 7$
Switch x and y, and solve for y.
$$x = -4y - 7$$
$$4y = -x - 7$$
$$y = -\frac{1}{4}x - \frac{7}{4}$$
$$f^{-1}(x) = -\frac{1}{4}x - \frac{7}{4}$$

60. $f(x) = \dfrac{x-3}{x^2 - 2x - 35} = \dfrac{x-3}{(x-7)(x+5)}$
The domain is the set of all real numbers except 7 and -5 since these values make the denominator equal zero.

61. The sequence is geometric with a common ratio of 4.
$$a_n = a_1 r^{n-1}$$
$$a_{10} = 2(4)^{10-1} = 524{,}288$$

62. This is an arithmetic sequence with a common difference of $d = 4$. Since $a_1 = -86$ and $a_n = 170$, we have:
$$a_n = a_1 + d(n-1)$$
$$170 = -86 + 4(n-1)$$
$$256 = 4(n-1)$$
$$64 = n - 1$$
$$65 = n$$
The last term in the sequence is term 65.

63. This is a geometric series with $r = \dfrac{1}{2}$ and $a_1 = 98{,}304$.
$$a_n = a_1 r^{n-1}$$
$$3 = 98{,}304\left(\frac{1}{2}\right)^{n-1}$$
$$\frac{3}{98{,}304} = \left(\frac{1}{2}\right)^{n-1}$$
$$\ln\left(\frac{3}{98{,}304}\right) = (n-1)\ln\left(\frac{1}{2}\right)$$
$$\frac{\ln\left(\dfrac{3}{98{,}304}\right)}{\ln\left(\dfrac{1}{2}\right)} = n - 1$$
$$15 = n - 1$$
$$16 = n$$
There are 16 terms in the series.
$$S_{16} = \frac{a_1(1 - r^{16})}{1 - r} = \frac{98{,}304\left(1 - \left(\dfrac{1}{2}\right)^{16}\right)}{1 - \dfrac{1}{2}} = 196{,}605$$
The sum of the series is 196,605.

64. This is an arithmetic series with $d = 3$ and $a_1 = 11$.
$$a_n = a_1 + (n-1)d$$
$$182 = 11 + (n-1)3$$
$$171 = 3(n-1)$$
$$57 = n - 1$$
$$58 = n$$
There are 58 terms in the series.
$$S_n = \frac{n(a_1 + a_n)}{2}$$
$$S_{58} = \frac{58(11 + 182)}{2} = 29(193) = 5597$$

65. Let x be number of liters of 15% acid solution.
Let y be number of liters of 30% acid solution.

$$x + y = 6 \qquad \text{Equation (1)}$$
$$0.15x + 0.30y = (0.25)6 \quad \text{Equation (2)}$$

Multiply both sides of equation (2) by 100.
$$15x + 30y = 150$$

To use elimination, multiply both sides of equation (1) by -15 to eliminate x when added to equation 2.

$$\begin{aligned} x + y = 6 &\quad \rightarrow \quad -15x - 15y = -90 \\ 15x + 30y = 150 &\quad \rightarrow \quad \underline{15x + 30y = 150} \\ & \qquad\qquad\qquad\quad 15y = 60 \\ & \qquad\qquad\qquad\quad\;\; y = 4 \end{aligned}$$

Substitute $y = 4$ into $x + y = 6$ to find x.
$$x + 4 = 6$$
$$x = 2$$

The chemist needs to mix 2 liters of the 15% acid solution with 4 liters of the 30% acid solution to create 6 liters of the 25% acid solution.

66. a. Create a scattergram of the data.

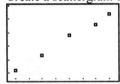

Use linear regression to find an equation.

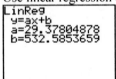

$$f(t) = 29.38t + 532.59$$

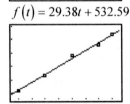

The equation fits the points well.

b. $n = f(t) = 29.38t + 532.59$

Solve for t.
$$n = 29.38t + 532.59$$
$$n - 532.59 = 29.38t$$
$$t = \frac{n - 532.59}{29.38}$$
$$t = 0.034n - 18.13$$
$$f^{-1}(n) = 0.034n - 18.13$$

c. $f(18) = 29.38(18) + 532.59 = 1061.43$

There will be about 1061 thousand slot machines and video poker machines in 2018.

d. $f^{-1}(1100) = 0.034(1100) - 18.13 = 19.27$

There will be 1100 thousand (1.1 million) slot machines and video poker machines in 2019.

e. The slope of $f(t) = 29.38t + 532.59$ is 29.38. The slope means that the number of slot machines and video poker machines is increasing by 29.38 thousand machines per year.

67. a. We are given two data points for India's population: $(0, 0.687)$ and $(31, 1.210)$

The y-intercept is $(0, 0.687)$ so $b = 0.687$.
The slope of the line can be found by using the two given points.
$$m = \frac{1.210 - 0.687}{31 - 0} \approx 0.0169$$
The linear model is
$$L(t) = 0.0169t + 0.687.$$

b. We now fit the model $y = a(b)^x$. Since the y-intercept is $(0, 0.687)$, we have $a = 0.687$.

Substitute the point $(31, 1.210)$ into the equation $y = 0.687(b)^x$ and solve for b.
$$1.210 = 0.687(b)^{31}$$
$$b^{31} = \frac{1.210}{0.687}$$
$$b = \sqrt[31]{\frac{1.210}{0.687}} \approx 1.0184$$

The model is $E(t) = 0.687(1.0184)^t$.

c. $L(70) = 0.0169(70) + 0.687 = 1.870$

$E(70) = 0.687(1.0184)^{70} \approx 2.462$

In 2050, India's population will be 1.870 billion according to the linear model and 2.462 billion according to the exponential model.

d. $(E-L)(70) = 2.462 - 1.870 = 0.592$

The difference in India's population in 2050 between the models is 592 million, which is greater than the predicted U.S. population of 439 million for 2050.

e. $L(t) = 0.0169t + 0.687$

$1.424 = 0.0169t + 0.687$

$0.737 = 0.0169t$

$t \approx 44$

$E(t) = 0.687(1.0184)^t$

$1.424 = 0.687(1.0184)^t$

$\dfrac{1.424}{0.687} = 1.0184^t$

$\log\left(\dfrac{1.424}{0.687}\right) = \log\left(1.0184^t\right)$

$\log\left(\dfrac{1.424}{0.687}\right) = t\log(1.0184)$

$\dfrac{\log\left(\dfrac{1.424}{0.687}\right)}{\log(1.0184)} = t$

$40 \approx t$

According to the linear model, India's population will reach 1.424 billion about 44 years after 1980, in 2024.

According to the exponential model, India's population will reach 1.424 billion about 40 years after 1980 in 2020.

68. a. Create a scattergram of the data.

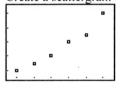

Use exponential regression to find an equation.

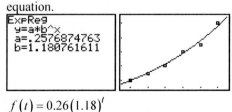

$f(t) = 0.26(1.18)^t$

Use quadratic regression to find an equation.

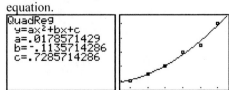

$f(t) = 0.0179t^2 - 0.114t + 0.73$

Each fits most of the data well.
Answers may vary.

b. The exponential model describes the situation better for years before 2005. Answers may vary. Example:

By zooming out on the graph, we see that the quadratic model is decreasing before 2005.

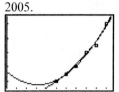

c.
$f(t) = 0.26(1.18)^t$

$5 = 0.26(1.18)^t$

$\dfrac{5}{0.26} = 1.18^t$

$\log\left(\dfrac{5}{0.26}\right) = \log\left(1.18^t\right)$

$\log\left(\dfrac{5}{0.26}\right) = t\log(1.18)$

$\dfrac{\log\left(\dfrac{5}{0.26}\right)}{\log(1.18)} = t$

$17.86 \approx t$

According to the exponential model, the annual sales will be 5 million 9-liter cases in 2018.

d. $f(t) = 0.0179t^2 - 0.114t + 0.73$

$5 = 0.0179t^2 - 0.114t + 0.73$

$0 = 0.0179t^2 - 0.114t - 4.27$

$x = \dfrac{-b \pm \sqrt{b^2 - 4ac}}{2a}$

$= \dfrac{-(-0.114) \pm \sqrt{(-0.114)^2 - 4(0.0179)(-4.27)}}{2(0.0179)}$

$= \dfrac{0.114 \pm \sqrt{0.318728}}{0.0358}$

≈ -12.59 or 18.95

Discard the negative solution in this situation.

According to the quadratic model, the annual sales will be 5 million 9-liter cases in 2019.

e. Answers may vary. Example:
Exponential growth functions tend to grow
much more rapidly than quadratic ones.

69. a. Create a scattergram of the data.

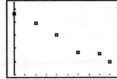

Use a linear regression to find $B(t)$.

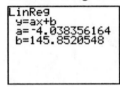

$$B(t) = -4.04t + 145.85$$

b. Create a scattergram of the data.

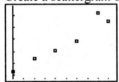

Use a linear regression to find $R(t)$.

LinReg
y=ax+b
a=102.3616438
b=1849.252055

$$R(t) = 102.36t + 1849.25$$

c. $P(t) = \dfrac{B(t)}{R(t)} \cdot 100$

$$= \frac{-4.04t + 145.85}{102.36t + 1849.25} \cdot 100$$

$$= \frac{-404t + 14,585}{102.36t + 1849.25}$$

d.

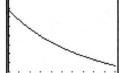

P is decreasing for t-values between 0 and
20. This means that between 2000 and
2020, the percentage of total recreational
expenditures that consists of book
purchases is decreasing.

e.
$$2 = \frac{-404t + 14,585}{102.36t + 1849.25}$$

$$2(102.36t + 1849.25) = -404t + 14,585$$

$$204.72t + 3698.5 = -404t + 14,585$$

$$608.72t = 10,886.5$$

$$t \approx 17.88$$

The model predicts that 2% of recreational
expenditures will consist of book
purchases in $2000 + 17.88 \approx 2018$.

Chapter 11
Additional Topics

1. $|x| = 7$

$x = 7$ or $x = -7$

3. $|x| = -3$

Since $|x|$ is always nonnegative, the solution set for $|x| = -3$ is the empty set.

5. $5|p| - 3 = 15$

$\qquad 5|p| = 18$

$\qquad |p| = \dfrac{18}{5}$

$p = -\dfrac{18}{5}$ or $p = \dfrac{18}{5}$

7. $|x + 2| = 5$

$x + 2 = -5$ or $x + 2 = 5$

$\quad x = -7$ or $\qquad x = 3$

9. $|x - 5| = 0$

$x - 5 = 0$

$\quad x = 5$

11. $|3t - 1| = 11$

$3t - 1 = -11$ or $3t - 1 = 11$

$\quad 3t = -10$ or $\qquad 3t = 12$

$\quad t = -\dfrac{10}{3}$ or $\qquad t = 4$

13. $|2x + 9| = -6$

Since $|2x + 9|$ is always nonnegative, the solution set for $|2x + 9| = -6$ is the empty set.

15. $|4x| + 1 = 9$

$\quad |4x| = 8$

$4x = -8$ or $4x = 8$

$\quad x = -2$ or $\quad x = 2$

17. $2|a + 5| = 8$

$\quad |a + 5| = 4$

$a + 5 = -4$ or $a + 5 = 4$

$\quad a = -9$ or $\qquad a = -1$

19. $|2x - 5| - 4 = -3$

$\qquad |2x - 5| = 1$

$2x - 5 = -1$ or $2x - 5 = 1$

$\quad 2x = 4$ or $\qquad 2x = 6$

$\quad x = 2$ or $\qquad x = 3$

21. $|4x - 5| = |3x + 2|$

$4x - 5 = -(3x + 2)$ or $4x - 5 = 3x + 2$

$4x - 5 = -3x - 2$ or $\qquad 4x = 3x + 7$

$\quad 7x = 3$ or $\qquad$ or $x = 7$

$\quad x = \dfrac{3}{7}$ or $\qquad$ or $x = 7$

23. $|5w + 1| = |3 - w|$

$5w + 1 = -(3 - w)$ or $5w + 1 = 3 - w$

$5w + 1 = -3 + w$ or $\quad 5w = 2 - w$

$\quad 4w = -4$ or $\qquad 6w = 2$

$\quad w = -1$ or $\qquad w = \dfrac{1}{3}$

25. $\left|\dfrac{4x + 3}{2}\right| = 5$

$\dfrac{4x + 3}{2} = -5$ or $\dfrac{4x + 3}{2} = 5$

$4x + 3 = -10$ or $4x + 3 = 10$

$\quad 4x = -13$ or $\qquad 4x = 7$

$\quad x = -\dfrac{13}{4}$ or $\qquad x = \dfrac{7}{4}$

27. $\left|\dfrac{1}{2}x - \dfrac{5}{3}\right| = \dfrac{7}{6}$

$\dfrac{1}{2}x - \dfrac{5}{3} = -\dfrac{7}{6}$ or $\dfrac{1}{2}x - \dfrac{5}{3} = \dfrac{7}{6}$

$\dfrac{1}{2}x = -\dfrac{7}{6} + \dfrac{5}{3}$ or $\dfrac{1}{2}x = \dfrac{7}{6} + \dfrac{5}{3}$

$\dfrac{1}{2}x = \dfrac{1}{2}$ or $\dfrac{1}{2}x = \dfrac{17}{6}$

$\quad x = 1$ or $\qquad x = \dfrac{17}{3}$

29. $\left|\dfrac{2}{3}k + \dfrac{4}{9}\right| = \left|\dfrac{5}{6}k - \dfrac{1}{3}\right|$

$\dfrac{2}{3}k + \dfrac{4}{9} = -\left(\dfrac{5}{6}k - \dfrac{1}{3}\right)$ or $\dfrac{2}{3}k + \dfrac{4}{9} = \dfrac{5}{6}k - \dfrac{1}{3}$

$\dfrac{2}{3}k + \dfrac{5}{6}k = \dfrac{1}{3} - \dfrac{4}{9}$ or $\dfrac{2}{3}k - \dfrac{5}{6}k = -\dfrac{1}{3} - \dfrac{4}{9}$

$\dfrac{3}{2}k = -\dfrac{1}{9}$ or $-\dfrac{1}{6}k = -\dfrac{7}{9}$

$k = -\dfrac{2}{27}$ or $k = \dfrac{14}{3}$

31. $4.7|x| - 3.9 = 8.8$

$4.7|x| = 12.7$

$|x| \approx 2.70$

$x \approx -2.70$ or $x \approx 2.70$

33. $|2.1x + 5.8| - 9.7 = 10.2$

$|2.1x + 5.8| = 19.9$

$2.1x + 5.8 = -19.9$ or $2.1x + 5.8 = 19.9$

$2.1x = -25.7$ or $2.1x = 14.1$

$x \approx -12.24$ or $x \approx 6.71$

35.
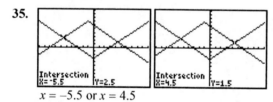

$x = -5.5$ or $x = 4.5$

37.
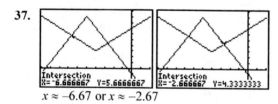

$x \approx -6.67$ or $x \approx -2.67$

39. $x = \pm 4$

41. $x = 2$ or $x = -3$

43. $f(x) = 2|x| - 11$

$f(-5) = 2|-5| - 11 = 2(5) - 11 = -1$

45. $f(x) = 2|x| - 11$

$-5 = 2|x| - 11$

$6 = 2|x|$

$3 = |x|$

$x = -3$ or $x = 3$

47. $f(x) = |4x + 7| - 9$

$f(-3) = |4(-3) + 7| - 9$

$f(-3) = |-12 + 7| - 9$

$f(-3) = |-5| - 9$

$f(-3) = 5 - 9$

$f(-3) = -4$

49. $f(x) = |4x + 7| - 9$

$-3 = |4x + 7| - 9$

$6 = |4x + 7|$

$4x + 7 = -6$ or $4x + 7 = 6$

$4x = -13$ or $4x = -1$

$x = -\dfrac{13}{4}$ or $x = -\dfrac{1}{4}$

51. $|x| < 4$

$-4 < x < 4$

Interval: $(-4, 4)$

53. $|x| \geq 3$

$x \leq -3$ or $x \geq 3$

Interval: $(-\infty, -3] \cup [3, \infty)$

55. $|r| < -3$

Since $|r|$ is nonnegative, the inequality $|r| < -3$ has an empty set solution.

57. $|x| > 0$

$x < 0$ or $x > 0$

Interval: $(-\infty, 0) \cup (0, \infty)$

59. $|x - 6| \geq 7$

$x - 6 \leq -7$ or $x - 6 \geq 7$

$x \leq -1$ or $x \geq 13$

Interval: $(-\infty, -1] \cup [13, \infty)$

61. $|2x + 5| < 15$

$-15 < 2x + 5 < 15$

$-20 < 2x < 10$

$-10 < x < 5$

Interval: $(-10, 5)$

63. $|7x + 15| > -4$

Since $|7x + 15|$ is always nonnegative, the solution set for the inequality $|7x + 15| > -4$ is the set of all real numbers.

Interval: $(-\infty, \infty)$

65. $|0.25t - 1.3| \geq 1.1$

$\quad 0.25t - 1.3 \leq -1.1 \quad$ or $\quad 0.25t - 1.3 \geq 1.1$

$\qquad 0.25t \leq 0.2 \quad$ or $\qquad 0.25t \geq 2.4$

$\qquad\qquad t \leq 0.8 \quad$ or $\qquad\qquad t \geq 9.6$

Interval: $(-\infty, 0.8] \cup [9.6, \infty)$

67. $2|x| - 5 > 3$

$\quad 2|x| > 8$

$\quad |x| > 4$

$x < -4$ or $x > 4$

Interval: $(-\infty, -4) \cup (4, \infty)$

69. $2 - 5|p| \leq -8$

$\quad -5|p| \leq -10$

$\quad |p| \geq 2$

$p \leq -2$ or $p \geq 2$

Interval: $(-\infty, -2] \cup [2, \infty)$

71. $7 - |x + 3| \leq 2$

$\quad -|x + 3| \leq -5$

$\quad |x + 3| \geq 5$

$x + 3 \leq -5$ or $x + 3 \geq 5$

$\quad x \leq -8$ or $\qquad x \geq 2$

Interval: $(-\infty, -8] \cup [2, \infty)$

73. $\left|\dfrac{x + 4}{3}\right| \geq 2$

$\dfrac{x + 4}{3} \leq -2 \quad$ or $\quad \dfrac{x + 4}{3} \geq 2$

$\quad x + 4 \leq -6 \quad$ or $\quad x + 4 \geq 6$

$\qquad x \leq -10 \quad$ or $\qquad x \geq 2$

Interval: $(-\infty, -10] \cup [2, \infty)$

75. $\left|\dfrac{2x}{5} + \dfrac{3}{2}\right| \leq \dfrac{9}{20}$

$-\dfrac{9}{20} \leq \dfrac{2x}{5} + \dfrac{3}{2} \leq \dfrac{9}{20}$

$-\dfrac{39}{20} \leq \dfrac{2x}{5} \leq -\dfrac{21}{20}$

$-\dfrac{39}{4} \leq 2x \leq -\dfrac{21}{4}$

$-\dfrac{39}{8} \leq x \leq -\dfrac{21}{8}$

Interval: $\left[-\dfrac{39}{8}, -\dfrac{21}{8}\right]$

77. $|mx + b| + c = k$

$\quad |mx + b| = k - c$

$mx + b = -(k - c) \quad$ or $mx + b = k - c$

$\quad mx = -b - (k - c)$ or $mx = -b + (k - c)$

$\qquad x = \dfrac{-b \pm (k - c)}{m}$

79. Answers may vary. Example: The student failed to use the absolute value property for equations. Instead, the student tried to take the absolute value of $x - 5$ directly, but this leads to the wrong answer. Also, the student incorrectly interchanges $x - 5$ with $x + 5$ in the second line.

$|x - 5| = 7$

$x - 5 = -7 \quad$ or $\quad x - 5 = 7$

$\quad x = -2 \quad$ or $\qquad x = 12$

81. Answers may vary. Example: The student attempted to use the absolute value property for equations, but he or she should have used the absolute value property for inequalities.

$|x + 3| < 10$

$-10 < x + 3 < 10$

$-13 < x < 7$

83. a. $|2x + 3| = 13$

$2x + 3 = -13 \quad$ or $2x + 3 = 13$

$\quad 2x = -16 \quad$ or $\qquad 2x = 10$

$\qquad x = -8 \quad$ or $\qquad x = 5$

b. $|2x + 3| < 13$

$-13 < 2x + 3 < 13$

$-16 < 2x < 10$

$-8 < x < 5$

c. $|2x+3| > 13$

$$2x+3 < -13 \quad \text{or} \quad 2x+3 > 13$$
$$2x < -16 \quad \text{or} \quad 2x > 10$$
$$x < -8 \quad \text{or} \quad x > 5$$

d.

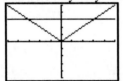

Answers may vary. Example:
The three different graphs each use –8 and 5, either as included points or as non-included endpoints. Between the three graphs, every part of the number line is covered.

85. Answers may vary. Example:

The graph of $y = |x|$ and the graph of $y = 3$ intersect at $(-3, 3)$ and $(3, 3)$. The graph of $y = |x|$ is below the graph of $y = 3$ for $-3 < x < 3$. Therefore $|x| < 3$ is equivalent to $-3 < x < 3$.

87. Answers may vary. Example:
The statement "$|a + b| = |a| + |b|$ for all real numbers a and b" is false. For a counterexample, let $a = -1$ and $b = 3$.

$$|-1+3| \overset{?}{=} |-1| + |3|$$
$$|2| \overset{?}{=} 1 + 3$$
$$2 \overset{?}{=} 4 \quad \text{false}$$

89. $|x - 5| = 4$

$$x - 5 = -4 \quad \text{or} \quad x - 5 = 4$$
$$x = 1 \quad \text{or} \quad x = 9$$

91. $|2^y - 5| = 4$

$$2^y - 5 = -4 \quad \text{or} \quad 2^y - 5 = 4$$
$$2^y = 1 \quad \text{or} \quad 2^y = 9$$
$$\log(2^y) = \log(1) \quad \text{or} \quad \log(2^y) = \log(9)$$
$$y\log(2) = \log(1) \quad \text{or} \quad y\log(2) = \log(9)$$
$$y = \frac{\log(1)}{\log(2)} \quad \text{or} \quad y = \frac{\log(9)}{\log(2)}$$
$$y = 0 \quad \text{or} \quad y \approx 3.1699$$

93. $\left| \dfrac{2x+3}{x-2} - 5 \right| = 4$

$$\frac{2x+3}{x-2} - 5 = -4 \quad \text{or} \quad \frac{2x+3}{x-2} - 5 = 4$$
$$\frac{2x+3}{x-2} = 1 \quad \text{or} \quad \frac{2x+3}{x-2} = 9$$
$$2x+3 = x-2 \quad \text{or} \quad 2x+3 = 9x-18$$
$$x = -5 \quad \text{or} \quad -7x = -21$$
$$x = -5 \quad \text{or} \quad x = 3$$

95. $3(2x) - 5 \le 7$

$$6x \le 12$$
$$x \le 2$$

Interval: $(-\infty, 2]$

97. $3|2x| - 5 \le 7$

$$3|2x| \le 12$$
$$|2x| \le 4$$
$$-4 \le 2x \le 4$$
$$-2 \le x \le 2$$

Interval: $[-2, 2]$

99.

$$y = 3(x-2) + 1 = 3x - 5$$

This is a linear function.

101. $3|x - 2| + 1 = 7$

$$3|x - 2| = 6$$
$$|x - 2| = 2$$
$$x - 2 = -2 \quad \text{or} \quad x - 2 = 2$$
$$x = 0 \quad \text{or} \quad x = 4$$

This is an absolute value equation in one variable.

103. If the line contains the points $(-4, 2)$ and $(5, -3)$, the slope is $\dfrac{-3-2}{5-(-4)} = -\dfrac{5}{9}$.

$$y = mx + b$$
$$2 = -\frac{5}{9}(-4) + b$$
$$2 = \frac{20}{9} + b$$
$$-\frac{2}{9} = b$$

The equation is $y = -\dfrac{5}{9}x - \dfrac{2}{9}$.

This is a linear equation in two variables.

11.1 Quiz

1. $3|t| - 4 = 11$
$$3|t| = 15$$
$$|t| = 5$$
$$t = 5 \text{ or } t = -5$$

2. $5|6r - 5| = 15$
$$|6r - 5| = 3$$
$$6r - 5 = -3 \text{ or } 6r - 5 = 3$$
$$6r = 2 \text{ or } 6r = 8$$
$$r = \frac{1}{3} \text{ or } r = \frac{4}{3}$$

3. $|7x + 1| = -3$

Since $|7x + 1|$ is always nonnegative, the solution set for $|7x + 1| = -3$ is the empty set.

4. $|5x - 2| = |3x + 6|$
$$5x - 2 = -(3x + 6) \text{ or } 5x - 2 = 3x + 6$$
$$5x = -3x - 4 \text{ or } 5x = 3x + 8$$
$$8x = -4 \text{ or } 2x = 8$$
$$x = -\frac{1}{2} \text{ or } x = 4$$

5. $\left|\dfrac{3}{4}x - \dfrac{1}{2}\right| = \dfrac{7}{8}$
$$\frac{3}{4}x - \frac{1}{2} = -\frac{7}{8} \text{ or } \frac{3}{4}x - \frac{1}{2} = \frac{7}{8}$$
$$\frac{3}{4}x = -\frac{3}{8} \text{ or } \frac{3}{4}x = \frac{11}{8}$$
$$x = -\frac{1}{2} \text{ or } x = \frac{11}{6}$$

6. The statement "$|a - b| = |a| - |b|$ for all real numbers a and b" is false.
Answers may vary. Example:
For a counterexample, let $a = -1$ and $b = 3$.
$$|-1 - 3| \neq |-1| - |3|$$
$$|-4| \neq 1 - 3$$
$$4 \neq -2$$

7. $3|k| - 4 \geq 2$
$$3|k| \geq 6$$
$$|k| \geq 2$$
$$k \leq -2 \text{ or } k \geq 2$$

Interval: $(-\infty, -2] \cup [2, \infty)$

8. $|4c - 8| > 12$
$$4c - 8 < -12 \text{ or } 4c - 8 > 12$$
$$4c < -4 \text{ or } 4c > 20$$
$$c < -1 \text{ or } c > 5$$

Interval: $(-\infty, -1) \cup (5, \infty)$

9. $7|3x - 2| \leq 42$
$$|3x - 2| \leq 6$$
$$-6 \leq 3x - 2 \leq 6$$
$$-4 \leq 3x \leq 8$$
$$-\frac{4}{3} \leq x \leq \frac{8}{3}$$

Interval: $\left[-\dfrac{4}{3}, \dfrac{8}{3}\right]$

10. $|x - 5| < -7$

Since $|x - 5|$ is always nonnegative, the solution set for $|x - 5| < -7$ is the empty set.

Homework 11.2

1. $(4 - 7i) + (3 + 10i) = 4 - 7i + 3 + 10i$
$$= 4 + 3 - 7i + 10i$$
$$= 7 + 3i$$

3. $\left(5 - \sqrt{-9}\right) + \left(2 - \sqrt{-25}\right) = 5 - 3i + 2 - 5i$
$$= 5 + 2 - 3i - 5i$$
$$= 7 - 8i$$

5. $(6-5i)-(2-13i) = 6-5i-2+13i$
$$= 6-2-5i+13i$$
$$= 4+8i$$

7. $\left(6-\sqrt{-49}\right)-\left(1+\sqrt{-81}\right) = 6-7i-1-9i$
$$= 6-1-7i-9i$$
$$= 5-16i$$

9. $2i \cdot 9i = 18i^2 = 18(-1) = -18$

11. $-10i(-5i) = 50i^2 = 50(-1) = -50$

13. $\sqrt{-4}\sqrt{-25} = 2i \cdot 5i = 10i^2 = 10(-1) = -10$

15. $\sqrt{-3}\sqrt{-5} = i\sqrt{3} \cdot i\sqrt{5}$
$$= i^2\sqrt{15}$$
$$= (-1)\sqrt{15}$$
$$= -\sqrt{15}$$

17. $(8i)^2 = 64i^2 = 64(-1) = -64$

19. $5i(3-2i) = 5i \cdot 3 - 5i \cdot 2i$
$$= 15i - 10i^2$$
$$= 15i - 10(-1)$$
$$= 15i + 10$$
$$= 10 + 15i$$

21. $20 - 3i(2-7i) = 20 - 3i \cdot 2 + 3i \cdot 7i$
$$= 20 - 6i + 21i^2$$
$$= 20 - 6i + 21(-1)$$
$$= 20 - 6i - 21$$
$$= -1 - 6i$$

23. $(2+5i)(3+4i) = 2 \cdot 3 + 2 \cdot 4i + 5i \cdot 3 + 5i \cdot 4i$
$$= 6 + 8i + 15i + 20i^2$$
$$= 6 + 23i + 20(-1)$$
$$= 6 + 23i - 20$$
$$= -14 + 23i$$

25. $(3-6i)(5+2i) = 3 \cdot 5 + 3 \cdot 2i - 6i \cdot 5 - 6i \cdot 2i$
$$= 15 + 6i - 30i - 12i^2$$
$$= 15 - 24i - 12(-1)$$
$$= 15 - 24i + 12$$
$$= 27 - 24i$$

27. $(-6+4i)(-2+7i) = 12 - 42i - 8i + 28i^2$
$$= 12 - 28 - 42i - 8i$$
$$= -16 - 50i$$

29. $(5+4i)(5-4i) = 5^2 - (4i)^2$
$$= 25 - 16i^2$$
$$= 25 - 16(-1)$$
$$= 25 + 16$$
$$= 41$$

31. $(2-9i)(2+9i) = 2^2 - (9i)^2$
$$= 4 - 81i^2$$
$$= 4 - 81(-1)$$
$$= 4 + 81$$
$$= 85$$

33. $(1+i)(1-i) = 1^2 - i^2 = 1 - (-1) = 1 + 1 = 2$

35. $(2+7i)^2 = 2^2 + 2(2)(7i) + (7i)^2$
$$= 4 + 28i + 49i^2$$
$$= 4 - 49 + 28i$$
$$= -45 + 28i$$

37. $(4-5i)^2 = 4^2 - 2(4)(5i) + (5i)^2$
$$= 16 - 40i + 25i^2$$
$$= 16 - 25 - 40i$$
$$= -9 - 40i$$

39. $(-4+3i)^2 = (-4)^2 + 2(-4)(3i) + (3i)^2$
$$= 16 - 24i + 9i^2$$
$$= 16 - 9 - 24i$$
$$= 7 - 24i$$

41. $\dfrac{3}{2+5i} = \dfrac{3}{2+5i} \cdot \dfrac{2-5i}{2-5i}$
$$= \dfrac{6-15i}{4-25i^2}$$
$$= \dfrac{6-15i}{4-25(-1)}$$
$$= \dfrac{6-15i}{4+25}$$
$$= \dfrac{6-15i}{29}$$
$$= \dfrac{6}{29} - \dfrac{15}{29}i$$

43. $\dfrac{3i}{7-2i} = \dfrac{3i}{7-2i} \cdot \dfrac{7+2i}{7+2i}$

$= \dfrac{21i + 6i^2}{49 - 4i^2}$

$= \dfrac{-6 + 21i}{53}$

$= -\dfrac{6}{53} + \dfrac{21}{53}i$

45. $\dfrac{2+3i}{7+i} = \dfrac{2+3i}{7+i} \cdot \dfrac{7-i}{7-i}$

$= \dfrac{14 + 21i - 2i - 3i^2}{49 - i^2}$

$= \dfrac{17 + 19i}{50}$

$= \dfrac{17}{50} + \dfrac{19}{50}i$

47. $\dfrac{3+4i}{3-4i} = \dfrac{3+4i}{3-4i} \cdot \dfrac{3+4i}{3+4i}$

$= \dfrac{9 + 12i + 12i + 16i^2}{9 - 16i^2}$

$= \dfrac{9 + 24i + 16(-1)}{9 - 16(-1)}$

$= \dfrac{-7 + 24i}{25}$

$= -\dfrac{7}{25} + \dfrac{24}{25}i$

49. $\dfrac{3-5i}{2-9i} = \dfrac{3-5i}{2-9i} \cdot \dfrac{2+9i}{2+9i}$

$= \dfrac{6 + 27i - 10i - 45i^2}{4 - 81i^2}$

$= \dfrac{51 + 17i}{85}$

$= \dfrac{51}{85} + \dfrac{17}{85}i$

$= \dfrac{3}{5} + \dfrac{1}{5}i$

51. $\dfrac{5+7i}{4i} = \dfrac{5+7i}{4i} \cdot \dfrac{i}{i}$

$= \dfrac{5i + 7i^2}{4i^2}$

$= \dfrac{-7 + 5i}{-4}$

$= \dfrac{7}{4} - \dfrac{5}{4}i$

53. $\dfrac{7}{5i} = \dfrac{7}{5i} \cdot \dfrac{i}{i} = \dfrac{7i}{5i^2} = -\dfrac{7}{5}i$

55. Student 2's work is correct
Answers may vary. Example:
If a radical has a negative radicand, you must rewrite the radical using i before performing any operations. Student 1 did not perform this step, and therefore got the wrong answer.

57. a. Answers may vary. Example:
$a = 3$, $b = -2$, $c = -8$, and $d = 5$.
$(a + bi) + (c + di) = (3 - 2i) + (-8 + 5i)$
$= 3 - 8 - 2i + 5i$
$= -5 + 3i$

b. Answers may vary. Example:
$a = -2$, $b = 5$, $c = -3$, and $d = -5$.
$(a + bi) + (c + di) = (-2 + 5i) + (-3 - 5i)$
$= -2 - 3 + 5i - 5i$
$= -5$

c. Answers may vary. Example:
$a = 7$, $b = -4$, $c = -7$, and $d = -2$.
$(a + bi) + (c + di) = (7 - 4i) + (-7 - 2i)$
$= 7 - 7 - 4i - 2i$
$= -6i$

59. Answers may vary. Example:
If a radical has a negative radicand and you perform any operations before rewriting the radical using i, you might simplify incorrectly.
For example, $\sqrt{-3}\sqrt{-3} = \sqrt{-3(-3)} = \sqrt{9} = 3$, which is incorrect. The correct simplification is $\sqrt{-3}\sqrt{-3} = i\sqrt{3}\left(i\sqrt{3}\right) = 3i^2 = -3$.

61. The square of a pure imaginary number will always be a negative real number.
Answers may vary. Example:
A pure imaginary number has two parts: the coefficient and i. When the coefficient is squared, it always becomes a positive real number. When i is squared, it always becomes -1. Multiplying a positive real number by -1 always gives a negative real product.

63.
$$\frac{4}{3+2\sqrt{x}} = \frac{4}{3+2\sqrt{x}} \cdot \frac{3-2\sqrt{x}}{3-2\sqrt{x}}$$
$$= \frac{12-8\sqrt{x}}{9-\left(2\sqrt{x}\right)^2}$$
$$= \frac{12-8\sqrt{x}}{9-4x}$$

65.
$$\frac{4}{3+2i} = \frac{4}{3+2i} \cdot \frac{3-2i}{3-2i}$$
$$= \frac{12-8i}{9-4i^2}$$
$$= \frac{12-8i}{13}$$
$$= \frac{12}{13} - \frac{8}{13}i$$

67. $3x^2 - 2x + 3 = 0$
$$x = \frac{2 \pm \sqrt{(-2)^2 - 4(3)(3)}}{2(3)}$$
$$= \frac{2 \pm \sqrt{-32}}{6}$$
$$= \frac{2 \pm 4i\sqrt{2}}{6}$$
$$= \frac{1 \pm 2i\sqrt{2}}{3}$$

69. $5x^2 - 4x = -1$
$$5x^2 - 4x + 1 = 0$$
$$x = \frac{4 \pm \sqrt{(-4)^2 - 4(5)(1)}}{2(5)}$$
$$= \frac{4 \pm \sqrt{-4}}{10}$$
$$= \frac{4 \pm 2i}{10}$$
$$= \frac{2 \pm i}{5}$$

71.
$$(p-3)(2p+1) = -10$$
$$2p^2 + p - 6p - 3 + 10 = 0$$
$$2p^2 - 5p + 7 = 0$$
$$p = \frac{5 \pm \sqrt{(-5)^2 - 4(2)(7)}}{2(2)}$$
$$= \frac{5 \pm \sqrt{-31}}{4}$$
$$= \frac{5 \pm i\sqrt{31}}{4}$$

73.
$$x(3x-2) = 2 + 2(x-3)$$
$$3x^2 - 2x = 2 + 2x - 6$$
$$3x^2 - 4x + 4 = 0$$
$$x = \frac{4 \pm \sqrt{(-4)^2 - 4(3)(4)}}{2(3)}$$
$$= \frac{4 \pm \sqrt{-32}}{6}$$
$$= \frac{4 \pm 4i\sqrt{2}}{6}$$
$$= \frac{2 \pm 2i\sqrt{2}}{3}$$

75.
$$(5w+3)^2 = -20$$
$$25w^2 + 30w + 9 = -20$$
$$25w^2 + 30w + 29 = 0$$
$$w = \frac{-30 \pm \sqrt{30^2 - 4(25)(29)}}{2(25)}$$
$$= \frac{-30 \pm \sqrt{900 - 2900}}{50}$$
$$= \frac{-30 \pm \sqrt{-2000}}{50}$$
$$= \frac{-30 \pm 20i\sqrt{5}}{50}$$
$$= \frac{-3 \pm 2i\sqrt{5}}{5}$$

77. $4x^2 - 2x + 3 = 0$
$$x = \frac{2 \pm \sqrt{(-2)^2 - 4(4)(3)}}{2(4)}$$
$$= \frac{2 \pm \sqrt{-44}}{8}$$
$$= \frac{2 \pm 2i\sqrt{11}}{8}$$
$$= \frac{1 \pm i\sqrt{11}}{4}$$

This is a quadratic equation in one variable.

79. $10x^2 - 19x + 6 = (5x-2)(2x-3)$

This is a quadratic (or second-degree) polynomial in one variable.

81. $(3i-7)(4i+6) = 12i^2 + 18i - 28i - 42$
$$= -12 - 42 + 18i - 28i$$
$$= -54 - 10i$$

This is an imaginary number.

11.2 Quiz

1. $(6-2i)+(3-4i) = 6-2i+3-4i$
$$= 6+3-2i-4i$$
$$= 9-6i$$

2. $(3+7i)-(8-2i) = 3+7i-8+2i$
$$= 3-8+7i+2i$$
$$= -5+9i$$

3. $-4i \cdot 3i = -12i^2 = -12(-1) = 12$

4. $\sqrt{-2}\sqrt{-7} = i\sqrt{2} \cdot i\sqrt{7} = i^2\sqrt{14} = -\sqrt{14}$

5. $(5-3i)(7+i) = 5\cdot7+5\cdot i-3i\cdot7-3i\cdot i$
$$= 35+5i-21i-3i^2$$
$$= 35-16i-3(-1)$$
$$= 35-16i+3$$
$$= 38-16i$$

6. $(4-3i)^2 = (4)^2-2(4)(3i)+(3i)^2$
$$= 16-24i+9i^2$$
$$= 16-24i+9(-1)$$
$$= 16-24i-9$$
$$= 7-24i$$

7. $(8+5i)(8-5i) = (8)^2-(5i)^2$
$$= 64-25i^2$$
$$= 64-25(-1)$$
$$= 64+25$$
$$= 89$$

8. $\dfrac{3+2i}{5-4i} = \dfrac{3+2i}{5-4i}\cdot\dfrac{5+4i}{5+4i}$
$$= \dfrac{15+12i+10i+8i^2}{25-16i^2}$$
$$= \dfrac{15+22i+8(-1)}{25-16(-1)}$$
$$= \dfrac{7+22i}{41}$$
$$= \dfrac{7}{41}+\dfrac{22}{41}i$$

9. $\dfrac{5-7i}{6i} = \dfrac{5-7i}{6i}\cdot\dfrac{i}{i}$
$$= \dfrac{5i-7i^2}{6i^2}$$
$$= \dfrac{7+5i}{-6}$$
$$= -\dfrac{7}{6}-\dfrac{5}{6}i$$

10. False. Answers may vary. Example: The number $3i$ (or $0+3i$) is a complex number and i is a pure imaginary number. Then $(i)(3i) = 3i^2 = -3$. Since -3 is not an imaginary number, the statement is false.

Homework 11.3

1. $c^2 = a^2+b^2$
$$c^2 = 5^2+12^2$$
$$c^2 = 25+144$$
$$c^2 = 169$$
$$c = 13$$

3. $c^2 = a^2+b^2$
$$c^2 = 4^2+5^2$$
$$c^2 = 16+25$$
$$c^2 = 41$$
$$c = \sqrt{41}$$

5. $a^2+b^2 = c^2$
$$3^2+b^2 = 8^2$$
$$9+b^2 = 64$$
$$b^2 = 55$$
$$b = \sqrt{55}$$

7. $a^2+b^2 = c^2$
$$a^2+5^2 = 7^2$$
$$a^2+25 = 49$$
$$a^2 = 24$$
$$a = \sqrt{24} = 2\sqrt{6}$$

9. $c^2 = a^2+b^2$
$$c^2 = \left(\sqrt{2}\right)^2+\left(\sqrt{5}\right)^2$$
$$c^2 = 2+5$$
$$c^2 = 7$$
$$c = \sqrt{7}$$

11. The lengths of the two legs are given, so let $a = 11$ and $b = 7$.

$$c^2 = a^2 + b^2$$
$$c^2 = 11^2 + 7^2$$
$$c^2 = 121 + 49$$
$$c^2 = 170$$
$$c = \sqrt{170}$$

13. The lengths of a leg and the hypotenuse are given, so let $a = 10$ and $c = 12$.

$$a^2 + b^2 = c^2$$
$$10^2 + b^2 = 12^2$$
$$100 + b^2 = 144$$
$$b^2 = 44$$
$$b = 2\sqrt{11}$$

15. $a = 5, b = 20, c = $ length of ladder

$$c^2 = a^2 + b^2$$
$$c^2 = 5^2 + 20^2$$
$$c^2 = 25 + 400$$
$$c^2 = 425$$
$$c = \sqrt{425} = 5\sqrt{17} \approx 20.6$$

The ladder must be approximately 20.6 feet long.

17. $a = 13, b = $ width of screen, $c = 20$

$$a^2 + b^2 = c^2$$
$$13^2 + b^2 = 20^2$$
$$169 + b^2 = 400$$
$$b^2 = 231$$
$$b = \sqrt{231} \approx 15.2$$

The screen must be approximately 15.2 inches wide.

19. $a = 2.8$, $b = $ distance across lake, $c = 3.4$

$$a^2 + b^2 = c^2$$
$$2.8^2 + b^2 = 3.4^2$$
$$7.84 + b^2 = 11.56$$
$$b^2 = 3.72$$
$$b = \sqrt{3.72} \approx 1.9$$

The distance across the lake must be approximately 1.9 miles.

21. $a = 465, c = 964, b = $ distance

$$a^2 + b^2 = c^2$$
$$465^2 + b^2 = 964^2$$
$$b^2 = 713071$$
$$b = \sqrt{713071} \approx 844.4$$
$$a + b + c = 465 + 844.4 + 964 = 2273.4$$

The total distance of the road trip would be approximately 2273.4 miles.

23. $(2, 9)$ and $(8, 1)$

$$d = \sqrt{(8-2)^2 + (1-9)^2}$$
$$= \sqrt{6^2 + (-8)^2}$$
$$= \sqrt{36 + 64}$$
$$= \sqrt{100}$$
$$= 10$$

25. $(-3, 5)$ and $(4, 2)$

$$d = \sqrt{(4-(-3))^2 + (2-5)^2}$$
$$= \sqrt{7^2 + (-3)^2}$$
$$= \sqrt{49 + 9}$$
$$= \sqrt{58}$$

27. $(-6, -3)$ and $(-4, 1)$

$$d = \sqrt{(-4-(-6))^2 + (1-(-3))^2}$$
$$= \sqrt{2^2 + 4^2}$$
$$= \sqrt{4 + 16}$$
$$= \sqrt{20}$$
$$= 2\sqrt{5}$$

29. $(-4, -5)$ and $(-8, -9)$

$$d = \sqrt{(-8-(-4))^2 + (-9-(-5))^2}$$
$$= \sqrt{(-4)^2 + (-4)^2}$$
$$= \sqrt{16 + 16}$$
$$= \sqrt{32}$$
$$= 4\sqrt{2}$$

31. $(2.1, 8.9)$ and $(5.6, 1.7)$

$$d = \sqrt{(5.6-2.1)^2 + (1.7-8.9)^2}$$
$$= \sqrt{3.5^2 + (-7.2)^2}$$
$$= \sqrt{12.25 + 51.84}$$
$$= \sqrt{64.09} \approx 8.01$$

33. $(-2.18, -5.74)$ and $(3.44, 6.29)$

$$d = \sqrt{(3.44 - (-2.18))^2 + (6.29 - (-5.74))^2}$$
$$= \sqrt{5.62^2 + 12.03^2}$$
$$= \sqrt{31.5844 + 144.7209}$$
$$= \sqrt{176.3053} \approx 13.28$$

35. $C(0,0)$ and $r = 7$

$$(x - h)^2 + (y - k)^2 = r^2$$
$$(x - 0)^2 + (y - 0)^2 = 7^2$$
$$x^2 + y^2 = 49$$

37. $C(0,0)$ and $r = 6.7$

$$(x - h)^2 + (y - k)^2 = r^2$$
$$(x - 0)^2 + (y - 0)^2 = 6.7^2$$
$$x^2 + y^2 = 44.89$$

39. $C(5,3)$ and $r = 2$

$$(x - h)^2 + (y - k)^2 = r^2$$
$$(x - 5)^2 + (y - 3)^2 = 2^2$$
$$(x - 5)^2 + (y - 3)^2 = 4$$

41. $C(-2,1)$ and $r = 4$

$$(x - h)^2 + (y - k)^2 = r^2$$
$$(x - (-2))^2 + (y - 1)^2 = 4^2$$
$$(x + 2)^2 + (y - 1)^2 = 16$$

43. $C(-7, -3)$ and $r = \sqrt{3}$

$$(x - h)^2 + (y - k)^2 = r^2$$
$$(x - (-7))^2 + (y - (-3))^2 = (\sqrt{3})^2$$
$$(x + 7)^2 + (y + 3)^2 = 3$$

45. $x^2 + y^2 = 25$

The equation has the form $x^2 + y^2 = r^2$.
Therefore, $C = (0,0)$ and

$$r^2 = 25$$
$$r = \sqrt{25} = 5$$

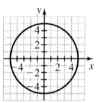

47. $x^2 + y^2 = 8$

The equation has the form $x^2 + y^2 = r^2$.
Therefore, $C = (0,0)$ and

$$r^2 = 8$$
$$r = \sqrt{8} = 2\sqrt{2}$$

49. $(x - 3)^2 + (y - 5)^2 = 16$

The equation is in the form
$(x - h)^2 + (y - k)^2 = r^2$.
The center is (h,k) or $C(3,5)$ and

$$r^2 = 16$$
$$r = \sqrt{16} = 4$$

51. $(x + 6)^2 + (y - 1)^2 = 7$

$$(x - (-6))^2 + (y - 1)^2 = (\sqrt{7})^2$$

The equation is in the form
$(x - h)^2 + (y - k)^2 = r^2$. The center is
$C(-6, 1)$ and the radius is $r = \sqrt{7}$.

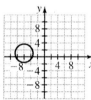

53. $(x+3)^2 + (y+2)^2 = 1$

$(x-(-3))^2 + (y-(-2))^2 = 1^2$

The equation is in the form

$(x-h)^2 + (y-k)^2 = r^2$. The center is

$C(-3,-2)$ and the radius is $r = 1$.

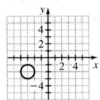

55. $C(0,0)$ and $r = 3$.

$(x-h)^2 + (y-k)^2 = r^2$

$(x-0)^2 + (y-0)^2 = 3^2$

$x^2 + y^2 = 9$

57. $C(-3,2)$ and $r = 2$.

$(x-h)^2 + (y-k)^2 = r^2$

$(x-(-3))^2 + (y-2)^2 = 2^2$

$(x+3)^2 + (y-2)^2 = 4$

59. The radius is the distance from the center to any point on the circle. The distance between $C(3,2)$ and $(5,6)$ is given by:

$d = \sqrt{(5-3)^2 + (6-2)^2}$

$= \sqrt{2^2 + 4^2}$

$= \sqrt{4 + 16}$

$= \sqrt{20}$

The radius is $r = \sqrt{20}$.
The equation of the circle is

$(x-h)^2 + (y-k)^2 = r^2$

$(x-3)^2 + (y-2)^2 = (\sqrt{20})^2$

$(x-3)^2 + (y-2)^2 = 20$

61. Answers may vary. Example:

$(x-2)^2 + (y-3)^2 = 9$

$(x-5)^2 + (y-6)^2 = 9$

63. a. $x^2 + y^2 = 16$

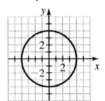

b. Answers may vary. Example:

x	y
-4	0
-2	$2\sqrt{3} \approx 3.46$
-2	$-2\sqrt{3} \approx -3.46$
0	4
0	-4
2	$2\sqrt{3} \approx 3.46$
2	$-2\sqrt{3} \approx -3.46$
4	0

c. For each input-output pair, the sum of the square of the input and the square of the output is 16.

65. Find the equation of the circle that has center $C(3,2)$ and $r = 4$.

$(x-h)^2 + (y-k)^2 = r^2$

$(x-3)^2 + (y-2)^2 = 4^2$

$(x-3)^2 + (y-2)^2 = 16$

Find the coordinates of five points, (x,y), that satisfy this equation.
Points may vary. Example:

$(3,6), (3,-2), (7,2), (-1,2), (5,2+\sqrt{12})$

67. No. Answers may vary. Example:
The graph of the relation is a circle with radius 7 and centered at the origin. The graph fails the vertical line test.

69. a. The square root of a nonnegative number is a nonnegative real number and the square root of a negative number is an imaginary number. Therefore, $y \geq 0$ for real number values of y.

b.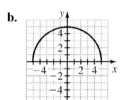

71. a. Sketches may vary. Example:

b. $a = k, b = k$
$$c^2 = a^2 + b^2$$
$$c^2 = k^2 + k^2$$
$$c^2 = 2k^2$$
$$c = \sqrt{2k^2} = \sqrt{2}\sqrt{k^2} = k\sqrt{2}$$

c. $c = k\sqrt{2} = 3\sqrt{2}$

d. $c = k\sqrt{2}$
$5 = k\sqrt{2}$
$$k = \frac{5}{\sqrt{2}} = \frac{5}{\sqrt{2}} \cdot \frac{\sqrt{2}}{\sqrt{2}} = \frac{5\sqrt{2}}{2}$$

73. $x + y = 4$

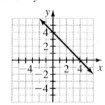

75. $x^2 + y^2 = 4$

77. $2^x + y = 0$

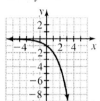

79. $f(x) = 2(x-4)^2 - 3$

This is a quadratic function.

81. $6x^2 - 16x + 8 = 2\left(3x^2 - 8x + 4\right)$
$$= 2(3x - 2)(x - 2)$$

This is a quadratic (or second-degree) polynomial in one variable.

83. $x(5x - 3) = 3(x + 1)$
$$5x^2 - 3x = 3x + 3$$
$$5x^2 - 6x - 3 = 0$$
$$x = \frac{-b \pm \sqrt{b^2 - 4ac}}{2a}$$
$$= \frac{6 \pm \sqrt{(-6)^2 - 4(5)(-3)}}{2(5)}$$
$$= \frac{6 \pm \sqrt{36 + 60}}{10}$$
$$= \frac{6 \pm 4\sqrt{6}}{10}$$
$$= \frac{3 \pm 2\sqrt{6}}{5}$$

This is a quadratic equation in one variable.

11.3 Quiz

1. $a = 4, c = 8$
$$a^2 + b^2 = c^2$$
$$4^2 + b^2 = 8^2$$
$$16 + b^2 = 64$$
$$b^2 = 48$$
$$b = \sqrt{48} = 4\sqrt{3} \approx 6.9$$

The other leg is about 6.9 inches.

2. $b = 16, c = 19$

$$a^2 + b^2 = c^2$$
$$a^2 + 16^2 = 19^2$$
$$a^2 + 256 = 361$$
$$a^2 = 105$$
$$a = \sqrt{105} \approx 10.2$$

The height of the screen is about 10.2 inches.

3. $(-2, -5)$ and $(3, -1)$

$$d = \sqrt{(3 - (-2))^2 + (-1 - (-5))^2}$$
$$= \sqrt{5^2 + 4^2}$$
$$= \sqrt{25 + 16}$$
$$= \sqrt{41}$$

4. $(-3, 2)$ and $(-7, -2)$

$$d = \sqrt{(-7 - (-3))^2 + (-2 - 2)^2}$$
$$= \sqrt{(-4)^2 + (-4)^2}$$
$$= \sqrt{16 + 16}$$
$$= \sqrt{32}$$
$$= 4\sqrt{2}$$

5. $C(-3, 2)$ and $r = 6$

$$(x - h)^2 + (y - k)^2 = r^2$$
$$(x - (-3))^2 + (y - 2)^2 = 6^2$$
$$(x + 3)^2 + (y - 2)^2 = 36$$

6. $C(0, 0)$ and $r = 2.8$

$$(x - h)^2 + (y - k)^2 = r^2$$
$$(x - 0)^2 + (y - 0)^2 = 2.8^2$$
$$x^2 + y^2 = 7.84$$

7. $x^2 + y^2 = 12$

The equation is in the form $x^2 + y^2 = r^2$. The center is $C(0, 0)$ and

$$r^2 = 12$$
$$r = \sqrt{12} = 2\sqrt{3}$$

8. $(x + 4)^2 + (y - 3)^2 = 25$

$$(x - (-4))^2 + (y - 3)^2 = 5^2$$

The equation is in the form $(x - h)^2 + (y - k)^2 = r^2$. The center is $C(-4, 3)$ and the radius is $r = 5$.

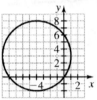

9. $C(2, -1)$

The radius is the distance from the center, $C(2, -1)$, to the point $(4, 7)$ that lies on the circle.

$$d = \sqrt{(4 - 2)^2 + (7 - (-1))^2}$$
$$= \sqrt{2^2 + 8^2}$$
$$= \sqrt{4 + 64}$$
$$= \sqrt{68}$$

The equation of the circle is

$$(x - h)^2 + (y - k)^2 = r^2$$
$$(x - 2)^2 + (y - (-1))^2 = (\sqrt{68})^2$$
$$(x - 2)^2 + (y + 1)^2 = 68$$

10. Answers may vary. Example:

$$(x + 2)^2 + y^2 = 4$$
$$(x - 3)^2 + y^2 = 9$$

Homework 11.4

1. $\dfrac{x^2}{36} + \dfrac{y^2}{9} = 1$

$a^2 = 36, a = 6$

x-intercepts: $(-6, 0), (6, 0)$

$b^2 = 9, b = 3$

y-intercepts: $(0, -3), (0, 3)$

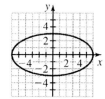

3. $\dfrac{x^2}{4} + \dfrac{y^2}{36} = 1$

$a^2 = 4, a = 2$

x-intercepts: $(-2, 0), (2, 0)$

$b^2 = 36, b = 6$

y-intercepts: $(0, -6), (0, 6)$

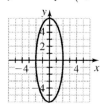

5. $\dfrac{x^2}{100} + \dfrac{y^2}{16} = 1$

$a^2 = 100, a = 10$

x-intercepts: $(-10, 0), (10, 0)$

$b^2 = 16, b = 4$

y-intercepts: $(0, -4), (0, 4)$

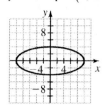

7. $25x^2 + 4y^2 = 100$

$\dfrac{25x^2}{100} + \dfrac{4y^2}{100} = \dfrac{100}{100}$

$\dfrac{x^2}{4} + \dfrac{y^2}{25} = 1$

$a^2 = 4, a = 2$

x-intercepts: $(-2, 0), (2, 0)$

$b^2 = 25, b = 5$

y-intercepts: $(0, -5), (0, 5)$

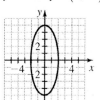

9. $9x^2 + 100y^2 = 900$

$\dfrac{9x^2}{900} + \dfrac{100y^2}{900} = \dfrac{900}{900}$

$\dfrac{x^2}{100} + \dfrac{y^2}{9} = 1$

$a^2 = 100, a = 10$

x-intercepts: $(-10, 0), (10, 0)$

$b^2 = 9, b = 3$

y-intercepts: $(0, -3), (0, 3)$

11. $x^2 + y^2 = 36$

$\dfrac{x^2}{36} + \dfrac{y^2}{36} = \dfrac{36}{36}$

$\dfrac{x^2}{36} + \dfrac{y^2}{36} = 1$

$a^2 = 36, a = 6$

x-intercepts: $(-6, 0), (6, 0)$

$b^2 = 36, b = 6$

y-intercepts: $(0,-6),(0,6)$

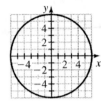

13. $x^2 + 25y^2 = 25$

$\dfrac{x^2}{25} + \dfrac{25y^2}{25} = \dfrac{25}{25}$

$\dfrac{x^2}{25} + \dfrac{y^2}{1} = 1$

$a^2 = 25, a = 5$

x-intercepts: $(-5,0),(5,0)$

$b^2 = 1, b = 1$

y-intercepts: $(0,-1),(0,1)$

15. $5x^2 + 16y^2 = 80$

$\dfrac{5x^2}{80} + \dfrac{16y^2}{80} = \dfrac{80}{80}$

$\dfrac{x^2}{16} + \dfrac{y^2}{5} = 1$

$a^2 = 16, a = 4$

x-intercepts: $(-4,0),(4,0)$

$b^2 = 5, b = \sqrt{5}$

y-intercepts: $\left(0,-\sqrt{5}\right),\left(0,\sqrt{5}\right)$

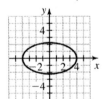

17. The x-intercepts are $(5,0)$ and $(-5,0)$, so $a = 5$. The y-intercepts are $(3,0)$ and $(-3,0)$, so $b = 3$. Therefore:

$\dfrac{x^2}{5^2} + \dfrac{y^2}{3^2} = 1$

$\dfrac{x^2}{25} + \dfrac{y^2}{9} = 1$

19. $\dfrac{x^2}{16} - \dfrac{y^2}{4} = 1$

$a^2 = 16, a = 4$

x-intercepts: $(-4,0),(4,0)$

$b^2 = 4, b = 2$

Sketch a dashed rectangle that contains the points $(-4,0),(4,0),(0,-2)$, and $(0,2)$, and then sketch the inclined asymptotes.

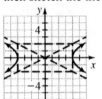

21. $\dfrac{y^2}{16} - \dfrac{x^2}{25} = 1$

$b^2 = 16, b = 4$

y-intercepts: $(0,-4),(0,4)$

$a^2 = 25, a = 5$

Sketch a dashed rectangle that contains the points $(-5,0),(5,0),(0,-4)$, and $(0,4)$, and then sketch the inclined asymptotes.

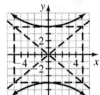

23. $\dfrac{x^2}{25} - \dfrac{y^2}{81} = 1$

$a^2 = 25, b = 5$

x-intercepts: $(5,0),(-5,0)$

$b^2 = 81, b = 9$

Sketch a dashed rectangle that contains the points $(-5,0),(5,0),(0,-9)$, and $(0,9)$, and then sketch the inclined asymptotes.

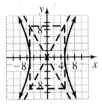

25. $16x^2 - 4y^2 = 64$

$$\frac{16x^2}{64} - \frac{4y^2}{64} = \frac{64}{64}$$

$$\frac{x^2}{4} - \frac{y^2}{16} = 1$$

$a^2 = 4, a = 2$

x-intercepts: $(-2, 0), (2, 0)$

$b^2 = 16, b = 4$

Sketch a dashed rectangle that contains the points $(-2, 0), (2, 0), (0, -4)$, and $(0, 4)$, and then sketch the inclined asymptotes.

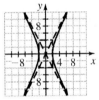

27. $x^2 - 9y^2 = 9$

$$\frac{x^2}{9} - \frac{9y^2}{9} = \frac{9}{9}$$

$$\frac{x^2}{9} - \frac{y^2}{1} = 1$$

$a^2 = 9, a = 3$

x-intercepts: $(-3, 0), (3, 0)$

$b^2 = 1, b = 1$

Sketch a dashed rectangle that contains the points $(-3, 0), (3, 0), (0, -1)$, and $(0, 1)$, and then sketch the inclined asymptotes.

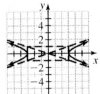

29. $y^2 - x^2 = 4$

$$\frac{y^2}{4} - \frac{x^2}{4} = \frac{4}{4}$$

$$\frac{y^2}{4} - \frac{x^2}{4} = 1$$

$b^2 = 4, b = 2$

y-intercepts: $(0, -2), (0, 2)$

$a^2 = 4, a = 2$

Sketch a dashed rectangle that contains the points $(-2, 0), (2, 0), (0, -2)$, and $(0, 2)$, and then sketch the inclined asymptotes.

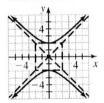

31. $16y^2 - x^2 = 16$

$$\frac{16y^2}{16} - \frac{x^2}{16} = \frac{16}{16}$$

$$\frac{y^2}{1} - \frac{x^2}{16} = 1$$

$b^2 = 1, b = 1$

y-intercepts: $(0, -1), (0, 1)$

$a^2 = 16, a = 4$

Sketch a dashed rectangle that contains the points $(-4, 0), (4, 0), (0, -1)$, and $(0, 1)$, and then sketch the inclined asymptotes.

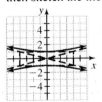

33. $25x^2 - 7y^2 = 175$

$$\frac{25x^2}{175} - \frac{7y^2}{175} = \frac{175}{175}$$

$$\frac{x^2}{7} - \frac{y^2}{25} = 1$$

$a^2 = 7, a = \sqrt{7}$

x-intercepts: $(-\sqrt{7}, 0), (\sqrt{7}, 0)$

$b^2 = 25, b = 5$

Sketch a dashed rectangle that contains the points $\left(-\sqrt{7},0\right),\left(\sqrt{7},0\right),(0,-5)$, and $(0,5)$, and then sketch the inclined asymptotes.

35. $\dfrac{x^2}{64}+\dfrac{y^2}{4}=1$

This is an ellipse.

$a^2=64, a=8$

x-intercepts: $(-8,0),(8,0)$

$b^2=4, b=2$

y-intercepts: $(0,-2),(0,2)$

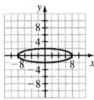

37. $x^2-y^2=1$

$\dfrac{x^2}{1}-\dfrac{y^2}{1}=1$

This is a hyperbola.

$a^2=1, a=1$

x-intercepts: $(-1,0),(1,0)$

$b^2=1, b=1$

Sketch a dashed rectangle that contains the points $(-1,0),(1,0),(0,-1)$, and $(0,1)$, and then sketch the inclined asymptotes.

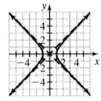

39. $81x^2+49y^2=3969$

$\dfrac{81x^2}{3969}+\dfrac{49y^2}{3969}=\dfrac{3969}{3969}$

$\dfrac{x^2}{49}+\dfrac{y^2}{81}=1$

This is an ellipse.

$a^2=49, a=7$

x-intercepts: $(-7,0),(7,0)$

$b^2=81, b=9$

y-intercepts: $(0,-9),(0,9)$

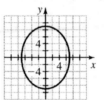

41. $x^2+y^2=1$

$\dfrac{x^2}{1}+\dfrac{y^2}{1}=1$

This is a circle.

$a^2=1, a=1$

x-intercepts: $(-1,0),(1,0)$

$b^2=1, b=1$

y-intercepts: $(0,-1),(0,1)$

43. $9y^2-4x^2=144$

$\dfrac{9y^2}{144}-\dfrac{4x^2}{144}=\dfrac{144}{144}$

$\dfrac{y^2}{16}-\dfrac{x^2}{36}=1$

This is a hyperbola.

$b^2=16, b=4$

y-intercepts: $(0,-4),(0,4)$

$a^2=36, a=6$

Sketch a dashed rectangle that contains the points $(-6,0),(6,0),(0,-4)$, and $(0,4)$, and then sketch the inclined asymptotes.

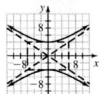

45. $\dfrac{x^2}{25} - \dfrac{y^2}{25} = 1$

This is a hyperbola.

$a^2 = 25, a = 5$

x-intercepts: $(-5, 0), (5, 0)$

$b^2 = 25, b = 5$

Sketch a dashed rectangle that contains the points $(-5, 0), (5, 0), (0, -5),$ and $(0, 5)$, and then sketch the inclined asymptotes.

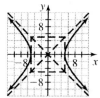

47. $x^2 + y^2 = 16$

$\dfrac{x^2}{16} + \dfrac{y^2}{16} = 1$

This is a circle.

$a^2 = 16, a = 4$

x-intercepts: $(-4, 0), (4, 0)$

$b^2 = 16, b = 4$

y-intercepts: $(0, -4), (0, 4)$

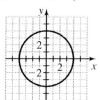

49. $9x^2 + 16y^2 = 144$

$\dfrac{9x^2}{144} + \dfrac{16y^2}{144} = 1$

$\dfrac{x^2}{16} + \dfrac{y^2}{9} = 1$

This is an ellipse.

$a^2 = 16, a = 4$

x-intercepts: $(-4, 0), (4, 0)$

$b^2 = 9, b = 3$

y-intercepts: $(0, -3), (0, 3)$

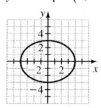

51. $\dfrac{x^2}{16} + \dfrac{y^2}{16} = 1$

This is a circle.

$a^2 = 16, a = 4$

x-intercepts: $(-4, 0), (4, 0)$

$b^2 = 16, b = 4$

y-intercepts: $(0, -4), (0, 4)$

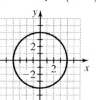

53. a. i. $\dfrac{x^2}{c} + \dfrac{y^2}{d} = 1$

$\dfrac{x^2}{4} + \dfrac{y^2}{16} = 1$

This is an ellipse.

$a^2 = 4, a = 2$

x-intercepts: $(-2, 0), (2, 0)$

$b^2 = 16, b = 4$

y-intercepts: $(0, -4), (0, 4)$

ii. $\dfrac{x^2}{c} + \dfrac{y^2}{d} = 1$

$\dfrac{x^2}{4} - \dfrac{y^2}{16} = 1$

This is a hyperbola.

$a^2 = 4, a = 2$

x-intercepts: $(-2, 0), (2, 0)$

$b^2 = 16, b = 4$

Sketch a dashed rectangle that contains the points $(-2, 0), (2, 0),$ $(0, -4),$ and $(0, 4)$, and then sketch the inclined asymptotes.

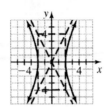

iii. $\dfrac{x^2}{c} + \dfrac{y^2}{d} = 1$

$\dfrac{y^2}{16} - \dfrac{x^2}{4} = 1$

This is a hyperbola.

$b^2 = 16, b = 4$

y-intercepts: $(0, -4), (0, 4)$

$a^2 = 4, a = 2$

Sketch a dashed rectangle that contains the points $(-2, 0), (2, 0), (0, -4),$ and $(0, 4)$, and then sketch the inclined asymptotes.

iv. $\dfrac{x^2}{c} + \dfrac{y^2}{d} = 1$

$\dfrac{x^2}{4} + \dfrac{y^2}{4} = 1$

This is a circle.

$a^2 = 4, a = 2$

x-intercepts: $(-2, 0), (2, 0)$

$b^2 = 4, b = 2$

y-intercepts: $(0, -2), (0, 2)$

b. If $c > 0$ and $d > 0$ and $c \neq d$, then the graph is an ellipse.
If $c > 0$ and $d < 0$, then the graph is a hyperbola with x-intercepts.

If $c < 0$ and $d > 0$, then the graph is a hyperbola with y-intercepts.
If $c = d$ and $c > 0$, then the graph is a circle.

55. a. $4x^2 + 25y^2 = 100$

$\dfrac{4x^2}{100} + \dfrac{25y^2}{100} = 1$

$\dfrac{x^2}{25} + \dfrac{y^2}{4} = 1$

This is an ellipse.

$a^2 = 25, a = 5$

x-intercepts: $(-5, 0), (5, 0)$

$b^2 = 4, b = 2$

y-intercepts: $(0, -2), (0, 2)$

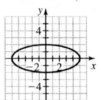

b. Answers may vary. Example:

x	y
-5	0
-2	$\sqrt{3.36} \approx 1.83$
-2	$-\sqrt{3.36} \approx -1.83$
0	2
0	-2
2	$\sqrt{3.36} \approx 1.83$
2	$-\sqrt{3.36} \approx -1.83$
5	0

c. Four times the square of x plus 25 times the square of y equals 100.

57. a. $y = \dfrac{5}{2}\sqrt{4 - x^2}$

Answers may vary. Example:
The square root of a nonnegative number is a nonnegative real number and the square root of a negative number is an imaginary number. Since $\dfrac{5}{2}$ is positive, $\dfrac{5}{2}\sqrt{4 - x^2}$ is nonnegative for real values of y. Thus, $y \geq 0$ for real number values of y.

b.

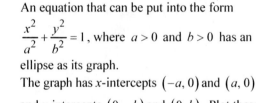

59. Answers may vary. Example:

$$\frac{x^2}{a^2} + \frac{y^2}{b^2} = 1$$

Substitute 0 for x and solve for y.

$$\frac{0^2}{a^2} + \frac{y^2}{b^2} = 1$$

$$\frac{y^2}{b^2} = 1$$

$$y^2 = b^2$$

$$y = \pm\sqrt{b^2}$$

$$y = \pm b$$

The y-intercepts are $(0, -b)$ and $(0, b)$.

61.
$$x^2 + y^2 = r^2$$

$$\frac{x^2}{r^2} + \frac{y^2}{r^2} = \frac{r^2}{r^2}$$

$$\frac{x^2}{r^2} + \frac{y^2}{r^2} = 1$$

This is an ellipse.

Since $a^2 = b^2 = r^2$, it is also a circle. (Note that all circles are ellipses just as all squares are rectangles.

63. Answers may vary. Example:
An equation that can be put into the form
$$\frac{x^2}{a^2} + \frac{y^2}{b^2} = 1$$, where $a > 0$ and $b > 0$ has an
ellipse as its graph.
The graph has x-intercepts $(-a, 0)$ and $(a, 0)$
and y-intercepts $(0, -b)$ and $(0, b)$. Plot these
intercepts and sketch an ellipse that contains
them.

65. a.

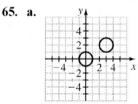

To get the graph of $(x-3)^2 + (y-2)^2 = 1$,
translate the graph of $x^2 + y^2 = 1$ by
3 units to the right and 2 units up.

b.

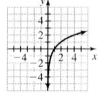

To get the graph of $(x+3)^2 + (y+2)^2 = 1$,
translate the graph of $x^2 + y^2 = 1$ by
3 units to the left and 2 units down.

c. Translate the graph of $x^2 + y^2 = r^2$ by h
units to the right if $h > 0$ or by $|h|$ units to
the left if $h < 0$, then by k units up if
$k > 0$ or by $|k|$ units down if $k < 0$.

d.

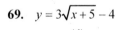

e.

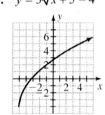

67. $y = \log_2(x)$

69. $y = 3\sqrt{x+5} - 4$

71. No, the relation is not a function.
Answers may vary. Example:

The graph of $\dfrac{x^2}{4} + \dfrac{y^2}{81} = 1$ is an ellipse, which

does not pass the vertical line test.

73. $y = 2x^2 - 8x + 3$

$y = 2\left(x^2 - 4x\right) + 3$

$y = 2\left(x^2 - 4x + 4\right) + 3 - 8$

$y = 2\left(x - 2\right)^2 - 5$

This is a quadratic function.

75. $2x^2 - 8x + 3 = 0$

$x = \dfrac{8 \pm \sqrt{\left(-8\right)^2 - 4(2)(3)}}{2(2)}$

$= \dfrac{8 \pm \sqrt{64 - 24}}{4}$

$= \dfrac{8 \pm 2\sqrt{10}}{4}$

$= \dfrac{4 \pm \sqrt{10}}{2}$

This is a quadratic equation in one variable.

77. $-5x\left(2x - 1\right)\left(3x - 1\right)$

$= -5x\left(2x \cdot 3x - 1(2x) - 1(3x) - 1(-1)\right)$

$= -5x\left(6x^2 - 5x + 1\right)$

$= -30x^3 + 25x^2 - 5x$

This is a cubic (or third-degree) polynomial in one variable.

11.4 Quiz

1. $\dfrac{x^2}{9} + \dfrac{y^2}{25} = 1$

$a^2 = 9, a = 3$

x-intercepts: $\left(-3, 0\right), \left(3, 0\right)$

$b^2 = 25, b = 5$

y-intercepts: $\left(0, -5\right), \left(0, 5\right)$

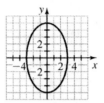

2. $\dfrac{y^2}{49} - \dfrac{x^2}{9} = 1$

$b^2 = 49, b = 7$

y-intercepts: $\left(0, -7\right), \left(0, 7\right)$

$a^2 = 9, a = 3$

Sketch a dashed rectangle that contains the points $\left(-3, 0\right), \left(3, 0\right), \left(0, -7\right)$, and $\left(0, 7\right)$, and then sketch the inclined asymptotes.

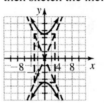

3. $4x^2 - y^2 = 16$

$\dfrac{4x^2}{16} - \dfrac{y^2}{16} = \dfrac{16}{16}$

$\dfrac{x^2}{4} - \dfrac{y^2}{16} = 1$

$a^2 = 4, a = 2$

x-intercepts: $\left(-2, 0\right), \left(2, 0\right)$

$b^2 = 16, b = 4$

Sketch a dashed rectangle that contains the points $\left(-2, 0\right), \left(2, 0\right), \left(0, -4\right)$, and $\left(0, 4\right)$, and then sketch the inclined asymptotes.

4. $16x^2 + 3y^2 = 48$

$\dfrac{16x^2}{48} + \dfrac{3y^2}{48} = \dfrac{48}{48}$

$\dfrac{x^2}{3} + \dfrac{y^2}{16} = 1$

$a^2 = 3, a = \sqrt{3}$

x-intercepts: $\left(-\sqrt{3}, 0\right), \left(\sqrt{3}, 0\right)$

$b^2 = 16, b = 4$

y-intercepts: $(0, -4), (0, 4)$

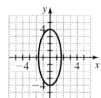

5. $x^2 - 9y^2 = 81$

$$\frac{x^2}{81} - \frac{9y^2}{81} = 1$$

$$\frac{x^2}{81} - \frac{y^2}{9} = 1$$

$a^2 = 81, a = 9$

x-intercepts: $(-9, 0), (9, 0)$

$b^2 = 9, b = 3$

Sketch a dashed rectangle that contains the points $(-9, 0), (9, 0), (0, -3)$, and $(0, 3)$, and then sketch the inclined asymptotes.

6. $4y^2 - 4x^2 = 16$

$$\frac{4y^2}{16} - \frac{4x^2}{16} = \frac{16}{16}$$

$$\frac{y^2}{4} - \frac{x^2}{4} = 1$$

$b^2 = 4, b = 2$

y-intercepts: $(0, -2), (0, 2)$

$a^2 = 4, a = 2$

Sketch a dashed rectangle that contains the points $(-2, 0), (2, 0), (0, -2)$, and $(0, 2)$, and then sketch the inclined asymptotes.

7. $\dfrac{x^2}{5} + \dfrac{y^2}{14} = 1$

$a^2 = 5, a = \sqrt{5}$

x-intercepts: $\left(-\sqrt{5}, 0\right), \left(\sqrt{5}, 0\right)$

$b^2 = 14, b = \sqrt{14}$

y-intercepts: $\left(0, -\sqrt{14}\right), \left(0, \sqrt{14}\right)$

8. $\dfrac{x^2}{8} + \dfrac{y^2}{3} = 1$

$a^2 = 8, a = 2\sqrt{2}$

x-intercepts: $\left(-2\sqrt{2}, 0\right), \left(2\sqrt{2}, 0\right)$

$b^2 = 3, b = \sqrt{3}$

y-intercepts: $\left(0, -\sqrt{3}\right), \left(0, \sqrt{3}\right)$

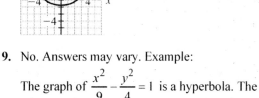

9. No. Answers may vary. Example:

The graph of $\dfrac{x^2}{9} - \dfrac{y^2}{4} = 1$ is a hyperbola. The graph fails the vertical line test.

10. Answers may vary. Example:

The points $(0, 3)$ and $(0, -3)$ are y-intercepts.

An ellipse with these y-intercepts has the form

$$\frac{x^2}{a^2} + \frac{y^2}{9} = 1.$$

Let $a = 1, 2,$ and 4.

$$\frac{x^2}{1} + \frac{y^2}{9} = 1$$

$$\frac{x^2}{4} + \frac{y^2}{9} = 1$$

$$\frac{x^2}{16} + \frac{y^2}{9} = 1$$

Homework 11.5

1. $x^2 + y^2 = 25$
   $4x^2 + 25y^2 = 100$

   The two intersection points $(-5, 0)$ and $(5, 0)$

   are the solutions to the system.
   Solve using elimination:

   $-4x^2 - 4y^2 = -100$

   $\underline{4x^2 + 25y^2 = 100}$

   $\qquad 21y^2 = 0$

   $\qquad y^2 = 0$

   $\qquad y = 0$

   Let $y = 0$ in $x^2 + y^2 = 25$ and solve for x.

   $x^2 + 0^2 = 25$

   $x^2 = 25$

   $x = \pm 5$

   The solutions are $(-5, 0)$ and $(5, 0)$.

3. $y = x^2 + 1$
   $y = -x + 3$

   The two intersection points $(-2, 5)$ and $(1, 2)$

   are solutions of the system.
   Solve using substitution.
   Substitute $x^2 + 1$ for y in the second equation.

   $y = -x + 3$

   $x^2 + 1 = -x + 3$

   $x^2 + x - 2 = 0$

   $(x + 2)(x - 1) = 0$

   $x + 2 = 0 \quad$ or $\quad x - 1 = 0$

   $x = -2 \quad$ or $\qquad x = 1$

   Let $x = -2$ and $x = 1$ in $y = -x + 3$ and solve
   for y.

   $\begin{aligned} y &= -(-2) + 3 & y &= -(1) + 3 \\ &= 2 + 3 & &= -1 + 3 \\ &= 5 & &= 2 \end{aligned}$

   The solutions are $(-2, 5)$ and $(1, 2)$.

5. $y = x^2 - 2$
   $y = -x^2 + 6$

   The two intersection points $(2, 2)$ and $(-2, 2)$

   are solutions of the system.
   Solve using substitution.

Substitute $x^2 - 2$ for y in the second equation.

$y = -x^2 + 6$

$x^2 - 2 = -x^2 + 6$

$2x^2 = 8$

$x^2 = 4$

$x = \pm 2$

Let $x = -2$ and $x = 2$ in $y = x^2 - 2$ and solve
for y.

$\begin{aligned} y &= (-2)^2 - 2 & y &= (2)^2 - 2 \\ &= 4 - 2 & &= 4 - 2 \\ &= 2 & &= 2 \end{aligned}$

The solutions are $(-2, 2)$ and $(2, 2)$.

7. $x^2 + y^2 = 49$
   $x^2 + y^2 = 16$

   The graphs do not intersect. The solution set is
   the empty set.
   Solve using elimination.

   $x^2 + y^2 = 49$

   $\underline{-x^2 - y^2 = -16}$

   $\qquad 0 = 33 \quad$ false

   There is no solution.

9. $x^2 + y^2 = 25$
   $\qquad y = -x - 1$

   The two intersection points $(-4, 3)$ and $(3, -4)$

   are solutions of the system.
   Solve using substitution.
   Substitute $-x - 1$ for y in the first equation.

   $x^2 + y^2 = 25$

   $x^2 + (-x - 1)^2 = 25$

   $x^2 + x^2 + 2x + 1 = 25$

   $2x^2 + 2x - 24 = 0$

   $x^2 + x - 12 = 0$

   $(x + 4)(x - 3) = 0$

   $x + 4 = 0 \quad$ or $\quad x - 3 = 0$

   $x = -4 \quad$ or $\qquad x = 3$

   Let $x = -4$ and $x = 3$ in $y = -x - 1$ and solve
   for y.

   $\begin{aligned} y &= -(-4) - 1 & y &= -(3) - 1 \\ &= 4 - 1 & &= -3 - 1 \\ &= 3 & &= -4 \end{aligned}$

   The solutions are $(-4, 3)$ and $(3, -4)$.

11. $y^2 - x^2 = 16$

$y + x^2 = 4$

The three intersection points $(-3, -5)$, $(0, 4)$,

and $(3, -5)$ are the solutions to the system.

Solve using elimination.

$y^2 - x^2 = 16$

$\underline{y + x^2 = 4}$

$y^2 + y = 20$

$y^2 + y - 20 = 0$

$(y + 5)(y - 4) = 0$

$y + 5 = 0$　or　$y - 4 = 0$

$y = -5$　or　　$y = 4$

Let $y = -5$ and $y = 4$ in $y + x^2 = 4$ and

solve for x.

$-5 + x^2 = 4$　　　　$4 + x^2 = 4$

$x^2 = 9$　　　　　$x^2 = 0$

$x = \pm 3$　　　　　$x = 0$

The solutions are $(-3, -5)$, $(3, -5)$, and

$(0, 4)$.

13. $25x^2 - 9y^2 = 225$

$4x^2 + 9y^2 = 36$

The two intersection points $(-3, 0)$ and $(3, 0)$

are solutions of the system.

Solve using elimination.

$25x^2 - 9y^2 = 225$

$\underline{4x^2 + 9y^2 = 36}$

$29x^2 = 261$

$x^2 = 9$

$x = \pm 3$

Let $x = -3$ and $x = 3$ in $4x^2 + 9y^2 = 36$ and

solve for y.

$4(-3)^2 + 9y^2 = 36$　　$4(3)^2 + 9y^2 = 36$

$36 + 9y^2 = 36$　　　$36 + 9y^2 = 36$

$9y^2 = 0$　　　　$9y^2 = 0$

$y^2 = 0$　　　　$y^2 = 0$

$y = 0$　　　　$y = 0$

The solutions are $(-3, 0)$ and $(3, 0)$.

15. $9x^2 + y^2 = 9$

$y = 3x + 3$

The two intersection points $(-1, 0)$ and $(0, 3)$

are solutions of the system.

Solve using substitution.

Substitute $3x + 3$ for y in the first equation.

$9x^2 + y^2 = 9$

$9x^2 + (3x + 3)^2 = 9$

$9x^2 + 9x^2 + 18x + 9 = 9$

$18x^2 + 18x = 0$

$18x(x + 1) = 0$

$18x = 0$　or　$x + 1 = 0$

$x = 0$　or　　$x = -1$

Let $x = 0$ and $x = -1$ in $y = 3x + 3$ and solve

for y.

$y = 3(0) + 3$　　　　$y = 3(-1) + 3$

$= 0 + 3$　　　　　$= -3 + 3$

$= 3$　　　　　　$= 0$

The solutions are $(0, 3)$ and $(-1, 0)$.

17. $4x^2 + 9y^2 = 36$

$16x^2 + 25y^2 = 225$

The graphs do not intersect. The solution set is

the empty set.

Solve using elimination.

$-16x^2 - 36y^2 = -144$

$\underline{16x^2 + 25y^2 = 225}$

$-11y^2 = 81$

$y^2 = -\dfrac{81}{11}$

Since y^2 cannot be negative (in the real

number system), there is no solution.

19. $y = \sqrt{x} - 3$

$y = -x - 1$

The intersection point $(1, -2)$ is the solution to

the system.

Solve using substitution.

Substitute $\sqrt{x} - 3$ for y in the second

equation.

$$y = -x - 1$$
$$\sqrt{x} - 3 = -x - 1$$
$$\sqrt{x} = -x + 2$$
$$\left(\sqrt{x}\right)^2 = \left(-x + 2\right)^2$$
$$x = x^2 - 4x + 4$$
$$x^2 - 5x + 4 = 0$$
$$\left(x - 4\right)\left(x - 1\right) = 0$$
$$x - 4 = 0 \quad \text{or} \quad x - 1 = 0$$
$$x = 4 \quad \text{or} \quad x = 1$$

Let $x = 4$ and $x = 1$ in $y = \sqrt{x} - 3$ and solve
for y.

$$y = \sqrt{4} - 3 \qquad\qquad y = \sqrt{1} - 3$$
$$= 2 - 3 \qquad\qquad\quad = 1 - 3$$
$$= -1 \qquad\qquad\qquad = -2$$

Check each result in $y = -x - 1$.

$$-1 \overset{?}{=} -(4) - 1 \qquad -2 \overset{?}{=} -1 - 1$$
$$-1 \overset{?}{=} -5 \text{ false} \qquad -2 \overset{?}{=} -2 \text{ true}$$

The only solution is $(1, -2)$.

21. $y = 2x^2 - 5$
  $y = x^2 - 2$

The two intersection points $\left(-\sqrt{3}, 1\right)$ and

$\left(\sqrt{3}, 1\right)$ are solutions of the system.

Solve using substitution.

Substitute $2x^2 - 5$ for y in the second
equation.

$$y = x^2 - 2$$
$$2x^2 - 5 = x^2 - 2$$
$$x^2 = 3$$
$$x = \pm\sqrt{3}$$

Let $x = -\sqrt{3}$ and $x = \sqrt{3}$ in $y = x^2 - 2$ and
solve for y.

$$y = \left(-\sqrt{3}\right)^2 - 2 \qquad y = \left(\sqrt{3}\right)^2 - 2$$
$$= 3 - 2 \qquad\qquad\quad = 3 - 2$$
$$= 1 \qquad\qquad\qquad = 1$$

The solutions are $\left(-\sqrt{3}, 1\right)$ and $\left(\sqrt{3}, 1\right)$.

23. $25y^2 - 4x^2 = 100$
  $9x^2 + y^2 = 9$

The four intersection points $\left(-0.74, -2.02\right)$,

$\left(-0.74, 2.02\right)$, $\left(0.74, -2.02\right)$, and $\left(0.74, 2.02\right)$

are solutions of the system.
Solve using substitution.

$$9x^2 + y^2 = 9$$
$$y^2 = 9 - 9x^2$$

Substitute $9 - 9x^2$ for y^2 in the first
equation.

$$25y^2 - 4x^2 = 100$$
$$25\left(9 - 9x^2\right) - 4x^2 = 100$$
$$225 - 225x^2 - 4x^2 = 100$$
$$-229x^2 = -125$$
$$x^2 = \frac{125}{229}$$
$$x = \pm\sqrt{\frac{125}{229}} \approx \pm 0.74$$

Let $x = -0.74$ and $x = 0.74$ in $9x^2 + y^2 = 9$
and solve for y.

$$9\left(-0.74\right)^2 + y^2 = 9$$
$$y^2 = 4.07$$
$$y = \pm\sqrt{4.07} \approx \pm 2.02$$
$$9\left(0.74\right)^2 + y^2 = 9$$
$$y^2 = 4.07$$
$$y = \pm\sqrt{4.07} \approx \pm 2.02$$

The solutions are $\left(-0.74, -2.02\right)$,

$\left(-0.74, 2.02\right)$, $\left(0.74, -2.02\right)$, and $\left(0.74, 2.02\right)$.

25. $25x^2 + 9y^2 = 225$
  $x^2 + y^2 = 16$

The four intersection points $\left(-2.25, -3.31\right)$,

$\left(-2.25, 3.31\right)$, $\left(2.25, -3.31\right)$, and $\left(2.25, 3.31\right)$

are solutions of the system.
Solve using elimination.

$$25x^2 + 9y^2 = 225$$
$$\underline{-9x^2 - 9y^2 = -144}$$
$$16x^2 = 81$$
$$x^2 = \frac{81}{16}$$
$$x = \pm\frac{9}{4} = \pm 2.25$$

Let $x = -2.25$ and $x = 2.25$ in $x^2 + y^2 = 16$
and solve for y.

$$\left(-2.25\right)^2 + y^2 = 16$$
$$y^2 = 10.9375$$
$$y = \pm 3.31$$

$$(2.25)^2 + y^2 = 16$$
$$y^2 = 10.9375$$
$$y = \pm 3.31$$

The solutions are $(-2.25, -3.31)$,
$(-2.25, 3.31)$, $(2.25, -3.31)$, and $(2.25, 3.31)$.

27. $9x^2 + y^2 = 85$
$2x^2 - 3y^2 = 6$

Solve using elimination.

$$27x^2 + 3y^2 = 255$$
$$\underline{2x^2 - 3y^2 = 6}$$
$$29x^2 = 261$$
$$x^2 = 9$$
$$x = \pm 3$$

Let $x = -3$ and $x = 3$ in $9x^2 + y^2 = 85$ and solve for y.

$$9(-3)^2 + y^2 = 85 \qquad 9(3)^2 + y^2 = 85$$
$$81 + y^2 = 85 \qquad\quad 81 + y^2 = 85$$
$$y^2 = 4 \qquad\qquad\quad y^2 = 4$$
$$y = \pm 2 \qquad\qquad\quad y = \pm 2$$

The four solutions are $(-3, -2)$, $(-3, 2)$, $(3, -2)$, and $(3, 2)$. Each result satisfies both equations.

29. $x^2 + 4y^2 = 25$
$y = -x + 5$

Solve using substitution.
Substitute $-x + 5$ for y in the first equation.

$$x^2 + 4y^2 = 25$$
$$x^2 + 4(-x + 5)^2 = 25$$
$$x^2 + 4(x^2 - 10x + 25) = 25$$
$$x^2 + 4x^2 - 40x + 100 = 25$$
$$5x^2 - 40x + 75 = 0$$
$$x^2 - 8x + 15 = 0$$
$$(x - 5)(x - 3) = 0$$
$$x - 5 = 0 \text{ or } x - 3 = 0$$
$$x = 5 \text{ or } \quad x = 3$$

Let $x = 5$ and $x = 3$ in $y = -x + 5$ and solve for y.

$$y = -(5) + 5 \qquad\quad y = -(3) + 5$$
$$= -5 + 5 \qquad\qquad = -3 + 5$$
$$= 0 \qquad\qquad\qquad = 2$$

The solutions are $(3, 2)$ and $(5, 0)$. Each result satisfies both equations.

31. $y = x^2 - 3x + 2$
$y = 2x - 4$

Solve using substitution.
Substitute $2x - 4$ for y in the first equation.

$$y = x^2 - 3x + 2$$
$$2x - 4 = x^2 - 3x + 2$$
$$x^2 - 5x + 6 = 0$$
$$(x - 3)(x - 2) = 0$$
$$x - 3 = 0 \text{ or } x - 2 = 0$$
$$x = 3 \text{ or } \quad x = 2$$

Let $x = 3$ and $x = 2$ in $y = 2x - 4$ and solve for y.

$$y = 2(3) - 4 \qquad\quad y = 2(2) - 4$$
$$= 6 - 4 \qquad\qquad\quad = 4 - 4$$
$$= 2 \qquad\qquad\qquad = 0$$

The solutions are $(3, 2)$ and $(2, 0)$. Each result satisfies both equations.

33. $x^2 + y^2 = 25$
$4x^2 - 25y^2 = 100$
$4x^2 + 25y^2 = 100$

The two intersection points for all three graphs are $(-5, 0)$ and $(5, 0)$. These are the solutions to the system. Each result satisfies all three equations.

35. Answers may vary. Example:
$$x^2 + y^2 = 16$$
$$x^2 - y^2 = 16$$

37. $y = x^2$
$y = -x^2 + c$

Solve the system using substitution.

$$x^2 = -x^2 + c$$
$$2x^2 = c$$
$$x^2 = \frac{c}{2}$$
$$x = \pm\sqrt{\frac{c}{2}} = \pm\frac{\sqrt{c}}{\sqrt{2}} \cdot \frac{\sqrt{2}}{\sqrt{2}} = \pm\frac{\sqrt{2c}}{2}$$

a. This system has two solutions if $c > 0$.

b. This system has one solution if $c = 0$.

c. This system has no solutions if $c < 0$.

39. $2x^2 + cy^2 = 82$

$$y = x^2 + dx + 5$$

Substitute $(1, 4)$ into each equation.

$$
\begin{array}{ll}
2(1)^2 + c(4)^2 = 82 & 4 = (1)^2 + d(1) + 5 \\
2 + 16c = 82 & 4 = 1 + d + 5 \\
16c = 80 & d = -2 \\
c = 5 &
\end{array}
$$

41. Answers may vary. Example:
Graph each of the equations on the same coordinate plane, then find the points of intersection. These points are the solutions to the system.

43. $y = 2^x$

$$y = 4\left(\frac{1}{2}\right)^x$$

Solve by substitution.

Substitute 2^x for y in the second equation.

$$2^x = 4\left(\frac{1}{2}\right)^x$$

$$2^x = 2^2\left(2^{-1}\right)^x$$

$$2^x = 2^2\left(2^{-x}\right)$$

$$2^x = 2^{2-x}$$

$$x = 2 - x$$

$$x = 1$$

Let $x = 1$ in $y = 2^x$ and solve for y.

$$y = 2^1 = 2$$

The solution is $(1, 2)$

45. Answers may vary. Example:
$y = -2x + 9$

47. Answers may vary. Example:

$$\frac{5x - 7}{2x + 2} = 4$$

$$5x - 7 = 4(2x + 2)$$

$$5x - 7 = 8x + 8$$

$$-3x = 15$$

$$x = -5$$

49. Answers may vary. Example:
$y = 3^x$

51. Answers may vary. Example:
$y = x^2 - 6x + 7$

53. Answers may vary. Example:

$$2x^2 + 8x - 10 = 0$$

$$2\left(x^2 + 4x - 5\right) = 0$$

$$2(x + 5)(x - 1) = 0$$

$$x + 5 = 0$$

$$x = -5$$

$$x - 1 = 0$$

$$x = 1$$

The solutions are -5 and 1.

11.5 Quiz

1. $9x^2 + y^2 = 81$

$$x^2 + y^2 = 9$$

The two intersection points $(-3, 0)$ and $(3, 0)$ are the solutions to the system.

Solve using elimination.

$$
\begin{array}{l}
9x^2 + y^2 = 81 \\
\underline{-9x^2 - 9y^2 = -81} \\
-8y^2 = 0 \\
y^2 = 0 \\
y = 0
\end{array}
$$

Let $y = 0$ in $x^2 + y^2 = 9$ and solve for x.

$$x^2 + y^2 = 9$$

$$x^2 + 0^2 = 9$$

$$x^2 = 9$$

$$x = \pm 3$$

The solutions are $(-3, 0)$ and $(3, 0)$.

2. $y = x^2 - 2$

$$y = -2x + 1$$

The two intersection points $(1, -1)$ and $(-3, 7)$ are the solutions to the system.

Solve using substitution.

Substitute $x^2 - 2$ for y in the second equation.

$$y = -2x + 1$$

$$x^2 - 2 = -2x + 1$$

$$x^2 + 2x - 3 = 0$$

$$(x + 3)(x - 1) = 0$$

$$x + 3 = 0 \quad \text{or} \quad x - 1 = 0$$

$$x = -3 \quad \text{or} \quad x = 1$$

Let $x = -3$ and $x = 1$ in $y = -2x + 1$ and solve for y.

$$y = -2(-3) + 1 \qquad y = -2(1) + 1$$
$$= 6 + 1 \qquad\qquad = -2 + 1$$
$$= 7 \qquad\qquad\quad = -1$$

The solutions are $(-3, 7)$ and $(1, -1)$.

3. $y = x^2 + 3$

$y = x^2 - 6x + 9$

The intersection point $(1, 4)$ is the solution to the system.

Solve using substitution.

Substitute $x^2 + 3$ for y in the second equation.

$$y = x^2 - 6x + 9$$
$$x^2 + 3 = x^2 - 6x + 9$$
$$6x - 6 = 0$$
$$x - 1 = 0$$
$$x = 1$$

Let $x = 1$ in $y = x^2 + 3$ and solve for y.

$$y = (1)^2 + 3 = 1 + 3 = 4$$

The solution is $(1, 4)$.

4. $25x^2 - 4y^2 = 100$

$9x^2 + y^2 = 9$

The graphs do not intersect. There are no solutions to the system.

Solve using elimination.

$$225x^2 - 36y^2 = 900$$
$$\underline{-225x^2 - 25y^2 = -225}$$
$$-61y^2 = 675$$
$$y^2 = -\frac{675}{61}$$

Since y^2 must be nonnegative (in the real number system), there are no solutions to the system. The solution set is the empty set.

5. $x^2 - y^2 = 16$

$x^2 + y^2 = 16$

$y = (x + 4)^2$

The intersection point of all three graphs is $(-4, 0)$. This is the solution of the system.

6. Answers may vary. Example:

$$x^2 + y^2 = 25$$
$$y = x^2 + 5$$

Appendix A
Reviewing Prerequisite Material

Section A.1

1–8.

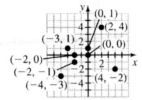

Section A.2

1. 4 and 85 are counting numbers.

2. 4, 0, –2, and 85 are integers.

3. $\dfrac{2}{9}$, 4, –7.19, 0, –2, and 85 are rational numbers.

4. $\sqrt{17}$ is an irrational number.

5. $\dfrac{2}{9}$, 4, –7.19, 0, –2, $\sqrt{17}$, and 85 are real numbers.

Section A.3

1. $|-3| = 3$

2. $|4.69| = 4.69$

3. $|0| = 0$

4. $|-\pi| = \pi$

Section A.4

1. $-3(7) = -21$

2. $5(-6) = -30$

3. $-9(-4) = 36$

4. $-8(-2) = 16$

5. $-(-4) = (-1)(-4) = 4$

6. $-(-(-9)) = -((-1)(-9)) = (-1)(-1)(-9) = -9$

7. $\dfrac{8}{-2} = (-1)\left(\dfrac{8}{2}\right) = -4$

8. $\dfrac{-6}{-2} = \dfrac{6}{2} = 3$

9. $-3 + (-5) = -8$

10. $-6 + (-7) = -13$

11. $2 + (-8) = -6$

12. $-7 + 3 = -4$

13. $-1 + 6 = 5$

14. $8 + (-3) = 5$

15. $-4 + (-6) = -10$

16. $-2 + (-3) = -5$

17. $3 - 7 = 3 + (-7) = -4$

18. $2 - 8 = 2 + (-8) = -6$

19. $5 - (-3) = 5 + (-(-3)) = 5 + 3 = 8$

20. $9 - (-4) = 9 + (-(-4)) = 9 + 4 = 13$

21. $-4 - 9 = -4 + (-9) = -13$

22. $-2 - 4 = -2 + (-4) = -6$

23. $-1 - (-1) = -1 + (-(-1)) = -1 + 1 = 0$

24. $-10 - (-6) = -10 + (-(-6)) = -10 + 6 = -4$

Section A.5

1. $7^2 = 7 \cdot 7 = 49$

2. $9^2 = 9 \cdot 9 = 81$

3. $6^3 = 6 \cdot 6 \cdot 6 = 216$

4. $5^4 = 5 \cdot 5 \cdot 5 \cdot 5 = 625$

5. $(-2)^4 = (-2)(-2)(-2)(-2) = 16$

6. $(-3)^3 = (-3)(-3)(-3) = -27$

7. $-2^4 = -(2 \cdot 2 \cdot 2 \cdot 2) = -16$

8. $-3^3 = -(3 \cdot 3 \cdot 3) = -27$

Section A.6

1. $3 + 5 \cdot 2 = 3 + 10 = 13$

2. $2(8) - 4 = 16 - 4 = 12$

3. $2 + 10 \div (-5) = 2 + (-2) = 0$

4. $14 \div (-2) - 1 = -7 - 1 = -8$

5. $2(1 - 3) + 4 \cdot 2 = 2(-2) + 8 = -4 + 8 = 4$

6. $\begin{aligned} 10(2 - 7) + 5 \cdot 4 &= 10(-5) + 20 \\ &= -50 + 20 \\ &= -30 \end{aligned}$

7. $\begin{aligned} (5 - 9)(4 + 2) \div 8 + 2 &= (-4)(6) \div 8 + 2 \\ &= -24 \div 8 + 2 \\ &= -3 + 2 \\ &= -1 \end{aligned}$

8. $\begin{aligned} (3 + 5)(2 - 6) \div 4 + 1 &= (8)(-4) \div 4 + 1 \\ &= -32 \div 4 + 1 \\ &= -8 + 1 \\ &= -7 \end{aligned}$

9. $4(3)^2 = 4(9) = 36$

10. $-3(2)^3 = -3(8) = -24$

11. $-3^2 + (-3)^2 - (-3)^2 = -9 + 9 - 9 = -9$

12. $2^3 - (-2)^3 + (-2)^3 = 8 - (-8) + (-8) = 8$

13. $5 - 4^2 + (-8) \div (-2) = 5 - 16 + 4 = -7$

14. $2^3 - 10 \div (-5) + 1 = 8 - (-2) + 1 = 11$

15. $6 - (3 - 1)^3 + 8 = 6 - (2)^3 + 8 = 6 - 8 + 8 = 6$

16. $\begin{aligned} 10 - (9 - 6)^3 + 5 &= 10 - (3)^3 + 5 \\ &= 10 - 27 + 5 \\ &= -12 \end{aligned}$

Section A.7

1. $y = mx + b$ is an equation.

2. $3x^2 - 5x + 4 = 8$ is an equation.

3. $2x - 5\pi + 1$ is an expression.

4. $x^3 - 8$ is an expression.

Section A.8

1. $2(x + 4) = 2x + 2(4) = 2x + 8$

2. $4(x + 7) = 4x + 4(7) = 4x + 28$

3. $6(2t - 3) = 6(2t) + 6(-3) = 12t - 18$

4. $5(4w - 6) = 5(4w) + 5(-6) = 20w - 30$

5. $(x + 8)(-3) = -3x + 8(-3) = -3x - 24$

6. $(x + 5)(-4) = -4x + 5(-4) = -4x - 20$

7. $(2x - 9)(-5) = -10x - 9(-5) = -10x + 45$

8. $(3x - 1)(-6) = -18x - 1(-6) = -18x + 6$

9. $2.8(p + 4.1) = 2.8p + 2.8(4.1) = 2.8p + 11.48$

10. $\begin{aligned} -5.2(b + 3.9) &= -5.2b - 5.2(3.9) \\ &= -5.2b - 20.28 \end{aligned}$

Section A.9

1. $4x + 3x = (4 + 3)x = 7x$

2. $7x - 2x = (7 - 2)x = 5x$

3. $\begin{aligned} 5x - 9y - 3x + 2y &= 5x - 3x - 9y + 2y \\ &= (5 - 3)x + (-9 + 2)y \\ &= 2x - 7y \end{aligned}$

4. $8x - 4y - 6x + 5y = 8x - 6x - 4y + 5y$
$$= (8 - 6)x + (-4 + 5)y$$
$$= 2x + y$$

5. $7a - 4 + b - 9a - 3b + 2$
$$= 7a - 9a + b - 3b - 4 + 2$$
$$= (7 - 9)a + (1 - 3)b - 2$$
$$= -2a - 2b - 2$$

6. $4t - 2w + 5 + t - 1 - 8w$
$$= 4t + t - 2w - 8w + 5 - 1$$
$$= (4 + 1)t + (-2 - 8)w + 4$$
$$= 5t - 10w + 4$$

7. $4(2x + 3) + 5(4x - 1) = 8x + 12 + 20x - 5$
$$= 8x + 20x + 12 - 5$$
$$= (8 + 20)x + 7$$
$$= 28x + 7$$

8. $5(3x + 2) + 2(3x + 6) = 15x + 10 + 6x + 12$
$$= 15x + 6x + 10 + 12$$
$$= (15 + 6)x + 22$$
$$= 21x + 22$$

9. $2(5x - y) - 3(4x + y)$
$$= 10x - 2y - 12x - 3y$$
$$= 10x - 12x - 2y - 3y$$
$$= (10 - 12)x + (-2 - 3)y$$
$$= -2x - 5y$$

10. $3(4x - y) - 5(2x + y)$
$$= 12x - 3y - 10x - 5y$$
$$= 12x - 10x - 3y - 5y$$
$$= (12 - 10)x + (-3 - 5)y$$
$$= 2x - 8y$$

11. $10 - (3m - 2n) + 4m - 7n$
$$= 10 - 3m + 2n + 4m - 7n$$
$$= 10 - 3m + 4m + 2n - 7n$$
$$= 10 + (-3 + 4)m + (2 - 7)n$$
$$= 10 + m - 5n$$
$$= m - 5n + 10$$

12. $6 - (6a - 3b) - 5b + 2a$
$$= 6 - 6a + 3b - 5b + 2a$$
$$= 6 - 6a + 2a + 3b - 5b$$
$$= 6 + (-6 + 2)a + (3 - 5)b$$
$$= 6 - 4a - 2b$$
$$= -4a - 2b + 6$$

Section A.10

1. $\quad x + 5 = 9$
$$x + 5 - 5 = 9 - 5$$
$$x = 4$$

2. $\quad x - 3 = 4$
$$x - 3 + 3 = 4 + 3$$
$$x = 7$$

3. $\quad 4x = 12$
$$\frac{4x}{4} = \frac{12}{4}$$
$$x = 3$$

4. $\quad -3x = 21$
$$\frac{-3x}{-3} = \frac{21}{-3}$$
$$x = -7$$

5. $\quad 5(w - 3) = 13$
$$5w - 15 = 13$$
$$5w - 15 + 15 = 13 + 15$$
$$\frac{5w}{5} = \frac{28}{5}$$
$$w = \frac{28}{5}$$

6. $\quad -2(k - 4) = 5$
$$-2k + 8 = 5$$
$$-2k + 8 - 8 = 5 - 8$$
$$\frac{-2k}{-2} = \frac{-3}{-2}$$
$$k = \frac{3}{2}$$

7. $\quad 2x + 5 = 6x - 3$
$$2x + 5 - 5 = 6x - 3 - 5$$
$$2x - 6x = 6x - 6x - 8$$
$$\frac{-4x}{-4} = \frac{-8}{-4}$$
$$x = 2$$

8. $\quad 4x - 7 = 9x + 3$
$$4x - 7 + 7 = 9x + 3 + 7$$
$$4x - 9x = 9x - 9x + 10$$
$$\frac{-5x}{-5} = \frac{10}{-5}$$
$$x = -2$$

9. $5 - 4(2x - 3) = 13$

$5 - 8x + 12 = 13$

$-8x + 17 - 17 = 13 - 17$

$\dfrac{-8x}{-8} = \dfrac{-4}{-8}$

$x = \dfrac{1}{2}$

10. $7 - 2(3x + 5) = 19$

$7 - 6x - 10 = 19$

$-6x - 3 + 3 = 19 + 3$

$\dfrac{-6x}{-6} = \dfrac{22}{-6}$

$x = -\dfrac{11}{3}$

11. $\dfrac{2}{3}t + \dfrac{1}{4} = \dfrac{5}{12}$

$12\left(\dfrac{2}{3}t + \dfrac{1}{4}\right) = 12\left(\dfrac{5}{12}\right)$

$12 \cdot \dfrac{2}{3}t + 12 \cdot \dfrac{1}{4} = 12 \cdot \dfrac{5}{12}$

$8t + 3 = 5$

$8t + 3 - 3 = 5 - 3$

$8t = 2$

$\dfrac{8t}{8} = \dfrac{2}{8}$

$t = \dfrac{1}{4}$

12. $\dfrac{5}{9}w + \dfrac{1}{2} = \dfrac{7}{6}$

$\dfrac{5}{9}w + \dfrac{1}{2} - \dfrac{1}{2} = \dfrac{7}{6} - \dfrac{1}{2}$

$\dfrac{5}{9}w = \dfrac{7}{6} - \dfrac{3}{6}$

$\dfrac{5}{9}w = \dfrac{2}{3}$

$\dfrac{5}{9}w\left(\dfrac{9}{5}\right) = \dfrac{2}{3}\left(\dfrac{9}{5}\right)$

$w = \dfrac{6}{5}$

13. $\dfrac{5}{2}x - \dfrac{7}{4} = \dfrac{3}{8}x$

$\dfrac{5}{2}x - \dfrac{5}{2}x - \dfrac{7}{4} = \dfrac{3}{8}x - \dfrac{5}{2}x$

$-\dfrac{7}{4} = \dfrac{3}{8}x - \dfrac{20}{8}x$

$-\dfrac{7}{4} = -\dfrac{17}{8}x$

$-\dfrac{7}{4}\left(-\dfrac{8}{17}\right) = -\dfrac{17}{8}x\left(-\dfrac{8}{17}\right)$

$\dfrac{14}{17} = x$

14. $\dfrac{5}{3}x - \dfrac{7}{2} = \dfrac{11}{6}x$

$\dfrac{5}{3}x - \dfrac{5}{3}x - \dfrac{7}{2} = \dfrac{11}{6}x - \dfrac{5}{3}x$

$-\dfrac{7}{2} = \dfrac{11}{6}x - \dfrac{10}{6}x$

$-\dfrac{7}{2} = \dfrac{1}{6}x$

$-\dfrac{7}{2}(6) = \dfrac{1}{6}x(6)$

$-21 = x$

Section A.11

1. $2x + y = 8$

$y = -2x + 8$

2. $3x - y = 5$

$-y = -3x + 5$

$y = 3x - 5$

3. $3x - 5y = 15$

$3x = 15 + 5y$

$x = \dfrac{5}{3}y + 5$

4. $3x - 5y = 15$

$-5y = 15 - 3x$

$y = \dfrac{3}{5}x - 3$

5. $ax - by = c$

$-by = c - ax$

$y = \dfrac{ax - c}{b}$

6. $ax - by = c$

$$ax = by + c$$

$$x = \frac{by + c}{a}$$

7. $-4x + 3y = 2x + 9$

$$-6x + 3y = 9$$

$$-6x = -3y + 9$$

$$\frac{-6x}{-6} = \frac{-3y}{-6} + \frac{9}{-6}$$

$$x = \frac{1}{2}y - \frac{3}{2}$$

8. $-4x + 3y = 2x + 9$

$$3y = 6x + 9$$

$$\frac{3y}{3} = \frac{6x}{3} + \frac{9}{3}$$

$$y = 2x + 3$$

9. $\frac{1}{2}x - \frac{3}{4}y = \frac{5}{8}$

$$8\left(\frac{1}{2}x - \frac{3}{4}y\right) = 8\left(\frac{5}{8}\right)$$

$$8 \cdot \frac{1}{2}x - 8 \cdot \frac{3}{4}y = 8 \cdot \frac{5}{8}$$

$$4x - 6y = 5$$

$$-6y = -4x + 5$$

$$y = \frac{2}{3}x - \frac{5}{6}$$

10. $\frac{3}{4}x - \frac{2}{3}y = \frac{1}{4}$

$$12\left(\frac{3}{4}x - \frac{2}{3}y\right) = 12\left(\frac{1}{4}\right)$$

$$12 \cdot \frac{3}{4}x - 12 \cdot \frac{2}{3}y = 12 \cdot \frac{1}{4}$$

$$9x - 8y = 3$$

$$-8y = -9x + 3$$

$$y = \frac{9}{8}x - \frac{3}{8}$$

11. $\frac{x}{a} - \frac{y}{a} = 1$

$$x - y = a$$

$$x = y + a$$

12. $\frac{x}{a} - \frac{y}{a} = 1$

$$x - y = a$$

$$-y = a - x$$

$$y = x - a$$

Section A.12

1. The expressions are equivalent.
Answers may vary. Example:
$$5(x - 4) = 5 \cdot x + 5(-4) = 5x - 20$$

2. The equations are equivalent.
Answers may vary. Example:
$$x + 8 = 0$$
$$x + 8 - 8 = 0 - 8$$
$$x = -8$$

3. The expressions are not equivalent.
Answers may vary. Example:
$$4x - 3x + 8 = x + 8$$
$$-12x + 8 \neq x + 8$$

4. The expressions are not equivalent.
Answers may vary. Example:
$$3(x + 1) + 7 = 3x + 3 + 7 = 3x + 10$$
$$3x + 8 \neq 3x + 10$$

5. The pair is neither.
Answers may vary. Example:
An expression and an equation cannot be equivalent to each other.

6. The expressions are equivalent.
Answers may vary. Example:
$$-3(2x - 5) = -3 \cdot 2x + (-3)(-5) = -6x + 15$$

7. The equations are equivalent.
Answers may vary. Example:
$$3x + 1 = 16$$
$$3x + 1 - 1 = 16 - 1$$
$$3x = 15$$

8. The equations are not equivalent.
Answers may vary. Example:
$$2(x - 3) + 5 = 25$$
$$2x - 6 + 5 = 25$$
$$2x = 26$$

9. The equations are not equivalent.
Answers may vary. Example:
$$-3(x - 4) = -18$$
$$-3x + 12 = -18$$
$$-3x = -30$$
$$x = 10$$

10. The expressions are not equivalent.
Answers may vary. Example:
$$3x + 4x - 2 = 7x - 2$$
$$2x + x - 2 + 5x = 8x - 2$$
$$7x - 2 \neq 8x - 2$$